International Series in Operations Research & Management Science

Volume 352

The book series **International Series in Operations Research and Management Science** encompasses the various areas of operations research and management science. Both theoretical and applied books are included. It describes current advances anywhere in the world that are at the cutting edge of the field. The series is aimed especially at researchers, advanced graduate students, and sophisticated practitioners.

The series features three types of books:

- Advanced expository books that extend and unify our understanding of particular areas.
- Research monographs that make substantial contributions to knowledge.
- Handbooks that define the new state of the art in particular areas. Each handbook will be edited by a leading authority in the area who will organize a team of experts on various aspects of the topic to write individual chapters. A handbook may emphasize expository surveys or completely new advances (either research or applications) or a combination of both.

The series emphasizes the following four areas:

Mathematical Programming: Including linear programming, integer programming, nonlinear programming, interior point methods, game theory, network optimization models, combinatorics, equilibrium programming, complementarity theory, multiobjective optimization, dynamic programming, stochastic programming, complexity theory, etc.

Applied Probability: Including queuing theory, simulation, renewal theory, Brownian motion and diffusion processes, decision analysis, Markov decision processes, reliability theory, forecasting, other stochastic processes motivated by applications, etc.

Production and Operations Management: Including inventory theory, production scheduling, capacity planning, facility location, supply chain management, distribution systems, materials requirements planning, just-in-time systems, flexible manufacturing systems, design of production lines, logistical planning, strategic issues, etc.

Applications of Operations Research and Management Science: Including telecommunications, health care, capital budgeting and finance, economics, marketing, public policy, military operations research, humanitarian relief and disaster mitigation, service operations, transportation systems, etc.

This book series is indexed in Scopus.

Eduardo Souza de Cursi

Uncertainty Quantification with R

Bayesian Methods

Springer

Eduardo Souza de Cursi
Department Mechanics / Civil Engineering
INSA Rouen Normandie
Saint-Etienne du Rouvray, France

ISSN 0884-8289 ISSN 2214-7934 (electronic)
International Series in Operations Research & Management Science
ISBN 978-3-031-48207-6 ISBN 978-3-031-48208-3 (eBook)
https://doi.org/10.1007/978-3-031-48208-3

This Springer imprint is published by the registered company Springer Nature Switzerland AG
The registered company address is: Gewerbestrasse 11, 6330 Cham, Switzerland

Introduction

This book is an independent companion volume to *Uncertainty Quantification with R*, complementing certain of its topics and taking up others from a different angle – the Bayesian Standpoint.

This is a fairly general book, covering a range of methods and techniques, each of which can be further developed in the literature specific to each of the fields covered – and the reader can find scientific articles that develop them in depth.

This book targets the use of R, which is a GNU project to develop a tool for language and environment for statistical computing and graphics. An IDE is proposed by RStudio. The popularity of R and RStudio make that the reader will find on the web many sites and information about it. A wide literature can also be found about this software. The community of the users of R proposes a large choice of packages to extend the possibilities of R. You will find repositories containing them, such as, for instance, https://cran.r-project.org/web/packages/.

The book contains many programs. If you are an expert in R, you will find certainly a large number of improvements to the programs presented. Analogously, the community of the users of R proposes many packages to solve a large number of practical problems and implementing the methods considered in this book. We cite many of them, but – probably, even certainly – not all the existing contributions. The author apologizes in advance to any forgotten contributors, whose works are not cited in the book, but who have made the effort and been kind enough to make their work product available to the community. As mentioned above, we recommend that R users search software repositories such as https://cran.r-project.org/.

If you are not familiarized with Uncertainty Quantification (UQ), you can consider it as a collection of methods for the analysis of numerical data, namely when uncertainty or variability is involved, having as general objective the determination of probability distributions. The general aim of UQ is to characterize the observed variability in a quantity X by using another random variable U, by using the available information about (X, U) to construct an explanation of X by U, into a form which will be useful for use in numerical calculations involving X. The information may be, for instance, an equation, a numerical problem involving both the variables, or samples. UQ applies to a wide range of situations.

In the book *Uncertainty Quantification with R*, we tried to illustrate the practical use of UQ techniques under R. In this book, the philosophy is the same: we focus on practical aspects, and the theoretical arguments are reduced to the strictly minimal amount necessary to the understanding of the practical methods introduced. We ask for your indulgence on this point – as indicated, we are more concerned with the practical aspects and do not deal with the theoretical aspects in this book – the reader should refer to the texts in the literature to study the mathematical arguments underlying the methods discussed.

Contents

Chapter 1
Basic Bayesian Probabilities

Bayesian probabilities can be seen as a particular point of view on probabilities. In this chapter, we recall the basic elements of probability theory that will be used in the book, and we give some elements about the Bayesian point of view about probabilities.

1.1 A Historical Perspective

The fundamental text for Bayesian probabilities is the article "An Essay Towards Solving a Problem in the Doctrine of Chances," conceived by Thomas Bayes and published posthumously by his friend Richard Price in *Philosophical Transactions of the Royal Society* (Bayes and Price 1763). In his text, Price says that Bayes formulates the following problem:

> **P R O B L E M.**
>
> ***Given*** **the number of times in which an unknown event has happened and failed:** ***Required*** **the chance that the probability of its happening in a ſingle trial lies ſomewhere between any two degrees of probability that can be named.**

E. Souza de Cursi, *Uncertainty Quantification with R*, International Series in Operations Research & Management Science 352,
https://doi.org/10.1007/978-3-031-48208-3_1

In terms of modern probability, Bayes' problem reads as

Problem 1.1 Let be given a sample $\mathcal{X} = (X_1, \ldots, X_n)$ of n variates from a Bernoulli distribution $B(p)$ $(P(X_i = 1) = p, P(X_i = 0) = 1 - p)$. Let be given $0 \leq a < b \leq 1$. Assuming that $\sum_{i=1}^{n} X_i = k$, determine the probability of the event $p \in (a, b)$, conditionally to the observed data.

A complete analysis of Bayes' solution can be found in Stigler (1982). It is based on two propositions: on the one hand,

PROP. 3.

The probability that two ſubſequent events will both happen is a ratio compounded of the probability of the 1ſt, and the probability of the 2d on ſuppoſition the 1ſt happens.

i.e., in terms of modern probability:

$$P(E_1 \cap E_2) = P(E_2|E_1)P(E_1). \tag{1.1}$$

On the other hand,

PROP. 5.

If there be two ſubſequent events, the probability of the 2d $\frac{b}{N}$ and the probability of both together $\frac{P}{N}$, and it being 1ſt diſcovered that the 2d event has happened, from hence I gueſs that the 1ſt event has alſo happened, the probability I am in the right is $\frac{P}{b}$†.

i.e., in terms of modern probability (see Sect. 1.3):

$$P(E_1|E_2) = \frac{P(E_1 \cap E_2)}{P(E_2)}. \tag{1.2}$$

Combining these two expressions, we obtain the classical equality known as ***Bayes' formula*** (see Sect. 1.3):

$$P(E_1|E_2) = \frac{P(E_2|E_1)}{P(E_2)} P(E_1). \tag{1.3}$$

In modern probability, Bayes' formula is used to show that the solution for Problem 1.1 verifies

$$P(p \in (a,b)|\mathcal{X}) = \int_a^b P(p|\mathcal{X})dp, \tag{1.4}$$

with

$$P(p|\mathcal{X}) = \frac{P(\mathcal{X}|p)P(p)}{P(\mathcal{X})}, \quad P(\mathcal{X}) = \int_0^1 P(\mathcal{X}|p)P(p)dp. \tag{1.5}$$

The practical use of Eq. (1.5) requires the knowledge of $P(\mathcal{X}|p)$ (usually referred to as ***Likelihood***) and $P(p)$ (usually referred to as ***Prior Distribution***) . In practice, Likelihood yields from a model, while the Prior must be chosen by the user, according to his assumptions or beliefs about p – except in particular situations. In the case studied by Bayes, an analogy – often called *Bayes' billiard* – is used to conclude that $P(p) = 1$, which corresponds to a uniform distribution, while the model for the Likelihood is binomial (Bayes and Price 1763). As a consequence, we have, in modern probabilistic language,

$$P(p|\mathcal{X}) = \frac{\binom{k}{n} p^k (1-p)^{n-k}}{\int_0^1 \binom{k}{n} p^k (1-p)^{n-k} dp}. \tag{1.6}$$

This expression can be written in a simplified form as

$$P(p|\mathcal{X}) = A\, p^k (1-p)^{n-k}\ , \tag{1.7}$$

where A is a constant such that

$$A \int_0^1 p^k (1-p)^{n-k} dp = 1 \Rightarrow \mathrm{A} = \frac{(n+1)!}{k!(n-k)!}. \tag{1.8}$$

The work of Bayes remained ignored for a long time (a complete history can be found in (Dale 1999)). In 1774, Pierre-Simon Laplace published a memoir entitled *Mémoire sur la probabilité des causes par les événements* (Laplace 1774, 1891). In this work, Laplace considers the following problem:

> Dans le Problème ſuivant, *une urne étant ſuppoſée renfermer un nombre donné de billets blancs & noirs dans un rapport inconnu, ſi l'on tire un billet & qu'il ſoit blanc, déterminer la probabilité que le rapport des billets blancs aux noirs eſt celui de* p *à q;* l'évènement eſt connu & la cauſe inconnue.

S.M. Stiegler proposes the following translation (Laplace 1986):

> In the following problem, *an urn is supposed to contain a given number of white and black tickets in an unknown ratio; if one draws a ticket and finds it white, determine the probability that the ratio of white to black tickets is that of p to q.* The event is known and the cause is unknown.

The problem considered by Laplace is analogous to the problem introduced by Bayes. Nevertheless, Laplace did not make any reference to Bayes – we can suppose that he was not aware of Bayes' work. Laplace states a principle for the solution:

> Si un évènement peut être produit par un nombre *n* de cauſes différentes, les probabilités de l'exiſtence de ces cauſes priſes de l'évènement, ſont entre elles comme les probabilités de l'évènement priſes de ces cauſes, & la probabilité de l'exiſtence de chacune d'elles, eſt égale à la probabilité de l'évènement priſe de cette cauſe, diviſée par la ſomme de toutes les probabilités de l'évènement priſes de chacune de ces cauſes.

S.M. Stiegler proposes the following translation (Laplace 1986):

> If an event can be produced by a number *n* of different causes, the probabilities of these causes given the event are to each other as the probabilities of the event given the causes, and the probability of the existence of each of these is equal to the probability of the event given that cause, divided by the sum of all the probabilities of the event given each of these causes.

In modern probability writing: Laplace considers n possible distinct causes C_i, $1 \leq i \leq n$ and an event : $C_i \cap C_j = \varnothing$, for $i \neq j$ and $\sum_{j=1}^{n} P(C_j) = 1$. Then, for an event E,

$$\frac{P(C_i|\mathrm{E})}{P(C_j|\mathrm{E})} = \frac{P(E|C_i)}{P(E|C_j)} \tag{1.9}$$

and (see Sect. 1.3)

$$P(C_i|\mathrm{E}) = \frac{P(E|C_i)}{\sum\limits_{j=1}^{n} P(E|C_j)}. \tag{1.10}$$

Notice that $P(E) = \sum\limits_{j=1}^{n} P(E|C_j)P(C_j)$ and $P(E \cap C_i) = P(E| C_i)P(C_i)$, so that Eq. (1.10) is equivalent to Eq. (1.2) when all the causes are equiprobable, i.e., $P(C_j) = 1/n, \forall j$. Indeed, Laplace assumes the equiprobability of the causes implicitly. A complete analysis of Laplace's work can be found in Dale (1982) or Dale (1999). A generalization to the actual form (1.10) appears in the works of Augustus de Morgan, William Fishburn Donkin, William Allen Whitworth, and Mathieu Paul Hermann Lauren (see Dale (1999)).

By 1920, the work of Bayes was rediscovered, namely, by Karl Pearson and Ronald Aylmer Fisher. The rediscovery generated a wide and polemical controversy about the use of a uniform distribution for p, justified by Bayes using the analogy of the billard.

One of the main criticisms was that such a choice is not invariant by transformation, so that it may lead to contradictions, if applied without precaution. Indeed, the criticism was not so much directed at Bayes' work as at the formulation that, in the absence of any information except the possible values of a quantity, one must consider a uniform distribution on these values. If such a principle is applied without adequate care, it can lead to contradictions, illustrated by a simple example: if a quantity X is uncertain and we do not have any information other than $X \in (1/_a, a)$, with $a > 1$, then the principle leads to

$$P(X \leq 1) = \frac{1 - \frac{1}{a}}{a - \frac{1}{a}} = \frac{1}{1 + a}.$$

However, $Y = \frac{1}{X}$ is also uncertain and $Y \in (1/_a, a)$. Thus, the same principle leads to

$$P(Y \geq 1) = \frac{a - 1}{a - \frac{1}{a}} = \frac{a}{1 + a} > \frac{1}{1 + a} = P(X \leq 1).$$

These results may appear to be contradictory, since $X \leq 1 \Leftrightarrow Y \geq 1$. Such a difficulty generated a controversy, which turned into a philosophical dispute about the meaning of probability, with two extreme standpoints: on the one hand, the "frequentist" one, where probabilities are the limits of relative frequencies or the result of enumerations; on the other hand, the "subjectivist" one, where probabilities are measurements degrees of belief. A highlight in this debate is the sentence of Bruno de Finetti (see (De Finetti 2017), p. xv) :

> *My thesis, paradoxically, and a little provocatively, but nonetheless genuinely, is simply this:*

PROBABILITY DOES NOT EXIST

The abandonment of superstitious beliefs about the existence of Phlogiston, the Cosmic Ether, Absolute Space and Time, . . ., or Fairies and Witches, was an essential step along the road to scientific thinking. Probability, too, if regarded as something endowed with some kind of objective existence, is no less a misleading misconception, an illusory attempt to exteriorize or materialize our true probabilistic beliefs.

Nowadays, many researchers (but not all!) regard the controversy as mainly obsolete and consider that they have at their disposal tools having as origin Bayes, Fisher, Pearson, etc. For these researchers, these tools form a toolbox from which they can choose what is necessary for their studies according to the circumstances.

Such a point de view is not universal, and the discussion goes on about the philosophical aspects and interpretations of probability – see, for instance, Howson and Urbach (2006), Jeffreys (1939), Savchuk and Tsokos (2011), Press and Tanur (2001), and Jaynes (1989, 2003).

The reader will certainly have his own point of view or will build it from his readings and his own thinking. We will not discuss these aspects in the following, and we will concentrate on the methods and their practical use, namely, with R.

1.2 Probabilities

Uncertainty can be modeled by different ways. For instance, we can use fuzzy variables to classify objects or individuals according to classes having imprecise limits. We can also model uncertain numerical variables as intervals and use interval analysis to manipulate them.

A classical model for uncertain quantities is the model of random variables, based on the classical definition of probabilities:

Definition 1.1 Let Ω be a nonempty set and $\mathcal{P}(\Omega) = \{E : E \subset \Omega\}$ be the set of the parts of Ω. A ***probability*** defined on Ω is an application $P : \mathcal{P}(\Omega) \longrightarrow \mathbb{R}$ such that

1. $P(E) \geq 0, \forall E \in \mathcal{P}(\Omega)$;
2. $P(\Omega) = 1$;
3. $P\left(\bigcup_{n \in \mathbb{N}} E_n\right) = \sum_{n \in \mathbb{N}} P(E_n), \forall \{E_n : n \in \mathbb{N}\}$ *such that* $E_i \cap E_j = \varnothing$, *if* $i \neq j$. ∎

In the context of probability, Ω is the ***universe***, and (Ω, P) is referred to as a ***probability space***. A subset $E \subset \Omega$ is an ***event***. From this definition, we can state the standard properties of probabilities, namely,

$$P(\varnothing) = 0; \tag{1.11}$$

$$P(E - F) = P(E) - P(E \cap F); \tag{1.12}$$

$$P(A \cup B) = P(A) + P(B) - P(A \cap B); \tag{1.13}$$

$$A \subset B \Rightarrow P(A) \leq P(B). \tag{1.14}$$

Definition 1.2 We say that $E \subset \Omega$ is ***almost sure (a.s.)*** if and only if $P(E) = 1$. We say that $E \subset \Omega$ is ***almost impossible (a.i.)*** or ***negligible*** if and only if $P(E) = 0$. ∎

Probabilities can be defined by **probability mass functions** on finite or enumerable universes, while **probability mass densities** are used on intervals of real numbers:

Definition 1.3

1. Let Ω be finite or enumerable. A **probability mass function** $\boldsymbol{\mu}$ on Ω is an application $\mu: \Omega \longrightarrow \mathbb{R}$ such that $\forall \omega \in \Omega$: $\mu(\omega) \geq 0$ and $\sum_{\omega \in \Omega} \mu(\omega) = 1$. The probability generated by μ is given by $P(\{\omega\}) = \mu(\omega)$.
2. Let $\Omega \subset \mathbb{R}$. A **probability mass density** $\boldsymbol{\mu}$ on Ω is an application $\mu: \Omega \longrightarrow \mathbb{R}$ such that $\mu(\omega) \geq 0$ on Ω and $\int_\Omega \mu(\omega) d\omega = 1$. We define a probability on Ω by considering $P((a, b)) = \int_a^b \mu(\omega) d\omega, \forall (a, b) \subset \Omega$. ∎

The extension to other subsets of Ω is made using condition 3 in Definition 1.1, the classical properties of probabilities, namely, Eqs. (1.12 and 1.13). For instance, when $\Omega = \{\omega_1, \ldots, \omega_n\}$ and

$$\mu(\omega_i) = p_i \geq 0, \quad \sum_{i=1}^{n} p_i = 1. \tag{1.15}$$

Then,

$$P(\{\omega_{i_1}, .., \omega_{i_k}\}) = \sum_{j=1}^{k} \mu(\omega_{i_j}). \tag{1.16}$$

R proposes functions for the manipulation of mass functions and mass densities. For instance, sum and integrate. The package pracma emulates Matlab® and proposes functions quad, romberg for the evaluation of integrals.

Example 1.1 Let us consider $\Omega = \{\omega_1, \ldots, \omega_n\}$ and $\mu(\omega_i) = \beta e^{-\alpha i}$. Let us determine the possible values of α and β. We have

$$\mu(\omega_i) \geq 0 \Leftrightarrow \beta \geq 0$$

and

$$\sum_{i=1}^{n} \mu(\omega_i) = 1 \Leftrightarrow \beta \sum_{i=1}^{n} e^{-\alpha i} = 1.$$

Thus,

$$\beta = \frac{1}{\sum_{i=1}^{n} e^{-\alpha i}}$$

We can plot the possible values as follows:

```
n = 6 # number of elements of Omega
Omega <- c(seq(1,n)) # universe
alfa = seq(-5,5,length.out = 1001)  # values of alpha
betta = rep(0, length(alfa))
for (i in seq(1,length(alfa))){betta[[i]]= 1/sum(exp(-alfa
[[i]]*Omega))}

str2 = substitute(paste("possible values of ( ", alpha,",
",beta," ), for n = ",v),list(v = n))
plot(alfa, betta,type = 'n', xlab = expression(alpha),ylab
= expression(beta), main = str2)
lines(alfa,betta,col='blue',lwd=2)
```

(continued)

Example 1.1 (continued)

The results are shown in Fig. 1.1. ■

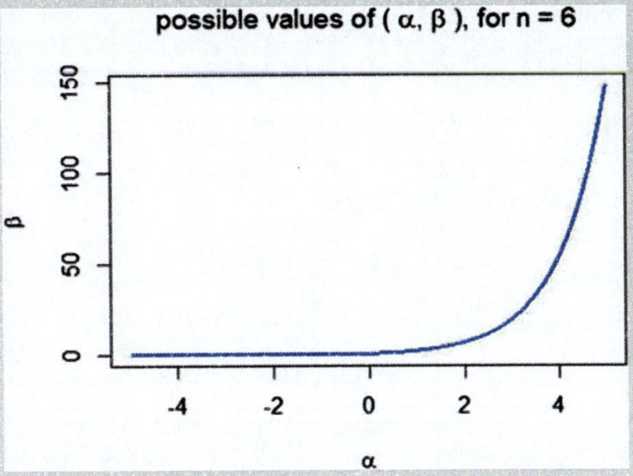

Fig. 1.1 Determination of the coefficients in the mass function

Example 1.2 Let us consider $\Omega = (0, 1) \subset \mathbb{R}$ and $\mu(\omega) = \beta e^{-\alpha\omega}$. Let us determine the possible values of α and β. We have

$$\mu(\omega) \geq 0 \Leftrightarrow \beta \geq 0$$

And, for $\alpha \neq 0$:

$$\int_0^1 \mu(\omega) d\omega = 1 \Leftrightarrow \beta \frac{1 - e^{-\alpha}}{\alpha} = 1.$$

Thus,

$$\beta = \frac{\alpha}{1 - e^{-\alpha}} \quad (\alpha \neq 0).$$

If $\alpha = 0$, then $\mu(\omega) = \beta$, so that we must have $\beta = 1$. We can plot the possible values as follows:

(continued)

Example 1.2 (continued)

```
alfa = seq(-5,5,length.out = 1001)  # values of alpha
betta = rep(1, length(alfa))
for (i in seq(1,length(alfa))){if (alfa[[i]] != 0) {betta[
[i]]= alfa[[i]]/(1 - exp(-alfa[[i]]))}}

str2 = expression(paste("possible values of ( ", alpha,",
",beta," )"))
plot(alfa, betta,type = 'n', xlab = expression(alpha),ylab
= expression(beta), main = str2)
lines(alfa,betta,col='blue',lwd=2)
```

The results are shown in Fig. 1.2. ∎

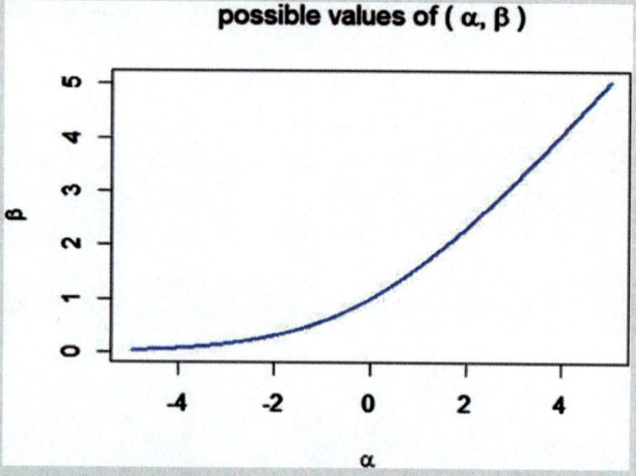

Fig. 1.2 Determination of the coefficients in the mass density

When $\Omega = \{\omega_1, \ldots, \omega_n\}$ is finite, a popular mass function is the uniform mass function: $\mu(\omega) = 1/n, \forall\, \omega \in \Omega$. In such a situation, probabilities are directly given by the number of elements of the subsets: $P(A) = |A|/n$, where $|A|$ is the number of elements of A. Its analogous to $\Omega = (a, b) \subset \mathbb{R}$ is the uniform mass density $\mu(\omega) = \frac{1}{b-a}, \forall \omega \in (a, b)$. Then, for $a \le \alpha \le \beta \le b$, $P((\alpha, \beta)) = \frac{\beta - \alpha}{b-a}$.

Example 1.3 In a 52-card deck of playing cards, you draw simultaneously three at random. Let us evaluate the probability of getting three diamonds: here, the universe $\Omega = \{(C_1, C_2, C_3)$: C_i: *card drawn at* $i\}$. Assuming a

(continued)

Example 1.3 (continued)

uniform mass function on Ω, we must evaluate, on the one hand, the number of elements of Ω, which is $n = 52 \times 51 \times 50$. On the other hand, we must evaluate the number of elements in $A = \{(C_1, C_2, C_3)$: C_i: *card drawn at i is a diamond*$\}$, which is $|A| = 13 \times 12 \times 11$, so that

$$P(3\ diamonds\ ) = \frac{13 \times 12 \times 11}{52 \times 51 \times 50.} = \frac{1716}{132600} \approx 0.01294$$

The evaluation can be made by R as follows:

```
library(arrangements)
k = npermutations(13,3)
n = npermutations(52,3)
pA = k/n
print(paste("P(A) = ",pA))
```

The result is

```
## [1] "P(A) =  0.0129411764705882"
```

Now let us evaluate the probability of getting three cards of the same color: in this case, we must evaluate the number of elements in $= \bigcup_{i=1}^{4} A_i\ A_i = \{(C_1,$ $C_2, C_3)$: C_j: *card drawn at j is* $c_i\}$, c_i is one of the four colors: clubs, diamonds, hearts, and spades. Thus, $|B| = 4 \times |A| = 4 \times 13 \times 12 \times 11$ and

$$P(3\ of\ \ same\ \ color) = \frac{4 \times 13 \times 12 \times 11}{52 \times 51 \times 50.} = \frac{6864}{132600} \approx 0.05176.$$

Using R:

```
k                        =                    4*npermutations(13,3)
n                        =                      npermutations(52,3)
pB                                   =                          k/n
print(paste("P(B) = ",pB))
```

The result is

```
## [1] "P(B) =  0.0517647058823529"
```

Example 1.4 Let $\Omega = (-1, 1)$ with the uniform mass density. Then, $\mu(\omega) = \frac{1}{2}$. Let $A_\alpha = \{\omega \in \Omega\colon \log |\omega| \leq -\alpha\}$. Since

$$A_\alpha = \{\omega \in \Omega : |\omega| \leq e^{-\alpha}\} = \begin{cases} \Omega, & if\ \alpha \leq 0 \\ (-e^{-\alpha}, e^{-\alpha}), & if\ \alpha \geq 0 \end{cases},$$

We have

$$P(\mathrm{A}_\alpha) = \begin{cases} 1, & if\ \alpha \leq 0 \\ e^{-\alpha}, & if\ \alpha \geq 0 \end{cases}.$$

We can evaluate these probabilities using R:

```
a <<- -1    # global variable
b <<- 1   # global variable
mu = function(omega){
  aux = 1/(b-a)
  return(aux)
}
alfa = seq(0,2, length.out = 201)
p_ex = exp(-alfa)
pcalc = numeric(length = length(alfa))
library(pracma)
for (i in seq(1,length(alfa))){
  om = exp(-alfa[[i]])
  pcalc[[i]] = quad(mu,-om,om)
}
str1 = expression(paste('P(A'[alpha],')'))
str2 = expression('probability of A'[alpha])
plot(alfa,pcalc,type='n',xlab = expression(alpha),ylab = s
tr1,
     main = str2)
lines(alfa,p_ex,col='blue',lwd = 2)
lines(alfa,pcalc,lwd=5,lty=2,col='red')
legend(x = "topright",legend = c("exact", "calculated"),
       lty = c(1, 2),col = c('blue', 'red'),lwd = c(2,3))
```

(continued)

Example 1.4 (continued)

The result appears in Fig. 1.3.

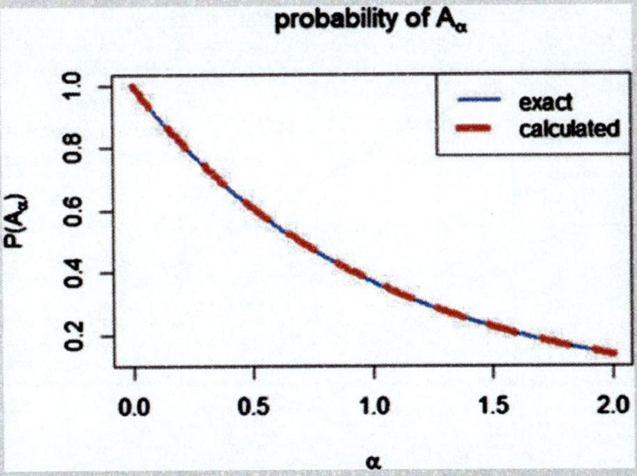

Fig. 1.3 Determination of the probabilities $P(A_\alpha)$

Now, let us consider $B_\alpha = \{\omega \in \Omega: \omega^2 - (\alpha + 1)\omega + \alpha \leq 0\}$. Since the roots of the polynomial are α and 1, we have

$$B_\alpha = \begin{cases} \Omega, \text{if } \alpha \leq -1 \\ (\alpha, 1), \text{if } -1 \leq \alpha \leq 1, \\ \varnothing, \text{if } \alpha > 1 \end{cases}$$

so that

$$P(B_\alpha) = \begin{cases} 1, \text{if } \alpha \leq -1 \\ \dfrac{1-\alpha}{2}, \text{if } -1 \leq \alpha \leq 1. \\ 0, \text{if } \alpha > 1 \end{cases}$$

Again, we can evaluate these probabilities using R:

```
alfa = seq(-1,1, length.out = 201)
p_ex = (1 - alfa)/2
pcalc = numeric(length = length(alfa))
library(pracma)
for (i in seq(1,length(alfa))){
  pcalc[[i]] = quad(mu,alfa[[i]],1)
}
```

(continued)

Example 1.4 (continued)

The result appears in Fig. 1.4.

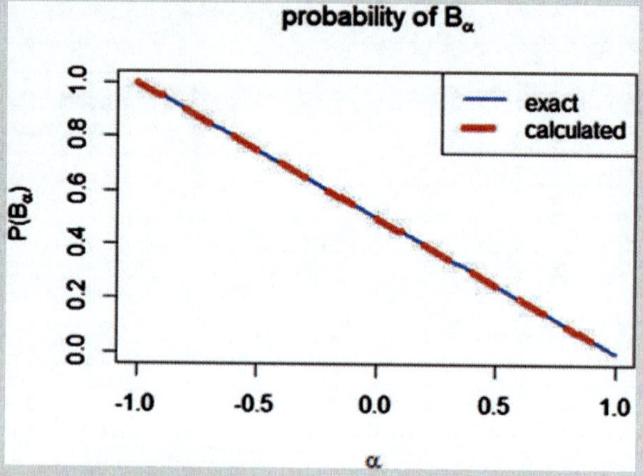

Fig. 1.4 Determination of the probabilities $P(B_\alpha)$

Exercises

1. Consider $\Omega = (0, 1)$ and the mass density $\mu(\omega) = \alpha \ln(2 + \omega) + \beta$. What are the possible values of α and β? Plot the admissible region.
2. Consider $\Omega = \{\omega_1, \omega_2, \omega_3\}$ and the mass function $\mu(\omega_i) = \alpha \ln(1 + i) + \beta$. What are the possible values of α and ? Plot the admissible region.
3. Consider $\Omega = (0, 1)$ and the mass density $\mu(\omega) = \alpha\omega^2 + \beta$. What are the possible values of α and ? Plot the admissible region.
4. Consider $\Omega = \{\omega_1, \omega_2, \omega_3, \omega_4\}$ and the mass function $\mu(\omega_i) = \alpha i + \beta$. What are the possible values of α and ? Plot the admissible region.
5. Consider $\Omega = (0, 1)$ and the mass density $\mu(\omega) = 3\omega^2$. Find the probabilities of the following events:

   (a) $A = \left\{\omega \in \Omega : \omega < \frac{1}{2}\right\}$.
   (b) $B = \left\{\omega \in \Omega : \omega^2 < \frac{1}{2}\right\}$.

(continued)

(c) $C = \{\omega \in \Omega: e^{\omega} < 2\}$.
(d) $D = \{\omega \in \Omega: 5\omega^2 - 9\omega + 1 \leq 0\}$.
(e) Write programs in R to find these probabilities.

6. Consider $\Omega = \{\omega_1, \ldots, \omega_n\}$ and $\mu(\omega_i) = \frac{2i}{n(n+1)}$.

   (a) Find $P(\{\omega_1, \omega_n\})$.
   (b) Find $P(\{\omega_2, \omega_{n-1}\})$.
   (c) Write programs in R to determine these probabilities.

7. Consider $\Omega = \{\omega_1, \omega_2, \omega_3, \omega_4\}$ and the events $A = \{\omega_1\}$, $B = \{\omega_1, \omega_2\}$, $C = \{\omega_1, \omega_2, \omega_4\}$. Let $P(A) = \frac{1}{5}$, $P(B) = \frac{3}{10}$, $P(C) = \frac{7}{10}$.

   (a) Find $P(\{\omega_2\})$, $P(\{\omega_3\})$, $P(\{\omega_4\})$.
   (b) Find $P(\{\omega_1, \omega_2, \omega_3\})$.

8. Consider $\Omega = \{\omega_1, \omega_2, \omega_3\}$ and the events $A = \{\omega_1\}$, $B = \{\omega_1, \omega_2\}$. Let $P(A) = \frac{1}{4}$ and $P(B) = \frac{1}{2}$.

   (a) Find $P(\{\omega_2\})$.
   (b) Find $P(\{\omega_3\})$.
   (c) Find $P(\{\omega_1, \omega_3\})$.

9. In a 52-card deck of playing cards, you draw sequentially four at random. At each draw, the card is put back in the deck, which is reshuffled. Find the probabilities of getting

   (a) Four spades.
   (b) Four cards of the same color.
   (c) Four aces.
   (d) At least three aces.

10. An urn contains 40 numbered balls of three colors: 12 red, 20 green, and 8 blue. We draw at random two balls simultaneously. Find the probability of getting

   (a) Two green balls.
   (b) Two balls of the same color.
   (c) Two balls of different colors.

1.3 Representing a Probability Space in R

When Ω is finite, we can represent the probability space using the class `dps` (discrete probability space) below:

```
library(R6)
#
dps=R6Class("dps",
            public=list(
              m = NULL,
              names = NULL,
              size = NULL,
              initialize=function(n,mu,nams){
                if (missing(n) == TRUE){
                  self$size = 2
                }else{self$size <<- n}
                if (missing(mu) == TRUE){
                  self$m <<- rep(1/self$size,self$size)
                }else{self$m <<- mu}
                if (missing(nams) == TRUE){
                  self$names <<- as.character(seq(1:self$siz
e))
                }else{self$names <<- nams}
              },
              prob= function(ev,ps){
                if (length(ev) == 0){p = 0}
                else{
                  ev1 = unique(ev)
                  p = 0
                  if (missing(ps) == TRUE){
                    for (i in ev1){p = p + self$m[[i]]}
                  }else{
                    for (i in ev1){p = p + ps$m[[i]]}
                  }
                  return(p)
                }
              },
              print_events_values = function(events,values,v
arnames, valname){
                nev = length(events)
                mp = matrix(nrow = nev, ncol = 2)
                for (i in 1:nev){
                  aux = events[[i]]
                  ele = length(aux)
                  nam = paste("{",varnames[[aux[[1]]]])
                  if (ele > 1){
                    for (j in 2:ele)
                      nam = paste(nam,",",varnames[[aux[[j]]
]])
```

```
          }
          nam = paste(nam,"}")
          mp[ i,1] = nam
          mp[ , 2] = values[[i]]
        }
        colnames(mp) = c("event",valname)
        mpd = as.data.frame(mp)
        print(mpd)
      },
      show = function(ps){
        if (missing(ps) == TRUE){
          events= seq(1:self$size)
          values = self$m
          varnames = self$names
        }else{
          events= seq(1:ps$size)
          values = ps$m
          varnames = ps$names
        }
        valname = "prob"
        self$print_events_values(events,values,varna
mes, valname)
      }
    )
)
```

These contents must be saved to a file `dps.R`, to be sourced before use. For this class, $\Omega = \{1, 2, ..., n\}$, but you can give names to the elements. If the mass function mu is not given, the mass is uniformly distributed on the elements of Ω. If the names are not given, they are set to "1," "2," ... R proposes intrinsic functions `intersect`, `union`, `setdiff` which can be used to determine the probabilities associated with intersections, unions, etc.

Example 1.1 Let us consider the result when a coin is tossed once, which can be Head (H) or Tail (T). Then, $\Omega = \{H, T\}$. Let μ be a mass function on Ω and $p = \mu(H)$, $q = \mu(T)$. Then: $p \geq 0, q \geq 0, p + q = 1$. The coin is fair when $p = q$, i.e., when $p = q = 1/2$.

We can create a probability space for a fair coin as follows:

```
source("dps.R")
ps = dps$new(2)
ps$names = c("H","T")
```

(continued)

Example 1.1 (continued)

We can see the probabilities of the elements of Ω as follows:

```
ps$show()
##    event prob
## 1 { H }  0.5
## 2 { T }  0.5
T = 2
ps$prob(T)
## [1] 0.5
```

```
a = c(1,2)
ps$prob(a)
## [1] 1
ps$prob(intersect(a,T))
## [1] 0.5
ps$prob(union(a,T))
## [1] 1 ■
```

Example 1.2 Let us consider the result when a dice having n faces is rolled in a single die. Here, $\Omega = \{1, 2, \ldots, n\}$. Let μ be a mass function on Ω and $p_i = \mu(i)$, $1 \leq i \leq n$. Then: $p_i \geq 0$, $1 \leq i \leq n$; $p_1 + \ldots + p_n = 1$. The dice is balanced when all the sides have the same probability – all the p_i are equal, i.e., when $p_i = {}^1\!/_n$, $1 \leq i \leq n$. An example of determination of probabilities using R is:

```
source("dps.R")
n = 16 # number of faces
ps = dps$new(n)
```

```
A = c(1,6, 12)
pA = ps$prob(A)
B = c(1, 3, 5, 7)
pB = ps$prob(B)
```

The results are

```
print(paste("P(A) = ",pA," ; P(B) = ", pB))
## [1] "P(A) =  0.1875  ; P(B) =  0.25" ■
```

Alternative representations of probability spaces are the binary and hexadecimal representations.

In the binary representation, the events $E \subset \Omega$ correspond to binary vectors of dimension n having all the components equal to zero, except those corresponding to the elements of E. For instance, let $n = 5$: the event $\{\omega_1, \omega_3\}$ is represented by the binary vector $(1, 0, 1, 0, 0)$, the event $\{\omega_2, \omega_3, \omega_5\}$ is represented by the binary vector $(0, 1, 1, 0, 1)$, and so on. The universe Ω is represented by a vector entirely formed by ones and the empty set ∅ by a vector of zeros. Under the binary representation, the intersection of two events corresponds to the product term by term of the binary vectors or the logical operation "AND." The union of two subsets can be achieved by considering the term-by-term maximum of the vectors or the logical operator "OR."

The difference $A - B$ can be evaluated by subtracting $A \cap B$ from A – the complementary set $\Omega - A$ can be evaluated by subtracting A from the universe vector formed by ones. These operations can be implemented in a class `bds` (*binary discrete set*), defined as follows:

```
library(R6)
#
bds=R6Class("bds",
 public=list(
 initialize=function(){
 },
 binter= function(ev1,ev2){
  result = pmin(ev1,ev2)
  return(result)
 },
 bunion= function(ev1,ev2){
  result = pmax(ev1,ev2)
  return(result)
 },
 bdiff= function(ev1,ev2){
  result = ev1 - pmin(ev1, ev2)
  return(result)
 },
 bcomplement = function(ev1){
  result = rep(1,length(ev1)) - ev1
  return(result)
 },
 bflip = function(ev){
  ind = (length(ev) + 1) - seq(1:length(ev))
  aux = ev[ind]
  return(aux)
 },
 isequal = function(ev1,ev2){
   # TRUE if ev1 is equal to ev2
   aux = sum(abs(ev1-ev2))
   if (aux > 0){r = FALSE}
   else{r = TRUE}
   return(r)
 },
 iscontained = function(ev1,ev2){
   # TRUE if ev1 is part of ev2
   ev = self$binter(ev1,ev2)
   r = self$isequal(ev,ev1)
   return(r)
 },
 contains = function(ev1,ev2){
   # TRUE if ev1 contains ev2
   ev = self$bunion(ev1,ev2)
   r = self$isequal(ev,ev1)
   return(r)
 }
 )
)
```

Example 1.3 Let us consider $n = 5$ and the events

$$A = \{\omega_1, \omega_3\}, B = \{\omega_2, \omega_3, \omega_5\}, C = \{\omega_1, \omega_3, \omega_4\},$$

Their binary representation is

```
A = c(1,0,1,0,0)
B = c(0,1,1,0,0)
```

```
C = c(1,0,1,1,0)
```

Let us evaluate $A \cap B$, $A \cup B$, $A - B$, $B - A$, $\Omega - A$, $\Omega - B$, and verify the inclusions in C.

```
source("bds.R")

bs = bds$new()
A
## [1] 1 0 1 0 0

bs$bflip(A)
## [1] 0 0 1 0 1

B
## [1] 0 1 1 0 0

bs$bflip(B)
## [1] 0 0 1 1 0

bs$binter(A,B)
## [1] 0 0 1 0 0

bs$bunion(A,B)
## [1] 1 1 1 0 0

bs$bdiff(A,B)
```

```
## [1] 1 0 0 0 0

bs$bdiff(B,A)
## [1] 0 1 0 0 0

bs$bcomplement(A)
## [1] 0 1 0 1 1

bs$bcomplement(B)
## [1] 1 0 0 1 1

bs$contains(C,A)
## [1] TRUE

bs$contains(C,B)
## [1] FALSE

bs$iscontained(A,C)
## [1] TRUE

bs$iscontained(B,C)
## [1] FALSE
```

■

In the binary representation, each $\omega_i \in \Omega$ is represented by a one-hot vector, i.e., a binary vector having a single element equal to one. Thus, we can bring the mass function to an Identity Matrix: each line represents one element of Ω. The probability space corresponding to the binary representation is implemented in a class bps (binary probability space), defined as follows – notice that class bps inherits from class bds:

```
library(R6)
source("bds.R")
#
bps=R6Class("bps",
  inherit=bds,
  public=list(
   m = NULL, names = NULL,
   size = NULL,frame = NULL,
   initialize=function(n,mu,nams){
     super$initialize()
     if (missing(n) == TRUE){
       self$size = 2
     }else{self$size <- n}
     if (missing(mu) == TRUE){
       self$m <- rep(1/self$size,self$size)
     }else{self$m <- mu}
    if (missing(nams) == TRUE){
       self$names <- as.character(seq(1:self$size))
     }else{self$names <- nams}
        self$frame <- diag(self$size)
   },
  bprob= function(ev,ps){
   # evaluates the probability of ev
   # in the binary probability space ps
   if (missing(ps) == TRUE){ps = self}
   ind = which(ev > 0)
   if (length(ind) == 0){p = 0}
   else{
     p = 0
     for (i in ind){p = p + ps$m[[i]]}
   }
   return(p)
  },
  bevent = function(nbs,ps){
   # creates an event in the binary probability
   # space ps using the positions given in nbs
   if (missing(ps) == TRUE){ps = self}
   res = rep(0,ps$size)
   if (length(nbs) > 0){res[nbs] = 1}
   return(res)
  },
```

```
  print_events_values = function(events,values,varnames, val
name){
    n_om = dim(events)[[1]]
    nvar = dim(events)[[2]]
    mp = matrix(nrow = n_om, ncol = nvar + 1)
    mp[ ,1:nvar] = events
    mp[ , nvar+1] = values
    colnames(mp) = c(varnames,valname)
    mpd = as.data.frame(mp)
    print(mpd)
  },
  show = function(ps){
    if (missing(ps) == TRUE){ps = self}
    events= ps$frame
    values = ps$m
    varnames = ps$names
    valname = "prob"
    self$print_events_values(events,values,varnames,valname)
  }
 )
)
```

Example 1.4 Let us consider the fair coin presented in Example 1.1. We create the corresponding binary probability space as follows:

```
source("bps.R")
ps = bps$new(2)
ps$names = c("H","T")
```

We can see the probabilities of the elements of Ω as follows:

```
ps$show()
##   H T prob
## 1 1 0  0.5
## 2 0 1  0.5
H = ps$bevent(1) # create an event : Head
print(H)
## [1] 1 0
T = c(0,1) #Tail
ps$bprob(T)
## [1] 0.5
a = ps$bunion(H,T) # Head or Tail
ps$bprob(a)
## [1] 1
ps$bprob(ps$binter(a,T)) #Tail
## [1] 0.5
ps$bflip(T)
## [1] 1 0  ■
```

Example 1.5 Let us consider the situation described in Example 1.2 (rolling a dice with n faces). We have:

```
source("bps.R")
n = 16 # number of faces
ps = bps$new(n)

A = ps$bevent(c(1,6, 12))
A
## [1] 1 0 0 0 0 1 0 0 0 0 0 1 0 0 0 0
ps$bflip(A)
## [1] 0 0 0 0 1 0 0 0 0 0 1 0 0 0 0 1
pA = ps$bprob(A)

B = ps$bevent(c(1, 3, 5, 7))
B
## [1] 1 0 1 0 1 0 1 0 0 0 0 0 0 0 0 0
ps$bflip(B)
## [1] 0 0 0 0 0 0 0 0 0 1 0 1 0 1 0 1
pB = ps$bprob(B)

print(paste("P(A) = ",pA," ; P(B) = ", pB))
## [1] "P(A) =  0.1875  ; P(B) =  0.25"
```
■

The hexadecimal representation compacts the binary representation into hexadecimal numbers. The set operations are defined by a class hds:

```
library(R6)
source("bds.R")
#
hds=R6Class("hds",
  inherit = bds,
  public=list(
  size = NULL,
  initialize=function(n){
   super$initialize()
   self$size = n
  },
  b2h = function(x){
   library(BMS)
   result = bin2hex(super$bflip(x))
   return(result)
  },
  h2b = function(x){
   library(BMS)
   aux = hex2bin(x)
   n1 = self$size-length(aux)
   if (n1 > 0){
     result= cbind(rep(0,n1),0)
```

```
    }else{
      if (n1 < 0){
        result = aux[(1-n1):length(aux)]
      }
      else{result = aux}
    }
    return(super$bflip(result))
  },
  hinter = function(x,y){
    aux = super$binter(self$h2b(x), self$h2b(y))
    aux2 = self$b2h(aux)
    return(aux2)
  },
  hunion = function(x,y){
    aux = super$bunion(self$h2b(x), self$h2b(y))
    aux2 = self$b2h(aux)
    return(aux2)
  },
  hdiff = function(x,y){
    aux = super$bdiff(self$h2b(x), self$h2b(y))
    aux2 = self$b2h(aux)
    return(aux2)
  },
 hc = function(x){
    aux = super$bcomplement(self$h2b(x))
    aux2 = self$b2h(aux)
    return(aux2)
 },
 isequal = function(ev1,ev2){
    # TRUE if ev1 is equal to ev2
    r = super$isequal(self$h2b(x),self$h2b(y))
    return(r)
 },
 iscontained = function(ev1,ev2){
    # TRUE if ev1 is part of ev2
    r = super$iscontained(self$h2b(x),self$h2b(y))
    return(r)
 },
 contains = function(ev1,ev2){
    # TRUE if ev1 contains ev2
    r = super$contains(self$h2b(x),self$h2b(y))
    return(r)
 }
)
)
```

Example 1.6 Let us consider $n = 5$ and the events of Example 1.1

$$A = \{\omega_1, \omega_3\}, B = \{\omega_2, \omega_3, \omega_5\}, C = \{\omega_1, \omega_2, \omega_3\}.$$

Their hexadecimal representation is

```
A = "05"
B = "16"
C = "07"
```

Again, let us evaluate $A \cap B, A \cup B, A - B, B - A, \Omega - A, \Omega - B$ and verify the inclusions of A, B in C:

```
source("hds.R")
hs = hds$new(5)
hs$h2b(A)
## [1] 1 0 1 0 0
hs$h2b(B)
## [1] 0 1 1 0 1
hs$h2b(C)
## [1] 1 1 1 0 0
E1 = hs$hinter(A,B)
hs$h2b(E1)
## [1] 0 0 1 0 0
E2 = hs$hunion(A,B)
E2
## [1] "17"
hs$h2b(E2)
## [1] 1 1 1 0 1
E3 = hs$hdiff(A,B)
E3
## [1] "01"
hs$h2b(E3)
## [1] 1 0 0 0 0
E4 = hs$hdiff(B,A)
E4
## [1] "12"
hs$h2b(E4)
## [1] 0 1 0 0 1
E5=hs$hc(A)
E5
## [1] "1a"
hs$h2b(E5)
## [1] 0 1 0 1 1
E6=hs$hc(B)
E6
## [1] "09"
hs$h2b(E6)
## [1] 1 0 0 1 0
E7=hs$hc(C)
E7
## [1] "18"
hs$h2b(E7)
## [1] 0 0 0 1 1
hs$iscontained(A,C)
## [1] TRUE
hs$contains(C,A)
## [1] TRUE
hs$iscontained(B,C)
## [1] FALSE
hs$contains(C,B)
## [1] FALSE
```
■

The probability space is created by the class hps:

```
library(R6)
source("hds.R")
#
hps=R6Class("hps",
 inherit = hds,
  public=list(
   m = NULL,
   names = NULL,
   size = NULL,
   initialize=function(n,mu,nams){
   if (missing(n) == TRUE){
     self$size = 2
   }else{self$size <- n}
   if (missing(mu) == TRUE){
     self$m <- rep(1/self$size,self$size)
   }else{self$m <- mu}
   if (missing(nams) == TRUE){
     self$names <- as.character(seq(1:self$size))
   }else{self$names <- nams}
     super$initialize(self$size)
   },
   hprob= function(hev,ps){
     x = super$h2b(hev)
     ind = which(x > 0)
     p = 0
     if (length(ind) > 0){
       if (missing(ps) == TRUE){p = sum(self$m[ind])}
       else{p = sum(psf$m[ind])}}
      return(p)
   },
   prob= function(ev,ps){
     p = 0
     if (length(ev) > 0) {
       if (missing(ps) == TRUE){p = sum(self$m[ev])}
       else{p = sum(ps$m[ev])}
     }
```

```
      return(p)
    },
    hevent = function(nbs){
      resa = rep(0,self$size)
      if (length(nbs) > 0){
        resa[nbs] = 1
      }
      res = super$b2h(resa)
      return(res)
    },
    print_events_values = function(events,values,varnames, va
lname){
      nev = length(events)
      mp = matrix(nrow = nev, ncol = 2)
      for (i in 1:nev){
        aux = events[[i]]
        ele = length(aux)
        nam = paste("{",varnames[[aux[[1]]]])
        if (ele > 1){
          for (j in 2:ele)
            nam = paste(nam,",",varnames[[aux[[j]]]])
        }
        nam = paste(nam,"}")
        mp[ i,1] = nam
        mp[ , 2] = values[[i]]
      }
      colnames(mp) = c("event",valname)
      mpd = as.data.frame(mp)
      print(mpd)
    },
    show = function(ps){
      if (missing(ps) == TRUE){ps = self}
      events= seq(1:ps$size)
      values = ps$m
      varnames = ps$names
      valname = "prob"
      self$print_events_values(events,values,varnames, valnam
e)
    }
  )
)
```

Example 1.7 Let us consider the fair coin presented in Example 1.1. We create the corresponding hexadecimal probability space as follows:

```
source("hps.R")
ps = hps$new(2)
ps$names = c("H","T")
```

We can see the probabilities of the elements of Ω as follows:

```
ps$show()
##    event prob
## 1 { H }  0.5
## 2 { T }  0.5
H = ps$hevent(1) # create an event : Head
print(paste("T = (1,0) in hex = ",H,", prob = ",ps$hprob(H
)))
## [1] "T = (1,0) in hex =  1 , prob =  0.5"
T = ps$hevent(2) #Tail
print(paste("H = (0,1) in hex = ",T,", prob = ",ps$hprob(T
)))
## [1] "H = (0,1) in hex =  2 , prob =  0.5"
a = ps$hunion(H,T) # Head or Tail
print(paste("(H,T) = (1,1) in hex = ",a,", prob = ",ps$hpr
ob(a)))
## [1] "(H,T) = (1,1) in hex =  3 , prob =  1"
b = ps$hinter(a,T)
print(paste("T = (0,1) in hex = ",b,", prob = ",ps$hprob(b
)))
## [1] "T = (0,1) in hex =  2 , prob =  0.5"
```
■

Example 1.8 Let us consider the situation described in Example 1.2 (rolling a dice with n faces). We have:

```
source("hps.R")
n = 16 # number of faces
ps = hps$new(n)

A = ps$hevent(c(1,6, 12))
A
## [1] "0821"
ps$h2b(A)
##  [1] 1 0 0 0 0 1 0 0 0 0 0 1 0 0 0 0

B = ps$hevent(c(1, 3, 5, 7))
B
## [1] "0055"
ps$h2b(B)
##  [1] 1 0 1 0 1 0 1 0 0 0 0 0 0 0 0 0

pA = ps$hprob(A)
pB = ps$hprob(B)
print(paste("P(A) = ",pA," ; P(B) = ", pB))
## [1] "P(A) =  0.1875  ; P(B) =  0.25"
```
■

To deal with the situation where Ω is an interval of real numbers, we need to evaluate integrals. A class that evaluates integrals is given below:

```
library(R6)
#
ints=R6Class("ints",
 public = list(
  initialize = function(){},
  int1 = function(f,a,b){
    library(pracma)
    aux = quad(Vectorize(f),a,b)
    return(aux)
  },
  int2 = function(f,a,b){
    library(pracma)
    aux = quad(Vectorize(function(x1) {
    quad(Vectorize(function(x2) {
    f(x1,x2) }), a[[2]],b[[2]]) }), a[[1]],b[[1]])
    return(aux)
  },
  int3 = function(f,a,b) {
    library(pracma)
    aux = quad(Vectorize(function(x1) {
    quad(Vectorize(function(x2) {
    quad(function(x3) { f(x1,x2,x3) }, a[[3]], b[[3]]) }), a
[[2]],b[[2]]) }), a[[1]],b[[1]])
    return(aux)
  }
 )
)
```

The probability space is defined by the class `cps` (continuous probability space):

```
library(R6)
source("ints.R")
#
cps=R6Class("cps",
 inherit = ints,
 public=list(
  m = NULL, name = NULL, a = NULL,
  b = NULL, dim = NULL, integ=NULL,
  initialize=function(a,b,mu,nam){
   super$initialize()
   self$dim = length(a)
   self$m = mu
   self$a = a
   self$b = b
   if (missing(nam) == TRUE){
     self$name = "X"}else{self$name = nam}
  },
  meas = function(ev){
   s = 1
   a = ev[[1]]
   b = ev[[2]]
   for (i in 1: length(a)){
     aa = a[[i]]
     bb = b[[i]]
     if (aa < bb){aux = bb-aa}else{aux=0}
     s = s*aux}
   return(s)
  },
  prob= function(ev,ps){
   integ = ints$new()
   dim = length(ev[[1]])
   if (self$meas(ev) == 0){p = 0}
   else{
     if (missing(ps) == TRUE){
       p = eval(parse(text=paste0('integ$int',dim,'(self$m,e
v[[1]],ev[[2]])')))
     }else{
       p = eval(parse(text=paste0('integ$int',dim,'(ps$m,ev[
[1]],ev[[2]])')))
     }
   }
   return(p)
   }
 )
)
```

Example 1.9 Let us consider $\Omega = (a, b) \subset \mathbb{R}$ and $\mu(\omega) = 1/(b - a)$. The probability of a subinterval $I = (\alpha, \beta) \subset \Omega$ $(\alpha \leq \beta)$ is $P(I) = (\beta - \alpha)/(b - a)$. We can evaluate this value using R as follows:

```
source("cps.R")
a = -1
b = 1
mu <- function(omega){1/(b-a)}
ps = cps$new(a,b,mu)
ev = c(0,0.5)
ps$prob(ev)
## [1] 0.25 ■
```

Example 1.10 Let us consider $\Omega = (-1, 1)^2 \subset \mathbb{R}^2$ and $\mu(\omega) = 1/4$. The probability of the region $I = \left\{0 \leq \ x_1 \leq \frac{1}{2}, \ -1 \leq x_2 \leq x_1\right\} \subset \Omega$ is $P(I) = 5/32$. We can evaluate this value using R as follows: (`int2var` is a modification of `int2` from class `ints` to take into account limits on `x2` that depend on `x1`).

```
source("cps.R")
a = c(-1,-1)
b = c(1,1)

mu <- function(om1,om2){1/prod(b-a)}

ps = cps$new(a,b,mu)

aa = list(NA,NA)
bb = list(NA,NA)
aa[[1]] = 0
aa[[2]] = a[[2]]
bb[[1]] = 0.5
bb[[2]] = function(x1){x1}
ps$int2var(mu,aa,bb)
## [1] 0.15625 ■
```

Exercises

1. Consider $\Omega = (0, 1)$ and the mass density $\mu(\omega) = 3\omega^2$. Represent the probability space with R and use it to determine the probabilities of the following events:

   (a) $A = \left\{\omega \in \Omega : \omega < \frac{1}{2}\right\}$.
   (b) $B = \left\{\omega \in \Omega : \omega^2 < \frac{1}{2}\right\}$.
   (c) $C = \{\omega \in \Omega\colon e^{\omega} < 2\}$.

2. Consider $\Omega = \{\omega_1, \ldots, \omega_n\}$ and $\mu(\omega_i) = \frac{2i}{n(n+1)}$. Represent the probability space for $n = 20$ in binary form and use it to determine the probabilities of

   (a) $A = \{\omega_1, \omega_{20}\}$.
   (b) $B = \{\omega_2, \omega_{19}\}$.
   (c) $C = \{\omega_1, \omega_2, \omega_5, \omega_{18}, \omega_{19}\}$.

3. Consider $\Omega = \{\omega_1, \ldots, \omega_n\}$ and $\mu(\omega_i) = \frac{2i}{n(n+1)}$. Represent the probability space for $n = 20$ in hexadecimal form and use it to determine the probabilities of

   (a) $A = \{\omega_1, \omega_{20}\}$.
   (b) $B = \{\omega_2, \omega_{19}\}$.
   (c) $C = \{\omega_1, \omega_2, \omega_5, \omega_{18}, \omega_{19}\}$.

4. In a 52-card deck of playing cards, you draw sequentially four at random. At each draw, the card is put back in the deck, which is reshuffled. Represent the corresponding probability space in R and find the probabilities of getting

   (a) Four spades.
   (b) Four cards of the same color.
   (c) Four aces.
   (d) At least three aces.

5. An urn contains 40 numbered balls of three colors: 12 red, 20 green, and 8 blue. We draw at random two balls simultaneously. Represent the corresponding probability space in R and find the probability of getting

   (a) Two green balls.
   (b) Two balls of the same color.
   (c) Two balls of different colors.

1.4 Independent Events and Bayes' Formula

Definition 1.4 Let (Ω, P) be a probability space. Let $E, F \subset \Omega$ be two events.

1. ***E* and *F* are independent** if and only if $P(E \cap F) = P(E)P(F)$.
2. Let $P(F) > 0$. The **probability of E conditional to F** is

$$P(E|F) = \frac{P(E \cap F)}{P(F)}. \quad \blacksquare \tag{1.17}$$

Thus, for $P(F) > 0$, E and F are independent if and only if $P(E|F) = P(E)$. If $P(E \cap F) > 0$, we have

$$P(E|F) = \frac{P(E \cap F)}{P(F)} \text{ and } P(F|E) = \frac{P(E \cap F)}{P(E)},$$

so that

$$P(E|F)P(F) = P(E \cap F) = P(F|E)P(E)$$

and we have the classical Bayes' formula:

$$P(F|E) = \frac{P(E|F)}{P(E)}P(F) \tag{1.18}$$

Let us consider a partition $\Omega = A_1 \cup A_2 \cup \ldots \cup A_n$, where $A_i \cap A_j = \emptyset$, for $i \neq j$, and $P(A_i) > 0$, for $1 \leq i \leq n$. Then

$$E = E \cap \Omega = (E \cap A_1) \cup (E \cap A_2) \cup \ldots \cup (E \cap A_n),$$

Since $(E \cap A_i) \cap (E \cap A_j) = \emptyset$, for $i \neq j$, we have

$$P(E) = P(E \cap A_1) + P(E \cap A_2) + \ldots + P(E \cap A_n),$$

so that

$$P(E) = P(E|A_1)P(A_1) + P(E|A_2)P(A_2) + \ldots + P(E|A_n)P(A_n),$$

and

$$P(F|E) = \frac{P(E|F)}{\sum_{i=1}^{n} P(E|A_i)P(A_i)} P(F). \quad (1.19)$$

Taking $F = A_k$, we obtain

$$P(A_k|E) = \frac{P(E|A_k)}{\sum_{i=1}^{n} P(E|A_i)P(A_i)} P(A_k). \quad (1.20)$$

Equation (1.20) is known as ***generalized Bayes' formula***.

Example 1.11 A fair coin has head and tail results equiprobable. It is tossed three times and the result is written. We have the information that one of the results is head. What is the probability that the two others are tails? Let us denote H a head and T a tail. Here, $\Omega = \{HHH, HHT, HTH, THH, TTH, THT, HTT, TTT\}$. Let us consider the events $E = \{HHH, HHT, HTH, THH, TTH, THT, HTT\}$ (one of them is a head) and the event $F = \{TTH, THT, HTT\}$ (two of them are tails). We have $E \cap F = \{TTH, THT, HTT\}$ and

$$P(F|E) = \frac{\mathrm{P(E \cap F)}}{P(E)} = \frac{\frac{3}{8}}{\frac{7}{8}} = \frac{3}{7}.$$

Example 1.12 A box contains 40% of red pins. Globally, 25% of the pins are damaged, but there is a difference between the colors: 40% of the red pins are damaged. What is the probability that a damaged pin is red? Let $E_1 =$ *the pin is red*; $E_2 =$ *the pin is not red*. We have $P(E_1) = 0.4$, $P(E_2) = 0.6$. Let $F =$ *the pin is damaged*. We have $\mathrm{P}(F| E_1) = 0.4$, $P(F) = 0.25$. Then

$$P(E_1|F) = \frac{P(E_1) \times P(F|E_1)}{P(F)} = \frac{0.4 \times 0.4}{0.25} = 0.64.$$

Example 1.13 Let us consider the events R: *weather is rainy* and $W=$ *weather is windy*. In a given location, the number of raining days is 10 %, and the number of windy days is 25 %. However, we observed that 80 % of the raining days are also windy. Then

$$P(R)=0.1, P(W)=0.25, P(W|R)=0.8.$$

We have $P(W \cap R) = P(R)P(W|R) = 0.08$. Analogously, let R^c be the complementary event of R: R^c: *weather is not rainy*. Then, $P(R^c) = 0.9$ and $P(W \cap R^c) = P(W) - P(W \cap R) = 0.25 - 0.08 = 0.17$. Finally, we have $P(W^c \cap R) = P(R) - P(W \cap R) = 0.1 - 0.08 = 0.02$, and $P(W^c \cap R^c) = P(R^c) - P(W \cap R^c) = 0.9 - 0.17 = 0.73$. Notice that $0.17 + 0.73 + 0.08 + 0.02 = 1$.

In addition, Bayes' formula furnishes

$$P(R|W)=\frac{P(R)\times P(W|R)}{P(W)}=\frac{0.1\times 0.8}{0.25}=0.32.$$

Thus, a windy day increases the probability of raining to 32 % – three times more than the standard. Similarly,

$$P(R^c|W)=\frac{P(W\cap R^c)}{P(W)}=\frac{P(W)-P(W\cap R)}{P(W)}=\frac{0.25-0.08}{0.25}=0.68.$$

Thus, a windy day decreases the probability of not raining from 0.9 to 0.68 – about a 26 % decrease. Notice that $P(R|W) + P(R^c|W) = 1$.

Exercises

1. Overall, it is estimated that 1% of the parts manufactured in a factory are defective. The quality control makes mistakes by accepting 1% of noncompliant parts and rejecting 2% of compliant parts. What is the probability that an accepted part is compliant?
2. Two boxes contain the same number of pins, divided between two colors: red and green. In box 1, 25 % are red and 75 % are green; in box 2, 50 % are red pins and 50 % are green. The boxes are equiprobable.

   (a) We take a red pin: what is the probability that it came from box 1?
   (b) We take a green pin: what is the probability that it came from box 2?
   (c) How are these values modified if box 1 is much larger than box 2, so that its probability is $\frac{3}{4}$?

(continued)

3. An urn contains 25 numbered balls of three colors: 1 to 10 in red, 11 to 15 in green, and 16 to 20 in blue. We draw a ball at random.
   (a) The number is odd: what is the probability that the color is green?
   (b) The number has two digits: what is the probability that the color is blue?
   (c) The number is prime: what is the probability that the color is red?
4. Three urns contain ten numbered balls in red and blue each, but the number of red balls is different in each one: six in urn 1, three in urn 2, and four in urn 3. The urns are equiprobable. A red ball is drawn.
   (a) What is the probability that it came from urn 2?
   (b) What is the probability that it did not come from urn 1?
   (c) What is the probability that it came from urn 3?
5. Determine if the events A and B are independent or not, when
   (a) $P(A) = \frac{1}{3}$, $P(B) = \frac{2}{3}$, $P(A \cap B) = \frac{1}{8}$.
   (b) $P(A) = \frac{1}{2}$, $P(B) = \frac{2}{3}$, $P(A \cap B) = \frac{1}{3}$.
   (c) $P(A) = \frac{3}{5}$, $P(B) = \frac{5}{6}$, $P(A \cap B) = \frac{1}{2}$.
   (d) $P(A) = \frac{3}{5}$, $P(B) = \frac{5}{6}$, $P(A \cap B) = \frac{1}{5}$.

1.5 Representation of Product Spaces

It is usual to manipulate events corresponding to vectors of events. For instance, if we toss a coin twice, the result is a vector (R_1, R_2) where R_i is the result of the toss number i. Such a situation can be studied by considering the cartesian product of two spaces: $\Omega = \Omega_1 \times \Omega_2$, where $\Omega_i = \{H, T\}$ is the result for the toss number i. The cartesian product of two probability spaces can be represented either by considering a traditional cartesian product or a binary representation where the binary vectors sequentially include the elements of each space – these binary numbers can be eventually compacted to hexadecimal representation. For instance, we can represent $\{H, T\} \times \{H, T\}$ by its cartesian product $\{(H, H), (H, T), (T, H), (T, T)\}$ or by the binary numbers $\{00,01,10,11\}$. If the spaces are independent in the sense that their events are independent, the generation of a product space can be easily made by taking the product of the mass functions or densities of each space. For instance, if the tosses are considered independent and the coin is fair, we have $\mu_{\Omega_1 \times \Omega_2} = \mu_{\Omega_1} \times \mu_{\Omega_2}$. If the two spaces are not independent, the mass function of $\Omega = \Omega_1 \times \Omega_2$ must be given – in such a situation, Ω is considered as a single probability space with its own mass function.

The generation of a representation of independent probability spaces can be made by using the class pps (product of probability spaces) below:

```
library(R6)
#
pps=R6Class("pps",
 public=list(
 spaces = NULL, nspaces = NULL,
 names = NULL,
 initialize=function(spaces){
  if (missing(spaces) == TRUE){
   self$spaces <- NA
   self$nspaces = 0
   self$names = NA
  }else{
   self$spaces = spaces
   self$nspaces = length(spaces)
   self$names = vector('list',length = self$nspaces)
   for (i in 1:self$nspaces){
     self$names[[i]] = self$spaces[[i]]$names
   }
  }
 },
 probsingle = function(ev){
  if (class(self)[[1]] == "pbpsb"){
    p = self$probb(ev)
    }else{p=self$probn(ev)}
  return(p)
 },
 probb =  function(evsingle){
  if (length(evsingle) == 0){p = 0}
  else{
   lev = self$ungroup(evsingle)
   p = 1
   for (i in 1:length(lev)){
     p = p*self$spaces[[i]]$bprob(lev[[i]])
   }
  }
  return(p)
 },
 probn = function(evsingle){
  if (length(evsingle) == 0){p = 0}
  else{
   p = 1
   for (i in 1:length(evsingle)){
    p = p*self$spaces[[i]]$prob(evsingle[[i]])
   }
  }
  return(p)
 },
```

```
  prob= function(ev){
   if (length(ev) == 0){p = 0}
   else{
     if (is.list(ev)){
       p = numeric(length = length(ev))
       for (i in 1:length(ev)){
         p[[i]] = self$probsingle(ev[[i]])}
       }else{
         if (is.matrix(ev)){

          p = numeric(length = dim(ev)[[1]])
          for (i in 1:dim(ev)[[1]]){
            p[[i]] = self$probsingle(ev[i, ])}
          }else{
            p = self$probsingle(ev)}
       }
   }
   return(p)
 },
 varnamsingle = function(evsingle){
  if (length(evsingle) == 0){aux = ""}
  else{
    aux = ""
    for (i in 1:length(evsingle)){
      aux = paste0(aux,self$names[[i]][[ evsingle[[i]] ]])
    }
  }
  return(aux)
 },
 varnams = function(ev){
  if (is.list(ev)){
    varnam = vector("list",length = length(ev))
    for (i in 1:length(ev)){
      varnam[[i]] = self$varnamsingle(ev[[i]])}
  }else{
     if (is.matrix(ev)){
       varnam = vector("list",nrow = dim(ev)[[1]])
       for (i in 1:dim(ev)[[1]]){
         varnam[[i]] = self$varnamsingle(ev[i,])}
     }else{
       varnam = self$varnamsingle(ev)
     }
  }
  return(varnam)
 }
 )
)
```

This class is used, for instance, to generate products of discrete probability spaces: the class pdps (product of discrete probability spaces) inherits from pps:

```
library(R6)
source("pps.R")
#
pdps=R6Class("pdps",
 inherit = pps,
 public=list(
  ps = NULL, size = NULL, names = NULL,
  m = NULL, frame = NULL,
  initialize=function(spaces){
    super$initialize(spaces)
    self$ps = pps$new(spaces)
    if (missing(spaces) == TRUE){
      self$size = 0
    }else{
      s = 1
      for (i in 1:self$ps$nspaces){
         s = s*spaces[[i]]$size
      }
      self$size = s
    }
    self$frame = self$complete_frame()
    self$m = NA
 },
 complete_frame = function(){
   if (self$ps$nspaces < 1){aux = NA}
   else{
     if (self$ps$nspaces < 2){
       aux = seq(1:self$ps$spaces[[1]]$size)}
     else{
       x = vector('list', length = self$ps$nspaces)
       x[[1]] = seq(1:self$ps$spaces[[1]]$size)
       ch = "expand.grid(x[[1]]"
       for (i in 2:self$ps$nspaces){
         x[[i]] = seq(1:self$ps$spaces[[1]]$size)
         aux2 = paste0(",x[[",i,"]]")
         ch = paste0(ch,aux2)
       }
       ch =paste0(ch,")")
       aux = eval(parse(text=ch))
     }
   }
   return(as.matrix(aux))
 },
 findm = function(){
   g = self$complete_frame()
```

```
    m = numeric(length = dim(g)[[1]])
    for (i in 1: dim(g)[[1]]){
      m[[i]]= self$ps$probsingle(g[i,])}
    self$m = m
    return(m)
  },
  print_events_values = function(events,values,varnames, valn
ame){
    if (is.list(events)){ nev = length(events)}
    else{if(is.matrix(events)){nev = dim(events)[[1]]}
      else{nev = length(events)}}
    mp = matrix(nrow = nev, ncol = 2)
    for (i in 1:nev){
      aux = events[i,]
      ele = length(aux)
      nam = paste0("(",varnames[[1]][[ aux[[1]] ]])
      if (ele > 1){
        for (j in 2:ele){
          nam = paste0(nam,",",varnames[[j]][[ aux[[j]] ]])}
      }
      nam = paste0(nam,")")
      mp[i, 1] = nam
      mp[ , 2] = values[[i]]
     }
     colnames(mp) = c("event",valname)
     mpd = as.data.frame(mp)
     print(mpd)
  },
  show = function(ps){
    if (missing(ps) == TRUE){
     events= self$complete_frame()
     values = self$findm()
     varnames = self$ps$names
    }else{
     events= seq(1:ps$size)
     values = ps$m
     varnames = ps$names
    }
    valname = "prob"
    self$print_events_values(events,values,varnames, valname)
  }
  )
)
```

Example 1.14 A fair coin has head and tail results equiprobable. It is tossed three times. We create the corresponding probability space as follows:

```
source("dps.R")

ps1 = dps$new(2)
ps1$names = c("H","T")

source("pdps.R")
lps = list(ps1,ps1,ps1)
pps1 = pdps$new(lps)

pps1$show()

##     event  prob
## 1 (H,H,H) 0.125
## 2 (T,H,H) 0.125
## 3 (H,T,H) 0.125
## 4 (T,T,H) 0.125
## 5 (H,H,T) 0.125
## 6 (T,H,T) 0.125
## 7 (H,T,T) 0.125
## 8 (T,T,T) 0.125

ev = rbind(c(1,1,1), c(1,2,1))  # matrix rows = events
pps1$prob(ev)

## [1] 0.125 0.125

ev = list(c(1,1,1), c(1,2,1))  # list of events
pps1$prob(ev)

## [1] 0.125 0.125
```

For the binary representation, we can use the class `pbps` (product of binary probability spaces) below – it also inherits from `pps`:

```
library(R6)
source("pps.R")
#
pbps=R6Class("pbps",
 inherit=pps,
 public=list(
 ps = NULL, m = NULL, index = NULL,
 frame = NULL, size = NULL,
 initialize=function(spaces){
  super$initialize(spaces)
  self$ps = pps$new(spaces)
  if (missing(spaces) == TRUE){
   self$m = NA
   self$frame = NA
   self$index = NA
   self$size = 0
  }else{
   ind = vector('list',length = self$ps$nspaces)
   ind[[1]] = c(1, self$ps$spaces[[1]]$size)
   s = self$ps$spaces[[1]]$size
   for (i in 2:self$ps$nspaces){
    ind[[i]] = c((ind[[i-1]][[2]]+1),(ind[[i-1]][[2]]+self$p
s$spaces[[i]]$size) )
    s = s + self$ps$spaces[[i]]$size
  }
  self$index = ind
  self$size = s
  self$complete_frame()
  }
 },
 complete_frame = function(){
  if (self$ps$nspaces < 1){
    self$frame = NA
    self$m = NA}
  else{
    if (self$ps$nspaces < 2){
      self$frame = self$spaces[[1]]$frame
      self$m = self$spaces[[1]]$m
    }
    else{
      x = vector('list', length = self$nspaces)
      x[[1]] = seq(1:self$spaces[[1]]$size)
```

```
      ch = "expand.grid(x[[1]]"
      for (i in 2:self$nspaces){
        x[[i]] = seq(1:self$spaces[[1]]$size)
        aux1 = paste0(",x[[",i,"]]")
        ch = paste0(ch,aux1)
      }
      ch =paste0(ch,")")
      chaux = as.matrix(eval(parse(text=ch)))
      frame = matrix(0,nrow = dim(chaux)[[1]],ncol = self$si
ze)
      mu = numeric(length = dim(chaux)[[1]])
      for (i in 1:dim(chaux)[[1]]){
        lev = vector('list',length = self$nspaces)
        for (j in 1: self$nspaces){
          aux2 = rep(0,self$spaces[[j]]$size)
          aux2[ chaux[[i,j]] ] = 1
          ev[[j]] = aux2
        }
        frame[i,] = self$group(lev)
        s = 1
        for (j in 1:dim(chaux)[[2]] ){
          s = s*self$ps$spaces[[j]]$m[[ chaux[[i,j]] ]]
        }
        mu[[i]] = s
      }
      self$m = mu
      self$frame = frame
    }
  }
 },
 group = function(lev){
   res = rep(0,self$size)
   for (i in 1:length(lev)){
     n = self$index[[i]]
     res[ n[[1]]:n[[2]] ] = lev[[i]]
   }
   return(res)
 },
 ungroup = function(ev){
   res = vector('list',length = self$nspaces)
   for (i in 1:self$nspaces){
     n = self$index[[i]]
     res[[i]] = ev[ n[[1]]:n[[2]] ]
   }
   return(res)
```

```
 },
 print_events_values = function(events,values,varnames, valn
ame){
   if (is.list(events)){ nev = length(events)}
   else{if(is.matrix(events)){nev = dim(events)[[1]]}
   else{nev = length(events)}}
   mp = matrix(nrow = nev, ncol = 2)
   for (i in 1:nev){
     aux = events[i,]
     auxd = self$ungroup(aux)
     ele = length(auxd)
     nam = paste0("(",varnames[[1]][[ which(auxd[[1]] > 0) ]
])
     if (ele > 1){
       for (j in 2:ele)
         {nam = paste0(nam,",",varnames[[j]][[ which(auxd[[j
]] > 0) ]])}
     }
     nam = paste0(nam,")")
     mp[i, 1] = nam
     mp[ , 2] = values[[i]]
     colnames(mp) = c("event",valname)
   }
   mpd = as.data.frame(mp)
   print(mpd)
 },
 show = function(ps){
   if (missing(ps) == TRUE){
     self$complete_frame()
     events= self$frame
     values = self$m
     varnames = self$ps$names
   }else{
     events= seq(1:ps$size)
     values = ps$m
     varnames = ps$names
   }
   valname = "prob"
   self$print_events_values(events,values,varnames, valname)
 }
 )
)
```

Example 1.15 Consider again three independent tosses of a fair coin. We create the corresponding binary probability space as follows:

```
source("bps.R")

ps = bps$new(2)
ps$names = c("H","T")

source("pbps.R")
ls = c(ps,ps,ps)
pps1 = pbps$new(ls)

pps1$show()

##     event  prob
## 1 (H,H,H) 0.125
## 2 (T,H,H) 0.125
## 3 (H,T,H) 0.125
## 4 (T,T,H) 0.125
## 5 (H,H,T) 0.125
## 6 (T,H,T) 0.125
## 7 (H,T,T) 0.125
## 8 (T,T,T) 0.125

TTT = c(0,1,0,1,0,1)
HTH = c(1,0,0,1,1,0)

ev = list(TTT,HTH)  # list of events
pps1$prob(ev)

## [1] 0.125 0.125

ev = rbind(TTT, HTH)  # matrix rows = events
pps1$prob(ev)

## [1] 0.125 0.125
```

The situation is analogous to continuous probability spaces:

```
library(R6)
source("pps.R")
source("ints.R")
#
pcps=R6Class("pcps",
 inherit = pps,
 public=list(
  ps = NULL, m = NULL, integ=NULL,
  initialize=function(spaces){
    super$initialize(spaces)
    self$ps = pps$new(spaces)
    self$integ = ints$new()
    mu = function(om1){self$spaces[[1]]$m(om1)}
    chvold = 'om1'
    chfold = 'self$spaces[[1]]$m(om1)'
    for (i in 2:self$nspaces){
      chvnew = paste0(chvold,",om",i)
      chfnew = paste0(chfold,'*self$spaces[[',i,']]$m(om',i,
')')
      chvold = chvnew
      chfold = chfnew
    }
    chf = paste0('function(',chvnew,'){ ',chfnew,'}')
    mu = eval(parse(text = chf))
    self$m = mu
  }
 )
)
```

Example 1.16 Consider again the situation of Example 1.10: $\Omega = (-1, 1)^2 \subset \mathbb{R}^2$ and $\mu(\omega) = 1/4$. Here, we consider this situation as a product of two probability spaces, created as follows:

```
source("cps.R")

a1 = -1
b1 = 1
mu1 <- function(om1){1/(b1-a1)}
ps1 = cps$new(a1,b1,mu1)

a2 = -1
b2 = 1
mu2 <- function(om2){1/(b2-a2)}
ps2 = cps$new(a2,b2,mu2)

source("pcps.R")
ls = c(ps1,ps2)
pcps1 = pcps$new(ls)
```

(continued)

Example 1.16 (continued)

The probability of the region $I = \left\{0 \le x_1 \le \frac{1}{2}, -1 \le x_2 \le x_1\right\} \subset \Omega$ is $P(I) = 5/32$:

```
source("cps.R")

a1 = -1
b1 = 1
mu1 <- function(om1){1/(b1-a1)}
ps1 = cps$new(a1,b1,mu1)

a2 = -1
b2 = 1
mu2 <- function(om2){1/(b2-a2)}
ps2 = cps$new(a2,b2,mu2)

source("pcps.R")
ls = c(ps1,ps2)
pps1 = pcps$new(ls)

aa = list(NA,NA)
bb = list(NA,NA)
aa[[1]] = 0
aa[[2]] = a2
bb[[1]] = 0.5
bb[[2]] = function(x1){x1}
pps1$integ$int2var(pps1$m,aa,bb)

## [1] 0.15625
```

We can also evaluate the probability of the event $E = (0.3, 0.5) \times (0.4, 0.6)$. We have $P(E) = \frac{0.04}{4} = 0.01$.

```
ev = list(c(0.3,0.5), c(0.4,0.6))
# event (0.3,0.5)x(0.4,0.6)
lev = list(ev)
pps1$prob(lev)

## [1] 0.01
```

Exercises

1. Consider $\Omega_1 = \{1, 2\}$ and $\Omega_2 = \{1, 2, 3\}$, both with a uniform mass.
   (a) Generate the product space using R.
   (b) Use the product space generated to find the probability of $(\{1\}, \{2, 3\})$.
2. Consider $\Omega = \{1, 2, \ldots, 16\}$ with a uniform mass.
   (a) Generate the corresponding space using R.
   (b) Generate the product space $\Omega \times \Omega \times \Omega$ using R.
   (c) Use the product space generated to find the probability of three odd numbers. TIP: generate the event "3 odd numbers" as $(2i + 1, 2j + 1, 2k + 1)$, $1 \leq i, j, k \leq 7$.
   (d) Use the product space generated to find the probability that the sum of the three is superior to 16. TIP: generate the event as (i, j, k), $i + j + k \leq 16$.
3. Let $\Omega = (0, 1)$, with the mass density $\mu(\omega) = 2\omega$.
   (a) Generate the corresponding space using R.
   (b) Generate the product space $\Omega \times \Omega$ using R.
   (c) Use the product space generated to find the probability of $E = (0.2, 0.4) \times (0.1, 0.3)$.
   (d) Use the product space generated to find the probability of $E = \left\{0 \leq x_1 \leq \frac{1}{2}, 0 \leq x_2 \leq x_1\right\}$.

1.6 Random Variables

As previously observed, uncertain quantities can be modeled by different ways, such as fuzzy variables. In the framework of probability, the classical model is the model of a random variable:

Definition 1.5 A random variable on (Ω, P) is an application $X: \Omega \longrightarrow \mathbb{R}$.
For $\subset \mathbb{R}$, $P(X \in A) = P(X^{-1}(A))$, where $X^{-1}(A) = \{\omega \in \Omega: X(\omega) \in A\}$. The **cumulative distribution function** (CDF) of X is

$$F(x) = P(X \leq x) = P(X \in]-\infty, x]).$$

F is called the **distribution** of X. If F is differentiable, we define the ***probability density function*** (PDF) of X as the derivative of F with respect to x: $f'(x) = F(x)$. ■

Let us denote

$$I_X(x)=X^{-1}((-\infty,x))=\{\omega\in\Omega:X(\omega)\leq x\}. \tag{1.21}$$

If the probability is defined by a mass function μ, then

$$F(x)=\sum_{\omega\in I(x)}\mu(\omega). \tag{1.22}$$

When the probability is defined by a mass density μ, we have

$$F(x)=\int_{I(x)}\mu(\omega)d\omega. \tag{1.23}$$

If Ω is finite, then its image $X(\Omega)=\{X_1,\ldots,X_n\}$ is finite too, and we have

$$p_i=P(X=X_i)=\sum_{\omega\in X^{-1}(X_i)}\mu(\omega). \tag{1.24}$$

Using ordered values $X_1<X_2<\ldots<X_{n-1}<X_n$, the CDF of X is

$$F(x)=\begin{cases}0,\text{if } x<X_1;\\ \sum_{j=1}^{i-1}p_j,\text{if } X_{i-1}\leq x<X_i, 1<i\leq n;\\ 1,\text{if } x\geq X_n\ .\end{cases} \tag{1.25}$$

In such a situation, it is usual to represent X by a table (Table 1.1):

Table 1.1 Representing X as a Table

Value	X_1	X_2	...	X_{n-1}	X_n
Probability	p_1	p_2	...	p_{n-1}	p_n

We define the mean of a random variable as follows:

Definition 1.6 Let $g: \mathbb{R} \longrightarrow \mathbb{R}$ be a function and $Y = g(X)$, where X is a random variable on (Ω, P). The ***mean*** or ***expectation*** of Y is $E(Y)$ given by

1. Let Ω be finite or enumerable and P be generated by the probability mass function μ. Then

$$E(Y) = E(g(X)) = \sum_{\omega \in \Omega} \mu(\omega) g(X(\omega)).$$

2. Let $\Omega \subset \mathbb{R}$. and P be generated by the probability mass density μ. Then $E(Y) = E(g(X)) = \int_{\Omega} g(X(\omega))\mu(\omega)d\omega$.

$E(Y)$ is the best approximation of Y by a constant, in the sense of least squares: let

$$K(E(Y)) \leq K(m), \forall m \in \mathbb{R}; \; K(s) = E\left((Y - s)^2\right) \; (s \in \mathbb{R}). \tag{1.26}$$

If Ω is finite, then its image $X(\Omega) = \{X_1, \ldots, X_n\}$ is finite too and we have

$$E(g(X)) = \sum_{i=1}^{n} p_i g(X_i) \; . \tag{1.27}$$

If $\Omega \subset \mathbb{R}$, then

$$E(g(X)) = \int_{\mathbb{R}} g(x) dF(x). \tag{1.28}$$

If the PDF exists, we have $dF(x) = f(x)dx$, so that

$$E(g(X)) = \int_{\mathbb{R}} g(x) f(x) dx. \tag{1.29}$$

Means have many useful properties. For instance, linearity and positivity:

$$E(\alpha f(X) + \beta g(X)) = \alpha E(f(X)) + \beta E(g(X)); \tag{1.30}$$

$$f(X) \geq 0 \Rightarrow E(f(X)) \geq 0. \tag{1.31}$$

There are also useful inequalities, such as Markov's inequality:

$$f(X) \geq 0 \Rightarrow \lambda P(f(X) \geq \lambda) \leq E(f(X)), \forall \lambda \in \mathbb{R}. \tag{1.32}$$

Another useful inequality is Jensen's inequality:

$$f \text{ convex} \Rightarrow f(E(X)) \leq E(f(X)). \tag{1.33}$$

The main statistics of X are as follows:

Definition 1.7

1. The **moment of order k of** X is $M_k(X) = E(X^k)$.
2. The **central moment of order k of** X is $CM_k(X) = E((X - E(X))^k)$.
3. The **variance of** X is $V(X) = CM_2(X) = E((X - E(X))^2)$.
4. The **standard deviation** of X is $\sigma(X) = \sqrt{V(X)}$.
5. A **median of** X is a number $med(X) = m$ such that $(X \leq m) \geq \frac{1}{2}$ and $P(X \geq m) \geq \frac{1}{2}$.
6. The **characteristic function** of X is $\varphi_X(t) = E(e^{itX})$.

We have

$$V(X) \geq 0, V(X) = E(X^2) - E(X)^2. \tag{1.34}$$

In addition,

$$V(X) = 0 \Leftrightarrow X = 0 \; a.s. \tag{1.35}$$

Chebyshev's inequality is derived from Markov's inequality by taking $f(X) = (X - E(X))^2$and $\lambda = \varepsilon^2$

$$P(|X - E(X)| \geq \varepsilon) \leq \frac{V(X)}{\varepsilon^2}, \forall \varepsilon > 0. \tag{1.36}$$

A discrete random variable is represented by a class `drv` as follows (these contents must be saved to a file "`drv.R`"):

```
library(R6)
#
drv=R6Class("drv",
  public=list(
  X=NULL,p=NULL,mean=NULL,m2=NULL,var=NULL,sd=NULL,cumul=NUL
L,med=NULL,phi=NULL,
  initialize= function(X,p,Xnames){
    self$X = X
    self$p = p
    self$mean = sum(p*X)
   self$m2 = sum(p*(X^2))
   self$var = self$m2 - self$mean^2
   self$sd = sqrt(self$var)
   aux = cumsum(p)
   self$cumul = aux
   medX = list(NA,NA)
   names(medX) = c("lower","upper")
   i1 = min(which( aux >= 0.5))
   if (i1 <= 1) {medX$lower = X[[1]]}
   else {
     if (aux[[i1-1]]>= 0.5) {medX$lower = X[[i1-1]]}
     else {medX$lower = X[[i1]]}}
     i2 = max(which( aux <= 0.5))
     if (i2 >= length(X)) {medX$upper = X[[length(X)]]}
     else {
       if (aux[[i2+1]]>= 0.5) {medX$upper = X[[i2+1]]}
       else {medX$upper = X[[i2]]}}
   self$med = medX
   self$phi = function(t){sum(p*exp(1i*t*X))}
  }
 )
)
```

Example 1.17 Let us consider a random variable X representing the result of the roll of a fair six-faced dice. It can be represented by Table 1.2.

Table 1.2 Result of fair six-faced dice

Value	1	2	3	4	5	6
Probability	$\frac{1}{6}$	$\frac{1}{6}$	$\frac{1}{6}$	$\frac{1}{6}$	$\frac{1}{6}$	$\frac{1}{6}$

Thus,

$$E(X) = \frac{1}{6}\sum_{i=1}^{6} i = \frac{7}{2},\ E(X^2) = \frac{1}{6}\sum_{i=1}^{6} i^2 = \frac{91}{6}, V(X) = \frac{35}{12}.$$

(continued)

Example 1.17 (continued)

Any number m such that $3 < m < 4$ is a median. The characteristic function of X is

$$\varphi_X(t) = \frac{1}{6}\sum_{k=1}^{6} e^{ikt} = \frac{1}{6}\sum_{k=1}^{6} (\cos(kt) + i\ \sin(kt)).$$

We can create this random variable in R as follows:

```
X = seq(1:6)  # values of X
p = rep(1/6, 6) # probabilities
source("drv.R")
va1 = drv$new(X,p)
```

Then, we use it as follows:

```
print(paste("mean = ",va1$mean,", exact = ", 7/2))

## [1] "mean =  3.5 , exact =  3.5"

print(paste("Moment M2 = ", va1$m2,", exact = ",91/6))

## [1] "Moment M2 =  15.1666666666667 , exact =  15.166666
6666667"

print(paste("variance = ",va1$var,", exact = ", 35/12))

## [1] "variance =  2.91666666666667 , exact =  2.91666666
666667"

print(paste("standard deviation = ", va1$sd,", exact = ",s
qrt(35/12)))

## [1] "standard deviation =  1.70782512765993 , exact =
1.70782512765993"

print(paste("median between ",va1$med$lower," and ",va1$me
d$upper))

## [1] "median between  3  and  4"

v1 = va1$phi(1)
c1 = sum(cos((X)))/6
s1 = sum(sin((X)))/6
print(paste("phi(1) = ",v1,", exact = ",c1 + 1i*s1))

## [1] "phi(1) =  -0.045941362671578-0.0172089747460912i ,
exact =  -0.045941362671578-0.0172089747460912i"
```

Analogously, we can create a class `crv` to deal with continuous random variables:

```
library(R6)
source("ints.R")
#
crv=R6Class("crv",
 inherit=ints,
  public=list(
   cdf = NULL, pdf = NULL, mean = NULL,
   m2 = NULL, var=NULL, sd = NULL,
   med = NULL, phi = NULL,
   initialize=function(F,f,a,b){
     super$initialize()
     self$cdf = F
     self$pdf = f
     h1 = function(x){x*f(x)}
     self$mean = super$int1(h1,a,b)
     h2= function(x){(x^2)*f(x)}
     self$m2 = super$int1(h2,a,b)
     self$var = self$m2 - self$mean^2
     self$sd = sqrt(self$var)
     fm = function(x){F(x) - 0.5}
     x0 = 0.5*(a+b)
     medX = fzero(fm,x0)
     self$med = medX$x
     self$phi = function(t){
       h3 = function(x){f(x)*exp(1i*t*x)}
       aux = super$int1(h3,a,b)
       return(aux)
     }
   }
  )
)
```

Example 1.18 Let us consider $\Omega = (0, 1)$ and the mass density $\mu(\omega) = 1$. Let $X(\omega) = \exp(\omega)$. Then $X(\omega) \leq x \Leftrightarrow \exp(\omega) \leq x \Leftrightarrow \omega \leq \log(x)$. Thus, the CDF of X is

$$F_X(x) = \begin{cases} 0, \text{if } x < 1; \\ \log(x), \text{if } 1 \leq x < \mathrm{e}; \\ 1, \text{if } x \geq \mathrm{e}. \end{cases}$$

(continued)

Example 1.18 (continued)

The PDF is

$$f_X(x) = \begin{cases} 0, \text{if } x < 1; \\ \frac{1}{x}, \text{if } 1 < x < e; \\ 0, \text{if } x > \text{e} \ . \end{cases}$$

The mean of X is

$$E(X) = \int_0^1 X(\omega)\mu(\omega)d\omega = \int_0^1 e^{\omega} d\omega = e - 1.$$

We have also

$$E(X) = \int_1^e xf(x)dx \int_1^e x\frac{1}{x}dx = \int_1^e 1dx = e - 1$$

In addition,

$$E(X^2) = \int_0^1 (X(\omega))^2\mu(\omega)d\omega = \int_0^1 e^{2\omega} d\omega = \frac{1}{2} \left(e^2 - 1\right),$$

which can be also determined as

$$E(X^2) = \int_1^e x^2 f(x)dx \int_1^e x^2\frac{1}{x}dx = \int_1^e xdx = \frac{e^2 - 1}{2}.$$

Thus

$$V(X) = 2e - \frac{1}{2}\left(e^2 + 3\right)$$

The median m verifies

$$F(m) = \frac{1}{2} \Leftrightarrow \log(m) = \frac{1}{2} \Leftrightarrow m = \exp\left(\frac{1}{2}\right)$$

The characteristic function is

$$\varphi_X(t) = \int_1^e \frac{e^{itx}}{x} dx = 1 + \sum_{k=1}^{+\infty} \frac{(it)^k}{k \times k!} \left(e^k - 1\right).$$

(continued)

Example 1.18 (continued)

We can find numerical approximations of these values by numerical integration: let us create the random variable

```
F = function(x){log(x)}
f = function(x){1/x}
a = 1 #Lower bound for x
b = exp(1) #Upper bound for x

source("crv.R")
va1 = crv$new(F,f,a,b)
```

Then, we have:

```
print(paste("mean of X = ",va1$mean,", exact = ",exp(1)-
1))
## [1] "mean of X =  1.71828182845905 , exact =
1.71828182845905"
print(paste("mean of X^2 = ",va1$m2,", exact =
",0.5*(exp(2)-1)))
## [1] "mean of X^2 =  3.19452804946532 , exact =
3.19452804946533"
print(paste("variance of X = ",va1$var,", exact =
",2*exp(1) - 0.5*(exp(2)+3)))
## [1] "variance of X =  0.242035607452765 , exact =
0.242035607452765"
print(paste("median of X = ",va1$med,", exact =
",exp(0.5)))
## [1] "median of X =  1.64872127070013 , exact =
1.64872127070013"
v1 = va1$phi(1)
s1 = 0 + 0i
s2 = 0 + 0i
kmax = 1000
aux1 = 1
aux2 = 1
for (k in 1:kmax){
  aux1 = 1i*aux1/k
  aux2 = 1i*exp(1)*aux2/k
  s1 = s1 + aux1/k
  s2 = s2 + aux2/k
}
s = s2 - s1 + 1
print(paste("phi(1) = ",v1,", exact = ",s))
  ## [1] "phi(1) =  -0.123445921541694+0.874957198781388i
, exact =  -0.123445921560589+0.874957198780384i"
```

(continued)

Example 1.18 (continued)

In practice, F and f can be given at a discrete set of values. For instance, assume the F is given at 51 points equally spaced on $(1, e)$:

```
Fex = function(x){log(x)}
fex = function(x){1/x}
aex = 1 #Lower bound for x
bex = exp(1) #Upper bound for x

xdata = seq(aex,bex,length.out = 51)
Fdata = Vectorize(Fex)(xdata)
```

Then, we can generate approximations using an SPH derivative and the function `approxfun`:

```
source("sph.R")
kern = "quintic"
h = xdata[[2]] - xdata[[1]]
sp = sph$new(kern,h)
fdata = sp$sph_derivation(xdata, Fdata, xdata, Fdata)

F = approxfun(xdata,Fdata,rule=2)
f = approxfun(xdata,fdata,rule=2)
```

These results can be used to generate a random variable:

```
source("crv.R")
a = min(xdata)
b = max(xdata)
va1 = crv$new(F,f,a,b)

print(paste("mean of X = ",va1$mean,", exact = ",exp(1)-1)
)
## [1] "mean of X =  1.71852294159437 , exact =  1.7182818
2845905"

print(paste("mean of X^2 = ",va1$m2,", exact = ",0.5*(exp(
2)-1)))

## [1] "mean of X^2 =  3.19528421997224 , exact =  3.19452
804946533"
```

(continued)

Example 1.18 (continued)

```
print(paste("variance of X = ",va1$var,", exact = ",2*exp(
1) - 0.5*(exp(2)+3)))
```

```
## [1] "variance of X =  0.241963119186066 , exact =  0.24
2035607452765"
```

```
print(paste("median of X = ",va1$med,", exact = ",exp(0.5)
))
```

```
## [1] "median of X =  1.64875999854769 , exact =  1.64872
127070013"
```

```
v1 = va1$phi(1)
print(paste("phi(1) = ",v1,", exact = ",s))
```

```
## [1] "phi(1) =  -0.123657295965049+0.874959132590891i ,
exact =  -0.123445921560589+0.874957198780384i"
```

The characteristic function can be considered as the Fourier transform of the PDF. Thus, the PDF f can be obtained from the characteristic function φ by inversion of the Fourier Transform. From the numerical standpoint, such an inversion can be made by the methods presented in Witkovský (2016). For instance, we can use the FFT:

```
invcf = function(cf,a,b,N){
k = 0:(N-1)
j = 0:(N-1)
dy = (b-a)/N
y = a + k * dy
u = (j - N/2)/(b-a)
phi = ((-1)^(-(2*a/(b-a))*j)) * cf(2*pi*u)
CC = ((-1)^((a/(b-a) + k/N)*N))/(b-a)
pdf = Re(CC * fft(phi))
resu = list(y,pdf)
names(resu) = c('x','f')
return(resu)
}
```

Alternatively, the reader can use the package `CharFun` (removed from the CRAN repository, but available in the archives (https://cran.r-project.org/src/contrib/Archive/CharFun/ , https://rdrr.io/cran/CharFun/) or `CharFunToolR` (https://github.com/gajdosandrej/CharFunToolR).A version in Matlab® is available at https://fr.mathworks.com/matlabcentral/fileexchange/64400-charfuntool-the-characteristic-functions-toolbox or https://github.com/witkovsky/CharFunTool). These packages propose other methods of numerical inversion.

Example 1.19 Let us consider $X \sim N(m, \sigma)$. Thus, $\varphi_X(t) = \exp(imt - t^2\sigma^2/2)$. Let us apply the method above to determine the PDF:

```
m = 1
sig = 2
cf = function(t){exp(1i*m*t-sig^2*t^2/2)}
N = 2^10 # Number of subintervals
a = -10 # Lower bound for X
b = 10 # Upper bound for X
r = invcf(cf,a,b,N)
y = dnorm(r$x,m,sig)
library(latex2exp)
plot(r$x,y,type='n',main = TeX(r'(Inversion of $\varphi(t)$)'),
     xlab = 't', ylab = 'f(t)')
lines(r$x,y,lwd=2,lty=1,col='black')
lines(r$x,r$f,lwd=2,lty=2,col='red')
legend('topleft',legend=c('exact','calculated'),
       col=c('black','red'),lty = c(1,2),lwd = c(2,2))
```

Alternatively, we can use the function `cf2DistGP` from `CharFun`, as shown at right.	`library(CharFun)` `library(latex2exp)r1` `= cf2DistGP(cf)`

The results obtained for $m = 2$, $\sigma = 1$ are shown in Fig. 1.5.

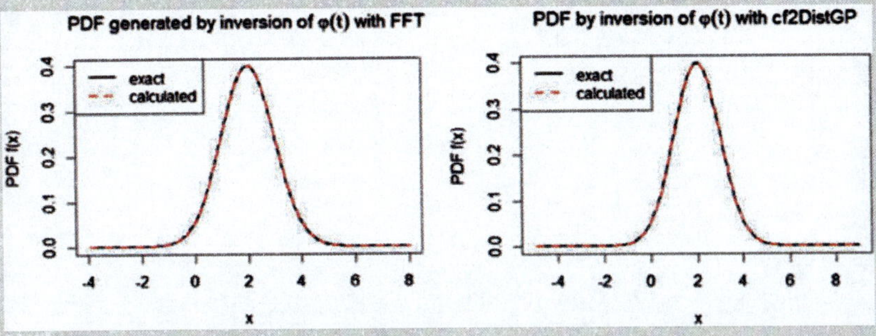

Fig. 1.5 Inversion of a characteristic function using the FFT and cf2DistGP

In practice, we may use data corresponding to observed values of φ. For instance, assume that we have the values $(t_i, \varphi(t_i))$, for $t_i = -10 + i\Delta t$,

(continued)

Example 1.19 (continued)

$\Delta t = 0.01$ and $i = 0, \ldots, 2000$. We can interpolate the data to generate a function:

```
m = 2
sig = 1
cf = function(t){exp(1i*m*t-sig^2*t^2/2)}
tc = seq(-10,10,by = 0.1)
cfc = cf(tc)
rcfc = Re(cfc)
icfc = Im(cfc)
rcf = approxfun(tc, rcfc,method="linear",
                , , rule = 2, f = 0, ties = mean)
icf = approxfun(tc, icfc,method="linear",
                , , rule = 2, f = 0, ties = mean)
cf = function(t){rcf(t) + 1i*icf(t)}
```

The results appear in Fig. 1.6.

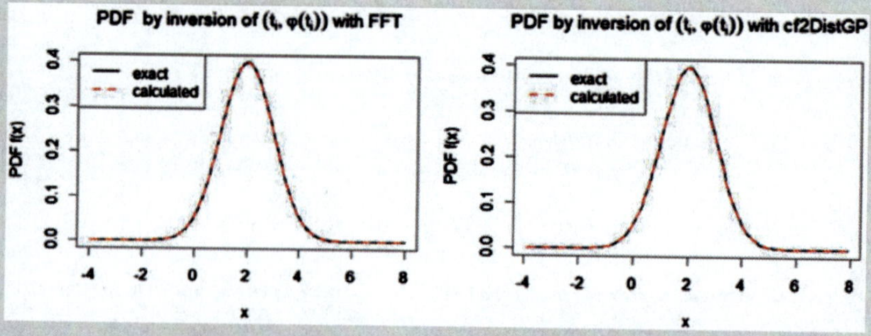

Fig. 1.6 Inversion of a characteristic function using cf2DistGP

Exercises

1. Consider $\Omega = (0, 1)$ and the mass density $\mu(\omega) = 3\omega^2$.
   (a) Find the distribution of $X(\omega) = e^{\omega}$.
   (b) Find the expression of $M_k(X)$.
   (c) Determine the mean and the variance of X.

(continued)

2. Consider $\Omega = (0, 1)$ and the mass density $\mu(\omega) = 1$.

   (a) Find the distribution of $X(\omega) = e^{\omega}$.
   (b) Find the expression of $M_k(X)$.
   (c) Determine the mean and the variance of X.

3. Consider $\Omega = (0, 1)$ and the mass density $\mu(\omega) = 2\omega$.

   (a) Find the distribution of $X(\omega) = \omega^3$.
   (b) Find the expression of $M_k(X)$.
   (c) Determine the mean and the variance of X.

4. Consider $\Omega = (0, 1)$ and the mass density $\mu(\omega) = 1$.

   (a) Find the distribution of $X(\omega) = \omega$.
   (b) Find the expression of $M_k(X)$.
   (c) Determine the mean and the variance of X.

5. Let $\Omega = \{0, 1, \ldots, n\}$ and $X(\omega) = \omega \; mod \; 2$ (remainder after division by 2). Assume the elements of Ω as equiprobable.

   (a) Show that $X(\Omega) = \{0, 1\}$.
   (b) Determine $p = P(X = 0)$ when $n = 2k$.

6. Determine the characteristic functions of the following random variables:

   (a) X is such that $P(X = 0) = 1$.
   (b) X is such that $P(X = 1) = p$, $P(X = 0) = 1 - p$.
   (c) X is Poisson distributed with parameter $\lambda : P(X = k) = \frac{e^{-\lambda}\lambda^k}{k!}$.
   (d) X is uniformly distributed on (a, b) – *its PDF is* : $f(x) = \frac{1}{b-a}$ on (a, b)
   (e) X is exponentially distributed –*its PDF is*: $f(x) = \lambda e^{-\lambda x}$ on $(0, +\infty)$.

7. Find the mean and the variance of the variable X defined by

Value of X	1	2	3	4	5	6
probability	0.1	0.25	0.05	0.2	0.15	0.25

8. Find the mean and the variance of the variable X defined by:

Value of X	1	2	3	4	5	6
probability	0.1	0.2	0.15	0.2	0.15	0.2

9. Find the mean and the variance of the variable X defined by:

Value of X	1	2	3	4	5	6
probability	0.2	0.15	0.2	0.15	0.2	0.1

10. Consider a random variable X having a mean $\frac{n}{2}$ and a variance $\frac{n}{4}$.

   (a) Determine an upper bound for the probability of obtaining a deviation of 10 % of the mean. TIP: Consider $P\left(\left|X - \frac{n}{2}\right| > \alpha\frac{n}{2}\right)$.
   (b) Find an upper bound for the probability $P(X > 0.9\ n)$. TIP: $P(|X - m| > a) = P(X < m - a) + P(X > m + a)$
   (c) Determine n such that $P\left(\left|X - \frac{n}{2}\right| > 0.01\frac{n}{2}\right) \leq 0.01$.

11. A company has a mean daily production of 500 units of a given product.

   (a) Use Markov's inequality to find an upper bound for the probabilities of productions of 510, 550, and 1000 unities.
   (b) Assume that the standard deviation of the daily production is 10. Use Chebishev's inequality to find an upper bound for these probabilities

   TIP: Consider $P(|X - 500| \geq x - 500)$.

12. Consider $\Omega = (-1, 1)$ and the mass density $\mu(\omega) = \alpha\omega^2$.

   (a) Show that $\alpha = 3/2$.
   (b) Find the distribution of $X(\omega) = \omega$.
   (c) Find the expression of $M_k(X)$.
   (d) Determine the mean and the variance of X.

13. Consider $\Omega = (-1, 1)$ and the mass density $\mu(\omega) = \alpha|\omega|$.

   (a) Show that $\alpha = 1$.
   (b) Find the distribution of $X(\omega) = \omega$.
   (c) Find the expression of $M_k(X)$.
   (d) Determine the mean and the variance of X.

14. Consider $\Omega = \{-1, 0, 1\}$ and the mass function $\mu(1) = \mu(-1) = \frac{1}{3}$. Let $X_n(\omega) = \omega^n$.

   (a) Determine $\mu(0)$.
   (b) Find the expression of $M_k(X_n)$.
   (c) Determine the mean and the variance of X_n.

15. Consider $\Omega = \{-1, 0, 1\}$ and the mass function $\mu(1) = \mu(-1) = \frac{1}{4}$. Let $X_n(\omega) = \omega^n$.

   (a) Determine $\mu(0)$.
   (b) Find the expression of $M_k(X_n)$.
   (c) Determine the mean and the variance of X_n.

16. Let $\Omega = \{0, 1, \ldots, n\}$ and $X(\omega) = \omega \bmod 2$ (remainder after division by 2). Assume the elements of Ω as equiprobable.

   (a) Show that $X(\Omega) = \{0, 1\}$.
   (b) Determine $p = P(X = 0)$ when $n = 2k$.
   (c) Determine $p = P(X = 0)$ when $n = 2k + 1$.

(continued)

17. Let X be a random variable having the CDF $F_X(x) = x^2$ for $0 < x < 1$
    (a) Show that $P(X \leq 0) = 0$.
    (b) Show that $P(X > 1) = 1$.
    (c) Find its PDF f_X.
    (d) Determine the mean and the variance of X.

1.7 The Exponential Family

Many of the usual random variables X belong to the exponential family, for which the PDF is defined by a vector of parameters $\boldsymbol{\theta}$, under the form

$$f_X(x|\boldsymbol{\theta}) = h(x)\exp(\boldsymbol{\eta}^t\boldsymbol{T}(x) - A(\boldsymbol{\eta})), \boldsymbol{\eta} = \boldsymbol{\psi}(\boldsymbol{\theta}). \tag{1.37}$$

$\boldsymbol{\eta}$ is called the canonical parameter, $\boldsymbol{T}$ is the sufficient statistic and A is the cumulant function. The family contains discrete and continuous variables, according to the possible values of x. One of the interesting properties of this family is that we can evaluate the means and covariances using the cumulant function A. Indeed,

$$\frac{\partial f_X}{\partial \eta_i} = \left(T_i(x) - \frac{\partial A}{\partial \eta_i}\right) f_X \,.$$

Since

$$\int_{\mathbb{R}} f_X(x|\boldsymbol{\theta})dx = 1 \Longrightarrow \frac{\partial}{\partial \eta_i}\int_{\mathbb{R}} f_X(x|\boldsymbol{\theta})dx = 0 \Longrightarrow \int_{\mathbb{R}} \frac{\partial f_X}{\partial \eta_i}dx = 0,$$

we have

$$E(T_i(X)) = \int_{\mathbb{R}} T_i(x) f_X(x|\boldsymbol{\theta})dx = \int_{\mathbb{R}} \frac{\partial A}{\partial \eta_i} f_X(x|\boldsymbol{\theta})dx = \frac{\partial A}{\partial \eta_i}\int_{\mathbb{R}} f_X(x|\boldsymbol{\theta})dx.$$

Thus:

$$E(T_i(X)) = \frac{\partial}{\partial \eta_i} A(\boldsymbol{\eta}). \tag{1.38}$$

Analogously,

$$\frac{\partial^2 f_X}{\partial \eta_i \partial \eta_j} = \left(T_i(x) - \frac{\partial A}{\partial \eta_i}\right)\left(T_j(x) - \frac{\partial A}{\partial \eta_j}\right) - f_X \frac{\partial^2 A}{\partial \eta_i \partial \eta_j},$$

so that

$$\mathrm{Cov}(T_i(X), T_j(X)) = \frac{\partial^2}{\partial \eta_i \partial \eta_j} A(\boldsymbol{\eta}). \tag{1.39}$$

To verify if a given density corresponds to a variable of the exponential family, it is usual to consider

$$\log f_X(x|\boldsymbol{\theta}) = \boldsymbol{\eta}^t \boldsymbol{T}(x) - A(\boldsymbol{\eta}) + w(x), \boldsymbol{\eta} = \boldsymbol{\psi}(\boldsymbol{\theta}). \tag{1.40}$$

If $\log f_X(x|\boldsymbol{\theta})$ can be put under the form (1.40), then the variable belongs to the exponential family and $h(x) = \exp(w(x))$. Otherwise, it is not a member of the family.

Example 1.20 Consider the Weibull distribution, which takes its values on $(0, +\infty)$ and whose PDF is

$$f(x) = \frac{\mathrm{a}}{b^a} x^{a-1} \exp\left(-\left(\frac{x}{b}\right)^2\right).$$

Consider $\boldsymbol{\theta} = (a, b)$. We have

$$\log f_X(x|\boldsymbol{\theta}) = \underbrace{(a-1)x - \left(\frac{x}{b}\right)^2}_{\boldsymbol{\eta}^t \boldsymbol{T}(x)} + \underbrace{\log a - a \log b}_{-A(\boldsymbol{\eta})}.$$

Let us consider $\boldsymbol{\eta} = (a - 1, -1/b^2)$, $\boldsymbol{T}(x) = (x, x^2)$. Then, $a = \eta_1 + 1$, $b = \sqrt{-1/\eta_2} = \sqrt{1/|\eta_2|}$, and

$$\log f_X(x|\boldsymbol{\theta}) = \boldsymbol{\eta}^t \boldsymbol{T}(x) + \underbrace{\log(\eta_1 + 1) + \frac{1}{2}(\eta_1 + 1)\log|\eta_2|}_{-A(\boldsymbol{\eta})} + \underbrace{0}_{w(x)}.$$

Thus, the variable is a member of the exponential family.

The reader will find in the literature many results about the exponential family – see, for instance, Brown (1986) or Sundberg (2019). We recall below some distributions of this family.

Classical Discrete Distributions of the Exponential Family

Some discrete distributions often used are recalled below.

Bernoulli Distribution it is denoted by $\mathcal{B}(p)$ and is the distribution of a binary variable, which can take only values in $\{0, 1\}$ (and, by extension any variable which can take only two values: Head or Tail, True or False, etc.). The value of p corresponds to the probability of the value 1: $p = P(X = 1)$. We have

$$P(X=0)=q=1-p, E(X)=p, V(X)=pq.$$

Thus, $f(x) = p^x q^{1-x}$. Considering $\theta = p$, we have

$$\log f_X(x|\theta) = \underbrace{x \log\left(\frac{p}{1-p}\right)}_{\eta} + \underbrace{q}_{-A(\eta)}.$$

Therefore, this variable is a member of the exponential family with

$$h(x)=1, \eta= \log\left(\frac{p}{1-p}\right), A(\eta)= \log(1+e^{\eta}), T(x)=x.$$

Binomial Distribution it is denoted by $\mathcal{B}(n, p)$ and is the distribution of the sum of n independent Bernoulli variables: $X_n = B_1 + \ldots + B_n$, with $B_i \sim \mathcal{B}(p)$, all independent. X_n is used to describe the number of successes (or failures) on n trials. We have $X_n \in \{0, \ldots, n\}$, and

$$P(X_n=k)=\binom{n}{k} p^k q^{n-k}, E(X_n)=np, V(X_n)=npq.$$

This variable is a member of the exponential family with $\theta = p$,

$$h(x)=\binom{n}{x}, \eta= \log\left(\frac{p}{q}\right), A(\eta)=n\log(1+e^{\eta}), T(x)=x.$$

Negative Binomial Distribution it is denoted by $\mathcal{NB}(n,p)$ and is the distribution of the number k of failures requested to obtain n successes, by making independent trials, when the probability of success in a trial is p. We have $X \in \mathbb{N}$, and

$$P(X=k)=\binom{k+n-1}{n-1}p^n q^k, E(X)=n\frac{q}{p}, V(X)=n\frac{q}{p^2}.$$

This variable is a member of the exponential family with $\theta = p$,

$$h(x)=\binom{x+n-1}{n-1}, \eta = \log q, A(\eta)=n\log(1-e^{\eta}), T(x)=x.$$

Poisson Distribution it is denoted by $\mathcal{P}(\lambda)$ and describes the number of events in a given time interval, when the events occur with a fixed probability and independently of the time elapsed since the previous event. We have $X \in \mathbb{N}$, and

$$P(X=k)=\frac{\lambda^k}{k!}e^{-\lambda}, E(X)=\lambda, V(X)=\lambda.$$

This variable is a member of the exponential family with $\theta = \lambda$,

$$h(x)=\frac{1}{x!}, \eta = \log \lambda, A(\eta)=e^{\eta}, T(x)=x.$$

Classical Continuous Distributions

Some continuous distributions often used are recalled below.

Normal Distribution it is denoted by $N(m,\sigma)$, which is the limiting distribution in the Central Limit Theorem. It takes its values on $\mathbb{R}$ and its PDF is:

$$f(x)=\frac{1}{\sigma\sqrt{2\pi}}\exp\left(-\frac{1}{2}\left(\frac{x-m}{\sigma}\right)^2\right).$$

Taking $\boldsymbol{\theta} = (m,\sigma)$, we have

(continued)

$$\log f_X(x|\theta) = \underbrace{-\frac{1}{2\sigma^2}x^2 + \frac{m}{\sigma}x}_{\eta^t T(x)} - \underbrace{\left(\frac{m^2}{2\sigma^2} + \log \sigma\right)}_{A(\eta)} + \underbrace{\log \frac{1}{\sqrt{2\pi}}}_{w(x)}$$

Therefore, this variable is a member of the exponential family with

$$h(x) = \frac{1}{\sqrt{2\pi}}, \boldsymbol{\eta} = \left(\frac{m}{\sigma}, -\frac{1}{2\sigma^2}\right),$$

$$A(\boldsymbol{\eta}) = -\frac{\eta_1^2}{4\eta_2} - \frac{1}{2}\log\left(-2\eta_2^2\right), \boldsymbol{T}(x) = \left(x, x^2\right).$$

We have

$$E(X) = m, \quad V(X) = \sigma^2.$$

The characteristic function of X is $\varphi_X(t) = \exp(imt - t^2\sigma^2/2)$.

Beta Distribution it is denoted by $\mathcal{Be}(\alpha, \beta)$. It takes its values on (0, 1) and its PDF is

$$f(x) = \frac{x^{\alpha-1}(1-x)^{\beta-1}}{B(\alpha, \beta)}, \quad B(\alpha, \beta) = \frac{\Gamma(\alpha)\Gamma(\beta)}{\Gamma(\alpha+\beta)}.$$

Here, B is the beta function, and Γ is the Gamma function:

$$\Gamma(x) = \int_0^{+\infty} t^{x-1}e^{-t}dt.$$

Taking $\boldsymbol{\theta} = (\alpha, \beta)$, we have

$$\log f_X(x|\theta) = \underbrace{\alpha \log x + \beta \log(1-x)}_{\eta^t T(x)} - \underbrace{(\log B(\alpha, \beta)}_{A(\eta)} \underbrace{- \log x(1-x)}_{w(x)}$$

Therefore, this variable is a member of the exponential family with

$$h(x) = \frac{1}{x(1-x)}, \boldsymbol{\eta} = (\alpha, \beta),$$

(continued)

$$A(\boldsymbol{\eta}) = \log B(\eta_1, \eta_2) \ , \boldsymbol{T}(x) = (\log x, \log(1-x)).$$

We have

$$E(X) = \frac{\alpha}{\alpha+\beta}, \quad V(X) = \frac{\alpha\beta}{(\alpha+\beta+1)(\alpha+\beta)^2}.$$

Exponential Distribution it is denoted by Exp(λ). It takes its values on (0, +∞) and its PDF is

$$f(x) = \lambda \exp(-\lambda x).$$

We have

$$E(X) = \frac{1}{\lambda}, \quad V(X) = \frac{1}{\lambda^2}.$$

This variable is a member of the exponential family with $\theta = \lambda$,

$$h(x) = 1, \eta = \lambda, A(\eta) = -\log \eta, T(x) = -x.$$

Exercises

1. Rayleigh's distribution takes its values on (0, +∞) and its PDF is

$$f(x) = \frac{x}{b^2} \exp\left(-\frac{x^2}{2b^2}\right).$$

   Show that Rayleigh's distribution with $\theta = b$ belongs to the exponential family. Determine its mean and variance.

2. The log-normal distribution takes its values on (0, +∞) and its PDF is

$$f(x) = \frac{1}{x\sigma\sqrt{2\pi}} \exp\left(-\frac{1}{2}\left(\frac{\log(x)-m}{\sigma}\right)^2\right).$$

   Show that the log-normal distribution with $\boldsymbol{\theta} = (m, \sigma)$ belongs to the exponential family: Determine its mean and variance.

(continued)

3. Chi-squared distribution takes its values on $(0, +\infty)$ and its PDF is.

$$f(x) = \frac{1}{2^{\frac{n}{2}}\Gamma\left(\frac{n}{2}\right)} x^{\frac{n}{2}-1} e^{-\frac{x}{2}}.$$

Show that the chi-squared distribution with $\theta = n$ belongs to the exponential family. Determine its mean and variance.

4. The Inverse Gaussian distribution takes its values on $(0, +\infty)$ and its PDF is

$$f(x) = \sqrt{\frac{\lambda}{2\pi x^3}} \exp\left(-\frac{\lambda(x-m)^2}{2m^2 x}\right).$$

Show that the Inverse Gaussian distribution with $\boldsymbol{\theta} = (m, \lambda)$ belongs to the exponential family: Determine its mean and variance.

5. The Gamma distribution takes its values on $(0, +\infty)$ and its PDF is

$$f(x) = \frac{x^{\alpha-1} e^{-\beta x} \beta^{\alpha}}{\Gamma(\alpha)}$$

Show that the Gamma distribution with $\boldsymbol{\theta} = (\alpha, \beta)$ belongs to the exponential family: Determine its mean and variance.

1.8 Distributions That Are Not in the Exponential Family

There are distributions that do not belong to the exponential family – generally distributions having the support dependent on the parameter $\boldsymbol{\theta}$, so that the function $w(x)$ (or $h(x)$) cannot be constructed. We must have $w(x, \boldsymbol{\theta})$ instead. For instance,

Dirac Distribution it is denoted by δ_x and describes a constant random variable $X(\omega) = x$, $\forall\ \omega \in \Omega$. Then, $P(X = x) = 1$

$$f_X(y) = \delta(y - x), E(X) = x, V(X) = 0.$$

Its parameter is $\theta = x$ and its support is $\text{supp}(f_X) = \{x\}$, so that it is not a member of the exponential family.

Discrete Uniform Distribution it is denoted by $\mathcal{U}(n)$ and describes a random choice among n objects, all equiprobable. We have $X \in \{1, \ldots, n\}$ and

$$P(X = k) = \frac{1}{n}, E(X) = \frac{n+1}{2}, V(X) = \frac{n^2 - 1}{12}.$$

Its parameter is $\theta = n$ and its support is $supp(f_X) = \{1, \ldots, n\}$, so that it is not a member of the exponential family.

Continuous Uniform Distribution it is denoted by $\mathcal{U}(a, b)$ and describes a random choice for a subinterval on the interval (a, b), with a probability proportional to the length of the subinterval. We have $X \in (a, b)$ and

$$f_X(x) = \frac{1}{b - a}, E(X) = \frac{a + b}{2}, V(X) = \frac{b^2 - a^2}{12}.$$

Its parameter is $\boldsymbol{\theta} = (a, b)$ and its support is $supp(f_X) = (a, b)$, so that it is not a member of the exponential family.

Exercises

1. Laplaces's distribution takes its values on $\mathbb{R}$ and its PDF is

$$f(x) = \frac{x}{\sigma\sqrt{2}} \exp\left(-\left|\frac{x - m}{\sigma/\sqrt{2}}\right|\right).$$

   Is Laplace's distribution member of the exponential family? Why?

2. Gumbel's distribution takes its values on $(\alpha, +\infty)$ and its PDF is

$$f(x) = \frac{1}{\beta} \exp\left(-\left(\frac{x - \alpha}{\beta} + \exp\left(-\frac{x - \alpha}{\beta}\right)\right)\right).$$

   Is Gumbel's distribution member of the exponential family? Why?

3. Pareto's distribution takes its values on $(\alpha, +\infty)$ and its PDF is

$$f(x) = \frac{\beta\alpha^\beta}{x^{\beta+1}}.$$

   Is Pareto's distribution member of the exponential family? Why?

1.9 Random Vectors

The notion of random variable generalizes to random vectors. Analogously to random variables (Ω, P) and X: $\Omega \longrightarrow \mathbb{R}^n$: we have Let us assume that X is defined for all the elements of Ω. Let $A \subset \mathbb{R}^n$: analogously to random variables.

Definition 1.8 A random vector of dimension n on the probability space (Ω, P) is an application X: $\Omega \longrightarrow \mathbb{R}^n$, i.e., $X = (X_1, \ldots, X_n)$, where X_i: $\Omega \longrightarrow \mathbb{R}$ is a random variable, for $1 \le i \le n$.

For $A \subset \mathbb{R}^n$, we consider $X^{-1}(A) = \{\omega \in \Omega\text{: } X(\omega) \in A\}$ and $P(X \in A) = P(X^{-1}(A))$.

Let $x = (x_1, \ldots, x_n) \in \mathbb{R}^n$. We use the notation $X \le x$ for the set of inequalities $X_i \le x_i$, for $1 \le i \le n$ (i.e., $X_i \in (-\infty, x_i]$, for $1 \le i \le n$). The **cumulative distribution function** (CDF) of X is

$$F(x) = P(X \le x) = P\left(X \in \prod_{i=1}^{n}]-\infty, x_i]\right).$$

F is the **distribution** of X. The ***probability density function*** (PDF) of X is

$$f = \frac{\partial^n}{\partial x_1 \partial x_2 \ldots \partial x_n} F.$$

The marginal distribution of X_i is

$$F_i(x_i) = P(X_i \le x_i) = P(X_i \in]-\infty, x_i]).$$

The marginal PDF of X_i is $f_i = F_i'$. ∎

Definition 1.9 Let g: $\mathbb{R}^n \longrightarrow \mathbb{R}$ be a function and $Y = g(X)$, where X is a random vector of dimension n on (Ω, P). The ***mean*** or ***expectation*** of Y is $E(Y)$ given by

1. Let Ω be finite or enumerable and P be generated by the probability mass function μ. Then

$$E(Y) = E(g(X)) = \sum_{\omega \in \Omega} \mu(\omega) g(X(\omega)).$$

(continued)

Definition 1.9 (continued)

2. Let $\Omega \subset \mathbb{R}$. and P be generated by the probability mass density μ. Then

$$E(Y) = E(g(\boldsymbol{X})) = \int_{\Omega} g(\boldsymbol{X}(\omega))\mu(\omega)d\omega.$$

If the PDF exists, we have

$$E(g(\boldsymbol{X})) = \int_{\mathbb{R}} g(\boldsymbol{x})f(\boldsymbol{x})d\boldsymbol{x}, d\boldsymbol{x} = dx_1 \ldots dx_n. \tag{1.41}$$

Definition 1.10 Let $\boldsymbol{X}$ be a random vector of dimension n on (Ω, P). The ***covariance matrix*** of $\boldsymbol{X}$ is

$$\boldsymbol{C} = (C_{ij} : 1 \le i, j \le n), C_{ij} = E((X_i - E(X_i))(X_j - E(X_j))).$$

The ***correlation matrix*** of $\boldsymbol{X}$ is
$\boldsymbol{R} = (R_{ij} : 1 \le i, j \le n), R_{ij} = \frac{C_{ij}}{\sigma(X_i)\sigma(X_j)}$. ■

We have, for $1 \le i, j \le n$:

$$C_{ij} = E(X_i X_j) - E(X_i)E(X_j), \tag{1.42}$$

$$C_{ii} = V(X_i); \tag{1.43}$$

$$|C_{ij}| \le \sqrt{C_{ii}C_{jj}} = \sqrt{V(X_i)\, V(X_j)} \tag{1.44}$$

$$\boldsymbol{C} \textit{ is positive} : \boldsymbol{a}^t\boldsymbol{C}\boldsymbol{a} = V(\boldsymbol{a}^t\boldsymbol{X}) \ge 0, \forall \boldsymbol{a} \in \mathbb{R}^n. \tag{1.45}$$

Considering (1.44), we have

$$|R_{ij}| \le 1, \text{for } 1 \le i, j \le n; R_{ii} = 1, \text{for } 1 \le i \le n. \tag{1.46}$$

Indeed, the correlation R_{ij} measures the linear dependence between $X_i$and X_j. Indeed, we can look for the best approximation of X_i as an affine function of X_j, in the sense of the least squares by minimizing $K(a, b) = E((X_i - aX_j - b)^2)$: the solution is $a_{ij}X_j + b_{ij}$, where

$$a_{ij} = \frac{C_{ij}}{\sigma(X_j)}, b_{ij} = E(X_i) - a_{ij}E(X_j). \tag{1.47}$$

Then,

$$E\left((X_i - a_{ij}X_j - b_{ij})^2\right) = \left(1 - |R_{ij}|^2\right)V(X_i), \tag{1.48}$$

so that $E((X_i - a_{ij}X_j - b_{ij})^2) = 0$, when $|R_{ij}| = 1$: in such a situation, $X_i = a_{ij}X_j + b_{ij}$ is almost sure.

Alternatively, we can look for the best approximation of X_i as a function of X_j, by minimizing

$$K_{ij}(\phi) = E\left((X_i - \phi(X_j))^2\right) \tag{1.49}$$

In this case, the solution is

$$\phi_i(X_j) = E(X_i|X_j), \tag{1.50}$$

the mean of X_i conditional to X_i. We can use ϕ_i to measure the nonlinear independence between X_i and X_j by considering $K_{ij}(\phi_i)$: its value is equal to zero if and only if $X_i = \phi(X_j)$. This observation suggests that we can use as measure

$$\eta(X_i|X_j) = \psi(K_{ij}(\phi_i), V(X_i)), \tag{1.51}$$

where ψ: $[0, +\infty) \times [0, +\infty) \longrightarrow [0, 1]$ is a strictly positive decreasing function such that $\psi(0, V(X_i)) = 1$ – for instance,

$$\psi_1(s_1, s_2) = 1 - \frac{s_1}{s_2}; \quad \psi_2(s_1, s_2) = \mathrm{e}^{-s_1}, \psi_3(s_1, s_2) = \frac{\mathrm{e}^{\psi_1(s_1, s_2)} - 1}{\mathrm{e} - 1}. \tag{1.52}$$

We have $K_{ij}(\phi_i) \leq K_{ij}(E(X_i)) = V(X_i)$, so that $0 \leq \psi_1(K_{ij}(\phi_i), V(X_i)) \leq 1$. In addition, $\eta\left(X_i|X_{j_1}, \ldots, X_{j_\ell}\right)$ can be evaluated (see Example 1.21). However, in general, $\eta(X_i| X_j) \neq \eta(X_j| X_i)$, so that we can prefer a ***generalized correlation***

$$\eta(X_i, X_j) = \frac{1}{2}\left(\psi(K_{ij}(\phi_i), V(X_i)) + \psi(K_{ji}(\phi_j), V(X_j))\right). \tag{1.53}$$

ϕ_i can be evaluated in a closed form using the marginal densities and conditional densities – we shall develop this point in the sequel (Eqs. 1.56, 1.57, 1.63, 1.64, 1.65, 1.66 and 1.73, 1.74, 1.75, 1.76).

Definition 1.11 Let $X = (X_1, \ldots, X_n)$ be a random vector of dimension n on (Ω, P). The ***marginal distribution of*** of X_i is

$$F_i(x_i) = P(X_i \leq x_i) = P\left(X_i \leq x_i \textit{ and } X_j < +\infty, \textit{for } j \neq i\right).$$

The ***marginal density*** of X_i is

$$f_i(x_i) = F_i'(x_i) = \frac{d}{dx_i} F_i(x_i).\blacksquare$$

We have

$$F_i(x_i) = F(\boldsymbol{X}_{\infty,i}), \quad \mathrm{f_i}(x_i) = \int_{-\infty}^{A_1} dx_1 \int_{-\infty}^{A_2} dx_2 \ldots \int_{-\infty}^{A_n} f(\boldsymbol{x}) dx_n, \tag{1.54}$$

with

$$\boldsymbol{X}_{\infty,i} = (A_1, \ldots, A_n); A_k = +\infty, \textit{if } k \neq i, A_i = x_i. \tag{1.55}$$

The density of X_i conditional to $(X_j, j \neq i)$ is

$$f(X_i | X_j, j \neq i) = \frac{f(\boldsymbol{x})}{\bar{f}(x_j, j \neq i)}, \bar{f}(x_j, j \neq i) = \int_{\mathbb{R}} f(\boldsymbol{x}) dx_i. \tag{1.56}$$

The mean of X_i conditional to $(X_j, j \neq i)$ is

$$E(X_i | X_j, j \neq i) = \int_{\mathbb{R}} x_i f(X_i | X_j, j \neq i) dx_i. \tag{1.57}$$

Definition 1.12 Let $X = (X_1, \ldots, X_n)$ be a random vector of dimension n on (Ω, P). The ***characteristic function of*** of X is

$$\varphi(t = (t_1, \ldots, t_n)) = E\left(\exp\left(i \sum_{j=1}^{n} t_j X_j\right)\right).\blacksquare$$

Example 1.21 Let us consider $\boldsymbol{X} = (X_1, X_2, X_3) \in (0, 1)^3$ such that

$$F(\boldsymbol{x}) = x_1 x_2^2 x_3^3 \ on \ (0, 1)^3$$

The PDF of $\boldsymbol{X}$ is

$$f(\boldsymbol{x}) = 6 x_2 x_3^2 \ on \ (0, 1)^3.$$

We have

$$E(X_1) = \frac{1}{2}, E(X_2) = \frac{2}{3}, E(X_3) = \frac{3}{4},$$

$$E(X_1^2) = \frac{1}{3}, E(X_2^2) = \frac{1}{2}, E(X_3^2) = \frac{3}{5}.$$

Using R:

```
source("ints.R")
integ = ints$new()

f = function(x1,x2,x3){6*x2*x3^2}
a = c(0,0,0)
b = c(1,1,1)

h1 = function(x1,x2,x3){x1*f(x1,x2,x3)}
EX1 = integ$int3(h1,a,b)
print(paste("E(X1) = ", EX1,", exact = ",0.5))
## [1] "E(X1) =  0.5 , exact =  0.5"
h2 = function(x1,x2,x3){x2*f(x1,x2,x3)}
EX2 = integ$int3(h2,a,b)
print(paste("E(X2) = ", EX2,", exact = ",2/3))
## [1] "E(X2) =  0.666666666666667 , exact =  0.6666666666
66667"
h3 = function(x1,x2,x3){x3*f(x1,x2,x3)}
EX3 = integ$int3(h3,a,b)
print(paste("E(X3) = ", EX3,", exact = ",0.75))
## [1] "E(X3) =  0.75 , exact =  0.75"
h12 = function(x1,x2,x3){x1^2*f(x1,x2,x3)}
EX12 = integ$int3(h12,a,b)
print(paste("E(X1^2) = ", EX12,", exact = ",1/3))
## [1] "E(X1^2) =  0.333333333333333 , exact =  0.33333333
3333333"
```

(continued)

Example 1.21 (continued)

```
h22 = function(x1,x2,x3){x2^2*f(x1,x2,x3)}
EX22 = integ$int3(h22,a,b)
print(paste("E(X2^2) = ", EX22,", exact = ",
## [1] "E(X2^2) =  0.5 , exact =  0.5"
h32 = function(x1,x2,x3){x3^2*f(x1,x2,x3)}
EX32 = integ$int3(h32,a,b)
print(paste("E(X3^2) = ", EX32,", exact = ",
## [1] "E(X3^2) =  0.6 , exact =  0.6"
```

The marginal CDFs are

$$F_1(x_1) = x_1, F_2(x_2) = x_2^2, F_3(x_3) = x_3^3 .$$

Under R, we determine the marginal distributions as follows:

```
F = function(x1,x2,x3){x1*(x2^2)*(x3^3)}
a = c(0,0,0) #Lower bound for each x
b = c(1,1,1) #Upper bound for each x

F1 = function(x1){F(x1,b[[2]],b[[3]])}
F2 = function(x2){F(b[[1]],x2,b[[3]])}
F3 = function(x3){F(b[[1]],b[[2]],x3)}
```

The results appear in Fig. 1.7.

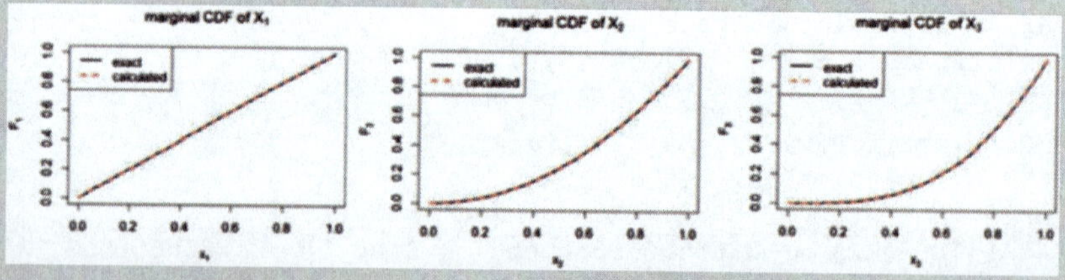

Fig. 1.7 Marginal CDFs

The marginal PDFs are

$$f_1(x_1) = 1, f_2(x_2) = 2x_2, f_3(x_3) = 3x_3^2 .$$

(continued)

Example 1.21 (continued)

Under R, we determine the marginal PDFs as follows:

```
source("ints.R")
integ = ints$new()
f = function(x1,x2,x3){6*x2*(x3^2)}
f1 = integ$int3marg1(f,a,b)
f2 = integ$int3marg2(f,a,b)
f3 = integ$int3marg3(f,a,b)
```

Function `int3margi` returns a function having x_i as an argument – the function evaluates the integral of f with respect to the other variables. For instance,

```
                int3marg1 = function(f,a,b) {
                  library(pracma)
                  aux = function(x1) { quad(Vectorize(functi
on(x2) {
                    quad(function(x3) { f(x1,x2,x3) }, a[[3]
], b[[3]]) }),
                    a[[2]],b[[2]]) }
                    return(aux)
                }
```

The results appear in Fig. 1.8.

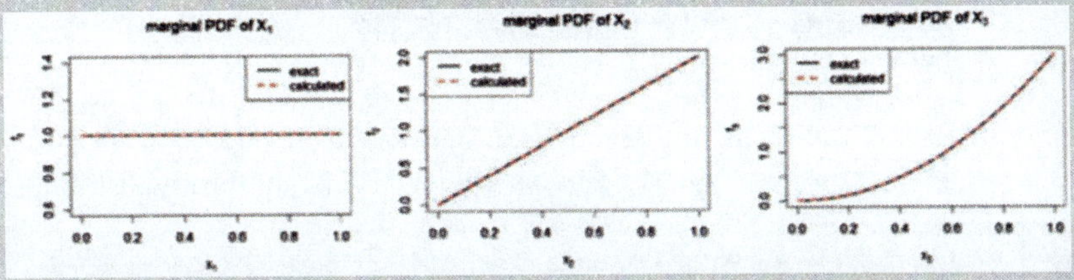

Fig. 1.8 Marginal PDFs

The conditional densities are

$$f(X_1|X_2, X_3) = 1,\ \ f(X_2|X_1, X_3) = 2x_2,\quad f(X_3|X_1, X_2) = 3x_3^2.$$

In this case, they coincide with the marginal densities (this indicates independency of the variables). The corresponding conditional means coincide with the means:

(continued)

Example 1.21 (continued)

$$E(X_1|X_2,X_3)=E(X_1),\quad E(X_2|X_1,X_3)=E(X_2),\quad E(X_3|X_1,X_2)=E(X_3).$$

Thus, using ψ_1 (see Eqs. 1.51 and 1.52):

$$\eta(X_1|X_2,X_3)=\eta(X_2|X_1,X_3)=\eta(X_3|X_1,X_2)=0.$$

These quantities and functions can be determined under R using the class `ints`. For instance, the evaluation of $f(X_3|X_1, X_2)$ (`fcond3`) and $E(X_3|X_1, X_2)$ (`Econd3`) is made as follows:

```
f12 = function(x1,x2){
  ff3 = function(x3){f(x1,x2,x3)}
  aux = integ$int1(ff3,a[[3]], b[[3]])
  return(aux)
}

zero = 1e-5

fcond3 = function(x1,x2,x3){
  aux = f12(x1,x2)
  if (abs(aux) > zero){v = f(x1,x2,x3)/aux}else{v=0}
  return(v)}

Econd3 = function(x1,x2){
      ff3 = function(x3){x3*fcond3(x1,x2,x3)}
      aux = integ$int1(ff3,a[[3]],b[[3]])
      return(aux)}
```

Using these functions, the comparison between the exact value $E(X_3|X_1, X_2)$ and those furnished by `Econd3` on a grid of 21×21 uniformly distributed points on $(0, 1)^2$ led to an absolute error of $2e-15$.

Let $g(\boldsymbol{X}) = X_1 + X_2 + X_3$. We have

$$E(g(\boldsymbol{X}))=6\int_{(0,1)^3}(x_1+x_2+x_3)x_2x_3^2dx_1dx_2dx_3=\frac{23}{12}.$$

(continued)

Example 1.21 (continued)

This result can be obtained using R as follows:

```
f = function(x1,x2,x3){6*x2*x3^2}
g = function(x1,x2,x3){x1+x2+x3}
h = function(x1,x2,x3){f(x1,x2,x3)*g(x1,x2,x3)}
a = c(0,0,0)
b = c(1,1,1)

mean_g = integ$int3(h,a,b)
print(paste("mean of g = ", mean_g,", exact = ",23/12))

## [1] "mean of g =  1.91666666666667 , exact =  1.9166666
6666667"
```

The covariance matrix is

$$C = \begin{pmatrix} 1/12 & 0 & 0 \\ 0 & 1/18 & 0 \\ 0 & 0 & 3/80 \end{pmatrix}.$$

We can use R to determine C as follows:

```
EX = c(EX1,EX2,EX3)
matCex = matrix(c(1/12,0,0,0,1/18,0,0,0,3/80), ncol=3,byro
w=TRUE) #exact value

matC = matrix(nrow = 3, ncol = 3)
for (i in 1:3){
  va1 = paste0("x",i,"*","x",i,"*f(x1,x2,x3)")
  h = function(x1,x2,x3){eval(parse(text=va1))}
  aux = integ$int3(h,a,b) - EX[[i]]*EX[[i]]
  matC[[i,i]] = aux
  if (i < 3){
    for (j in (i+1):3){
      va1 = paste0("x",i,"*","x",j,"*f(x1,x2,x3)")
      h = function(x1,x2,x3){eval(parse(text=va1))}
      aux = integ$int3(h,a,b) - EX[[i]]*EX[[j]]
      matC[[i,j]] = aux
      matC[[j,i]] = aux
    }
  }
}
print(matC)

##               [,1]          [,2]   [,3]
## [1,]  8.333333e-02 -5.551115e-17 0.0000
## [2,] -5.551115e-17  5.555556e-02 0.0000
## [3,]  0.000000e+00  0.000000e+00 0.0375
```

(continued)

Example 1.21 (continued)

```
print(matCex)

##            [,1]       [,2]   [,3]
## [1,] 0.08333333 0.00000000 0.0000
## [2,] 0.00000000 0.05555556 0.0000
## [3,] 0.00000000 0.00000000 0.0375

err = norm(matC-matCex)
print(paste("error: absolute = ",err,", relative = ",err/n
orm(matCex)))

  ## [1] "error: absolute = 1.31838984174237e-16 ,
relative = 1.58206781009085e-15"
```

The correlation matrix is

$$R = \begin{pmatrix} 1 & 0 & 0 \\ 0 & 1 & 0 \\ 0 & 0 & 1 \end{pmatrix}.$$

We can determine the correlation matrix R as follows:

```
matRex = diag(3) #exact value
VX = c(EX12-EX1^2,EX22 - EX2^2,EX32 - EX3^2)
matR = matrix(nrow = 3, ncol = 3)
for (i in 1:3){
    for (j in 1:3){
      matR[[i,j]] = matC[[j,i]]/(sqrt(VX[[i]]*VX[[j]]))
  }
}
print(matR)

##              [,1]         [,2] [,3]
## [1,]  1.00000e+00 -8.15844e-16    0
## [2,] -8.15844e-16  1.00000e+00    0
## [3,]  0.00000e+00  0.00000e+00    1

print(matRex)

##      [,1] [,2] [,3]
## [1,]    1    0    0
## [2,]    0    1    0
## [3,]    0    0    1

err = norm(matR-matRex)
print(paste("error: absolute = ",err,", relative = ",err/n
orm(matRex)))

## [1] "error: absolute =  8.15843973306311e-16 , relative
=  8.15843973306311e-16"
```

(continued)

Example 1.21 (continued)

Finally, we have

$$\varphi(t_1,t_2,t_3) = -\frac{6i(-1+e^{it_1})(-1+e^{it_2}(1-it_2))\left(-2i+e^{it_3}\left(2i+2t_3-it_3^2\right)\right)}{t_1 t_2^2 t_3^3},$$

and

$$\varphi(1,2,3) = -\frac{1}{18}i(-1+e^i)\left(-1+(1-2i)e^{2i}\right)\left(-2i+(6-7i)e^{3i}\right).$$

Thus, $\varphi(1,2,3) \approx -0.4025967431665892 - 0.6023362247523488\, i$. Let us evaluate it using R:

```
phi = function(t1,t2,t3){
  h = function(x1,x2,x3){f(x1,x2,x3)*exp(1i*(t1*x1+t2*x2+t
3*x3))}
  aux = integ$int3(h,a,b)
  return(aux)
}

aux = phi(1,2,3)
print(paste(expression(phi),"(1,2,3) = ",aux))

## [1] "phi (1,2,3) =  -0.402596743164089-0.60233622475644
3i"
```

For a couple of variables (X, Y) such that $X(\Omega) = \{X_1,\ldots,X_m\}$ and $Y(\Omega) = \{Y_1,\ldots,Y_n\}$, it is usual to introduce $p_{ij} = P(X = X_i, Y = Y_j)$ and to represent the couple by a Table analogous to Table 1.3.

In this case, the marginal distributions of X and Y are given by

$$P(X = X_i) = p_{i\blacksquare} = \sum_{j=1}^{n} p_{ij}, P(Y = Y_j) = p_{\blacksquare j} = \sum_{i=1}^{m} p_{ij}. \tag{1.58}$$

Table 1.3 Representation of a pair of random variables

Y			
X	Y_1	...	Y_n
X_1	p_{11}	...	p_{1n}
...	$\vdots$	$\ddots$	$\vdots$
X_m	p_{m1}	...	p_{mn}

We have also

$$E(g(X,Y)) = \sum_{i=1}^{m}\sum_{j=1}^{n} p_{ij} g(X_i, Y_j). \tag{1.59}$$

The characteristic function is

$$\varphi(t_1, t_2) = \sum_{i=1}^{m}\sum_{j=1}^{n} p_{ij} e^{it_1 X_i} e^{it_2 Y_j}. \tag{1.60}$$

The covariance between X and Y is

$$\text{cov}(X,Y) = E(XY) - E(X)E(Y), \tag{1.61}$$

And the coefficient of correlation between X and Y is

$$\rho(X,Y) = \frac{\text{cov}(X,Y)}{\sigma(X)\sigma(Y)}. \tag{1.62}$$

The distribution of X conditional to Y is

$$P(X = X_i | Y = Y_j) = \frac{P(X = X_i, Y = Y_j)}{P(Y = Y_j)} = \frac{p_{ij}}{p_{\blacksquare j}}, \tag{1.63}$$

and the distribution of Y conditional to X is

$$P(Y = Y_j | X = X_i) = \frac{P(X = X_i, Y = Y_j)}{P(X = X_i)} = \frac{p_{ij}}{p_{i\blacksquare}}. \tag{1.64}$$

The mean of X conditional to Y is

$$E(X | Y = Y_j) = \sum_{i=1}^{m} \frac{p_{ij}}{p_{i\blacksquare}} X_i. \tag{1.65}$$

Analogously, the mean of Y conditional to X is

$$E(Y | X = X_i) = \sum_{j=1}^{n} \frac{p_{ij}}{p_{\blacksquare j}} Y_j. \tag{1.66}$$

These elements can be determined using the class `drc` (discrete random couple) below:

```
library(R6)
source("drv.R")
#
drc=R6Class("drc",
 private= list(
  outer1d = function(x,y,f){
   v = matrix(nrow = length(x), ncol = length(y))
   for (i in 1:length(x)){
    for (j in 1:length(y)){
     v[i,j] = f(x[i],y[j])
    }
   }
   return(v)
  }
 ),
 public=list(
  X = NULL, Y =NULL, p=NULL, margin=NULL, va=NULL,
  cov=NULL, cond=NULL, condmean = NULL,
  corr = NULL, phi = NULL,
  initialize=function(X,Y,p,Xnames,Ynames){
   self$X = as.vector(X)
   self$Y = as.vector(Y)
   self$p = p
   pX <- as.vector(marginSums(p,1))
   pY <- as.vector(marginSums(p,2))
   m <- list(pX,pY)
   names(m) <- c("X","Y")
    self$margin = m
    vaX = drv$new(X,pX)
    vaY = drv$new(Y,pY)
    va = list(vaX,vaY)
    names(va) <- c("X","Y")
    self$va = va
    f = function(x,y){x*y}
    aux = private$outer1d(X,Y,f)
    mXY = sum(p * aux)
    mXYc = mXY - vaX$mean*vaY$mean
    covmat = matrix(c(vaX$var,mXYc,mXYc,vaY$var),nrow = 2, n
col= 2, byrow=TRUE)
    aux = list(mXY, mXYc,covmat)
    names(aux) <- c("notcenter","center","matrix")
    self$cov = aux
    rho = mXYc/sqrt(vaX$var*vaY$var)
    matR = matrix(c(1,rho,rho,1),ncol=2,byrow=TRUE)
    d <- dim(p)
    pX2Y = matrix(nrow = d[[2]], ncol = d[[1]])
```

```
    for (i in 1:d[[1]]){
      for (j in 1:d[[2]]){
        pX2Y[[j,i]] = p[[i,j]]/pY[[j]]
      }
    }
    pY2X = matrix(nrow = d[[1]], ncol = d[[2]])
    for (i in 1:d[[1]]){
      for (j in 1:d[[2]]){
        pY2X[[i,j]] = p[[i,j]]/pX[[i]]
      }
    }
    cxy <- list(pX2Y,pY2X)
    names(cxy)<- c("X2Y","Y2X")
    self$cond = cxy
    EX2Y = as.vector(cxy$X2Y %*% as.vector(X))
    EY2X = as.vector(cxy$Y2X %*% as.vector(Y))
    v = list(EX2Y,EY2X)
    names(v) <- c("X2Y","Y2X")
    self$condmean = v
    sX = 0
    for (i in 1:length(X)){
      for (j in 1: length(Y)){
        sX = sX + p[[i,j]]*(X[[i]] - EX2Y[[j]])^2
      }
    }
    sY = 0
   for (i in 1:length(X)){
     for (j in 1: length(Y)){
       sY = sY + p[[i,j]]*(Y[[j]] - EY2X[[i]])^2
     }
   }
   etaX2Y = 1 - sX/vaX$var
   etaY2X = 1 - sY/vaY$var
   eta = (etaX2Y + etaY2X)/2
   aux = list(rho,matR,etaX2Y, etaY2X, eta)
   names(aux) = c("coeff","matrix","X2Y","Y2X","gen")
   self$corr = aux
   phi = function(t1,t2){
      ff = function(x,y){exp(1i*(t1*x+t2*y))}
      aux = private$outer1d(X,Y,ff)
      aux2 = p*aux
      aux3 = sum(aux2)
      return(aux3)
   }
   self$phi = phi
  }
 )
)
```

Example 1.22 Let us consider the couple (X, Y) represented in Table 1.4.

Table 1.4 A couple (X, Y)

Y			
X	-1	0	1
0	0.2	0.15	0.1
1	0.1	0.25	0.2

The marginal distributions are shown in Table 1.5.

Table 1.5 Marginal distributions

Marginal distribution of X			Marginal distribution of Y			
Value	0	1	Value	-1	0	1
Probability	0.45	0.55	Probability	0.3	0.4	0.3

We have

$$E(X) = 0.55, E(Y) = 0, E(X^2) = 0.55, E(Y^2) = 0.6, E(XY) = 0.1.$$

We have

$$\varphi(t_1, t_2) = 0.15 + 0.25e^{it_1} + e^{-it_2}(0.2 + 0.1e^{it_1}) + e^{it_2}(0.1 + 0.2e^{it_1}).$$

Thus,

$$\varphi(1, 2) \approx 0.016263256769617096 + 0.06351490665058973\ \mathrm{i}.$$

We have also

$$cov(X, Y) = E(XY) - E(X)E(Y) = 0.1,$$

and the covariance matrix is

$$C = \begin{pmatrix} 0.2475 & 0.1 \\ 0.1 & 0.6 \end{pmatrix}.$$

(continued)

Example 1.22 (continued)

The correlation coefficient between X and Y is

$$\rho(X,Y)=\frac{cov(X,Y)}{\sigma(X)\sigma(Y)}=\frac{0.1}{\sqrt{0.2475\times 0.6}}\approx 0.2595$$

The conditional distributions are given in Table 1.6.

Table 1.6 Conditional distributions

Conditional distribution of X			Conditional distribution of Y			
Value	0	1	Value	−1	0	1
$P(X\mid Y=-1)$	2/3	1/3	$P(Y\mid X=0)$	4/9	1/3	2/9
$P(X\mid Y=0)$	0.375	0.625	$P(Y\mid X=1)$	2/11	5/11	4/11
$P(X\mid Y=1)$	1/3	2/3				

The conditional means are given in Table 1.7.

Table 1.7 Conditional means

	$E(X\mid Y)$				$E(Y\mid X)$	
Y	−1	0	1	X	0	1
$E(X\mid Y)$	1/3	0.625	2/3	$E(Y\mid X)$	−2/9	2/11

Then

$$E\left((X-E(X|Y))^2\right)=\sum_i\sum_j p_{ij}\left(X_i-E\left(X|Y=Y_j\right)\right)^2\approx 0.22708,$$

$$E\left((Y-E(Y|X))^2\right)=\sum_i\sum_j p_{ij}\left(Y_j-E\left(Y|X=X_j\right)\right)^2\approx 0.55960,$$

so that, using ψ_1 (see Eqs. (1.51 and 1.52)),

$$\eta(X|Y)\approx 0.08249, \eta(Y|X)\approx 0.06734, \eta(X,Y)\approx 0.07491$$

(continued)

Example 1.22 (continued)

We can evaluate these quantities using the class `drc` as follows:

```
X = c(0,1)
Y = c(-1,0,1)
p = matrix(c(0.2,0.15,0.1,0.1,0.25,0.2), nrow=length(X), n
col=length(Y),byrow=TRUE)

source("drc.R")
cva1 = drc$new(X,Y,p)

print("marginal distrib of X :")
## [1] "marginal distrib of X :"
print(cva1$margin$X)
## [1] 0.45 0.55
print("marginal distrib of Y :")
## [1] "marginal distrib of Y :"
print(cva1$margin$Y)
## [1] 0.3 0.4 0.3
print(paste("mean of X = ",cva1$va$X$mean,", mean of Y = "
,cva1$va$Y$mean))
## [1] "mean of X =  0.55 , mean of Y =  0"
print(paste("mean of X^2 = ",cva1$va$X$m2,", mean of Y^2 =
",cva1$va$Y$m2))
## [1] "mean of X^2 =  0.55 , mean of Y^2 =  0.6"
aux = cva1$phi(1,2)
print(paste("phi(1,2) = ",aux))
## [1] "phi(1,2) =  0.0162632567696171+0.0635149066505897i
"
print("Covariance Matrix C = ")
## [1] "Covariance Matrix C = "
print(cva1$cov$matrix)
##        [,1] [,2]
## [1,] 0.2475  0.1
## [2,] 0.1000  0.6
print("Correlation Matrix R = ")
## [1] "Correlation Matrix R = "
print(cva1$corr$matrix)
##           [,1]      [,2]
## [1,] 1.0000000 0.2594996
## [2,] 0.2594996 1.0000000
print("P(X | Y) :")
## [1] "P(X | Y) :"
print(cva1$cond$X2Y)
##           [,1]      [,2]
## [1,] 0.6666667 0.3333333
## [2,] 0.3750000 0.6250000
## [3,] 0.3333333 0.6666667
```

(continued)

Example 1.22 (continued)

```
print("P(Y | X) :")
## [1] "P(Y | X) :"
print(cva1$cond$Y2X)
##           [,1]      [,2]      [,3]
## [1,] 0.4444444 0.3333333 0.2222222
## [2,] 0.1818182 0.4545455 0.3636364
print("E(X | Y) :")
## [1] "E(X | Y) :"
print(cva1$condmean$X2Y)
## [1] 0.3333333 0.6250000 0.6666667
print("E(Y | X) :")
## [1] "E(Y | X) :"
print(cva1$condmean$Y2X)
## [1] -0.2222222  0.1818182
print(paste(expression(eta),"(X | Y) :",cva1$corr$X2Y))
## [1] "eta (X | Y) : 0.0824915824915824"
print(paste(expression(eta),"(Y | X) :",cva1$corr$Y2X))
## [1] "eta (Y | X) : 0.0673400673400676"
print(paste(expression(eta),"(X , Y) :",cva1$corr$gen))
## [1] "eta (X , Y) : 0.074915824915825"
```

For a couple of variables having a distribution $F(x, y) = P(X \leq x, Y \leq y)$ and a density $f(x, y) = \partial^2 F/\partial x \partial y$, the marginal distributions are

$$F_X(x) = F(x, +\infty), \ f_X(x) = \int_{\mathbb{R}} f(x, y) dy; \tag{1.67}$$

$$F_Y(y) = F(+\infty, y), \ f_Y(y) = \int_{\mathbb{R}} f(x, y) dx. \tag{1.68}$$

We have also

$$E(g(X, Y)) = \int_{\mathbb{R}} \int_{\mathbb{R}} g(x, y) f(x, y) dxdy \, . \tag{1.69}$$

The characteristic function is

$$\varphi(t_1, t_2 = \int_{\mathbb{R}} \int_{\mathbb{R}} f(x, y) e^{it_1 x} e^{it_2 y} dxdy. \tag{1.70}$$

The covariance between X and Y is

$$\mathrm{Cov}(X, Y) = E(XY) - E(X)E(Y), \tag{1.71}$$

and the correlation coefficient between X and Y is

$$\rho(X, Y) = \frac{\mathrm{Cov}(X, Y)}{\sigma(X)\sigma(Y)}. \tag{1.72}$$

We can also consider the events $E = X \in (x, x + dx)$ and $B = Y \in (y, y + dy)$. Then,

$$P(E|F) = \frac{P(E \cap F)}{P(F)} = \frac{f(x, y)dxdy}{f_y(y)dy} = \frac{f(x, y)}{f_Y(y)}dx$$

Then, the PDF of X conditional to Y is

$$f(x|Y = y) = \frac{f(x, y)}{f_Y(y)}, \tag{1.73}$$

Analogously, the PDF of Y conditional to X is

$$f(y|X = x) = \frac{f(x, y)}{f_X(x)}. \tag{1.74}$$

The conditional means are

$$E(X|Y = y) = \int_{\mathbb{R}} xf(x|Y = y)dx, \tag{1.75}$$

$$E(Y|X = x) = \int_{\mathbb{R}} yf(y|X = x)dy. \tag{1.76}$$

Example 1.23 Let us consider a couple $(X, Y) \in (0, 1)^2$ such that its CDF is

$$F(x, y) = \frac{1}{2}\left(x^2 y + xy^2\right) \quad \text{on } (0, 1)^2.$$

Thus, the PDF of the couple (X, Y) is

$$f(x, y) = x + y \text{ on } (0, 1)^2$$

(continued)

Example 1.23 (continued)

The marginal CDFs are

$$F_X(x) = \frac{1}{2}\left(x^2 + x\right) \; on \; (0,1); \; F_Y(y) = \frac{1}{2}\left(y^2 + y\right) \text{ on } (0,1).$$

The marginal PDFs are

$$f_X(x) = x + \frac{1}{2} \; on \; (0,1); \; f_Y(y) = y + \frac{1}{2} \text{ on } (0,1).$$

We have

$$E(X) = \int_0^1 \int_0^1 x(x+y)dxdy = \frac{7}{12}.$$

$$E\left(X^2\right) = \int_0^1 \int_0^1 x^2(x+y)dxdy = \frac{5}{12}.$$

Analogously:

$$E(Y) = \int_0^1 \int_0^1 y(x+y)dxdy = \frac{7}{12}.$$

$$E\left(X^2\right) = \int_0^1 \int_0^1 y^2(x+y)dxdy = \frac{5}{12}.$$

We have also

$$\varphi(t_1, t_2) = \frac{-i(-1+e^{it_1})(-1+e^{it_2})\mathrm{t}_2 + \mathrm{t}1(-i+e^{it_1}(i+\mathrm{t}_2)+e^{it_2}(i+\mathrm{t}_2-e^{it_1}(i+2\mathrm{t}_2)))}{\mathrm{t}_1^2\mathrm{t}_2^2}.$$

Thus,

$$\varphi(1,2) \approx -0.15516607141933914 + 0.819875172198004\,\mathrm{i}.$$

(continued)

Example 1.23 (continued)

We have

$$E(XY) = \int_0^1\int_0^1 xy(x+y)dxdy = \frac{2}{3}.$$

The covariance matrix is

$$C = \begin{pmatrix} 11/144 & -1/144 \\ -1/144 & 11/144 \end{pmatrix}.$$

The correlation between X and Y is

$$\rho(X,Y) = \frac{-1/144}{11/144} = -\frac{1}{11}$$

The conditional PDFs are, for $0 \leq x, y \leq 1$:

$$f(x|Y=y) = 2\frac{x+y}{2y+1}, f(y|X=x) = 2\frac{x+y}{2x+1}.$$

The conditional means are

$$E(X|Y=y) = \frac{2}{2y+1}\int_0^1 x(x+y)dx = \frac{2+3y}{3+6y},$$

$$E(Y|X=x) = \frac{2}{2x+1}\int_0^1 y(x+y)dx = \frac{2+3x}{3+6x}.$$

Then

$$E\left((X-E(X|Y))^2\right) = \int_0^1\int_0^1 (x-E(X|Y=y))^2 f(x,y)dxdy \approx 0.07570,$$

$$E\left((Y-E(Y|X))^2\right) = \int_0^1\int_0^1 (y-E(Y|X=x))^2 f(x,y)dxdy \approx 0.07570,$$

so that, using ψ_1 (see Eqs. 1.51 and 1.52),

$$\eta(X|Y) \approx 0.008965, \eta(Y|X) \approx 0.008965, \eta(X,Y) \approx 0.008965.$$

(continued)

Example 1.23 (continued)

We can evaluate these quantities using a class `crc` (continuous random couple) defined as follows:

```
library(R6)
source("ints.R")
source("crv.R")
#
crc=R6Class("crc",
  public=list(
  cdf = NULL, pdf = NULL, lower = NULL, upper=NULL,
  margin = NULL, va = NULL, cov =NULL, cond = NULL,
  condmean = NULL, corr=NULL, phi=NULL,
  initialize=function(F,f,a,b){
   integ = ints$new()
   self$cdf = F
   self$pdf = f
   self$lower = a
   self$upper = b
   m = list(NA,NA)
   names(m) = c("cdf","pdf")
   aux = list(NA, NA)
   names(aux) = c("X","Y")
   F1 = function(x){F(x,b[[2]])}
   aux$X = F1
   F2 = function(y){F(b[[1]],y)}
   aux$Y = F2
   m$cdf = aux
   f1 = function(x){
      ff = function(y){f(x,y)}
     aux = integ$int1(ff,a[[2]],b[[2]])
     return(aux)
  }
  aux$X = f1
  f2 = function(y){
     ff = function(x){f(x,y)}
     aux = integ$int1(ff,a[[1]],b[[1]])
     return(aux)
  }
  aux$Y = f2
  m$pdf = aux
  self$margin = m
  vaX = crv$new(F1,f1,a[[1]],b[[1]])
  vaY = crv$new(F2,f2,a[[2]],b[[2]])
  va = list(vaX,vaY)
  names(va) <- c("X","Y")
 self$va = va
 faux = function(x,y){x*y*f(x,y)}
 mXY = integ$int2(faux,a,b)
 mXYc = mXY - vaX$mean*vaY$mean
 covmat = matrix(c(vaX$var,mXYc,mXYc,vaY$var),nrow = 2, nc
ol= 2, byrow=TRUE)
```

(continued)

Example 1.23 (continued)

```
aux = list(mXY, mXYc,covmat)
names(aux) <- c("notcenter","center","matrix")
self$cov = aux
rho = mXYc/sqrt(vaX$var*vaY$var)
matR = matrix(c(1,rho,rho,1),ncol=2,byrow = TRUE)
fX2Y = function(x,y){
     f(x,y)/f2(y)
}
fY2X = function(x,y){
     f(x,y)/f1(x)
}
cxy <- list(fX2Y,fY2X)
names(cxy)<- c("X2Y","Y2X")
self$cond = cxy
EX2Y = function(y){
  ff = function(x){x*fX2Y(x,y)}
  aux = integ$int1(ff,a[[1]],b[[1]])
  return(aux)
}
EY2X = function(x){
  ff = function(y){y*fY2X(x,y)}
  aux = integ$int1(ff,a[[2]],b[[2]])
  return(aux)
 }
 v = list(EX2Y,EY2X)
 names(v) <- c("X2Y","Y2X")
 self$condmean = v
 ff = function(x,y){(x - EX2Y(y))^2*f(x,y) }
 sX = integ$int2(ff,a,b)
 ff = function(x,y){(y - EY2X(x))^2*f(x,y) }
 sY = integ$int2(ff,a,b)
 etaX2Y = 1 - sX/vaX$var
 etaY2X = 1 - sY/vaY$var
 eta = (etaX2Y + etaY2X)/2
 aux = list(rho,matR,etaX2Y, etaY2X, eta)
 names(aux) = c("coeff","matrix","X2Y","Y2X","gen")
 self$corr = aux
 phi = function(t1,t2){
   ff = function(x,y){f(x,y)*exp(1i*t1*x)*exp(1i*t2*y)}
   aux = integ$int2(ff,a,b)
   return(aux)
 }
 self$phi = phi
}
)
)
```

(continued)

Example 1.23 (continued)

These contents must be saved to a file named "`crc.R`." Then, they can be used as follows:

```
F = function(x,y){0.5*(x^2*y+y^2*x)}
f = function(x,y){x+y}
a = c(0,0) #Lower bounds for x and y
b = c(1,1) #Upper bounds for x and y

source("crc.R")
va1 = crc$new(F,f,a,b)

print(paste("mean of X = ",va1$va$X$mean,", exact = ",7/12
))
## [1] "mean of X =  0.583333333333333 , exact =  0.583333
333333333"
print(paste("mean of X^2 = ",va1$va$X$m2,", exact = ",5/12
))
## [1] "mean of X^2 =  0.416666666666667 , exact =  0.4166
66666666667"
print(paste("variance of X = ",va1$va$X$var,", exact = ",1
1/144))
## [1] "variance of X =  0.0763888888888889 , exact =  0.07
63888888888889"
aux = va1$phi(1,2)
print(paste("phi(1,2) = ", aux))
## [1] "phi(1,2) =  -0.155166071422716+0.819875172196479i"
matCex = matrix(c(11/144,-1/144,-1/144,11/144), ncol=2,byr
ow=TRUE) #exact value
print(va1$cov$matrix)
##              [,1]         [,2]
## [1,]  0.076388889 -0.006944444
## [2,] -0.006944444  0.076388889
print(matCex)
##              [,1]         [,2]
## [1,]  0.076388889 -0.006944444
## [2,] -0.006944444  0.076388889
err = norm(matCex - va1$cov$matrix)
matRex = matrix(c(1,-1/11,-1/11,1), ncol=2,byrow=TRUE) #ex
act value
print(va1$corr$matrix)
##             [,1]        [,2]
## [1,]  1.00000000 -0.09090909
## [2,] -0.09090909  1.00000000
```

(continued)

Example 1.23 (continued)

```
print(matRex)

##              [,1]        [,2]
## [1,]  1.00000000 -0.09090909
## [2,] -0.09090909  1.00000000

err = norm(matRex - va1$corr$matrix)
print(paste("error: absolute = ", err,", relative = ", err
/norm(matRex)))

## [1] "error: absolute =  1.12410081243297e-15 , relative
=  1.03042574473022e-15"
```

The figures below present the marginal CDF (Fig. 1.9), marginal PDF (Fig. 1.10), and conditional means (Fig. 1.11)

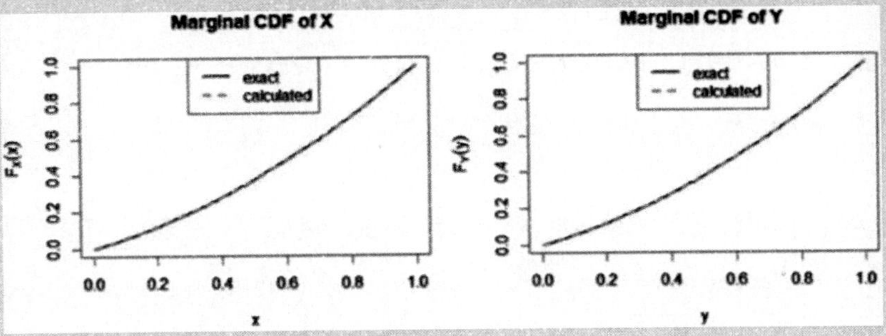

Fig. 1.9 Marginal CDFs

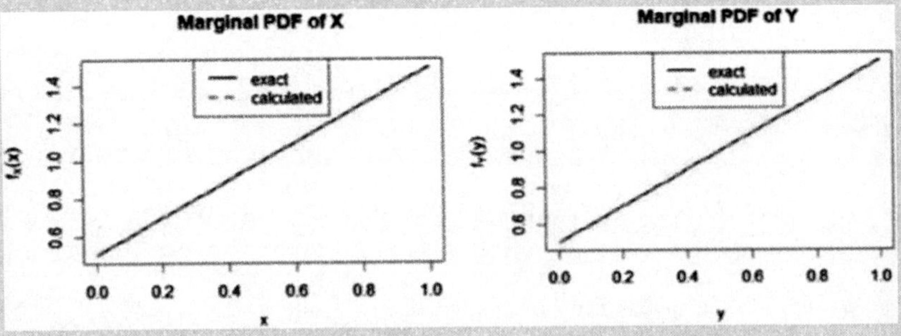

Fig. 1.10 Marginal PDFs

(continued)

Example 1.23 (continued)

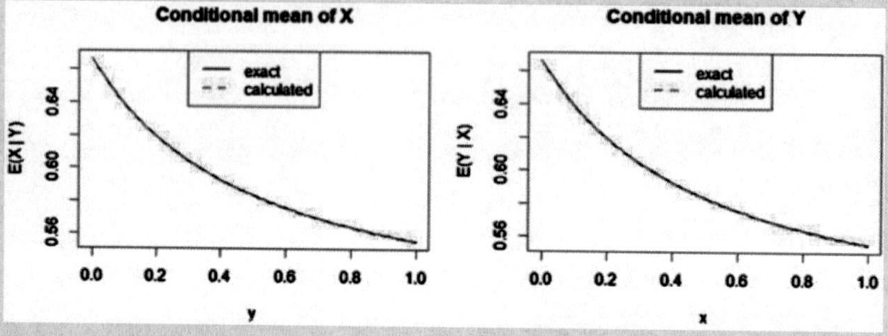

Fig. 1.11 Conditional means

The generalized correlation is evaluated as follows:

```
print(paste(expression(eta),"(X | Y) :",val$corr$X2Y))
## [1] "eta (X | Y) : 0.0089647534826226"
print(paste(expression(eta),"(Y | X) :",val$corr$Y2X))
## [1] "eta (Y | X) : 0.00896475364901683"
print(paste(expression(eta),"(X , Y) :",val$corr$gen))
## [1] "eta (X , Y) : 0.00896475356581972"
```

A particular case concerns Gaussian vectors:

Definition 1.13 $\boldsymbol{\mathcal{X}_n} = (X_1, \ldots, X_n)$ is a Gaussian vector of size n if and only if

$$\boldsymbol{a}^t \boldsymbol{\mathcal{X}_n} = \sum_{i=1}^{n} a_i X_i \text{ is gaussian}, \forall \boldsymbol{a} = (a_1, \ldots, a_n)^t \in \mathbb{R}^n.$$

We denote the distribution of $\boldsymbol{\mathcal{X}_n}$ by $N(\boldsymbol{m}, \boldsymbol{C})$, where

$$\boldsymbol{m} = (E(X_1), \ldots, E(X_n))^t, \boldsymbol{C} = \left(C_{ij} = cov\left(X_i, X_j\right), 1 \leq i, j \leq n\right).$$

(continued)

Definition 1.13 (continued)

The PDF of $\mathcal{X}_n$ is, for $\boldsymbol{x} = (x_1, \ldots, x_n)^t$,

$$f(\boldsymbol{x}) = \frac{1}{\sqrt{(2\pi)^n |\det \boldsymbol{C}|}} \exp\left(-\frac{1}{2}(\boldsymbol{x}-\boldsymbol{m})^t \boldsymbol{C}^{-1}(\boldsymbol{x}-\boldsymbol{m})\right),$$

The characteristic function of $\mathcal{X}_n$ is, for $\boldsymbol{t} = (t_1, \ldots, t_n)^t$,

$$\varphi_{X_n}(\boldsymbol{t}) = \exp\left(i\boldsymbol{m}^t\boldsymbol{t} - \frac{1}{2}\boldsymbol{t}^t\boldsymbol{C}^{-1}\boldsymbol{t}\right).$$

Exercises

1. Consider $\Omega = \{1, 2, \ldots, 12\}$ and the variables $X(\omega) = \omega \ mod\ 2$, $Y(\omega) = \omega \ mod\ 3$. Assume that $\mu(\omega) = 1/12$.
   (a) Verify that $X(\Omega) = \{0, 1\}$ and $(X = 0) = P(X = 1) = 1/2$.
   (b) Verify that $Y(\Omega) = \{0, 1, 2\}$ and $(Y = 0) = P(Y = 1) = P(Y = 2) = 1/3$.
   (c) Construct the Table 1.3 for the couple (X, Y).
   (d) Determine the covariance matrix of the couple.
2. Consider $\Omega = \{1, 2, \ldots, 12\}$ and the variables $X(\omega) = \omega \ mod\ 2$, $Y(\omega) = \omega \ mod\ 4$. Assume that $\mu(\omega) = 1/12$.
   (a) Verify that $X(\Omega) = \{0, 1\}$ and $(X = 0) = P(X = 1) = 1/2$.
   (b) Verify that $Y(\Omega) = \{0, 1, 2, 3\}$ and $(Y = 0) = P(Y = 1) = P(Y = 2) = P(Y = 3) = 1/4$.
   (c) Determine $P(X = i, Y = j)$ for all the values of (i, j).
3. Consider $\Omega = \{1, 2, \ldots, 12\}$ and the variables $X(\omega) = \omega \ mod\ 3$, $Y(\omega) = |\sin(\omega\pi/2)|$. Assume that $\mu(\omega) = 1/12$.
   (a) Verify that $X(\Omega) = \{0, 1, 2\}$ and $(X = 0) = P(X = 1) = P(X = 2) = 1/3$.
   (b) Verify that $Y(\Omega) = \{0, 1\}$ and $P(Y = 0) = P(Y = 1) = 1/2$.
   (c) Determine $P(X = i, Y = j)$ for all the values of (i, j).
4. Let $\Omega = (0, 1)$, with $P((a, b)) = b - a$ ($\mu(\omega) = 1$) and $\boldsymbol{X} = (X_1, X_2)$, $X_1(\omega) = \omega^2$, $X_2(\omega) = \omega^4$.
   (a) Find the CDF F of $\boldsymbol{X}$.
   (b) Determine the PDF f of $\boldsymbol{X}$.
   (c) Determine the marginal distributions and densities.
   (d) Find the distributions of $E(X_2 | X_1)$ and $E(X_1 | X_2)$.

(continued)

5. Let $\boldsymbol{X} = (X_1, X_2)$ be a pair of random variables having density

$$f(x_1, x_2) = \begin{cases} x_1 + x_2, \text{ if } 0 < x_1, x_2 < 1; \\ 0, \text{otherwise.} \end{cases}$$

(a) Determine the marginal distributions and densities.
(b) Find the distributions of $E(X_2 | X_1)$ and $E(X_1 | X_2)$.

6. Let $\boldsymbol{X} = (X_1, X_2)$ be a pair of random variables having density

$$f(x_1, x_2) = \begin{cases} \dfrac{2}{(x_2 + 1)^3}, \text{ if } 0 < x_1 < x_2; \\ 0, \text{otherwise.} \end{cases}$$

(a) Determine the marginal distributions and densities.
(b) Find the distributions of $E(X_2 | X_1)$ and $E(X_1 | X_2)$.

7. Let $\boldsymbol{X} = (X_1, X_2)$ be a pair of non-negative independent random variables such that

$$f_1(x_1) = \lambda e^{-\lambda x_1}, f_2(x_2) = \mu e^{-\mu x_2}.$$

Let $Y = X_1 + X_2$.

(a) Determine the distribution of Y.
(b) Determine the distribution of the couple (X_1, Y).
(c) Find the distribution of $E(X_1 | Y)$.

8. Let $\boldsymbol{X} = (X_1, \ldots, X_n)^t$ be a Gaussian vector $N(\boldsymbol{0}, \boldsymbol{Id})$. Let $\boldsymbol{A} = (A_{ij}, 1 \leq i \leq m, 1 \leq j \leq n)$ be a real matrix and $\boldsymbol{B} = (B_1, \ldots, B_m)^t$ be a vector of real numbers. Determine the distribution of $\boldsymbol{Y} = \boldsymbol{AX} + \boldsymbol{B}$.
9. Let $\boldsymbol{X} = (X_1, X_2)^t$ be a Gaussian vector $N(\boldsymbol{0}, \boldsymbol{Id})$. Let $X_3 = X_1 + X_2$, $X_4 = X_1 - X_2$. Determine the distribution of X_3 and X_4. Is the vector (X_3, X_4) Gaussian?
10. Let $\boldsymbol{X} = (X_1, X_2, X_3, X_4)^t$ be a Gaussian vector such that

$$m = \begin{pmatrix} 1 \\ -1 \\ 1 \\ 2 \end{pmatrix}, \quad C = \begin{pmatrix} 3 & 0 & 1 & -1 \\ 0 & 5 & 0 & 1 \\ 1 & 0 & 4 & 0 \\ -1 & 1 & 0 & 1 \end{pmatrix}$$

Which are the pairs (X_i, X_j) which are independent ? Determine α such that X_1 and $X_1 + X_2 + X_3 + \alpha X_4$ are independent.

1.10 Independent Random Variables

The notion of independence of events extends straightly to random vectors. Such a notion was criticized by many Bayesian researchers, namely, by De Finetti, which proposed an alternative approach, based on the notion of exchangeability. We shall examine the ideas of De Finetti in Sect. 1.12. Here, we consider the classical definition of independence and its consequences.

Definition 1.14 Let $\boldsymbol{X}$ be a random vector of dimension n on (Ω, P). We say that $\boldsymbol{X}$ is a vector of independent random variables if and only if its CDF is the product of its marginal CDFs, i.e.,

$$\forall \boldsymbol{x} = (x_1, \ldots, x_n) \in \mathbb{R}^n : \ F(\boldsymbol{x}) = \prod_{i=1}^{n} F_i(x_i). \blacksquare \tag{1.77}$$

Observe that we can say, equivalently, that $\boldsymbol{X}$ is a vector of independent random variables if and only if its PDF is the product of its marginal PDFs, i.e.,

$$\forall \boldsymbol{x} = (x_1, \ldots, x_n) \in \mathbb{R}^n: \ f(\boldsymbol{x}) = \prod_{i=1}^{n} f_i(x_i). \tag{1.78}$$

We can also characterize independency using the characteristic functions: $\boldsymbol{X}$ is a vector of independent random variables if and only if its characteristic function is the product of the individual characteristic functions of its components, i.e.,

$$\forall \boldsymbol{t} = (t_1, \ldots, t_n) \in \mathbb{R}^n: \ \varphi_X(\boldsymbol{t}) = \prod_{i=1}^{n} \varphi_i(t_i). \tag{1.79}$$

A couple (X, Y) given by Table 1.3 is independent if and only if

$$p_{ij} = p_{i\blacksquare} p_{\blacksquare j}, \forall i, j. \tag{1.80}$$

Then,

$$P(X = X_i | Y = Y_j) = P(X = X_i), \ P(Y = Y_j | X = X_i) = P(Y = Y_j).$$

Consequently,

$$E(X|Y) = E(X), \quad E(Y|X) = E(Y).$$

For a couple of variables having a distribution $F(x_1, x_2) = P(X_1 \leq x_1, X_2 \leq x_2)$ and a density $f(x_1, x_2) = \partial^2 F / \partial x_1 \partial x_2$, $\boldsymbol{X} = (X_1, X_2)$ is independent if and only if

$$F(x_1, x_2) = F_1(x_1) F_2(x_2), \forall \boldsymbol{x} = (x_1, x_2) \in \mathbb{R}^2. \tag{1.81}$$

Or, equivalently,

$$f(x_1, x_2) = f_1(x_1) f_2(x_2), \forall \boldsymbol{x} = (x_1, x_2) \in \mathbb{R}^2. \tag{1.82}$$

In this case,

$$f(x|Y=y) = f_X(x), f(y|X=x) = f_Y(y)$$

and we have $E(X| Y) = E(X)$, $E(Y| X) = E(Y)$.

If $\boldsymbol{X}$ is a vector of independent random variables, then

$$E(X_i X_j) = E(X_i) E(X_j), \text{if } i \neq j. \tag{1.83}$$

Then,

$$C_{ij} = E(X_i X_j) - E(X_i) E(X_j) = 0, \text{if } i \neq j. \tag{1.84}$$

and the covariance matrix is diagonal. The converse is, in general false, except for Gaussian variables: the expression of the characteristic function of a Gaussian vector (see Definition 1.13) shows that $C_{ij} = 0$, if $i \neq j \Longrightarrow$ (1.79) is satisfied.

Example 1.24 Let us come back to Example 1.21: $\boldsymbol{X} = (X_1, X_2, X_3) \in (0, 1)^3$ such that

$$F(\boldsymbol{x}) = x_1 x_2^2 x_3^3 \text{ on } (0, 1)^3$$

The PDF of $\boldsymbol{X}$ is

$$f(\boldsymbol{x}) = 6 x_2 x_3^2 \text{ on } (0, 1)^3.$$

(continued)

The marginal PDFs are

$$f_1(x_1) = 1, f_2(x_2) = 2x_2, f_3(x_3) = 3x_3^2 .$$

We have $f(\boldsymbol{x}) = f_1(x_1)f_2(x_2)f_3(x_3)$, so that the variables are independent. The conditional densities are

$$f(X_1|X_2, X_3) = 1, \ f(X_2|X_1, X_3) = 2x_2, \quad f(X_3|X_1, X_2) = 3x_3^2.$$

As noticed above, we have $f(X_1|X_2, X_3) = f_1$, $f(X_2|X_1, X_3) = f_2$, $f(X_3|X_1, X_2) = f_3$.

The covariance matrix is

$$C = \begin{pmatrix} 1/12 & 0 & 0 \\ 0 & 1/18 & 0 \\ 0 & 0 & 3/80 \end{pmatrix}.$$

As previously observed, C is diagonal. The characteristic function is

$$\varphi(t_1, t_2, t_3) = \\ -\frac{6\mathrm{i}(-1 + \mathrm{e}^{\mathrm{i}t_1})(-1 + \mathrm{e}^{\mathrm{i}t_2}(1 - \mathrm{i}t_2))(-2\mathrm{i} + \mathrm{e}^{\mathrm{i}t_3}(2\mathrm{i} + 2t_3 - \mathrm{i}t_3^2))}{t_1 t_2^2 t_3^3},$$

In this case, we have

$$\varphi(t_1, t_2, t_3) = \varphi_1(t_1)\varphi_2(t_2)\varphi_3(t_3)$$

Definition 1.15 Let (Ω, P) be a probability space and X be a random variable defined on (Ω, P). $X_n = (X_1, \ldots, X_n)$ is a sample vector of size n from X if and only all the components of X_n are independent and each X_i has the same distribution than X. Let $E(X) = m$ and $V(X) = \sigma^2$. The empirical mean $\overline{X}_n$, the squared deviation D_n^2, and the empirical variance d_n^2 of the sample vector are

(continued)

Definition 1.15 (continued)

$$\bar{X}_n = \frac{1}{n}\sum_{i=1}^{n} X_i, D_n^2 = \sum_{i=1}^{n} (X_i - m)^2, d_n^2 = \frac{D_n^2}{n}.$$

We have

$$E(\bar{X}_n) = m, V(\bar{X}_n) = \frac{\sigma^2}{n}, E(d_n^2) = \sigma^2, V(d_n^2) = \frac{1}{n}(CM_4(X) - \sigma^4).$$

By extension, we call sample a realization of X_n, i.e., a vector of n independent variates from X. The squared deviation S_n^2, the population variance $s_{n,p}^2$, and the empirical variance s_n^2 associated to X_n are

$$S_n^2 = \sum_{i=1}^{n} (X_i - \bar{X}_n)^2, s_{n,p}^2 = \frac{S_n^2}{n}, s_n^2 = \frac{S_n^2}{n-1}.$$

$s_{n,\ p}$, s_n are referred to as the population standard deviation and the empirical standard deviation of the sample, respectively.

Example 1.25 Let us consider $X \sim N(m, \sigma)$. Then,

$$\bar{X}_n \sim N\left(m, \frac{\sigma}{\sqrt{n}}\right), \frac{D_n^2}{\sigma^2} \sim \chi^2(n).$$

Thus,

$$E\left(\frac{D_n^2}{\sigma^2}\right) = n, V\left(\frac{D_n^2}{\sigma^2}\right) = 2n \Longrightarrow E(d_n^2) = \sigma^2, V(d_n^2) = \frac{2}{n}\sigma^2.$$

Indeed,

$$CM_4(X) = 3\sigma^4 \Longrightarrow \frac{1}{n}(CM_4(X) - \sigma^4) = \frac{2}{n}\sigma^4.$$

Let us generate $ns = 1E4$ samples of $n = 100$ variates of $N(m, \sigma)$, with $m = 2, \sigma = 1$:

(continued)

Example 1.25 (continued)

```
ns = 1e4
n=100
m = 2
sig = 1
xbar = numeric(length = ns)
d2 = numeric(length = ns)
for (i in 1: ns){
  x = rnorm(n,m,sig)
  aux = mean(x)
  xbar[[i]] = aux
  d2[[i]] = mean((x-aux)^2)}
mxbar = mean(xbar)
md2 = mean(d2)
sxbar = var(xbar)
vard2 = var(d2)
print(paste('mean of empirical means: ',mxbar,', exact: ',m))
## [1] "mean of empirical means:  1.9983664893426 , exact:  2"
print(paste('variance of empirical means: ',sxbar,', exact:
',sig^2/n))
## [1] "variance of empirical means:  0.00988890792457432 , ex-
act:  0.01"
print(paste('mean of empirical variances: ',sxbar,', exact:
',sig/n))
## [1] "mean of empirical variances:  0.00988890792457432 , ex-
act:  0.01"
print(paste('variance of empirical variances: ',vard2,', exact:
',2*sig/n))
## [1] "variance of empirical variances:  0.0196277806607795 ,
exact:  0.02"
```

The histograms appear in Fig. 1.12 – they can be plotted as follows (we use the package `latex2exp` for the titles):

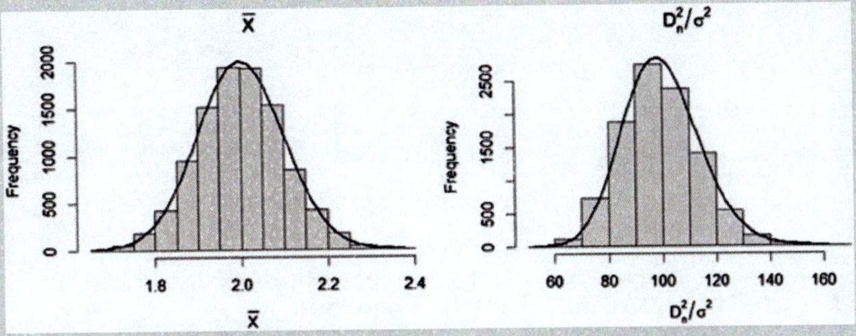

Fig. 1.12 Comparison between the exact distributions and the empirical ones

(continued)

Example 1.25 (continued)

```
library(latex2exp)
str = TeX(r'($\bar{X}$)')
hd <- hist(xbar,main = str , xlab = str)
xv <- seq(min(xbar), max(xbar), length = 100)
yv <- dnorm(xv, mean = m, sd = sqrt(sig/n))
yv <- yv * diff(hd$mids[1:2]) * length(xbar)
lines(xv, yv, lwd = 2)
str = TeX(r'($D^2_n / \sigma^2$)')
hd <- hist(D2,main = str,xlab=str )
xv <- seq(min(D2), max(D2), length = 100)
yv <- dchisq(xv, n)
yv <- yv * diff(hd$mids[1:2]) * length(D2)
lines(xv, yv, lwd = 2)
```

Exercises

1. Consider $\Omega = \{1, 2, \ldots, 12\}$ and the variables $X(\omega) = \omega \; mod \; 2$, $Y(\omega) = \omega \; mod \; 3$. Assume that $\mu(\omega) = 1/12$. Are these variables independent?
2. Consider $\Omega = \{1, 2, \ldots, 12\}$ and the variables $X(\omega) = \omega \; mod \; 2$, $Y(\omega) = \omega \; mod \; 4$. Assume that $\mu(\omega) = 1/12$. Are these variables independent?
3. Let $\Omega = (0, 1)$, with $P((a, b)) = b - a$ ($\mu(\omega) = 1$) and $\boldsymbol{X} = (X_1, X_2)$, $X_1(\omega) = \omega^2$, $X_2(\omega) = \omega^4$. Are these variables independent?
4. Let $\boldsymbol{X} = (X_1, X_2)$ be a pair of independent random variables taking their values on $\{1, \ldots, 8\}$. Let $Y_1 = X_1 \; mod \, 2$, $Y_2 = X_2 \; mod \, 2$, $Y_3 = (X_1 + X_2) \; mod \, 2$ Are these variables independent?
5. Let $\boldsymbol{X} = (X_1, X_2)$ be a sample from $N(0,1)$. Determine la distribution of $\boldsymbol{Y} = (Y_1, Y_2)$, where $Y_1 = X_1 + X_2$, $Y_2 = X_1 - X_2$. Are these variables independent?
6. Let $\boldsymbol{X} = (X_1, X_2)$ be a sample from $\mathcal{U}(a, b)$. Determine la distribution of $\boldsymbol{Y} = (Y_1, Y_2)$, where $Y_1 =$ min (X_1, X_2) , $Y_2 =$ max (X_1, X_2). Are these variables independent?
7. Let $\boldsymbol{X} = (X_1, X_2)$ be a sample from $N(0,1)$. Determine la distribution of $\boldsymbol{Y} = (Y_1, Y_2)$, where $Y_1 = X_1 + X_2$, $Y_2 = X_1 - X_2$. Are these variables independent?

1.11 Central Limit and Levy's Theorem

The Central Limit Theorem and Levy's Theorem are two of the main results concerning random variables. Both involve the notion of convergence in the distribution of a sequence of random variables:

Definition 1.16 Let (Ω, P) be a probability space and $\{X_n\colon n \in N\}$ be a sequence of random variables defined on (Ω, P). Let F_n be the CDF of X_n and F be the CDF of X.

We say that $X_n \rightarrow X$ in distribution if and only if $F_n(x) \longrightarrow F(x)$ at any point x where F is continuous, where. $X_n \longrightarrow X$.

Example 1.26 Let X_n be uniformly distributed on $\left(-\frac{1}{n}, \frac{1}{n}\right)$. Then,

$$F_n(x) = \begin{cases} 0, \mathit{if}\ x \leq -\dfrac{1}{n} \\ nx + 1, \mathit{if}\ x \in \left(-\dfrac{1}{n}, \dfrac{1}{n}\right) \\ 1, \text{otherwise} \end{cases}.$$

Observe that

$$F_n(x) \rightarrow F(x) = \begin{cases} 0, \text{if } x < 0 \\ 1, \text{if } x \geq 0 \end{cases}.$$

Thus, $X_n \rightarrow \delta_0$ in distribution.

Lévy's theorem (Lévy, 1922) reads as:

Theorem 1.1 (Lévy) Let $\{X_n\colon n \in N\}$ be a sequence of random variables such that the characteristic function of X_n is φ_n. Let X be a random variable having as characteristic function φ. Then, $X_n \longrightarrow X$ in distribution if and only if $\varphi_n(t) \longrightarrow \varphi(t)$ almost everywhere.

Remark 1.1 Lévy's theorem shows that we can approximate the distribution of X by approximating its moments. Indeed, $\varphi(t) = E(e^{itX})$ expands as

$$\varphi(t) = \sum_{n=0}^{+\infty} \frac{(it)^k}{k!} M_k(X) = 1 + itM_1(X) + \frac{(it)^2}{2} M_2(X) + \frac{(it)^3}{2} M_3(X) + \ldots$$

(continued)

Remark 1.1 (continued)

We may consider X_n as the variable has a characteristic function

$$\varphi_n(t) = \sum_{n=0}^{n} \frac{(it)^k}{k!} M_k(X) \approx E\left(e^{itX_n}\right).$$

Then, from Lévy's theorem, $X_n \longrightarrow X$ in distribution. In practice, we look for a variable verifying the equations:

$$M_k(X_n) = M_k(X), \quad \text{for } 1 \leq k \leq n. \tag{1.85}$$

This method of approximation is known as the Moments Matching Method. See the companion book (Souza de Cursi 2023) for examples.

A first consequence of Lévy's theorem is the following:

Corollary 1.2 Let $\{X_n: n \in \mathbb{N}\}$ be a sequence of normally distributed random variables, $X_n \sim N(m_n, \sigma_n)$. If $X_n \longrightarrow X$ in distribution, then X is normally distributed, $X \sim N(m, \sigma)$. In addition, $X_n \longrightarrow X$ in distribution if and only if $(m_n, \sigma_n) \longrightarrow (m, \sigma)$.

A second consequence of Lévy's theorem is the Central Limit Theorem:

Theorem 1.3 Let $\{X_k: k \in \mathbb{N}\}$ be a sequence of independent random variables having the same distribution of finite mean m and variance σ^2. Let

$$Z_n = \frac{\bar{X}_n - m}{\frac{\sigma}{\sqrt{n}}}.$$

Then Z_n converges in distribution to a Standard Gaussian $N(0, 1)$.

Remark 1.1 Other types of convergence exist, as, for instance:

1. $X_n \rightarrow X$: in quadratic mean if and only if $E((X_n - X)^2) \longrightarrow 0$, for $n \longrightarrow +\infty$;
2. $X_n \rightarrow X$ almost surely if and only if the event $E = \{X_n \longrightarrow X\}$ is almost sure, i.e., $P(E) = 1$;
3. $X_n \rightarrow X$ in probability if and only if: for any $\varepsilon > 0$, the event $E_n(\varepsilon) = \{\|X_n - X\| \geq \varepsilon\}$ verifies $P(E_n(\varepsilon)) \longrightarrow 0$, for $n \longrightarrow +\infty$.

Exercises

1. Let X_λ be a random variable such that its PDF is f_λ given by

$$f_\lambda(x) = 2\lambda x \exp\left(-\lambda x^2\right), \textit{if } x > 0; \quad 0, \text{otherwise.}$$

   (a) Determine the CDF F_λ of X_λ.
   (b) Show that, for $\lambda \to +\infty$, $X_\lambda \to \delta_0$ in distribution.

2. Let X_λ be a random variable such that its PDF is f_λ given by

$$f_\lambda(x) = \frac{\lambda}{\pi\left(1 + \lambda^2 x^2\right)}, \textit{if } x > 0; \quad 0, \text{otherwise.}$$

   (a) Determine the CDF F_λ of X_λ.
   (b) Show that, for $\lambda \to +\infty$, $X_\lambda \to \delta_0$ in distribution.

3. Let $\{X_n: n \in \mathbb{N}\}$ be a sequence of independent random variables uniformly distributed on (a, b). Their common CDF is

$$F(x) = \frac{x - a}{b - a}, \textit{if } a \le x \le b; \quad 0, \textit{if } x \le a; 1 \text{ otherwise.}$$

   (a) Determine the CDF of $Y_n = \min \{X_1, \ldots, X_n\}$ and $Z_n = \max \{X_1, \ldots, X_n\}$
   (b) Show that $Y_n \to \delta_a$ in distribution.
   (c) Show that $n(b - Z_n) \to T$ in distribution. Determine the distribution of T.

1.12 Exchangeability

The notion of independence of events extends straightly to random vectors, but it was an object of criticism by Bayesians, namely, by Bruno De Finetti – which introduced the notion of exchangeability and proved a theorem, usually referred to as De Finetti's representation theorem.

As an example, let us consider a coin, for which, in principle, the probability p of a Head (H) is equal to the probability $1 - p$ of a Tail (T), so that $p = 1/2$ (this may be considered as a prior). Suppose that you toss the coin three times, and you get the results *HHH*. Let us consider the fourth toss: under the concept of independence, the probability of the fourth toss is still $p = 1/2$, but one may consider that such reasoning neglects the information about the previous results – for instance, we can think that the coin is not fair – this is a prior assumption on the coin.

Some researchers consider that you must reevaluate p by using an estimation $\widehat{p}$ of the mean of a Bernoulli variable $X = 1$, *for* H; $X = 0$ *for* T – for instance, Wald's method with continuity correction furnishes, for a confidence 95%, the confidence interval $(0.83, 1)$ – so that the coin appears as unfair. With a thousand trials furnishing H as a result, Wilson's method with continuity correction furnishes, for a confidence 95%, the confidence interval (0.95, 1) – so that the coin appears as really unfair.

De Finetti introduced a different point of view, based on the prior assumption on p and proved a central result in the Bayesian Approach of probabilities, involving the notion of exchangeability instead of the notion of independence. We shall examine these elements in the sequel. Let us start with the classical definition of independence and its consequences.

Now, let us turn to the ideas of De Finetti, who considered that the notion of independence was inadequate. De Finetti introduced the idea of exchangeability (De Finetti 1930, 1937). We recall the definition of Bernardo and Smith (2000):

Definition 1.17 Let $\boldsymbol{X} = (X_1, \ldots, X_n)$ be a random vector on (Ω, P). We say that $\boldsymbol{X}$ is exchangeable if and only if the distribution of $\boldsymbol{X}_{\boldsymbol{\Pi}} = (X_{\Pi(1)}, \ldots, X_{\Pi(n)})$ coincides with the distribution of $\boldsymbol{X}$, for any permutation Π of $\{1, \ldots, n\}$.

Let $\boldsymbol{X} = \{X_i: i \in \mathcal{N}^*\}$. We say that $\boldsymbol{X}$ is finitely exchangeable if and only if $\boldsymbol{X}_{(n)} = (X_1, \ldots, X_n)$ is exchangeable, $\forall n \in \mathcal{N}^*$. We say that $\boldsymbol{X}$ is exchangeable if and only if $(X_{i_1}, \ldots, X_{i_n})$ is exchangeable, $\forall n \in \mathcal{N}^*, \{i_1 < i_2 < \ldots < i_n\} \in (\mathcal{N}^*)^n$. ∎

A first fundamental result shows that exchangeable sequences are ***identically distributed (i.d.)***:

Theorem 1.4 Let $\boldsymbol{X} = \{X_i: i \in \mathcal{N}^*\}$ be a sequence of real-valued random variables. If $\boldsymbol{X}$ is exchangeable then

(i) It is ***identically distributed***, i.e., all the variables X_i have the same marginal distribution: $\exists F: \mathbb{R} \rightarrow [0, 1]$ such that $P(X_i \leq x) = F(x), \forall i \in \mathcal{N}^*$.
(ii) It is **stationary**, i.e., the distribution of $(X_1, X_2, \ldots, X_n)$ coincides with the distribution of $(X_{k+1}, X_{k+2}, \ldots, X_{k+n}), \forall k \in \mathcal{N}$.
(iii) It is **conditionally identically distributed**, i.e., the distribution of $(X_1, X_2, \ldots, X_n, X_{n+1})$ and $(X_1, X_2, \ldots, X_n, X_{n+2})$ coincide $\forall n \in \mathcal{N}^*$. ∎

Proof: (i) Let us denote F the CDF of X_1. Let $\boldsymbol{X}_{(2)} = (X_1, X_2)$ and $F_{(2)} = F_2(x_1, x_2)$ be its CDF. Since the sequence is exchangeable, $F_{(2)}$ is also the distribution of (X_2, X_1). Thus,

$$P(X_1 \leq x, X_2 < +\infty) = F_{(2)}(x, +\infty) = P(X_2 \leq x, X_1 < +\infty),$$

(continued)

Theorem 1.4 (continued)

so that

$$P(X_2 \leq x) = P(X_1 \leq x) = F(x).$$

Thus, the CDF of X_2 is also F. Now, assume that the CDF of X_i is F, for $1 \leq i \leq n$, with $n \geq 2$. Let us consider $\boldsymbol{X}_{(n+1)} = (X_1, X_2, \ldots, X_n, X_{n+1})$ and its CDF $F_{(n+1)}$: exchangeability shows that the distribution of $(X_{n+1}, X_2, \ldots, X_n, X_1)$ is still F_{n+1}. Thus, analogously:

$$P(X_{n+1} \leq x) = F_{(n+1)}(x, +\infty, \ldots, +\infty) = P(X_1 \leq x).$$

Thus, the distribution of X_{n+1} is also F. By recurrence, we have the distribution of X_i is F, $\forall i \in \mathbb{N}^*$

(ii) Let us denote $\boldsymbol{X}_{(i,j,k)} = (X_{i+k}, \ldots, X_{j+k})$. If $k > n$, we have $\boldsymbol{X}_{(n+k)} = (\boldsymbol{X}_{(1,n,0)}, \boldsymbol{X}_{(n+1,k,0)}, \boldsymbol{X}_{(1,n,k)})$. Exchangeability shows that the distributions of $(\boldsymbol{X}_{(1,n,k)}, \boldsymbol{X}_{(n+1,k,0)}, \boldsymbol{X}_{(1,n,0)})$ and $\boldsymbol{X}_{(n+k)}$ coincide. Consequently, arguments analogous to those used for (i) show that $\boldsymbol{X}_{(1,n,k)}$ and $\boldsymbol{X}_{(1,n,0)}$ have the same distribution.

(iii) $\boldsymbol{X}_{(n+2)} = (\boldsymbol{X}_{(n)}, X_{n+1}, X_{n+2})$ and $(\boldsymbol{X}_{(n)}, X_{n+2}, X_{n+1})$ have the same distribution F_{n+2}. Thus, arguments analogous to those of (i) can be used:

$$P(X_1 \leq x_1, \ldots, X_n \leq x_n, X_{n+1} \leq x) = F_{(n+2)}(x_1, \ldots, x_n, x, +\infty);$$

$$P(X_1 \leq x_1, \ldots, X_n \leq x_n, X_{n+2} \leq x) = F_{(n+2)}(x_1, \ldots, x_n, x, +\infty).$$

Consequently, $(X_1, X_2, \ldots, X_n, X_{n+1})$ and $(X_1, X_2, \ldots, X_n, X_{n+2})$ have the same distribution. ∎

Conversely, a stationary and conditionally identically distributed sequence is exchangeable (Kallenberg 1988).

It is obvious that a vector of independent random variables having all the same distribution forms an exchangeable random vector. However, we can have dependent vectors that satisfy the condition of exchangeability.

Example 1.27 Let us consider an urn containing two balls, having colors red (R) and blue (B). We successively draw two balls without remitting the ball drawn in the urn: Let us consider the random variable $X_k = 1$,if the ball drawn at k is red, $X_k = 0$, otherwise. The couple (X_1, X_2) can take only two values: $(1, 0)$ or $(0, 1)$, with $P((0, 1)) = P((1, 0)) = 1/2$. The couple is exchangeable, since the couple (X_2, X_1) has the same distribution. However, the variables are not independent, since $P(X_2 = 0 | X_1 = 0) = 0 \neq P(X_2 = 0) = 1/2$. ∎

Example 1.28 A second example is furnished by *Polya's Urn* (Polya 1930): consider an urn that contains initially r_0 red balls and b_0 blue balls. When a ball is drawn, it is put back in the urn, and another ball of the same color is added to the urn: the number of balls in the urn increases unity by the addition of a ball having the same color as the ball drawn. Let $X_k = 1$, if the ball drawn at k is red, $X_k = 0$, otherwise. Let us denote by r_k, b_k, respectively, the number of red and blue balls at stage k. Let $n_k = b_k + r_k$ be the number of balls in the urns at stage k. We have

$$P(X_k = 1) = \frac{r_k}{n_k}, P(X_k = 0) = \frac{b_k}{n_k}.$$

Thus:

$$P(r_{k+1} = r_k) = P(b_{k+1} = b_k + 1) = \frac{b_k}{n_k},$$

$$P(r_{k+1} = r_k + 1) = P(b_{k+1} = b_k) = \frac{r_k}{n_k}.$$

In addition:

$$P(X_{k+1} = 1 | X_k = 0) = \frac{r_k}{n_k + 1}, P(X_{k+1} = 1 | X_k = 1) = \frac{r_k + 1}{n_k + 1}.$$

$$P(X_{k+1} = 0 | X_k = 0) = \frac{b_k + 1}{n_k + 1}, P(X_{k+1} = 0 | X_k = 1) = \frac{b_k}{n_k + 1}.$$

Consequently:

$$P(X_k = 0, X_{k+1} = 0) = \frac{b_k(b_k + 1)}{n_k(n_k + 1)},$$

$$P(X_k = 1, X_{k+1} = 0) = \frac{b_k r_k}{n_k(n_k + 1)}$$

$$P(X_k = 0, X_{k+1} = 1) = \frac{b_k r_k}{n_k(n_k + 1)},$$

$$P(X_k = 1, X_{k+1} = 1) = \frac{r_k(r_k + 1)}{n_k(n_k + 1)}$$

(continued)

Example 1.28 (continued)

and,

$$P(X_{k+1}=1)=\frac{r_k b_k+(r_k+1)r_k}{(n_k+1)n_k}=\frac{r_k(n_k+1)}{(n_k+1)n_k}=\frac{r_k}{n_k}=P(X_k=1),$$

$$P(X_{k+1}=0)=\frac{b_k r_k+(b_k+1)b_k}{(n_k+1)n_k}=\frac{b_k(n_k+1)}{(n_k+1)n_k}=\frac{b_k}{n_k}=P(X_k=0).$$

Thus, the distribution of all the variables X_k coincides with the distribution of X_0. In addition,

$$P(X_k=0|X_{k+1}=0)=\frac{P(X_k=0,X_{k+1}=0)}{P(X_{k+1}=0)}=\frac{b_k+1}{n_k+1},$$

$$P(X_k=1|X_{k+1}=0)=\frac{P(X_k=1,X_{k+1}=0)}{P(X_{k+1}=0)}=\frac{r_k}{n_k+1}$$

$$P(X_k=0|X_{k+1}=1)=\frac{P(X_k=0,X_{k+1}=1)}{P(X_{k+1}=1)}=\frac{b_k}{n_k+1},$$

$$P(X_k=1|X_{k+1}=1)=\frac{P(X_k=1,X_{k+1}=1)}{P(X_{k+1}=1)}=\frac{r_k+1}{n_k+1}$$

Consequently, the distribution of (X_k, X_{k+1}) coincides with the distribution of (X_{k+1}, X_k). Notice that the variables are not independent. This analysis can be generalized to establish that $(X_0, \ldots, X_n)$ is exchangeable (Polya 1930).∎

The fundamental result about Exchangeable random variables is called *De Finetti's representation theorem* or, simply, *De Finetti's theorem*:

Theorem 1.5 Let $X=\{X_i: i\in N^*\}$ be a sequence of binary random variables. If X is exchangeable, then there exists an unique distribution F such that

$$P(X_1=x_1,\ldots,X_n=x_n)=\int_0^1 \theta^{n\bar{X}_n}(1-\theta)^{n-n\bar{X}_n}dF(\theta).$$

In addition,

$$F(\theta)=\lim_{n\to\infty}F_n(\theta), F_n : \text{distribution } of\ \bar{X}_n.$$ ∎

De Finetti's theorem shows that a sequence of exchangeable Bernoulli variables $\mathcal{B}(p)$ can be represented using the distribution F of the parameter p, taken as the limit of the distributions of the empirical estimations of p on samples.

Example 1.29 Let us consider the situation where a coin is tossed n times and we set $X_k = 1$, if the result is H at toss no. k. Let us evaluate the probability $p_4 = P(X_4 = 1 | X_1 = X_2 = X_3 = 1)$.

As previously observed, under the assumption of a fair coin and independent tosses, we have $p_4 = 1/2$. If we make an estimation by Wald's method with continuity correction, and a confidence $\alpha = 0.95$, we obtain $0.83 \leq p_4 \leq 1$.

Let us use De Finetti's theorem: we must have a prior distribution F for p. If we consider that the coin is fair, we have $F(p) = 0$, if $p < 1/2$; $F(p) = 1$, if $p > 1/2$ (the distribution is concentrated at $p = 1/2$). In this case, De Finetti's theorem shows that:

$$P(X_1 = X_2 = X_3 = 1) = \left(\frac{1}{2}\right)^3, P(X_1 = X_2 = X_3 = X_4 = 1) = \left(\frac{1}{2}\right)^4,$$

so that

$$p_4 = \frac{P(X_1 = X_2 = X_3 = X_4 = 1)}{P(X_1 = X_2 = X_3 = 1)} = \frac{1}{2}.$$

Now, assume that we are not sure about the coin: we consider that $p_{\min} \leq p \leq p_{\max}$, uniformly distributed on this interval. In this case,

$$dF(p) = \frac{dp}{p_{\max} - p_{\min}},$$

and

$$P(X_1 = X_2 = X_3 = 1) = \frac{1}{p_{\max} - p_{\min}} \int_{p_{\min}}^{p_{\max}} p^3 dp = \frac{p_{\max}^4 - p_{\min}^4}{4(p_{\max} - p_{\min})}.$$

Analogously,

$$P(X_1 = X_2 = X_3 = X_4 = 1) = \frac{1}{p_{\max} - p_{\min}} \int_{p_{\min}}^{p_{\max}} p^4 dp = \frac{p_{\max}^5 - p_{\min}^5}{5(p_{\max} - p_{\min})}.$$

(continued)

Example 1.29 (continued)

Thus,

$$p_4 = \frac{P(X_1 = X_2 = X_3 = X_4 = 1)}{P(X_1 = X_2 = X_3 = 1)} = \frac{4}{5} p_{\max} \left(\frac{1 - a^5}{1 - a^4} \right), a = \frac{p_{\min}}{p_{\max}}.$$

For $p_{\min} = 0$, $p_{\max} = 1$, we have $p_4 = 0.8$. For $p_{\min} = 0.4$, $p_{\max} = 0.6$, we have $p_4 \approx 0.52$: our prior influences the result. If we consider n tosses of identical result H, we have

$$p_{n+1} = P(X_{n+1} = 1 | X_k = 1, 1 \leq k \leq n) = \frac{n+1}{n+2} p_{\max} \left(\frac{1 - a^{n+2}}{1 - a^{n+1}} \right).$$

Since $0 < a < 1$: $p_{n+1} \approx p_{\max}$, for large n. ■

The original De Finetti's theorem was extensively studied in the literature, and many extensions can be found, such as, for instance, versions for $\boldsymbol{X}$ finitely exchangeable (Diaconis 1977; Gavalakis and Kontoyiannis 2021), versions for general random variables and with converse conditions (Hewitt & Savage, 1955; Svante et al. 2016; Fritz et al. 2021), characterizations of exchangeable vectors (Kallenberg 1988; Hill et al. 1987), and extensions to arrays (Kallenberg 1989; Ivanoff and Weber 2004).

A second main result is an extension to general real-valued random variables (Hewitt and Savage 1955; Bernardo and Smith 2000):

Theorem 1.6 Let $\boldsymbol{X} = \{X_i: i \in N^*\}$ be a sequence of real-valued random variables. If $\boldsymbol{X}$ is exchangeable with probability measure P, then there exists an unique probability μ over the space of all the CDFs $F: \mathbb{R} \to [0, 1]$ such that

$$P(X_i \in A_i, 1 \leq i \leq n) = \int \mu(dF) \prod_{i=1}^{n} F(A_i).$$

In addition, μ is the distribution of

$$F(A) = \lim_{n \to \infty} F_n(A) \quad , F_n : \text{distribution of} \frac{1}{n} \sum_{i=1}^{n} \chi_A(X_i),$$

$\chi_A(s) = 1$, if $s \in A$; 0, otherwise. ■

If the distributions are characterized by a vector of parameters $\boldsymbol{\theta}$, we have

Theorem 1.7 Let $\boldsymbol{X} = \{X_i: i \in N^*\}$ be a sequence of real-valued random variables. Let $\boldsymbol{X}$ be exchangeable and having marginal density $f(x, \boldsymbol{\theta})$, defined by a vector of parameters $\boldsymbol{\theta}$. Then there exists an unique probability μ such that the density $f_{(n)}$ of $\boldsymbol{X}_{(n)} = (X_1, \ldots, X_n)$ verifies

$$f_{(n)}(x_1, \ldots, x_n) = \int \mu(d\boldsymbol{\theta}) \prod_{i=1}^{n} f(x_i, \boldsymbol{\theta}). \blacksquare$$

Notice that F_n is the empirical distribution of the samples $(x_1, \ldots, x_n) \in A_1 \times \ldots \times A_n$ and F is their limit. Indeed, if we consider $A = (-\infty, x]$, then $F_n(x)$ is the empirical value of $P(X \leq x)$: $F_n(x) = |A_n(x)|/n$, $A_n(x) = \{i: x_i \leq x\}$. In this general form, the priors must be defined on μ: you must define what are the possible choices for the CDF F of X – eventually by choosing the distribution of the parameters $\boldsymbol{\theta}$. If $\mu(d\boldsymbol{\theta}) = g(\boldsymbol{\theta})d\boldsymbol{\theta}$, we have

$$f_{(n)}(x_1, \ldots, x_n) = \int d\boldsymbol{\theta} g(\boldsymbol{\theta}) \prod_{i=1}^{n} f(x_i, \boldsymbol{\theta}). \tag{1.86}$$

Example 1.30 Let us consider a Gaussian variable having as real distribution $N(m, \sigma)$. We believe that the real values of m and σ verify $m \in (0.5, 1.7)$, $\sigma \in (0.4, 0.7)$. Assume that the data available is given in Table 1.8 – this data was generated by `rnorm` with $m = 1, \sigma = 0.5$.

Table 1.8 Data available

0.4015706	0.7331831	1.3658122	0.3789766	0.8516060

In this case, $\boldsymbol{\theta} = (m, \sigma)$. We are interested in the evaluation of $P(X \leq 1)$ and in determining the distribution of X. In such a situation, we can, for instance, determine the Maximum Likelihood Estimator (MLE) for $\boldsymbol{\theta}$ and use it to fit a distribution – alternatively, we can use the unbiased estimators $\widehat{m}$ and $\widehat{\sigma}$ given by

$$\widehat{m} = \bar{X}_n = \frac{1}{n} \sum_{i=1}^{n} X_i \;, \widehat{\sigma}^2 = s_n = \frac{S_n}{n-1}, S_n = \sum_{i=1}^{n} (X_i - \bar{X}_n)^2.$$

(continued)

Example 1.30 (continued)

The unbiased estimators are evaluated under R as follows:

```
X = c(0.4015706, 0.7331831, 1.3658122, 0.3789766, 0.851606
0)
m_u = mean(X)
s_u = sd(X)
print(paste('Unbiased: m = ',m_u,', ',expression(sigma),'
= ',s_u))

## [1] "Unbiased: m =  0.7462297 ,  sigma  =  0.4027473113
97581"
```

However, the distribution corresponding to these estimators does not furnish a good evaluation of $P(X \leq 1)$:

```
xc = 1
p_u = pnorm(xc,m_u,s_u)
p_ex = pnorm(xc,m_ex,s_ex)
print(paste('Unbiased: P(X <= 1) = ',p_u,', exact = ',p_ex
))

## [1] "Unbiased: P(X <= 1) =  0.735684785892781 , exact =
0.5"
```

The results can be improved if supplementary information is available on $\boldsymbol{\theta}$. Since we believe that $m \in (0.5, 1.7)$ and $\sigma \in (0.4, 0.7)$ – notice that the real values are not in the center of these intervals, we can use the generalized De Finetti's theorem to generate the PDF $f_{(1)}(x)$. If the density associated with μ is g, we have, from Eq. (1.86):

$$f_{(1)}(x) = \int f(x, m, \sigma) g(m, \sigma) dm d\sigma,$$

where f is the Gaussian density

$$f(x, m, \sigma) = \frac{1}{\sigma\sqrt{2\pi}} e^{-\frac{1}{2}\left(\frac{x-m}{\sigma}\right)^2}.$$

The CDF $F_{(1)}(x)$ can be generated by integrating $f_{(1)}$. We have $P(X \leq 1) = F_{(1)}(1)$.

For instance, let us assume that μ corresponds to a uniform distribution – which ***is not*** the distribution corresponding to De Finetti's theorem. We evaluate $P(X \leq 1)$ as follows (the methods `romb1`, `romb2` in class `ints` implement Romberg's method, using `pracma`'s `romberg` – they are analogous to `int1` and `int2`):

(continued)

Example 1.30 (continued)

```
m_lower = 0.5
m_upper = 1.7
s_upper = 0.7
s_lower = 0.4
surf1 = (m_upper - m_lower)*(s_upper - s_lower)
integrand_1 = function(x,m,s){
  aux = dnorm(x,m,s)/surf1
  return(aux)
}
source('ints.R')
integ = ints$new()
f1 = function(x,f){
  a = c(m_lower, s_lower)
  b = c(m_upper, s_upper)
  g = function(m,s){f(x,m,s)}
  aux = integ$romb2(g,a,b)
  return(aux)
}
pdf_1 = function(x){f1(x,integrand_1)}
cdf_1 = function(x){integ$romb1(pdf_1,minus_inf,x)}
minus_inf = -10
p_fin_1 = cdf_1(xc)
print(paste("P(X <= 1), De Finetti uniform = ",p_fin_1," e
xact = ",p_ex))

## [1] "P(X <= 1), De Finetti uniform =  0.439460250034702
exact =  0.5"
```

Although the result is not exactly the true value, it is greatly improved: we get the estimation 0.44 instead of 0.74 for the correct value 0.5. If the information is that $m \in (0.5, 1.6)$ and $\sigma \in (0.4, 0.7)$, we obtain:

```
## [1] "P(X <= 1), De Finetti uniform =  0.468738757647511
exact =  0.5"
```

For the information $m \in (0.5, 1.5)$ and $\sigma \in (0.4, 0.6)$, we obtain:

```
## [1] "P(X <= 1), De Finetti uniform =  0.500000000009039
exact =  0.5"
```

Notice that the quality improves even for bad estimations of the intervals. For instance, for $m \in (0.5, 1.3)$ and $\sigma \in (0.4, 0.8)$:

(continued)

Example 1.30 (continued)

```
## [1] "P(X <= 1), De Finetti uniform =  0.563151560994566
exact =  0.5"
```

Nevertheless, if we desire to remain as close as possible to the Theorem, we must consider the limit of the empirical distributions. The Central Limit Theorem shows that

$$Q_n^2 \sim \chi^2(n-1) \Longrightarrow \frac{Q_n^2 - n + 1}{\sqrt{2(n-1)}} \xrightarrow{n \to +\infty} N(0,1) \text{ in distribution,}$$

while

$$\frac{\bar{X}_n - m}{s_n/\sqrt{n}} \xrightarrow{n \to +\infty} N(0,1) \text{ in distribution,}$$

Since $Q_n^2 = S_n^2/\sigma^2 \sim \chi^2(n-1)$, we can consider that both m and S^2/σ^2 have as limit a Gaussian distribution and take g as the corresponding density:

$$g(m,\sigma) = f\left(m, \bar{X}_n, \frac{s_n}{\sqrt{n}}\right) f\left(\frac{S^2}{\sigma^2}, n-1, \sqrt{2(n-1)}\right). \tag{1.87}$$

Figure 1.13 compares the PDF and the CDF generated by these approximations to the exact value – here, $n = 5$.

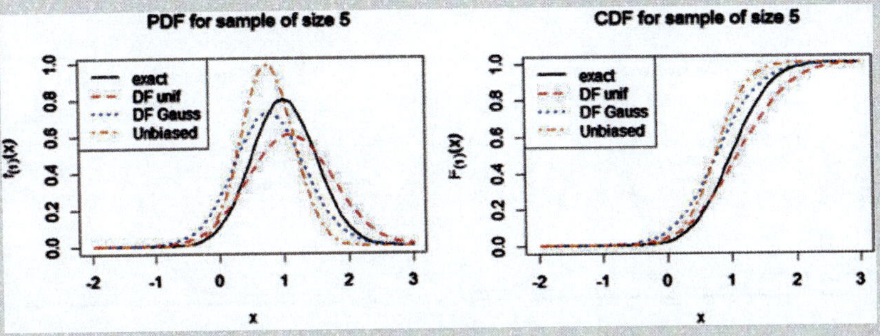

Fig. 1.13 Comparison between the exact distribution and the distributions furnished by De Finetti's theorem with two different priors and the unbiased estimators, using the sample of 5 values in Table 1.8

If more data is available, the estimations $\bar{X}_n, s_n$ are improved, so that the results improve also. For instance, let us consider the data formed by ten

(continued)

Example 1.30 (continued)

variates shown in Table 1.9 (also generated by `rnorm`, with $m = 1$, $\sigma = 0.5$). Figure 1.14 shows the results obtained for this data.

Table 1.9 A sample with ten variates

1.2624664	1.2348981	0.8707804	1.0677579	0.9279395
1.5103962	0.5496903	1.1348646	0.7162212	1.2706213

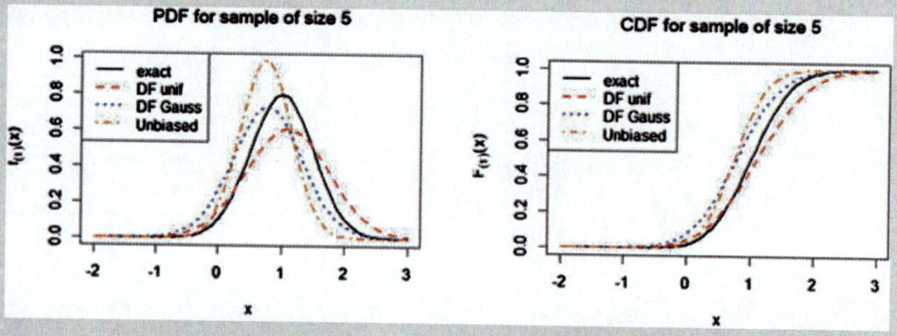

Fig. 1.14 Comparison between the exact distribution and the distributions furnished by De Finetti's theorem with two different priors, and the unbiased estimators, using the sample of ten variates in Table 1.9

If you have a large amount of data, both the unbiased estimators and De Finetti's theorem tend to furnish analogous results and become closer to the exact solution. For instance, Figs. 1.15 and 1.16 show the results for samples of sizes 20 and 100. The data of 20 variates (generated by `rnorm`, with $m = 1$, $\sigma = 0.5$) is shown in Table 1.10.

Fig. 1.15 Comparison between the exact distribution and the distributions furnished by De Finetti's theorem with two different priors, and the unbiased estimators, using a sample of 20 variates: unbiased estimator and Gaussian De Finetti's results become closer

(continued)

Example 1.30 (continued)

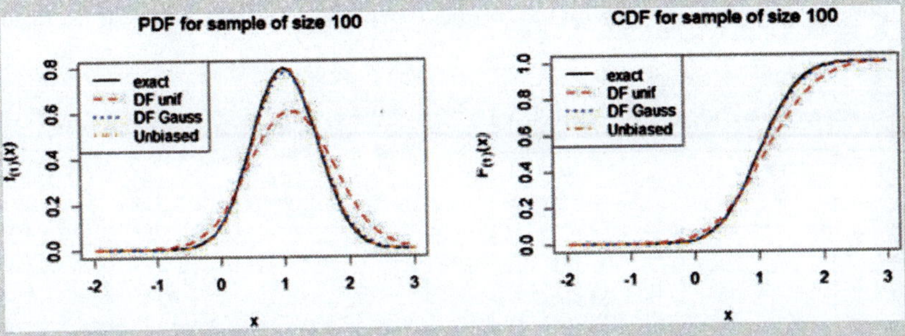

Fig. 1.16 Comparison between the exact distribution and the distributions furnished by De Finetti's theorem with two different priors, and the unbiased estimators, using a sample of 100 variates: unbiased estimator and Gaussian De Finetti's results almost coincide

Table 1.10 A sample of 20 variates

0.659774963	1.305985068	0.254624832	1.056616172	0.526085348
0.968261258	0.631565921	1.308790264	1.674394036	0.003233857
1.35247904	1.54011605	1.68689486	1.29357813	1.28123995
1.29226605	−0.12678400	0.76884859	1.06233044	0.856199880

In such a situation, both the Unbiased and De Finetti's estimations are close to the real one. Notice that the asymptotic Gaussian distribution must be used to get good results: the uniform prior does not furnish an approximation as close as the two others.

In practice, it may arise that we do not have any information about the values of θ. In such a situation, we can use a confidence interval. For instance, let us consider again the data in Table 1.8: the confidence interval for $\alpha = 0.05$ (confidence 95 %) is generated as follows:

```
m_u = mean(X)
s_u = sd(X)
S2 = sum((X-m_u)^2)
alfa = 0.05
ta=qt(1 - 0.5*alfa,ndf)
dm = ta*s_u/sqrt(nd)
```

```
m_lower = m_u - dm
m_upper = m_u + dm
A1 = qchisq(1 - 0.5*alfa,ndf)
A2 = qchisq(0.5*alfa,ndf)
s_lower = sqrt(S2/A1)
s_upper = sqrt(S2/A2)
```

We obtain m_u ≈ 0.75, s_u ≈ 0.40 and the confidence intervals:

$$m \in (0.25,1.25), \quad \sigma \in (0.24, 1.16).$$

(continued)

Example 1.30 (continued)

With these data, we obtain the results in Fig. 1.17. The results are inferior in quality.

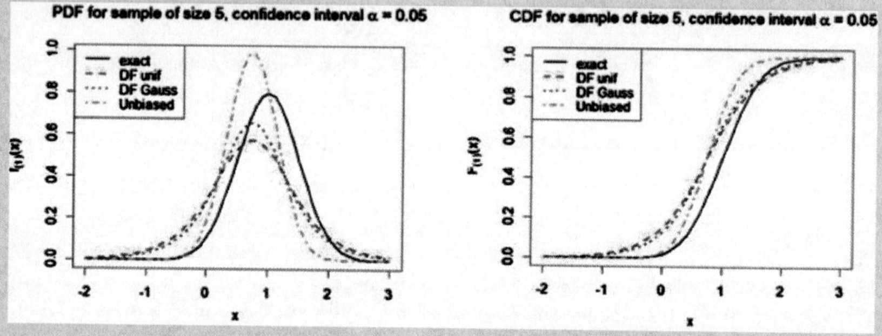

Fig. 1.17 Comparison between the exact distribution and the distributions furnished by De Finetti's theorem using the estimators and confidence intervals generated by the sample of five values in Table 1.8

Figure 1.13 compares the PDF and the CDF generated by these approximations to the exact value – here, $n = 5$.

If more data is available, the estimations $\bar{X}_n, s_n$ are improved, so that the results are expected to improve also. They become closer to the distributions furnished by the unbiased estimators – and this one becomes closer to the real one. For instance, let us consider the data formed by ten variates shown in Table 1.11 (also generated by `rnorm`, with $m = 1$, $\sigma = 0.5$). Figure 1.18 shows the results obtained for this data. For the samples formed by 20 and 100 variates, the results appear in Fig. 1.19 – they are close to the exact ones.

Table 1.11 A sample of ten variates

1.2624664	1.2348981	0.8707804	1.0677579	0.9279395
1.5103962	0.5496903	1.1348646	0.7162212	1.2706213

(continued)

Example 1.30 (continued)

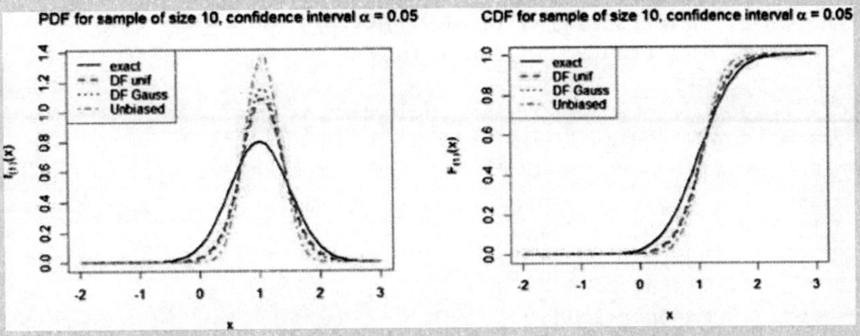

Fig. 1.18 Comparison between the exact distribution and the distributions furnished by De Finetti's theorem using the estimators and confidence intervals generated by 10 variates, given in Table 1.9

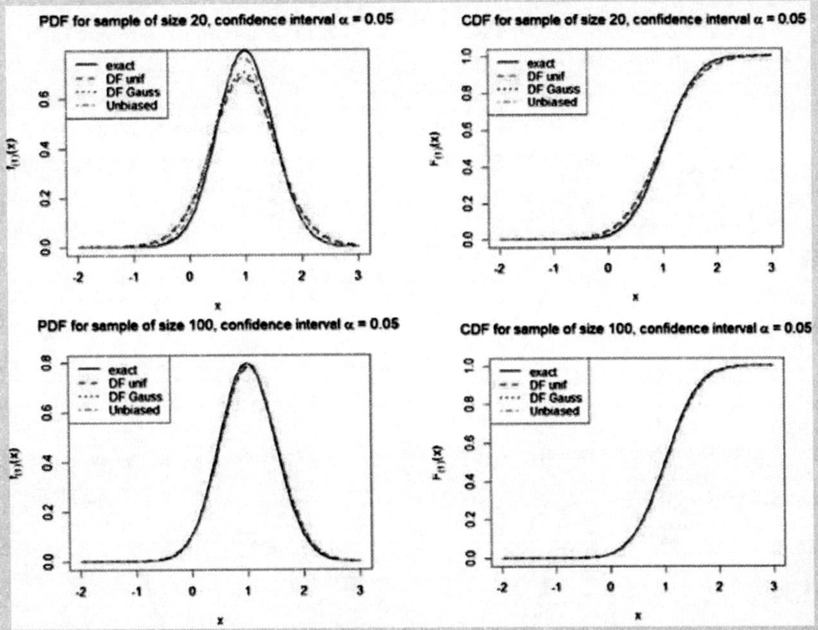

Fig. 1.19 Comparison between the exact distribution and the distributions furnished by De Finetti's theorem using the estimators and confidence intervals with larger samples

Eventually, we can use Bootstrap to improve the result: R proposes a package `boot` to generate bootstrap estimations and confidence intervals. Again, the results will be closer to those furnished by the unbiased estimators. For instance, using 100 bootstrap replicates:

(continued)

Example 1.30 (continued)

```
theta2 <- function(x,ind){mean(x[ind])}
theta3 <- function(x,ind){sd(x[ind])}
library(boot)
r2a = boot(X,theta2,100,sim='ordinary',stype='i')
r3a = boot(X,theta3,100,sim='ordinary',stype='i')
m2a=boot.ci(r2a, conf=0.95, index=1, type=c('perc'))
m3a=boot.ci(r3a, conf=0.95, index=1, type=c('perc'))
m_lower = m2a$percent[[4]]
m_upper = m2a$percent[[5]]
s_lower = m3a$percent[[4]]
s_upper = m3a$percent[[5]]
```

For the data under consideration, typical confidence intervals are

$$m \in (0.47, 1.16), \quad \sigma \in (0.12, 0.54).$$

Typical results appear in Fig. 1.20: we observe a rapprochement of the results, which are also improved.

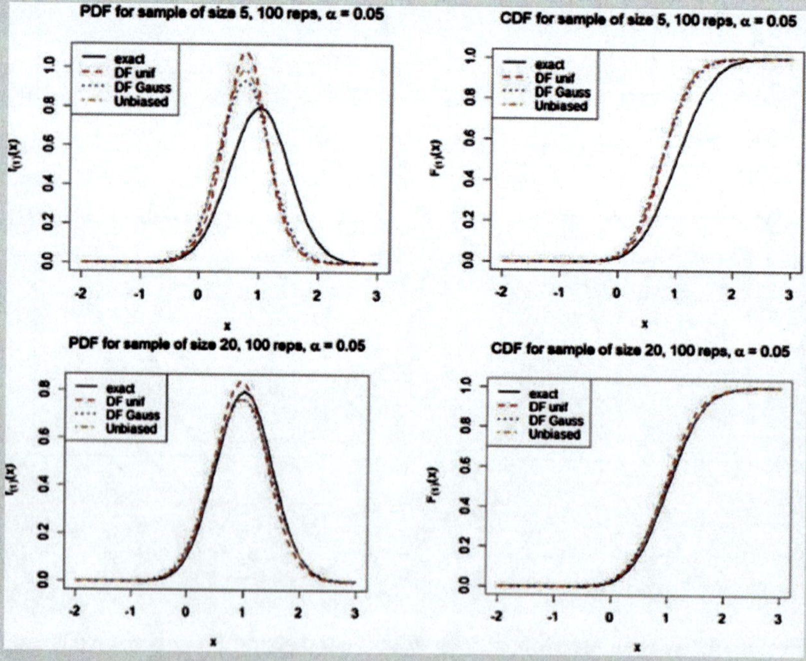

Fig. 1.20 Comparison between the exact distribution and the distributions furnished by De Finetti's theorem using the bootstrap estimators and associated confidence intervals, generated by 5 and 20 variates

(continued)

Example 1.30 (continued)

An alternative approach can combine uncertainty quantification (UQ) representation methods and De Finetti's theorem. For instance, we can consider generating *ns* samples $\mathcal{U}_5^k, k = 1, \ldots, ns$ of five variates from $N(0, 1)$ and apply the UQ representation approach: for each combination $\mathcal{U}_5^k$, we can determine a UQ representation $X \approx P^k X = \sum_{i=0}^{n} c_i^k \varphi_i(u)$ ($\{\varphi_i\colon i \in \mathbb{N}\}$ is a convenient Hilbert basis) – see, for instance, Souza de Cursi (2023). Each representation can be used to generate a large sample of *nv* variates from *X*, which furnishes an empirical mean $\widehat{m}^k = \overline{X_{nv}}^k$ and an empirical unbiased estimation $\widehat{\sigma}^k = s_{nv}^k$ of σ. These results generate samples from $\widehat{m} = \overline{X}_{nv}$ and $\widehat{\sigma} = s_{nv}$, which can be used to define μ. Such a definition can be made simply by considering that both the variables are approximately Gaussian, since *nv* is large: in such a situation, we use μ defined by the density given in Eq. (1.87) or simplify it to ($s(\overline{V})$ denotes the empirical unbiased estimation of the standard deviation of $\overline{V}$, *f* is the Gaussian density previously introduced):

$$g(m, \sigma) = f\left(m, \overline{\widehat{m}}, s\left(\widehat{m}\right)\right) f\left(s, \overline{\widehat{\sigma}}, s\left(\widehat{\sigma}\right)\right). \tag{1.88}$$

As an alternative, we can also use the UQ approach again and determine representations $P\widehat{m}$ of $\overline{X}_{nv}$ and $P\widehat{\sigma}$ of s_{nv}, which can be used to determine large samples formed by *nv* variates of each variable and use these samples to determine their marginal CDFs and PDFs. This approach furnishes an empirical PDF *g* by taking the product of the marginal distributions found.

As an example, let us generate $ns = 30$ samples $\mathcal{U}_5^k$ and $nv = 2E4$ variates. We eliminate the outliers of the bilateral quantile corresponding to $\alpha = 0.05$. The UQ expansion $P^k X$ uses a polynomial basis, with degree $n = 3$, and it is determined by moment matching with 4 moments – see, for instance, Souza de Cursi (2023). We apply De Finetti's theorem to four possible choices of μ: a uniform distribution on the rectangle defined by the extreme values of the values of $\widehat{m}^k$ and $\widehat{\sigma}^k$, g given by Eq. (1.86), g given by Eq. (1.87), and the empirical distribution generated by $P\widehat{m}$ and $P\widehat{\sigma}$. The results appear in Fig. 1.21 – the unbiased estimation corresponds to the distribution having as density $g\left(x, \overline{\widehat{m}}, \overline{\widehat{\sigma}}\right)$. Again, increasing the size of the sample makes that the results become closer to the exact one, as shown by the results for the sample of 20 variates in Table 1.10 exhibited in Fig. 1.21.

(continued)

Example 1.30 (continued)

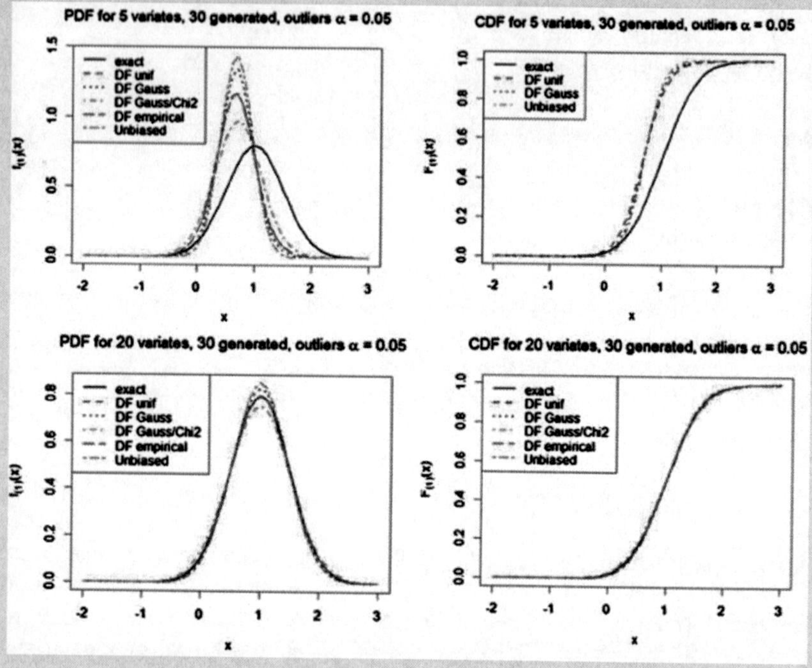

Fig. 1.21 Comparison between the exact distribution and the distributions furnished by De Finetti's theorem combined with the UQ approach

1.13 Stochastic Processes

Stochastic processes are presented in the companion book (Souza de Cursi, 2023). Here, we recall only the definition and some essential elements.

Definition 1.18 Let (Ω, P) be a probability space. A stochastic process indexed by $\tau \in \mathrm{T}$ is a family of random variables $\{X_\tau: \tau \in \mathrm{T}\}$, with X_τ defined on (Ω, P). We say that the process is discrete if T is a discrete set and that the process is continuous if T is an interval. τ is said to be the time.

For each time τ, $:X_\tau$ is a random variable, so that we can consider its statistics:

Definition 1.19 The **mean** of X is $m_X(t) = E(X_t)$.
The **variance** of X is $V_X(t) = V(X_t) = E(X_t^2) - (E(X_t))^2$
The **standard deviation** of X is $\sigma_X(t) = \sqrt{V_X(t)}$.
The **moment of order** k of X is $M_{X,k}(t) = E(X_t^k)$.
X is a **second-order process** if and only if $X_t \in L^2(\Omega, P)$, $\forall\ t$, i.e., $E(X_t^2) < \infty, \forall t$ ..

Stochastic processes are often characterized by their autocorrelations and autocovariances:

Definition 1.20 Let $X = \{X_t : t \in \mathcal{T}\}$ be a second order process.
The **autocorrelation** of X is $R_{XX}(t, s) = E(X_t X_s)$.
The **autocovariance** of X is $C_{XX}(t, s) = E(X_t X_s) - E(X_t)E(X_s)$.
The **crosscorrelation** of X and Y is $R_{XY}(t, s) = E(X_t Y_s)$.
The **crosscovariance** of X and Y is $C_{XY}(t, s) = E(X_t Y_s) - E(X_t)E(Y_s)$

Notice that

$$C_{XX}(t, s) = C_{XX}(s, t), R_{XX}(t, s) = R_{XX}(s, t).$$

Important classes of processes are defined by stationarity and ergodicity:

Definition 1.21

1. X is **weakly stationary** if and only if $V_X(t)$ is finite for any t; m_X is independent from t: $m_X(t) = m_0$, $\forall\ t$ and $R_{XX}(s, t) = R_X(|s - t|)$, $\forall\ s, t$. R_X is extended by symmetry to negative values: $R_X(a) = R_X(-a)$, so that $R_{XX}(s, t) = R_X(s - t) = R_X(t - s)$.
2. If X and Y are weakly stationary, $m_X(t) = m_0$, $m_Y(t) = n_0$, $\forall\ t$. Then, we can evaluate $R_{XY}(s) = E(X_{t+s} Y_t)$, $C_{XY}(s) = E(X_{t+s} Y_t) - m_0 n_0$.
3. X is **stationary** if and only if the distribution of $(X(t_1), \ldots, X(t_n))$ coincides with the distribution of $(X(t_1 + \tau), \ldots, X(t_n + \tau))$ for any n, any $(t_1, .., t_n)$, and any τ. If X is stationary, then X is weakly stationary.
4. Let X be weakly stationary. X is **ergodic in the mean** if and only if $m_0 = \lim_{T \to +\infty} \frac{1}{T} \int_0^T X(s)ds$. It is **ergodic in correlation** if and only if $R_X(\tau) = \lim_{T \to +\infty} \frac{1}{T} \int_0^T X(s)X(s + \tau)ds$.

For a weakly stationary process,

$$C_{XX}(t,s)=R_{XX}(t,s)-m_X^2=R_X(|s-t|)-m_0^2=C_X(|s-t|)$$

Exercises

1. Let X_t be a discrete variable Poisson distributed with parameter λt, i.e.,

$$P(X_t=n)=\frac{(\lambda t)^n}{n!}e^{-\lambda t}.$$

   (a) Show that $m_X(t) = \lambda t$.
   (b) Show that $V_X(t) = \lambda t$.
   (c) Show that $R_{XX}(s,t) = \lambda \min(s,t) + \lambda^2 st$.
   (d) Determine $C_{XX}(s,t)$.
   (e) Is this process stationary?

2. Let $X_t = A_1 \cos(\omega t) + A_2 \sin(\omega t)$, where $\omega \in \mathbb{R}$, while A_1, A_2 are independent random variables such that $E(A_i)=m_i$, $V(A_i)=\sigma_i^2$.

   (a) Show that $m_X(t) = m_1 \cos(\omega t) + m_2 \sin(\omega t)$.
   (b) Show that $V_X(t)=\sigma_1^2\cos^2(\omega t)+\sigma_2^2\sin^2(\omega t)$.
   (c) Show that $C_{XX}(s,t)=\sigma_1^2\cos(\omega t)\cos(\omega s)+\sigma_2^2\sin(\omega t)\sin(\omega s)$.
   (d) Determine $R_{XX}(s,t)$.
   (e) Is this process stationary?

3. Let $\{\varepsilon_n: n \in \mathbb{N}\}$ be a family of independent Gaussian variables of the same distribution $N(0,1)$ and $X_n = \varepsilon_n\varepsilon_{n+1}$

   (a) Show that $m_X(n) = 0$.
   (b) Show that $V_X(n) = 1$.
   (c) Show that $R_{XX}(m,n) = 0$ *if* $m \neq n$.
   (d) Determine $R_{XX}(m,m)$.
   (e) Is this process stationary?

4. Let $\{\varepsilon_n: n \in \mathbb{N}\}$ be a family of independent Gaussian variables of the same distribution $N(0,1)$ and $X_n = \varepsilon_n + \varepsilon_{n+1}$

   (a) Show that $m_X(n) = 0$.
   (b) Show that $V_X(n) = 2$.
   (c) Show that $R_{XX}(m,n) = 0$ *if* $|m-n| \geq 2$.
   (d) Determine $R_{XX}(m,m)$, $R_{XX}(m,m+1)$.
   (e) Is this process stationary?

References

Attiaoui, D., Doré, P., Martin, A., & Ben Yaghlane, B. (2012). A Distance between Continuous Belief Functions. In E. Hüllermeier, S. Link, T. Fober, & B. Seeger (Ed.), *Scalable Uncertainty Management: SUM 2012* (pp. 194–205). Marburg, Germany: Springer. doi:https://doi.org/10.1007/978-3-642-33362-0_15

Bayes, T., & Price, R. (1763, January 01). An essay towards solving a problem in the doctrine of chances. *Philosophical Transactions of the Royal Society of London,* 53, pp. 370–418. https://doi.org/10.1098/rstl.1763.0053, https://www.jstor.org/stable/105741?seq=3; https://royalsocietypublishing.org/doi/10.1098/rstl.1763.0053; consulted April 12th 2024.

Benavoli, A. (2014). Belief function and multivalued mapping robustness in statistical estimation. *International Journal of Approximate Reasoning, 55*, 311–329. doi:https://doi.org/10.1016/j.ijar.2013.04.014

Bernardo, J. M., & Smith, A. F. (2000). *Bayesian Theory.* Bafiins Lane, Chichester, England: John Wiley & Sons.

Boivin, C. (2022a, December 25). https://cran.r-project.org/web/packages/dst/vignettes/Captain_Example.html. Retrieved from Captain's Example: https://cran.r-project.org/web/packages/dst/vignettes/Captain_Example.html

Boivin, C. (2022b, 12 24). *Introduction to Belief Functions.* Retrieved from https://cran.r-project.org/web/packages/dst/vignettes/: https://cran.r-project.org/web/packages/dst/vignettes/Introduction_to_Belief_Functions.html#fn2

Boivin, C. (2022c, December 25). *Introduction to Belief Functions: The Monty Hall Game.* Retrieved 2022, from https://cran.microsoft.com/snapshot/2018-08-11/web/packages/dst/vignettes/Monty-hall-Example.html: https://cran.microsoft.com/snapshot/2018-08-11/web/packages/dst/vignettes/Monty-hall-Example.html

Brown, L. D. (1986). *Fundamentals of Statistical Exponential Families with Applications in Statistical Decision Theory* (Lecture Notes-Monograph Series, vol. 9 ed.). Hayward, California, USA: Institute of Mathematical Statistics. doi:https://doi.org/10.1214/lnms/1215466757

Campagner, A., Ciucci, D., & Denœux, T. (2022). Belief Functions and Rough Sets: Suvey and New Insights. *International Journal of Approximate Reasoning, 143*, 92–215. doi:https://doi.org/10.1016/j.ijar.2022.01.011

Cobb, B. R., & Shenoy, P. (2006, April). On the plausibility transformation method for translating belief function models to probability models. *Journal of Approximate Reasoning, 41*(3), pp. 314–330. doi:https://doi.org/10.1016/j.ijar.2005.06.008

Couso, I., Dubois, D., & Sanchez, L. (2014). *Random Sets and Random Fuzzy Sets as Ill-Perceived Random Variables.* Springer. doi:https://doi.org/10.1007/978-3-319-08611-8

Dale, A. I. (1982). Bayes or Laplace? An Examination of the Origin and Early Applications of Bayes' Theorem. *Archive for History of Exact Sciences, 27*(1), pp. 23–47.

Dale, A. I. (1999). *A History of Inverse Probability—From Thomas Bayes to Karl Pearson* (2nd. ed.). New York: Springer.

De Finetti, B. (1930). Funzione Caratteristica di un Fenomeno Aleatorio. *Memoria della Reale Accademia dei Lincei, IV*(5), 86–133. Retrieved 1 29, 2023, from http://www.brunodefinetti.it/Opere/funzioneCaratteristica.pdf

De Finetti, B. (1937). La prévision : ses lois logiques, ses sources subjectives. *Annales de l'Institut henri Poïncaré, 7*(1), 1–68. Retrieved 1 29, 2023, from http://www.numdam.org/item/AIHP_1937__7_1_1_0.pdf

De Finetti, B. (2017). *Theory of Probability—A Critical Introductory Treatment.* (A. M. Smith, Trans.) UK: John Wiley and Sons.

DeGroot, M. H. (2004). *Optimal Statistical Decisions.* New Jersey: John Wiley & Sons.

Dempster, A. (1967, April). Upper and Lower Probabilities Induced by a Multivalued Mapping. *Ann. Math. Statist., 38*(2), pp. 325–339,. doi:https://doi.org/10.1214/aoms/1177698950

Denoeux, T. (2019, June). Decision-making with belief functions: A review. *International Journal of Approximate Reasoning, 109*, 87–110. doi:https://doi.org/10.1016/j.ijar.2019.03.009

Denoeux, T. (2021). *Uncertainty Analysis using Belief Functions: Applications to Statistical Inference and Pattern Recognition.* Université de technologie de Compiègne, Department of Computer Science, Compiègne, France. Retrieved December 20, 2022, from https://www.hds.utc.fr/~tdenoeux/dokuwiki/_media/en/book_bf.pdf

Denœux, T., Younes, Z., & Abdallah, F. (2010). Representing uncertainty on set-valued variables using belief functions. *Artificial Intelligence, 174*(7–8), 479–499. doi:https://doi.org/10.1016/j.artint.2010.02.002

Dezert, J., & Smarandache, F. (2008). A new probabilistic transformation of belief mass assignment. *Fusion 2008: International Conference on Information Fusion*, (pp. 1410–1417). Retrieved December 19, 2022, from https://hal.archives-ouvertes.fr/hal-00304319/document

Dezert, J., & Smarandache, F. (2009). Transformations of belief masses into subjective probabilities. In J. Dezert, & F. Smarandache (Eds.), *Advances and Applications of DSmT for Information Fusion* (pp. 85–136). Rehoboth: American Research Press . Retrieved December 19, 2022, from https://www.researchgate.net/publication/306413841_Transformations_of_belief_masses_into_subjective_probabilities

Dezert, J., Smarandache, F., & Daniel, M. (2004). A Generalized Pignistic Transformation. In J. Dezert, & F. Smarandache (Eds.), *Advances and Applications of DSmT for Information Fusion* (pp. 143–153). Rehoboth: American Research Press. Retrieved December 19, 2022, from https://www.onera.fr/sites/default/files/297/C023-Dezert-Fusion2004Stockholm.pdf

Diaconis, P. (1977). Finite forms of de Finetti's theorem on exchangeability. *Synthese, 36*, 271–281. doi:https://doi.org/10.1007/BF00486116

Doré, P. E., Fiche, A., & Martin, A. (2010). Models of belief functions—Impacts for patterns recognitions. *13th International Conference on Information Fusion.* Edinburgh, UK. doi:https://doi.org/10.1109/ICIF.2010.5711936

Doré, P., Martin, A., Abi-Zeid, I., Jousselme, A., & Maupin, P. (2011a, January). Belief functions induced by multimodalprobability density functions, an application to the search and rescue problem. *RAIRO—Operations Research, 44*(4), 323–343. doi:https://doi.org/10.1051/ro/2011001

Doré, P., Osswald, C., Martin, A., Jousselme, A., & Maupin, P. (2011b). Continuous belief functions to qualify sensors performances. In W. Liu (Ed.), *Symbolic and Quantitative Approaches to Reasoning with Uncertainty. ECSQARU 2011* (pp. 350–361). Belfast, Ireland: Springer. doi:https://doi.org/10.1007/978-3-642-22152-1_30

Dubois, D., & Prade, H. (1986). A Set-Theoretic View of Belief Functions. *International Journal of General Systems, 12*(3), 193–226. doi:https://doi.org/10.1080/03081078608934937

Fetz, T., & Oberguggenberger, M. (2016, November). Imprecise random variables, random sets, and Monte Carlo simulation. *78*, 252–264. doi:https://doi.org/10.1016/j.ijar.2016.06.012

Fiche, A., Martin, A., Cexus, J.-C., & Khenchaf, A. (2010). Continuous belief functions and α-stable distributions. *13th Conference on Information Fusion.* Edinburgh. doi:https://doi.org/10.1109/ICIF.2010.5711934

Fox, C. R., & See, K. E. (2006). Belief and Preference in Decision Under Uncertainty. In D. Hardman, & L. Macchi (Eds.), *Thinking: Psychological Perspectives on Reasoning, Judgment and Decision Making.* John Wiley & Sons. doi:https://doi.org/10.1002/047001332X.ch14

Fritz, T., Gonda, T., & Perrone, P. (2021). De Finetti's Theorem in Caqtegorical Probability. *Journal of Stochastic Analysis, 2*(4). doi:https://doi.org/10.31390/josa.2.4.06

Gavalakis, L., & Kontoyiannis, I. (2021). An information-theoretic proof of a finite de Finetti theorem. *Electronic Communications in Probability, 26*, 1–5. doi:https://doi.org/10.1214/21-ECP428

Guan, J., & Bell, D. (1993). Discounting and Combination Operations in Evidential Reasoning. In D. Heckerman, & A. Mamdani (Ed.), *Ninth Conference on Uncertainty in Artificial Intelligence* (pp. 477–484). Washington, USA: Elsevier. doi:https://doi.org/10.1016/B978-1-4832-1451-1.50062-7

Halmos, P. (1978). *Measure theory.* Berlin, Heidelberg, New York: : Springer Verlag.

Halpern, J., & Fagin, R. (1992). Two views of belief: belief as generalized probability and belief as evidence. *Artificial Intelligence,, 54*, pp. 275–317.

He, Y. (2013). *Uncertainty Quantification and Data Fusion based on Dempster-Shafer Theory.* Florida State University, Mathematics. Florida State University Libraries.

Hewitt, E., & Savage, L. (1955). Symmetric Measures on Cartesian Products. *Transactions of The American Mathematical Society, 80*(2), 470–501. doi:https://doi.org/10.2307/1992999

Hill, B. M., Lane, D., & Sudderth, W. (1987). Exchangeable Urn Processes. *Ann. Probab., 15*(4), 1586–1592. doi:https://doi.org/10.1214/aop/1176991995

Howson, C., & Urbach, P. (2006). *Scientific reasoning : the Bayesian approach* . Chicago: Open Court.

Hüllermeier, E., Kruse, R., & Hoffmann, F. (Eds.). (2010). Consonant continuous belief functions conflicts calculation. *IPMU'10: Proceedings of the Computational intelligence for knowledge-based systems design, and 13th international conference on Information processing and management of uncertainty* (pp. 706–715). Dortmund, Germany: Springer.

Ivanoff, B. G., & Weber, N. (2004). Predictable sampling for partially exchangeable arrays. *Statistics & Probability Letters, 70*(1), 95–108. doi:https://doi.org/10.1016/j.spl.2004.08.005

Jaffray, J.-Y., & Wakker, P. (1993). Decision making with belief functions: Compatibility and incompatibility with the sure-thing principle. *Journal of Risk and Uncertainty, 7*(3), 255–271. doi:https://doi.org/10.1007/BF01079626

Jaynes, E. (1989). *Clearing up Mysteries—The Original Goal.* (J. Skilling, Ed.) Dordrecht: Springer. doi:https://doi.org/10.1007/978-94-015-7860-8_1

Jaynes, E. (2003). *Probability Theory: The Logic of Science.* Cambridge: Cambridge University Press.

Jeffreys, H. (1939). *Theory of probability.* Oxford: University Press.

Kallenberg, O. (1988). Spreading and Predictable Sampling in Exchangeable Sequences and Processes. *The Annals of Probability, 16*(2), 508–534. doi:https://doi.org/10.1214/aop/1176991771

Kallenberg, O. (1989). On the representation theorem for exchangeable arrays. *Journal of Multivariate Analysis, 30*(1), 137–154. doi:https://doi.org/10.1016/0047-259X(89)90092-4

Klopotek, M. A., & Wierzchon, S. T. (1998). A New Qualitative Rough-Set Approach to Modeling Belief Functions. In L. Polkowski, & A. Skowron (Ed.), *Rough Sets and Current Trends in Computing, First International Conference, RSCTC'98* (pp. 346–354). Warzsaw, Poland: Springer. doi:https://doi.org/10.1007/3-540-69115-4_47

Laghmara, H., Laurain, T., Cudel, C., & Lauffenburger, J. P. (2020). Heterogeneous sensor data fusion for multiple object association using belief functions. *Information Fusion, 57*, 44–58. doi:https://doi.org/10.1016/j.inffus.2019.11.002

Laplace, P.-S. (1774). Memoire sur la probabilité des causes par les événements. *Memoires de Mathématique et de Physique, Presentés à l'Académie Royale des Sciences par divers Savans & lus dans ses Assemblées*, pp. 621–656.

Laplace, P.-S. (1891). *Oeuvres complètes* (Vol. 8). (A. d. Paris, Ed.) Paris: Gauthiers-Villars.

Laplace, P.-S. (1986). Memoir on the Probability of the Causes of Events. *Statistical Science, 1*(3), pp. 364–78. Retrieved from http://www.jstor.org/stable/2245476

Lévy, P. (1922, November 13). Sur la détermination des lois de probabilité par leurs fonctions caractéristiques. *Comptes Rendus de l'Académie des Sciences de Paris*, pp. 854–856.

Lian, C. (2017). *Information Fusion and Decision Making using Belief Functions.* Compiègne, France: Université de Technologie de Compiègne.

Liu, W. (2006, August). Analyzing the degree of conflict among belief functions. *Artificial Intelligence, 170* (11), 909–924. doi:https://doi.org/10.1016/j.artint.2006.05.002

Martin, A. (2019). Conflict management in information fusion with belief functions. In E. Bossé, & G. Rogova (Eds.), *Information quality in information fusion and decision making* (pp. 79–97). Springer. doi:https://doi.org/10.1007/978-3-030-03643-0_4

Martin, R., Zhang, J., & Liu, C. (2010, June). Dempster–Shafer Theory and Statistical Inference with Weak Beliefs. *Statistical Science, 25*(1), 72–87. doi:https://doi.org/10.1214/10-STS322

Mercier, D., Quost, B., & Denœux, T. (2005). Contextual Discounting of Belief Functions. In L. Godo (Ed.), *ECSQARU 2005: Symbolic and Quantitative Approaches to Reasoning with Uncertainty* (pp. 552–562). Barcelona, Spain: Springer. doi:https://doi.org/10.1007/11518655_47

Miranda, E., Couso, I., & Gil, P. (2005, July). Random sets as imprecise random variables. *Journal of Mathematical Analysis and Applications, 307*(1), 32–47. doi:https://doi.org/10.1016/j.jmaa.2004.10.022

Nguyen, H. T. (1977). *On Random sets and Belief Functions.* University of California, Berkeley, EECS Department. Retrieved from http://www2.eecs.berkeley.edu/Pubs/TechRpts/1977/28879.html

Nguyen, H. T. (1978). On Random Sets and Belief Functions. *Journal of Mathematical Analysis and Applications, 65*, 531–542. doi:https://doi.org/10.1016/0022-247X(78)90161-0

Nguyen, H., & Wang, T. (1997). Belief Functions and Random Sets. In J. Goutsias, R. Mahler, & H. Nguyen (Eds.), *Random Sets* (pp. 243–255). Springer. doi:https://doi.org/10.1007/978-1-4612-1942-2_11

Ni, S., Lei, Y., & Tang, Y. (2020). Improved Base Belief Function-Based Conflict Data Fusion Approach Considering Belief Entropy in the Evidence Theory. *Entropy, 22*(8). doi:https://doi.org/10.3390/e22080801

Polya, G. (1930). Sur quelques points de la théorie des probabilités. *Annales de l'Institut Penri Poïncaré, 1*(2), 117–161. Retrieved 1 29, 2023, from http://www.numdam.org/article/AIHP_1930__1_2_117_0.pdf

Press, S. J., & Tanur, J. M. (2001). *The Subjectivity of Scientists and the Bayesian Approach.* New York: John Wiley & Sons.

Risti, B., & Smets, P. (2006). Belief function theory on the continuous space with an application to model based classification. In B. Bouchon-Meunier, G. Coletti, & R. R. Yager (Ed.), *Modern Information Processing: From Theory to Applications. IPMU'04* (pp. 11–24). Perugia, Italy.: Elsevier. doi:https://doi.org/10.1016/B978-044452075-3/50002-9

Savchuk, V. P., & Tsokos, C. P. (2011). *Bayesian Theory and Methods with Applications.* Paris: Atlantis Press.

Shafer, G. (1976). *A Mathematical Theory of Evidence* . New Jersey: Princeton University Press.

Shafer, G. (1990). Perspectives on the Theory and Practice of Belief Functions. *International Journal of Approximate Reasoning, 4*(5–6), 323–362. doi:https://doi.org/10.1016/0888-613X(90)90012-Q

Smets, P. (1990). Constructing the Pignistic Probability Function in a Context of Uncertainty. *Machine Intelligence and Pattern Recognition, 10*, 29–39. doi:https://doi.org/10.1016/B978-0-444-88738-2.50010-5

Smets, P. (2000). Data fusion in the transferable belief model. *Proceedings of the Third International Conference on Information Fusion. 1*, pp. 21–33. Paris, France: IEEE. doi:https://doi.org/10.1109/IFIC.2000.862713.

Smets, P. (2005). Belief functions on real numbers. *International Journal of Approximate Reasoning, 40*, 181–223. doi:https://doi.org/10.1016/j.ijar.2005.04.001

Souza de Cursi, E. (2023). *Uncertainty Quantification using R.* Springer Cham.

Stigler, S. M. (1982). Thomas Bayes's Bayesian Inference. *Journal of the Royal Statistical Society. Series A (General), 145*(2), pp. .250–258.

Strat, T. (1984). Continuous belief functions for evidential reasoning. *Proceedings of the 4th National Conference on Artificial Intelligence.* Austin, Texas. Retrieved December 18, 2022, from https://www.aaai.org/Papers/AAAI/1984/AAAI84-035.pdf

Strat, T. (1987). The Generation of Explanations within Evidential Reasoning Systems. In J. P. McDermott (Ed.), *10th. International Joint Conference on Artificial Intelligence (IJCAI)*, (pp. 1097–1104). Milan, Italy. Retrieved December 20, 2022, from https://www.ijcai.org/Proceedings/87-2/Papers/104.pdf

Strat, T. (1990). Decision analysis using belief functions. *International Journal of Approximate Reasoning, 4*(5–6), 391–417. doi:https://doi.org/10.1016/0888-613X(90)90014-S

Sundberg, R. (2019). *Statistical Modelling by Exponential Families.* Cambridge, UK: Cambridge University Press. doi:https://doi.org/10.1017/9781108604574

Svante, J., Takis, K., & Linglong, Y. (2016). On a representation theorem for finitely exchangeable random vectors. *Journal of Mathematical Analysis and Applications, 442*(2), 703–714. doi: https://doi.org/10.1016/j.jmaa.2016.04.070

Taillandier, P., & Therond, O. (2011). Use of the Belief Theory to formalize Agent DecisionMaking Processes : Application to cropping Plan Decision Making. *European Simulation and Modelling Conference*, (pp. 138–142). Guimaraes, Portugal. Retrieved 12 17, 2022, from https://hal.archives-ouvertes.fr/hal-00688405

Wasserman, L. A. (1990, September). Belief functions and statistical inference. *Canadian Journal of Statistics, 18*(3), 183–196. doi:https://doi.org/10.2307/3315449

Witkovský, V. (2016). Numerical inversion of a characteristic function: An alternative tool to form the probability distribution of output quantity in linear Measurement models. *ACTA IMEKO, 5*(3), 32–44. doi:https://doi.org/10.21014/acta_imeko.v5i3.382.

Wu, W.-Z., & Mi, J.-S. (2008). An Interpretation of Belief Functions on Infinite Universes in the Theory of Rough Sets. In C. Chan, J. Grzymala-Busse, & W. P. Ziarko (Ed.), *Rough Sets and Current Trends in Computing. RSCTC 2008* (pp. 71–80). Akron, OH, USA: Springer. doi: https://doi.org/10.1007/978-3-540-88425-5_8

Yao, Y. Y., & Lingras, P. J. (1998). Interpretations of Belief Functions in the Theory of Rough Sets. *Information Sciences, 104*(1–2), 81–106. doi:https://doi.org/10.1016/S0020-0255(97)00076-5

Zhang, H., & Deng, Y. (2020). Weighted belief function of sensor data fusion in engine fault diagnosis. *Soft Comput, 24*, 2329–2339. doi:https://doi.org/10.1007/s00500-019-04063-7

Zhou, K., Martin, A., & Pan, Q. (2018). A belief combination rule for a large number of sources. . *Journal of Advances in Information Fusion, 13*(2). Retrieved December 2022, 20, from https://www.researchgate.net/deref/https%3A%2F%2Fhal.archives-ouvertes.fr%2Fhal-01883239

Chapter 2
Beliefs

In the framework of Bayesian methods, the interpretation of probabilities in terms of degrees of belief has led some researchers to seek to formalize the notion of belief. These efforts resulted in the formalism of *Belief and Plausibility functions* (DeGroot 2004; Dempster 1967; Shafer 1976; Halpern and Fagin 1992). In this chapter, we present some elements from Belief theory, with illustrations using examples.

2.1 Beliefs

Definition 2.1 Let Ω be a non-empty set and $\mathcal{P}(\Omega) = \{E : E \subset \Omega\}$ be the set of the parts of Ω. A ***belief function*** defined on Ω is an application $Bel : \mathcal{P}(\Omega) \longrightarrow \mathbb{R}$ such that

1. $Bel(E) \geq 0, \forall E \in \mathcal{P}(\Omega)$;
2. $Bel(\Omega) = 1$;
3. $Bel(\varnothing) = 0$
4. $Bel\left(\bigcup_{i=1}^{k} E_i\right) \geq \sum_{\substack{I \subset \{1,\ldots,k\}, \\ I \neq \varnothing}} (-1)^{1+|I|} Bel\left(\bigcap_{i \in I} E_i\right)$. $|I|$ denotes the number of elements of I.

The ***plausibility*** of E is $Pl(E) = 1 - Bel(\Omega - E)$. ∎

In the context of belief theory, Ω is called *frame of discernment* – it corresponds to the universe in probabilities. It can also be simply referred to as *frame*. Frames can be

E. Souza de Cursi, *Uncertainty Quantification with R*, International Series in Operations Research & Management Science 352,
https://doi.org/10.1007/978-3-031-48208-3_2

represented analogously to probability spaces. For instance, we can use binary vectors or hexadecimal numbers as shown in Sect. 1.3.

Analogously to probabilities, beliefs can be defined using belief mass functions:

Definition 2.2

1. Let Ω be finite or enumerable. A **belief mass function** μ on Ω is an application $\mu : \mathcal{P}(\Omega) \longrightarrow \mathbb{R}$ such that $\mu(\varnothing) = 0$; $\forall A \in \mathcal{P}(\Omega) : \mu(A) \geq 0$ and $\sum_{A \in \mathcal{P}(\Omega)} \mu(A) = 1$. Then,

$$Bel(A) = \sum_{B \subset A} \mu(B), Pl(A) = \sum_{B \cap A \neq \varnothing} \mu(B) .$$

2. Let $\Omega \subset \mathbb{R}$. A **belief mass density** μ on Ω is an application $\mu : \Omega \times \Omega \longrightarrow \mathbb{R}$ such that, on the one hand, $\mu(a, b) \geq 0$, if $a \leq b$ *and* $(a, b) \subset \Omega$; $\mu(a, b) = 0$, otherwise; on the other hand, $\iint_{\Omega} \mu(x, y) dx dy = 1$. Then, for $a \leq b$ (Smets 2005):

$$Bel([a, b]) = \int_a^b dx \int_x^b \mu(x, y) dy, Pl([a, b]) = \int_{-\infty}^b dx \int_{\max(x, a)}^{+\infty} \mu(x, y) dy.$$

Notice that the mass function must be defined for all the subsets of Ω, so that a belief mass function is not the same object as a probability mass function, which can be defined for the elements of Ω. The theory of belief functions also considers an ***allocation mass function (or density)*** $\alpha(\omega, A)$ having properties analogous to those of a probability mass function, namely,

$$\mu(A) = \sum_{\omega \in A} \alpha(\omega, A) \quad \left(or \quad \mu(A) = \int_A \alpha(\omega, A) d\omega \right).$$

However, such an allocation mass function (or density) $\alpha(\omega, A)$ remains dependent on the part of $A \subset \Omega$ and is not from the single element $\omega \in \Omega$.

Notice also that the definition in the continuous situation considers a closed interval $[a, b] \subset \mathbb{R}$ as a vector $(a, b) \subset \mathbb{R}^2$ and gives to it a density $\mu(a, b)$. The definition above implies that $\mu(\varnothing) = 0$. At their origins, belief mass functions were intended to apply to finite or enumerable frames of discernment. Extensions to enumerable sets and to continuous situations came later – see, for instance, Strat (1984), Smets (2005), Risti and Smets (2006), Doré et al. (2010), Fiche et al. (2010), Hüllermeier et al. (2010), Doré et al. (2011a, b).

It is usual to describe mass functions by giving only the elements for which they are strictly positive. In such a situation, we implicitly extend the mass function by zero to the subsets not considered in the definition.

We say that A is a ***focal set*** for μ if and only if $\mu(A) > 0$. In the framework of belief functions, a belief mass function μ is said to be:

- **Vacuous** if and only if $\mu(\Omega) = 1$. A vacuous μ represents a total ignorance.
- **Logical** if and only if there exists a unique $A \subset \Omega$ such that $\mu(\mathrm{A}) > 0$. A logical μ represents a full certainty concerning A.
- **Bayesian** if and only if $\mu(A) > 0 \Longrightarrow A \text{ is a singleton } A = \{a\}, a \in \Omega$. A Bayesian μ represents a probability mass function.
- **Consonant** if and only if the focal sets are nested: $A_1 \subset A_2 \subset \ldots \subset A_n$. In this case,

$$Bel\ (A \cap B) = \min\{Bel\ (A), Bel(B)\}, \quad Pl(A \cup B) = \max\{Pl(A), Pl(B)\}.$$

Notice that beliefs and plausibilities are different from probabilities (see, for instance, Example 1.23).

We can transform plausibilities in probabilities by generating a probability mass function from the plausibilities: we determine the plausibilities of all the singletons in Ω, and we normalize their values to get a total sum equal to the unity (Cobb and Shenoy 2006).

A second way to transform plausibilities in probabilities is the ***pignistic transformation*** – see, for instance, Smets (1990, 2000), Dezert et al. (2004), Dezert and Smarandache (2008), Dezert and Smarandache (2009), Attiaoui et al. (2012):

$$P(A) = \sum_{B \neq \emptyset} \frac{|A \cap B|}{|B|} \mu(B)$$

Conversely, it is possible to define a mass function on $\mathbb{R}$, using the probability P and the multivalued map X: $\mu_X(A) = P(\{\omega \in \Omega : X(\omega) = A\})$ (Denoeux 2019).

Example 2.1 Let us consider 3 possible options for a decision: O_1, O_2, O_3. Assume that an expert furnishes the mass function:

$$\mu(\{O_1, O_2, O_3\}) = 0.4; \mu(\{O_2, O_3\}) = 0.2; \mu(\{O_2\}) = 0.4$$

In such a situation, we can extend the definition by taking the other masses to be null:

$$\mu(\{O_1\}) = \mu(\{O_3\}) = \mu(\{O_1, O_2\}) = \mu(\{O_1, O_3\}) = 0.$$

(continued)

Example 2.1 (continued)

Let us evaluate *Bel*:

$$\begin{aligned}
Bel(\varnothing) &= Bel(\{O_1\}) = 0,\\
Bel(\{O_2\}) &= \mu(\{O_2\}) = 0.4,\\
Bel(\{O_3\}) &= 0,\\
Bel(\{O_1, O_2\}) &= \mu(\{O_2\}) = 0.4,\\
Bel(\{O_1, O_3\}) &= 0,\\
Bel(\{O_2, O_3\}) &= \mu(\{O_2\}) + \mu(\{O_2, O_3\}) = 0.6,\\
Bel(\{O_1, O_2, O_3\}) &= \mu(\{O_2\}) + \mu(\{O_2, O_3\}) + \mu(\{O_1, O_2, O_3\}) = 1.
\end{aligned}$$

Let us evaluate *Pl*:

$$\begin{aligned}
Pl(\varnothing) &= 0,\\
Pl(\{O_1\}) &= \mu(\{O_1, O_2, O_3\}) = 0.4,\\
Pl(\{O_2\}) &= \mu(\{O_1, O_2, O_3\}) + \mu(\{O_2, O_3\}) + \mu(\{O_2\}) = 1,\\
Pl(\{O_3\}) &= \mu(\{O_1, O_2, O_3\}) + \mu(\{O_2, O_3\}) = 0.6,\\
Pl(\{O_1, O_2\}) &= \mu(\{O_2\}) + \mu(\{O_2, O_3\}) + \mu(\{O_1, O_2, O_3\}) = 1,\\
Pl(\{O_1, O_3\}) &= \mu(\{O_2, O_3\}) + \mu(\{O_1, O_2, O_3\}) = 0.6,\\
Pl(\{O_2, O_3\}) &= \mu(\{O_1, O_2, O_3\}) + \mu(\{O_2, O_3\}) + \mu(\{O_2\}) = 1,\\
Pl(\{O_1, O_2, O_3\}) &= \mu(\{O_1, O_2, O_3\}) + \mu(\{O_2, O_3\}) + \mu(\{O_2\}) = 1.
\end{aligned}$$

Let us transform the plausibilities in probabilities: since

$$Pl(\{O_1\}) = 0.4, Pl(\{O_2\}) = 1, Pl(\{O_3\}) = 0.6,$$

we have (the plausibilities which sum up to 2):

$$P(\{O_1\}) = 0.2, P(\{O_2\}) = 0.5, P(\{O_3\}) = 0.3.$$

Consequently,

$$P(\{O_1, O_2\}) = 0.7, P(\{O_2, O_3\}) = 0.8, P(\{O_1, O_3\}) = 0.5$$

(continued)

Example 2.1 (continued)

Now let us use the pignistic transformation: the results appear in Table 2.1:

Table 2.1 Pignistic transformation

A	B	$\lvert B\rvert$	$\lvert A \cap B\rvert$	$P(A)$
$\{O_1\}$	$\{O_1, O_2, O_3\}$	3	1	$\frac{4}{30}$
$\{O_2\}$	$\{O_1, O_2, O_3\}$ $\{O_2, O_3\}$ $\{O_2\}$	3 2 1	1 1 1	$\frac{19}{30}$
$\{O_3\}$	$\{O_1, O_2, O_3\}$ $\{O_2, O_3\}$	3 2	1 1	$\frac{7}{30}$
$\{O_1, O_2\}$	$\{O_1, O_2, O_3\}$ $\{O_2, O_3\}$ $\{O_2\}$	3 2 1	2 1 1	$\frac{23}{30}$
$\{O_2, O_3\}$	$\{O_1, O_2, O_3\}$ $\{O_2, O_3\}$ $\{O_2\}$	3 2 1	2 2 1	$\frac{26}{30}$
$\{O_1, O_3\}$	$\{O_1, O_2, O_3\}$ $\{O_2, O_3\}$	3 2	2 1	$\frac{11}{30}$

Example 2.2 Let us consider tossing a coin once. The result belongs to $\Omega = \{H, T\}$. We have $\mathcal{P}(\Omega) = \{\emptyset, \{H\}, \{T\}, \Omega\}$. In the context of belief functions, we can model our total ignorance of the result as a mass function:

$$\mu(\emptyset) = \mu(\{H\}) = \mu(\{T\}) = 0; \;\; \mu(\Omega) = 1.$$

In this case,

$$Bel(\emptyset) = Bel(\{H\}) = Bel(\{T\}) = 0; \;\; Bel(\Omega) = 1$$

$$Pl(\emptyset) = 0, Pl(\{H\}) = Pl(\{T\}) = Pl(\Omega) = 1.$$

Let us determine the probabilities by the plausibility transformation: since $Pl(\{H\}) = Pl(\{T\}) = 1$, which sum up to 2, we have

(continued)

Example 2.2 (continued)

$$P(\{H\}) = P(\{T\}) = \frac{1}{2}.$$

The result is the same by the pignistic transformation:

$$P(\{H\}) = \frac{|\{H\}|}{|\Omega|} Pl(\Omega) = \frac{1}{2}; P(\{T\}) = \frac{|\{T\}|}{|\Omega|} Pl(\Omega) = \frac{1}{2}.$$

Exercises

1. A six-faced dice is rolled. A gambler considers that the mass function verifies $\mu(\{6\}) = 0.1$ and $\mu(\{1\}) = 0.24$, while the other faces have equal masses.
   (a) Determine the masses of the other faces.
   (b) Determine the beliefs associated with the faces.
   (c) Determine the plausibilities of the faces.
   (d) Find the probabilities of the faces by the plausibility transformation.
2. A coin is tossed twice. A gambler considers that the mass function verifies $\mu(\{H, H\}) = 0.1$ and $\mu(\{T, T\}) = 0.3$. No other information is given.
   (a) Create a model frame for this situation.
   (b) Determine the beliefs associated with the results.
   (c) Determine the plausibilities of the results.
   (d) Find the probabilities of the results by the plausibility transformation.
   (e) Determine the probabilities $P(\{H\})$ and $P(\{T\})$ by the two methods presented.
3. An urn contains balls of three colors: blue (B), red (R), and white (W). Two balls are drawn sequentially, with remittance of the first ball in the urn. We consider that $\mu(\{B\}) = 0.2$, $\mu(\{R\}) = 0.3$, *and* $\mu(\{W\}) = 0.5$.
   (a) Create a model frame for this situation.
   (b) Determine the beliefs associated with the results.
   (c) Determine the plausibilities of the results.
   (d) Find the probabilities of the results by the plausibility transformation.
   (e) Determine the probabilities $P(\{B\})$, $P(\{R\})$, and $P(\{W\})$ by the two methods presented.

2.2 Working with Beliefs in R

Using R, we can create a class that generates a frame, beliefs, and plausibilities from a mass function, using a binary representation, as follows:

```
library(R6)
#
dfram=R6Class("dfram",
  public=list(
  ev = NULL, m = NULL, names = NULL, nev = NULL,
  nvar = NULL, bel = NULL, pl = NULL, Q = NULL,
  initialize=function(events,mu,varnames){
    if( missing(events) == TRUE){events = matrix(c(1,2),nrow
= 2)}
    self$ev = events
    d = dim(events)
    self$nev = d[[1]]
    self$nvar = d[[2]]
    if (missing(mu) == TRUE){mu = rep(1/self$nev, self$nev)}
    self$m = mu
    if (missing(varnames) == TRUE){
      varn = list(rep(NA,self$nvar))
      for (i in (1:self$nvar)){varn[[i]]=paste0("var",i,sepa
rator="")}
      self$names = varn
    }else{
      self$names = varnames
    }
    self$bel = self$eval_at_events(self$evalbel,events)
    self$pl = self$eval_at_events(self$evalpl,events)
  },
  evalbel = function(event,frame){
    if (missing(frame) == TRUE){frame = self}
    s = 0
    for (i in (1:frame$nev)){
      aux = sum(event*frame$ev[i,])
      if (aux == sum(frame$ev[i,])){
         s = s + frame$m[[i]]
      }
    }
    return(s)
  },
  eval_at_events= function(fonc,events,frame){
    if (missing(frame) == TRUE){frame = self}
    nev1 = dim(events)[[1]]
    result = numeric(length = nev1)
    for (i in 1: nev1){
      result[[i]] = fonc(events[i,],frame)
    }
      return(result)
```

```
    },
    complete_frame = function(nel){
      if (missing(nel) == TRUE){nel = self$nvar}
      nc = 2^nel
      ttn = matrix(nrow = nc, ncol=nel)
      k = 1
      ttn[k,]=c(rep(0,nel))
      for (i in (1:nel)){
        c = combn(1:nel,i)
        for ( j in 1:dim(c)[[2]]){
          ind = c[,j]
          ev = c(rep(0, nel))
          ev[ind] = 1
          k = k + 1
          ttn[k,] = ev
        }
      }
      return(ttn)
    },
    print_events_values = function(events,values,varnames, v
alname){
      n_om = dim(events)[[1]]
      nvar = dim(events)[[2]]
      mp = matrix(nrow = n_om, ncol = nvar + 1)
      mp[ ,1:nvar] = events
      mp[ , nvar+1] = values
      colnames(mp) = c(varnames,valname)
      mpd = as.data.frame(mp)
      print(mpd)
    },
  psingle = function(f){
    if (missing(f) == TRUE){f = self}
    resu = numeric(length = f$nvar)
    evsing = diag(f$nvar)
    resu = self$eval_at_events(self$evalpl,evsing,f)
    result = list(evsing,resu)
    names(result) = c("evs","pls")
    return(result)
  },
  probsingle = function(f){
    if (missing(f) == TRUE){f = self}
    resu = self$psingle(f)
    resu2 = resu$pls/sum(resu$pls)
    result = list(resu$evs,resu2)
    names(result) = c("evs","probs")
    return(result)
  },
```

```
probf = function(f,evs){
  if (missing(f) == TRUE){f = self}
  if (missing(evs) == TRUE){
    evs = self$complete_frame(f$nvar)}
  nev = dim(evs)[[1]]
  resu = self$probsingle(f)
  ps = resu$probs
  p = numeric(length = nev )
  for (i in 1: nev){
    ind = which(evs[i,] > 0)
    p[[i]] = sum(ps[ind])
  }
  result = list(evs, p)
  names(result) = c("evs","probs")
  return(result)
},
pig = function(f,evs){
  if (missing(f) == TRUE){f = self}
  if (missing(evs) == TRUE){
    evs = self$complete_frame(f)}
  nevs = dim(evs)[[1]]
  prob = numeric(length = nevs)
  for (k in 1:nevs){
    u = evs[k,]
    s = 0
    for (i in 1: f$nev){
      v = f$ev[i,]
      if (sum(v) > 0){
        aux = v*u
        s = s + (sum(aux)/sum(v))*f$m[[i]]
      }
    }
    prob[[k]] = s
  }
  resu = list(evs, prob)
  names(resu) = c("evs","probs")
  return(resu)
}
)
)
```

Example 2.3 Let us use `dfram.R` to make the evaluations shown in Example 2.1: we start by creating a table of events ev1 (each line corresponds to an event) and a vector mu1 with their masses. The names of the variables are stocked in op. Then, we create the frame. *Bel* and *Pl* are automatically evaluated for the events in Table ev1.

```
ev1 = matrix(nrow = 3, ncol = 3)
ev1[1,] = c(1, 1, 1)
ev1[2,] = c(0, 1, 1)
ev1[3,] = c(0, 1, 0)
mu1 = c(0.4,0.2,0.4)

source("dfram.R")
fr1 = dfram$new(ev1,mu1)

fr1$print_events_values(fr1$ev,fr1$bel,fr1$names,"Bel")

##    var1 var2 var3 Bel
## 1     1    1    1 1.0
## 2     0    1    1 0.6
## 3     0    1    0 0.4

fr1$print_events_values(fr1$ev,fr1$pl,fr1$names,"Pl")

##    var1 var2 var3 Pl
## 1     1    1    1  1
## 2     0    1    1  1
## 3     0    1    0  1
```

If we desire to determine the values of *Bel* and *Pl* for another event, we can call the methods evalbel and evalpl. To evaluate any of these functions on several events, we use the method `eval_at_events`. For instance, the method `complete_frame` generates all the elements of $\mathcal{P}(\Omega)$, i.e., all the events for the given variables. Then, we can determine *Bel* and *Pl* for all the possible events:

(continued)

Example 2.3 (continued)

```
evc = fr1$complete_frame(fr1$nvar)
belc = fr1$eval_at_events(fr1$evalbel,evc)
fr1$print_events_values(evc,belc,fr1$names, "Bel")
```

```
##    var1 var2 var3 Bel
## 1     0    0    0 0.0
## 2     1    0    0 0.0
## 3     0    1    0 0.4
## 4     0    0    1 0.0
## 5     1    1    0 0.4
## 6     1    0    1 0.0
## 7     0    1    1 0.6
## 8     1    1    1 1.0
```

```
plc = fr1$eval_at_events(fr1$evalpl,evc)
fr1$print_events_values(evc,plc,fr1$names, "Pl")
```

```
##    var1 var2 var3  Pl
## 1     0    0    0 0.0
## 2     1    0    0 0.4
## 3     0    1    0 1.0
## 4     0    0    1 0.6
## 5     1    1    0 1.0
## 6     1    0    1 0.6
## 7     0    1    1 1.0
## 8     1    1    1 1.0
```

Let us determine the probabilities associated with the singletons using the method of (Cobb and Shenoy 2006):

```
p1 = fr1$probsingle()
fr1$print_events_values(p1$evs,p1$probs,fr1$names, "Prob")
```

```
##    var1 var2 var3 Prob
## 1     1    0    0  0.2
## 2     0    1    0  0.5
## 3     0    0    1  0.3
```

(continued)

Example 2.3 (continued)

Let us use the pignistic transformation:

```
evsing = diag(fr1$nvar)
p1g = fr1$pig(,evsing)
fr1$print_events_values(p1g$evs,p1g$probs,fr1$names, "Prob
")

##    var1 var2 var3      Prob
## 1    1    0    0 0.1333333
## 2    0    1    0 0.6333333
## 3    0    0    1 0.2333333
```

Now, let us evaluate the probabilities for the whole frame, using the two methods:

```
p2 = fr1$probf()
fr1$print_events_values(p2$evs,p2$probs,fr1$names, "Prob")

##    var1 var2 var3 Prob
## 1    0    0    0  0.0
## 2    1    0    0  0.2
## 3    0    1    0  0.5
## 4    0    0    1  0.3
## 5    1    1    0  0.7
## 6    1    0    1  0.5
## 7    0    1    1  0.8
## 8    1    1    1  1.0

p2g = fr1$pig(,evc)
fr1$print_events_values(p2g$evs,p2g$probs,fr1$names, "Prob
")

##    var1 var2 var3      Prob
## 1    0    0    0 0.0000000
## 2    1    0    0 0.1333333
## 3    0    1    0 0.6333333
## 4    0    0    1 0.2333333
## 5    1    1    0 0.7666667
## 6    1    0    1 0.3666667
## 7    0    1    1 0.8666667
## 8    1    1    1 1.0000000
```

R proposes packages to deal with belief functions, such as ibelief and `dst`. Using these packages, you can generate belief functions using mass functions.

Example 2.4 Let us consider the situation presented in Example 2.3

$$\mu(\{O_1, O_2, O_3\}) = 0.4; \mu(\{O_2, O_3\}) = 0.2; \mu(\{O_2\}) = 0.4$$

Let us use package `dst`:

```
library(dst)
f1bis <-  bca(tt = ev1, m= mu1,cnames = op, varnames = "Op
tions", idvar = 1) # a single option, the variable is iden
tified as number 1
```

Function `bca` of package `dst` makes the basic chance attribution, which consists of creating the events given in ev1 and their masses. The package `dst` offers a function `belplau` for the evaluation of the events defined by `bca`.

```
bp = belplau(f1bis)
print(bp)

##          Belief Plausibility Plty Ratio
## frame       1.0            1        Inf
## O2 + O3     0.6            1   2.500000
## O2          0.4            1   1.666667
```

The function `plautrans` evaluates the probabilities using the method of (Cobb and Shenoy 2006):

```
plautrans(f1bis)

## Warning in addTobca(x = x, f = diag(1, nc)): Parameter
name 'f' is deprecated.
## Use 'tt' instead.

##        col1 col2 col3 trplau
## col1      1    0    0    0.2
## col2      0    1    0    0.5
## col3      0    0    1    0.3
```

Example 2.5 Let us consider tossing a coin once as in Example 2.4. The binary representation of the complete frame is formed by 4 events: $\varnothing = (0, 0)$, $\{H\} = (1, 0)$, $\{T\} = (0, 1)$, $\Omega = (1, 1)$. We can generate this frame either by

(continued)

Example 2.5 (continued)

giving all the events (ev1, mu1, fr1) or by giving just the single event having non-null mass (ev2, mu2, fr2):

```
source("dfram.R")

ev1 = dfram$public_methods$complete_frame(2)
nam1 = c("H","T")
mu1 = c(0,0,0,1)
fr1 = dfram$new(ev1,mu1,nam1)
b1 = fr1$bel
p1 = fr1$pl

ev2 = matrix(1,nrow=1,ncol=2)
mu2 = matrix(1,nrow=1,ncol=1)
fr2 = dfram$new(ev2,mu2,nam1)
b2 = fr2$eval_at_events(fr2$evalbel,ev1)
p2 = fr2$eval_at_events(fr2$evalpl,ev1)

fr1$print_events_values(ev1,b1,nam1, "Bel")

##   H T Bel
## 1 0 0   0
## 2 1 0   0
## 3 0 1   0
## 4 1 1   1

fr2$print_events_values(ev1,b2,nam1, "Bel")

##   H T Bel
## 1 0 0   0
## 2 1 0   0
## 3 0 1   0
## 4 1 1   1

fr1$print_events_values(ev1,p1,nam1, "Pl")

##   H T Pl
## 1 0 0  0
## 2 1 0  1
## 3 0 1  1
## 4 1 1  1

fr2$print_events_values(ev1,p2,nam1, "Pl")

##   H T Pl
## 1 0 0  0
## 2 1 0  1
## 3 0 1  1
## 4 1 1  1
```

(continued)

Example 2.5 (continued)

If we choose a different model, setting $\mu(1,0) = \mu(0,1) = 0.5$, as usual in probability theory, we have:

```
ev3 = matrix(nrow = 2, ncol = 2)
ev3[1,] = c(1,0)
ev3[2,] = c(0,1)
mu3 = c(0.5,0.5)
fr3 = dfram$new(ev3,mu3,nam1)
b3 = fr3$eval_at_events(fr3$evalbel,ev1)
p3 = fr3$eval_at_events(fr3$evalpl,ev1)
fr3$print_events_values(ev1,b3,nam1, "Bel")

##   H T Bel
## 1 0 0 0.0
## 2 1 0 0.5
## 3 0 1 0.5
## 4 1 1 1.0

fr2$print_events_values(ev1,p3,nam1, "Pl")

##   H T  Pl
## 1 0 0 0.0
## 2 1 0 0.5
## 3 0 1 0.5
## 4 1 1 1.0
```

Let us transform these plausibilities in probabilities, using the method described in (Cobb and Shenoy 2006):

```
r1 = fr1$probsingle()
r2 = fr2$probsingle()
fr1$print_events_values(r1$evs,r1$probs,nam1, "prob")

##   H T prob
## 1 1 0  0.5
## 2 0 1  0.5

fr2$print_events_values(r2$evs,r2$probs,nam1, "prob")

##   H T prob
## 1 1 0  0.5
## 2 0 1  0.5

r3 = fr3$probsingle()
fr3$print_events_values(r3$evs,r3$probs,nam1, "prob")

##   H T prob
## 1 1 0  0.5
## 2 0 1  0.5
```

(continued)

Example 2.5 (continued)

Now, let us use the pignistic transformation:

```
rp1 = fr1$pig()
rp2 = fr2$pig()
rp3 = fr3$pig()
fr1$print_events_values(rp1$evs,rp1$probs,nam1, "prob")

##   H T prob
## 1 0 0  0.0
## 2 1 0  0.5
## 3 0 1  0.5
## 4 1 1  1.0

fr2$print_events_values(rp2$evs,rp2$probs,nam1, "prob")

##   H T prob
## 1 0 0  0.0
## 2 1 0  0.5
## 3 0 1  0.5
## 4 1 1  1.0

fr2$print_events_values(rp3$evs,rp3$probs,nam1, "prob")

##   H T prob
## 1 0 0  0.0
## 2 1 0  0.5
## 3 0 1  0.5
## 4 1 1  1.0
```

Exercises

1. A six-faced dice is rolled. A gambler considers that the mass function verifies $\mu(\{6\}) = 0.1$ and $\mu(\{1\}) = 0.24$, while the other faces have equal masses.

   (a) Determine the masses of the other faces.
   (b) Use R to determine the beliefs associated with the faces.
   (c) Use R to determine the plausibilities of the faces.
   (d) Use R to find the probabilities of the faces.

2. A coin is tossed twice. A gambler considers that the mass function verifies $\mu((H, H)) = 0.1$ and $\mu((T, T)) = 0.3$. No other information is given.

   (a) Create a model frame for this situation.
   (b) Use R to determine the beliefs associated with the results.
   (c) Use R to determine the plausibilities of the results.

(continued)

(d) Use R to find the probabilities of the results by the plausibility transformation.
(e) Use R to determine les probabilities $P(\{H\})$ and $P(\{T\})$ by the two methods presented.

3. An urn contains balls of three colors: blue (B), red (R), and white (W). Two balls are drawn sequentially, with remittance of the first ball in the urn. We consider that $\mu(\{B\}) = 0.2$, $\mu(\{R\}) = 0.3$, $\mu(\{W\}) = 0.5$.

   (a) Create a model frame for this situation.
   (b) Use R to determine the beliefs associated with the results.
   (c) Use R to determine the plausibilities of the results.
   (d) Use R to determine les probabilities $P(\{B\})$, $P(\{R\})$, and $P(\{W\})$ by the two methods presented.

2.3 Commonality

Beliefs can also be represented by their commonality function:

$$Q(A) = \sum_{A \subset B} m(B). \tag{2.1}$$

Commonality functions furnish a simple way to combine beliefs from different sources – we shall examine this aspect later (see Sect. 2.7)

Example 2.6 Let us consider the situation described in Examples 2.3 and 2.4 :

$$\mu(\{O_1, O_2, O_3\}) = 0.4; \mu(\{O_2, O_3\}) = 0.2; \mu(\{O_2\}) = 0.4$$

Let us evaluate the commonality (results in Table 2.2):

Table 2.2 Commonality function $Q(A)$

A	B	Q
$\varnothing$	$\{O_2\}$, $\{O_2, O_3\}$, Ω	1
$\{O_1\}$	Ω	0.4
$\{O_2\}$	$\{O_2\}$, $\{O_2, O_3\}$, Ω	1
$\{O_3\}$	$\{O_2, O_3\}$, Ω	0.6
$\{O_1, O_2\}$	Ω	0.4
$\{O_1, O_3\}$	Ω	0.4
$\{O_2, O_3\}$	$\{O_2, O_3\}$, Ω	0.6
$\{O_1, O_2, O_3\}$	Ω	0.4

(continued)

Example 2.6 (continued)

Under R, we can evaluate these values using the function `evalcom` below, to be added to the class `dfram` (a new property Q is added too):

```
evalcom = function (event, frame){
  if (missing(frame) == TRUE){frame = self}
  s = 0
  for (i in (1:frame$nev)){
    aux = sum(event*frame$ev[i,])
    if (aux == sum(event)){
      s = s + frame$m[[i]]
    }
  }
  return(s)}
```

We can evaluate the commonality Q as follows (`fr1` and the other functions were defined in Example 2.3. Recall that a new property `Q` was added to `dfram`):

```
Q = fr1$eval_at_events(fr1$evalcom,evc)
fr1$Q = Q
fr1$print_events_values(evc,Q, fr1$names, "Q")

##   var1 var2 var3   Q
## 1    0    0    0 1.0
## 2    1    0    0 0.4
## 3    0    1    0 1.0
## 4    0    0    1 0.6
## 5    1    1    0 0.4
## 6    1    0    1 0.4
## 7    0    1    1 0.6
## 8    1    1    1 0.4
```

Exercises

1. A six-faced dice is rolled. We suppose that
$\mu(\{1\}) = 0.24$, $\mu(\{6\}) = 0.12$, $\mu(\{i\}) = 0.16$, $2 \leq i \leq 5$.
    (a) Create a model frame for this situation.
    (b) Determine the commonality associated with your model.
    (c) Determine the beliefs and plausibilities and probabilities of each result.
2. A coin is tossed twice. A gambler considers that the mass function verifies $\mu((H,H)) = 0.1$ and $\mu((T,T)) = 0.3$. No other information is given.
    (a) Create a model frame for this situation.
    (b) Determine the commonality associated with your model.
    (c) Determine the beliefs and plausibilities and probabilities of each result.
3. An urn contains balls of three colors: blue (B), red (R), and white (W). Two balls are drawn sequentially, with remittance of the first ball in the urn. We consider that $\mu(\{B\}) = 0.2$, $\mu(\{R\}) = 0.3$, $\mu(\{W\}) = 0.5$.
    (a) Create a model frame for this situation.
    (b) Determine the commonality associated with your model.
    (c) Determine the beliefs, plausibilities, and probabilities.

2.4 Retrieving Mass Functions from Other Representations

There are relations between the different representations of beliefs by μ, Bel, Pl, and Q. For instance, we have:

$$\mu(A) = \sum_{B \subset A} (-1)^{|A|-|B|} Bel(B) = \sum_{A \subset B} (-1)^{|B-A|} Q(B).$$

We also have

$$\mu(A) = \sum_{B \subset A} (-1)^{|A|-|B|} (1 - Pl(\Omega - B)).$$

Example 2.7 Let us consider the situation described in Example 2.3

$$\mu(\{O_1, O_2, O_3\}) = 0.4; \mu(\{O_2, O_3\}) = 0.2; \mu(\{O_2\}) = 0.4$$

We have

$$Bel(\{O_2\}) = Bel(\{O_1, O_2\}) = 0.4 \ ,$$
$$Bel(\{O_2, O_3\}) = 0.6, \ \ Bel(\{O_1, O_2, O_3\}) = 1$$

Let us retrieve μ from *Bel* (results in Table 1.9) (Table 2.3):

Table 2.3 Retrieving μ from *Bel*

A	B	μ
$\varnothing$	$\varnothing$	0
$\{O_1\}$	$\varnothing$	0
$\{O_2\}$	$\{O_2\}$	0.4
$\{O_3\}$	$\varnothing$	0
$\{O_1, O_2\}$	$\{O_2\}$, $\{O_1, O_2\}$	$0.4 - 0.4 = 0$
$\{O_1, O_3\}$	$\varnothing$	0
$\{O_2, O_3\}$	$\{O_2, O_3\}$, $\{O_2\}$	$0.6 - 0.4 = 0.2$
$\{O_1, O_2, O_3\}$	Ω	0.4

Under R, we can evaluate these values using the function `bel2mu` below, to be added to the class `dfram` :

```
bel2mu = function (event, frame){
  if (missing(frame) == TRUE){frame = self}
  n = sum(event)
  s = 0
  if ( n == 0){return(s)}
    for (i in (1: frame$nev)){
      aux = sum(event*frame$ev[i,])
      if ((aux == sum(frame$ev[i,])) & (aux > 0)){
        s = s + (-1)^(n-sum(frame$ev[i,]))*frame$bel[[i]]
      }
    }
  return(s) }
```

(continued)

Example 2.7 (continued)

The function `bel2mu` is used as follows (see Example 2.3):

```
ev2 = matrix(nrow = 4, ncol = 3)
ev2[1,] = c(1, 1, 1) # {O1,O2,O3}
ev2[2,] = c(0, 1, 1) # {O2,O3}
ev2[3,] = c(1, 1, 0) # {O1,O2}
ev2[4,] = c(0, 1, 0) # {O2}

source("dfram.R")
fr2 = dfram$new(ev2)
fr2$bel = c(1,0.6,0.4,0.4)
evc = fr2$complete_frame(fr2$nvar)
#
mu2 = fr2$eval_at_events(fr2$bel2mu,evc)
fr2$print_events_values(evc,mu2,fr2$names, "mu")

##   var1 var2 var3  mu
## 1    0    0    0 0.0
## 2    1    0    0 0.0
## 3    0    1    0 0.4
## 4    0    0    1 0.0
## 5    1    1    0 0.0
## 6    1    0    1 0.0
## 7    0    1    1 0.2
## 8    1    1    1 0.4
```

Alternatively, we can use all the values of *Bel*:

```
f1c = fr1
f1c$ev = evc
f1c$bel = c(0,0,0.4,0,0.4,0,0.6,1)
f1c$nev = length(f1c$bel)

mu = fr1$eval_at_events(fr$bel2mu,evc, f1c)
f1c$mu = mu
fr$print_events_values(evc,mu,fr1$names, "mu")

##    O1 O2 O3  mu
## 1  0  0  0 0.0
## 2  1  0  0 0.0
## 3  0  1  0 0.4
## 4  0  0  1 0.0
## 5  1  1  0 0.0
## 6  1  0  1 0.0
## 7  0  1  1 0.2
## 8  1  1  1 0.4
```

(continued)

Example 2.7 (continued)

To retrieve μ from Q, we can use the function below (to be added to the class dfram):

```
com2mu = function(event, frame){
  if( missing(frame) == TRUE){frame = self}
  s = 0
  for (i in (1:dim(frame$ev)[[1]])){
    aux = sum(event*frame$ev[i,])
    if (aux == sum(event)){
     s = s + (-1)^(sum(frame$ev[i,]-event))*frame$Q[[i]]
    }
  }
 return(s) }
```

The function com2mu is used as follows:

```
fr2$ev = evc
fr2$Q = c(1, 0.4,1,0.6,0.4,0.4,0.6, 0.4)
mu3 = fr2$eval_at_events(fr2$com2mu,evc)
fr2$print_events_values(evc,mu3,fr2$names, "mu")

##   var1 var2 var3  mu
## 1    0    0    0 0.0
## 2    1    0    0 0.0
## 3    0    1    0 0.4
## 4    0    0    1 0.0
## 5    1    1    0 0.0
## 6    1    0    1 0.0
## 7    0    1    1 0.2
## 8    1    1    1 0.4
```

(continued)

Example 2.7 (continued)

Finally, to determine μ from Pl, we can use the function below (to be added to the class `dfram`):

```
pl2mu = function(event, frame){
  if (missing(frame) == TRUE){frame = self}
  s = 0
  n = sum(event)
  if ( n == 0){return(s)}
  for (i in (1:dim(frame$ev)[[1]])){
    aux = sum(event*frame$ev[i,])
    nn = sum(frame$ev[i,])
    if (aux == nn){
      evaux = rep(1,frame$nvar) - frame$ev[i,]
      auxx = 0
      for (j in (1:dim(frame$ev)[[1]])){
        if (sum(abs(frame$ev[j,]-evaux)) == 0){
          s = s +(-1)^(n-sum(frame$ev[i,]))+(-1)^(n-sum(fr
ame$ev[i,])-1)*frame$pl[[j]]
        }
      }
    }
  }
  return(s) }
```

The function `pl2mu` is used as follows:

```
fr2$pl = c(0, 0.4, 1, 0.6, 1, 0.6, 1, 1)
mu4 = fr2$eval_at_events(fr2$pl2mu,evc)
fr2$print_events_values(evc,mu4,fr2$names, "mu")

##   var1 var2 var3  mu
## 1    0    0    0 0.0
## 2    1    0    0 0.0
## 3    0    1    0 0.4
## 4    0    0    1 0.0
## 5    1    1    0 0.0
## 6    1    0    1 0.0
## 7    0    1    1 0.2
## 8    1    1    1 0.4
```

Exercises

1. A six-faced dice is rolled. The commonality function associated is null, except for
$Q(\emptyset) = 1$, $Q(\{1\}) = 0.24$, $Q(\{6\}) = 0.12$, $Q(\{i\}) = 0.16$, $2 \leq i \leq 5$.
   (a) Create a model frame for this situation.
   (b) Determine the commonality associated with your model.
   (c) Use the commonality to determine the beliefs and plausibilities.
2. A coin is tossed twice. A gambler considers that the mass function verifies $\mu(\{H, H\}) = 0.1$ and $\mu(\{T, T\}) = 0.3$. No other information is given.
   (a) Create a model frame for this situation.
   (b) Determine the commonality associated with your model.
   (c) Use the commonality to determine the beliefs and plausibilities.
3. An urn contains balls of three colors: blue (B), red (R), and white (W). Two balls are drawn sequentially, with remittance of the first ball in the urn.
   We consider that Bel is null, except for $Bel(\{B\}) = 0.2$, $Bel(\{R\}) = 0.3$, $Bel(\{W\}) = 0.6$, $Bel(\{B, R\}) = 0.5$, $Bel(\{B, W\}) = 0.8$, $Bel(\{R, W\}) = 0.9$, $Bel(\Omega) = 1$.
   (a) Create a model frame for this situation.
   (b) Determine the mass function associated with your model.
   (c) Determine the commonalities, beliefs, and plausibilities.

2.5 Couples, Product Frames, and Marginality

Analogously to joint distributions of random variables, we can consider beliefs of couples of variables and look for marginal distributions. For instance, let $\boldsymbol{X} = (X_1, X_2)$ be a pair of variables defined on a product frame $\Omega = \Omega_1 \times \Omega_2$, where Ω_i is a frame for X_i. Assume that we have a mass function μ_X for the pair and we are interested in the margin μ_1 for X_1: for $A_1 \subset \Omega_1$, it is determined by the aggregation of the masses of all the elements of Ω that have A_1 as projection onto Ω_1, i.e., by considering

$$\mu_1(A) = \sum_{A \in S(A_1)} \mu(A), \quad S(A_1) = \{A \subset \Omega : proj_{\Omega_1}(A) = A_1\},$$

With

$$proj_{\Omega_1}(A) = A \downarrow \Omega_1 = \{x_1 \in \Omega_1 : \exists x_2 \in \Omega_2 \text{ such that } (x_1, x_2) \in A\}.$$

We can rewrite this definition as (recall that many parts of Ω can have masses equal to zero)

$$S(A_1) = \{A_1 \times A_2 \subset \Omega : A_2 \subset \Omega_2\},$$

so that

$$\mu_1(A) = \sum_{A_2 \subset \Omega_2} \mu(A_1 \times A_2).$$

Example 2.8 Let us consider two persons P_1, P_2 and three possible options O_1, O_2, O_3 for each person. We can consider $\Omega_1 = \{P_1, P_2\}$, $\Omega_2 = \{O_1, O_2, O_3\}$. Assume that we have the estimations:

$$\mu(\{P_1\}, \{O_1\}) = 0.1, (\{P_1\}, \{O_2\}) = 0.15, \mu(\{P_1\}, \{O_1, O_2\}) = 0.15$$
$$\mu(\{P_2\}, \{O_2\}) = 0.15, \mu(\{P_2\}, \{O_1, O_3\}) = 0.1$$
$$\mu(\{P_1\}, \{O_1, O_2, O_3\}) = 0.25, \mu(\{P_2\}, \{O_1, O_2, O_3\}) = 0.1$$

In this case,

$$\mu_1(\{P_1\}) = 0.1 + 0.15 + +0.15 + 0.25 = 0.65$$
$$\mu_1(\{P_2\}) = 0.15 + 0.1 + 0.1 = 0.35$$
$$\mu_1(\{P_1, P_2\}) = 0$$

Under R, we start by generating the events and their masses: each event is represented by a binary vector of length $|\Omega_1| + |\Omega_2| = 2 + 3 = 5$. The first $|\Omega_1|$ components correspond to Ω_1 and the last ones $|\Omega_2|$ correspond to Ω_2 : for instance, the event $(\{P_1\}, \{O_1\})$ is represented by (1, 0, 1, 0, 0) and the event $(\{P_2\}, \{O_1, O_3\})$ is represented by (0, 1, 1, 0, 1). Under R, we have

(continued)

Example 2.8 (continued)

```
source("dfram.R")

ev12 = matrix(nrow = 7, ncol = 5)
ev12[1,] = c(1,0,1,0,0) # ({P1}, {O1})
ev12[2,] = c(1,0,0,1,0) # ({P1}, {O2})
ev12[3,] = c(1,0,1,1,0) # ({P1}, {O1, O2})
ev12[4,] = c(0,1,0,1,0) # ({P2}, {O2})
ev12[5,] = c(0,1,1,0,1) # ({P2}, {O1, O3})
ev12[6,] = c(1,0,1,1,1) # ({P1}, {O1, O2, O3})
ev12[7,] = c(0,1,1,1,1) # ({P2}, {O1, O2, O3})

mu12 = c(0.1, 0.15, 0.15, 0.15, 0.1, 0.25, 0.1)
nam12 = c("P1","P2","O1","O2","O3")

f12 = dfram$new(ev12,mu12, nam12)
```

To evaluate μ_1, we use the method `margin1` (to be added to the class `dframe`):

```
margin1 = function(nel1,f){
  if (missing(f) == TRUE){f = self}
  ev1 = self$complete_frame(nel1)
  n1 = dim(ev1)[[1]]
  mu1 = numeric(length = n1)
  n_om = dim(f$ev)[[1]]
  for (k in 1:n1){
     u1 = ev1[k,]
     s = 0
     for (i in 1: n_om){
        v = f$ev[i,]
         v1 = v[1:nel1]
         aux = sum(abs(u1-v1))
         if (aux == 0){ s = s + f$m[[i]]}
     }
     mu1[[k]] = s
  }
  f1 = dfram$new(ev1,mu1,f$names[1:nel1])
  return(f1)}
```

(continued)

Example 2.8 (continued)

This method is used as follows:

```
f1 = f12$margin1(2)
f12$print_events_values(f1$ev,f1$m,f1$names,"mass")

##   P1 P2 mass
## 1  0  0 0.00
## 2  1  0 0.65
## 3  0  1 0.35
## 4  1  1 0.00
```

The package `dst` offers a function `elim` which allows the evaluation of marginal masses. To use this function, we must use `bcaRel` instead of `bca`:

```
library(dst)

s12 = matrix(nrow = 7, ncol = 2,dimnames = list(NULL, c("s
pecnb", "mass")))
s12[,1] = c(1:7)
s12[,2] = mu12

ifv = matrix(nrow = 2, ncol = 2)
ifv[1,] = c(1,2)
ifv[2,] = c(2,3)
colnames(ifv) = c("idvar", "size")

f12r = bcaRel(tt=ev12,spec=s12, infovar = ifv,
              varnames = c("person", "option"),
              valuenames = c("P1","P2","O1","O2","O3"),
              relnb = 1)
```

(continued)

Example 2.8 (continued)

Once the basic chance attribution is done, we can eliminate the second variable to get marginal μ_1

```
# evaluate the marginal mu1
f1 = elim(f12r,xnb = 2) # eliminates the second variable (
option)
f1$tt

##    P1 P2
## P1  1  0
## P2  0  1

t(f1$spec)

##         [,1] [,2]
## specnb  1.00 2.00
## mass    0.65 0.35
```

The marginal mass function μ_2 can be evaluated by the same procedure. We have:

$$\mu_2(\{O_1\}) = 0.1, \mu_2(\{O_2\}) = 0.15 + 0.15 = 0.3, \mu_2(\{O_3\}) = 0,$$

$$\mu_2(\{O_1, O_2\}) = 0.15, \mu_2(\{O_1, O_3\}) = 0.1, \mu_2(\{O_2, O_3\}) = 0,$$

$$\mu_2(\{O_1, O_2, O_3\}) = 0.25 + 0.1 = 0.35.$$

Under R, we can use the function `margin2`:

```
margin2 = function(nel1,f){
  if (missing(f) == TRUE){f = self}
  nel = dim(f$ev)[[2]]
  nel2 = nel - nel1
  tt2 = self$complete_frame(nel2)
  n2 = dim(tt2)[[1]]
  mu2 = numeric(length = n2)
  n_om = dim(f$ev)[[1]]
  for (k in 1:n2){
    u2 = tt2[k,]
    s = 0
    for (i in 1: n_om){
      v = f$ev[i,]
```

(continued)

Example 2.8 (continued)

```
    v2 = v[(nel1+1):nel]
    aux = sum(abs(u2-v2))
    if (aux == 0){ s = s + f$m[[i]]}
  }
  mu2[[k]] = s
}
f2 = dfram$new(tt2,mu2,f$names[(nel1+1):nel])
return(f2)}
```

This method is used as follows:

```
f2 = f12$margin2(2)
f12$print_events_values(f2$ev,f2$m,f2$names,"mass")

##   O1 O2 O3 mass
## 1  0  0  0 0.00
## 2  1  0  0 0.10
## 3  0  1  0 0.30
## 4  0  0  1 0.00
## 5  1  1  0 0.15
## 6  1  0  1 0.10
## 7  0  1  1 0.00
## 8  1  1  1 0.35
```

We can also use `elim`:

```
# evaluate the marginal mu2
f2 = elim(f12r,xnb = 1) # eliminates the first variable (p
erson)
f2$tt

##         O1 O2 O3
## O1       1  0  0
## O2       0  1  0
## O1 + O2  1  1  0
## O1 + O3  1  0  1
## frame    1  1  1

t(f2$spec)

##        [,1] [,2] [,3] [,4] [,5]
## specnb  1.0  2.0 3.00  4.0 5.00
## mass    0.1  0.3 0.15  0.1 0.35
```

Alternatively, we can use a class to generate product spaces:

```
library(R6)
source("dfram.R")
#
pdf=R6Class("pdf",
 inherit = dfram,
 public=list(
  spaces = NULL, nspaces = NULL, index = NULL, ev = NULL,
  names = NULL,size = NULL, m = NULL, nvar = NULL, nev = NUL
L,
  frc = NULL,
  initialize=function(listdata,ltype){
    super$initialize()
    if (ltype == 'events'){
      self$use_events(listdata)
    }else{
      if (ltype == 'cartesian'){
        self$use_spaces(listdata)
      }else{
        if (ltype == 'relations'){
          self$use_rels(listdata)}
        else{return}
      }
    }
    self$frc = dfram$new(self$ev,self$m,self$names)
  },
  genindex = function(){
    ind = vector('list',length = self$nspaces)
    ind[[1]] = c(1, self$nvar[[1]])
    for (i in 2:self$nspaces){
      ind[[i]] = c((ind[[i-1]][[2]]+1),(ind[[i-1]][[2]]+self
$nvar[[i]]) )
    }
    self$index = ind
  },
```

```
use_spaces = function(spaces){
  self$spaces = spaces
  self$nspaces = length(spaces)
  aux = rep(0,self$nspaces)
  for (i in 1:self$nspaces){
    aux[[i]] = spaces[[i]]$nvar
  }
  self$nvar = aux
  self$size = sum(aux)
  self$genindex()
  nams = rep("0",self$size)
  for (i in 1:self$nspaces){
    ind = self$index[[i]]
    nams[ ind[[1]]: ind[[2]] ] = spaces[[i]]$names
  }
  self$names = nams
  x = vector('list', length = self$nspaces)
  x[[1]] = seq(1:self$spaces[[1]]$nev)
  ch = "expand.grid(x[[1]]"
  for (i in 2:self$nspaces){
    x[[i]] = seq(1:self$spaces[[i]]$nev)
    aux1 = paste0(",x[[",i,"]]")
    ch = paste0(ch,aux1)
  }
  ch =paste0(ch,")")
  chaux = as.matrix(eval(parse(text=ch)))
  ev1 = matrix(nrow = dim(chaux)[[1]], ncol = self$size)
  mu = rep(0,dim(chaux)[[1]])
  for (i in (1:dim(chaux)[[1]])){
    s = 1
    idev = chaux[i,]
    for (j in (1:dim(chaux)[[2]])){
      ind = self$index[[j]]
      ev1[i,ind[[1]]: ind[[2]]] = self$spaces[[j]]$ev[ ide
v[[j]], ]
      s = s*self$spaces[[j]]$m[[idev[[j]] ]]
    }
    mu[[i]] = s
  }
  self$m = mu
  self$ev = ev1
  self$nev = dim(ev1)[[1]]
},
use_rels = function(ldata){
  self$spaces = ldata$spaces
```

```
    self$nspaces = length(ldata$spaces)
    aux = rep(0,self$nspaces)
    for (i in 1:self$nspaces){
      aux[[i]] = self$spaces[[i]]$nvar
    }
    self$nvar = aux
    self$size = sum(aux)
    self$genindex()
    nams = rep("0",self$size)
    for (i in 1:self$nspaces){
      ind = self$index[[i]]
      nams[ ind[[1]]: ind[[2]] ] = self$spaces[[i]]$names
    }
    self$names = nams
    self$nev = dim(ldata$rels)[[1]]
    nn = dim(ldata$rels)[[2]]
    ev1 = matrix(0, nrow = self$nev, ncol = self$size)
    mu = rep(0,self$nev)
    for (i in (1:self$nev)){
      ind = which(ldata$rels[i,] > 0 )
      for (j in (1:length(ind))){
        indj = self$index[[ ind[[j]] ]]
        ev1[i,indj[[1]]: indj[[2]]] = self$spaces[[ ind[[j]]
]]$ev[ldata$rels[i,ind[[j]] ], ]
      }
      mu[[i]] = ldata$m[[i]]
    }
    self$m = mu/sum(mu)
    self$ev = ev1
  },
  use_events = function(events){
    self$nspaces = length(events$nvars)
    self$nvar = events$nvars
    self$size = sum(events$nvars)
    self$ev = events$ev
    self$nev = dim(events$ev)[[1]]
    if (is.null(events$m)){
      self$m = rep(1/self$nev,self$nev)}
    else{self$m = events$m}
    if (is.null(events$names))
    { varn = list(rep(NA,self$size))
      for (i in (1:self$size)){varn[[i]]=paste0("var",i,sepa
rator="")}
      self$names = varn
    }else{self$names = events$names}
```

```
      self$genindex()
      ls = vector("list",length = self$nspaces)
      aux = 1/self$nspaces
      mu = self$m^(aux)
      for (i in 1:self$nspaces){
        ind = self$index[[i]]
        ev1 = self$ev[ ,ind[[1]]:ind[[2]]]
        nams = self$names[ ind[[1]]:ind[[2]] ]
        aux = dfram$new(ev1,mu,nams)
        ls[[i]] = aux
      }
      self$spaces = ls
    }
  )
)
```

This class proposes three ways to generate a product frame. The first consists of giving a list of events containing fields `ev`, `nvars`, names, and m, which are used to generate the product frame. The second gives a list of frames: the events will be generated by a cartesian product of the events in each frame. The last one consists of giving two frames and the events `rels` to be connected, using the masses mu.

Example 2.9 Let us consider again the situation presented in Example 2.8. We can generate the product frame by giving the events as follows:

```
nvars12 = c(2,3)
events = list(ev12,mu12,nams12,nvars12)
names(events) = c("ev","m","names","nvars")

source("pdf.R")
pf1 = pdf$new( ,events)
```

(continued)

Example 2.9 (continued)

Then, we use the inheritage to evaluate the margins:

```
f1 = pf1$margin1(self$nvar[[1]])
pf1$print_events_values(f1$ev,f1$m,f1$names,"mass")

##   P1 P2 mass
## 1  0  0 0.00
## 2  1  0 0.65
## 3  0  1 0.35
## 4  1  1 0.00

f2 = pf1$margin2(self$nvar[[1]])
pf1$print_events_values(f2$ev,f2$m,f2$names,"mass")

##   O1 O2 O3 mass
## 1  0  0  0 0.00
## 2  1  0  0 0.10
## 3  0  1  0 0.30
## 4  0  0  1 0.00
## 5  1  1  0 0.15
## 6  1  0  1 0.10
## 7  0  1  1 0.00
## 8  1  1  1 0.35
```

Example 2.10 Let us consider again the situation presented in Example 2.8. We can generate the product frame by using relations as follows:

```
source("dfram.R")

tt1 = matrix(nrow = 5, ncol = 3)
tt1[1,] = c(1, 0, 0)  # {O_1}
tt1[2,] = c(0, 1, 0)  # {O_2}
tt1[3,] = c(1, 1, 0)  # {O_1 , O_2}
tt1[4,] = c(1, 0, 1)  # {O_1 , O_3}
tt1[5,] = c(1, 1, 1)  # {O_1 , O_2, O_3}

nams1 <- c( "O1", "O2", "O3")
eval_expert = c(0.45,0.18,0.45*0.18,0.45*0.32,0.45*0.32*0.
18)
mu1 = eval_expert/sum(eval_expert)
```

(continued)

Example 2.10 (continued)

```
f1 = dfram$new(tt1,mu1,nams1)

tt2 = matrix(nrow = 2, ncol = 2)
tt2[1,] = c(1, 0)
tt2[2,] = c(0, 1)
nams2 <- c( "P1", "P2")

mu2=c(0.3,0.7)
f2 = dfram$new(tt2,mu2,nams2)

rels = matrix(nrow = 7, ncol = 2)
rels[1,] = c(1,1) # 1 of f2 with 1 of f1
rels[2,] = c(1,2) # 1 of f2 with 2 of f1
rels[3,] = c(1,3) # 1 of f2 with 3 of f1
rels[4,] = c(2,2) # 2 of f2 with 2 of f1
rels[5,] = c(2,4) # 2 of f2 with 4 of f1
rels[6,] = c(1,5) # 1 of f2 with 5 of f1
rels[7,] = c(2,5) # 2 of f2 with 5 of f1

murels = c(0.1, 0.15, 0.15, 0.15, 0.1, 0.25, 0.1)

ls = list(f2,f1)
ldata = list(ls,rels,murels)
names(ldata) = c("spaces","rels","m")

source("pdf.R")
pf1 = pdf$new(ldata,'relations')
```

Again, we use the inheritage to evaluate the margins:

```
fr1 = pf1$margin1(pf1$nvar[[1]])
fr1$print_events_values(fr1$ev,fr1$m,fr1$names,"mass")

##    P1 P2 mass
## 1  0  0 0.00
## 2  1  0 0.65
## 3  0  1 0.35
## 4  1  1 0.00
```

(continued)

Example 2.10 (continued)

```
fr2 = pf1$margin2(pf1$nvar[[1]])
fr2$print_events_values(fr2$ev,fr2$m,fr2$names,"mass")

##   O1 O2 O3 mass
## 1  0  0  0 0.00
## 2  1  0  0 0.10
## 3  0  1  0 0.30
## 4  0  0  1 0.00
## 5  1  1  0 0.15
## 6  1  0  1 0.10
## 7  0  1  1 0.00
## 8  1  1  1 0.35
```

Example 2.11 To illustrate the creation of the product frame using a list of frames, let us consider the situation where a fair coin is tossed three times:

```
source("dfram.R")

ev1 = dfram$public_methods$complete_frame(2)
nam = c("H","T")
fr1 = dfram$new(ev1, ,nam)

source("pdf.R")
ls = c(fr1,fr1,fr1)
pf1 = pdf$new(ls)

pf1$ev[2, ]

## [1] 1 0 0 0 0 0

pf1$m[[2]]

## [1] 0.015625

pf1$ev[60, ]

## [1] 1 1 0 1 1 1

pf1$m[[60]]

## [1] 0.015625
```

(continued)

Example 2.11 (continued)

You can get the complete frame of 64 events using the instruction

```
pf1$print_events_values(pf1$ev,pf1$m,pf1$names, "mass")
```

Example 2.12 Assume that an expert gives its opinion about 3 options O_1, O_2, and O_3. The expert considers that O_1 is possible at 50 % that $\{O_1, O_2\}$ is possible at 70 % and that $\{O_1, O_3\}$ is possible at 90 %. These values cannot correspond to probabilities on $\Omega = \{O_1, O_2, O_3\}$: if they do, we have $P(O_2) = P(\{O_1, O_2\}) - P(\{O_1\}) = 0.2$, so that $P(\{O_1, O_2, O_3\}) = P(\{O_1, O_3\}) + P(\{O_2\}) = 0.9 + 0.2 = 1.1 > 1$. Analogously, $P(O_3) = P(\{O_1, O_3\}) - P(\{O_1\}) = 0.4$, so that $P(\{O_1, O_2, O_3\}) = P(\{O_1, O_2\}) + P(\{O_3\}) = 0.7 + 0.4 = 1.1 > 1$. Of course, we can normalize the data to have a global probability of 1, by dividing all the values by 1.1, which furnishes

$$P(\{O_1\}) \approx 0.45, P(\{O_2\}) \approx 0.18, P(\{O_3\}) \approx 0.36$$

However, if the expert did not evaluate probabilities – what is confirmed by the analysis above – it appears as more convenient to consider these values as belief masses.

Let us use package `dst` to analyze the expert's evaluations. Analogously to Example 2.3, we start by creating a matrix containing the available information, where each column corresponds to an option and the lines are vectors defining the subsets given. Here,

```
tt = matrix(nrow = 3, ncol = 3)
tt[1,] = c(1, 0, 0)  # {O_1}
tt[2,] = c(1, 1, 0)  # {O_1 , O_2}
tt[3,] = c(1, 0, 1)  # {O_1 , O_3}
print(tt)

##      [,1] [,2] [,3]
## [1,]    1    0    0
## [2,]    1    1    0
## [3,]    1    0    1
```

(continued)

Example 2.12 (continued)

Now, let us create the mass function associated with the evaluations given by the expert. Since μ must sum up to 1, we normalize the evaluations. Then we make the basic chance assignment `bca` to the elements of Ω.

```
eval_expert = c(0.5,0.7,0.9)
mu = eval_expert/sum(eval_expert)

library(dst)
expert <-  bca(tt = ett, m= mu,cnames =c("O1", "O2", "O3")
, varnames = "Options", idvar = 1) # a single option, the
variable is identified as number 1
```

Once done the `bca`, we can get the beliefs and plausibilities corresponding to the data:

```
bp = belplau(expert);
print(bp)

##               Belief Plausibility Plty Ratio
## O1         0.2380952            1   1.312500
## O1 + O2 0.5714286            1   2.333333
## O1 + O3 0.6666667            1   3.000000
```

We can transform these results in probabilities by the function `plautrans`, which uses the method in (Cobb and Shenoy 2006) – see (Boivin 2022a, b, c)

```
prob = plautrans(expert)

## Warning in addTobca(x = x, f = diag(1, nc)): Parameter
name 'f' is deprecated.
## Use 'tt' instead.

print(prob)

##    O1 O2 O3    trplau
## O1  1  0  0 0.5675676
## O2  0  1  0 0.1891892
## O3  0  0  1 0.2432432
```

(continued)

Example 2.12 (continued)

We see that $P(O_1) \approx 0.57$, $P(0_2) \approx 0.19$, $P(0_3) \approx 0.24$. Now, assume that there are two persons that are susceptible of choosing an option. Person P_1 prefers O_2 at the level 90 % and is indifferent to the other options. Person P_2 prefers O_3 at the level 80 % and rejects completely O_1.

Let us analyze the relations between persons and options: we start by creating the frame of discernment concerning the persons – we give to each person an equal mass:

```
ptt = matrix(nrow = 2, ncol = 2)
ptt[1,] = c(1, 0)
ptt[2,] = c(0, 1)

person <- bca(tt = ptt, m=c(0.5,0.5), cnames=c("P1", "P2")
, varnames = "Persons", idvar = 1)
```

As an alternative, we can create the frame including $\Omega = \{P_1, P_2\}$ as follows – we give the mass 1 to Ω (total ignorance about the person):

```
ptt = matrix(nrow = 3, ncol = 2)
ptt[1,] = c(1, 0)
ptt[2,] = c(0, 1)
ptt[3,] = c(1, 1)
person <- bca(tt = ptt, m=c(0,0,1), cnames=c("P1", "P2"), varnames =
"Persons", idvar = 1)
```

Then, we create the relations between the persons and the options: each possibility is represented by a vector having 5 components. The first two components represent the person and the three final ones represent the option. All the values are zero except those corresponding to the couple (person, option) under consideration – these two elements are taken as the unity.

```
rels <- matrix(nrow = 5, ncol = 5)
rels[1,] = c(1,0,1,0,0) # P1 with O1
rels[2,] = c(1,0,0,1,0) # P1 with O2
rels[3,] = c(1,0,0,0,1) # P1 with O3
rels[4,] = c(0,1,0,1,0) # P2 with O2
rels[5,] = c(0,1,0,0,1) # P2 with O3
colnames(rels) <- c("P1", "P2", "O1", "O2", "O3")
rownames(rels) <- nameRows(rels)
```

(continued)

Example 2.12 (continued)

Then, we generate a Table with the information about the variables:

```
infovars <- matrix(nrow = 2, ncol = 2 )
infovars[1, ] = c(1,2) # Variable 1 : persons, with 2 possible
values
infovars[2, ] = c(2,3) # Variable 2 : options, with 3 possible
values
colnames(infovars) = c("idvar", "size")
```

In the next step, we define the masses of the lines of the Table of relations. We normalize the preferences of the individuals to get a sum equal to 1:

```
specP = matrix(nrow = 5, ncol = 2,dimnames = list(NULL, c("specnb",
"mass")))
specP[1, ] = c(1, 0.45)
specP[2, ] = c(2, 0.025)
specP[3, ] = c(3, 0.025)
specP[4, ] = c(4, 0.4)
specP[5, ] = c(5, 0.1)
```

The function `bcaRel` generates the `bca` corresponding to the relations considered – the number `relnb` is an identification of the relation under analysis.

```
person_opts <- bcaRel(tt = rels, spec = specP, infovar = infovars, var-
names = c("Persons", "Options"), relnb = 1)
```

Assume that we know that person 1 will make the choice: we modify the masses according to this information and we extend the data to the product space $\{P_1, P_2\} \times \{O_1, O_2, O_3\}$:

```
person$spec[ ,2] = c(1, 0) # or c(1,0,0) in the alternative
person_extend <- extmin(person, person_opts)
```

(continued)

Example 2.12 (continued)

Then, we combine the relations

```
comb_rel <- nzdsr(dsrwon(person_extend, person_opts))
```

and we eliminate the first variable (person)

```
option_eval <- elim(comb_rel, xnb = 1)
```

Finally, we generate the singletons of the options

```
option_eval_singleton <- addTobca(option_eval, tt  = matrix(c(1,0,0),
ncol = 3))
```

and we evaluate the probabilities of each option: as expected, we get the probabilities corresponding to person 1: $P(O_1) = 0.9$, $P(0_2) = 0.05$, $P(0_3) = 0.05$.

```
plautrans(option_eval_singleton)

##      O1 O2 O3 trplau
## O1    1  0  0   0.90
## O2    0  1  0   0.05
## O3    0  0  1   0.05
```

If we know that the person making the choice will be person 1 with a probability 30 % and person 2 with a probability 70 %, we take person$spec[,2] = c(0.3, 0.7) and we get $P(O_1) = 0.27$, $P(0_2) = 0.575$, $P(0_3) = 0.155$:

```
##      O1 O2 O3 trplau
## O1    1  0  0  0.270
## O2    0  1  0  0.575
## O3    0  0  1  0.155
```

The reader can find other examples in (Boivin 2022a, b, c).

Exercises

1. A six-faced dice is rolled twice. A gambler considers that the mass function for a single roll verifies $\mu(\{6\}) = 0.12$ and $\mu(\{1\}) = 0.24$, while $\mu(\{i\}) = 0.16, 2 \leq i \leq 5$.
   (a) Create the frame for a single roll.
   (b) Create the frame for two rolls as the cartesian product of two single rolls.
   (c) Determine the plausibilities of the results of two rolls.
   (d) Find the probabilities of the results of two rolls.
   (e) Determine $P(\{i\}), 1 \leq i \leq 6$.
2. A coin is tossed four times. A gambler considers that the mass function verifies $\mu((H, H, H, H)) = 0.05$ and $\mu((T, T, T, T)) = 0.05$, $\mu((H, T, H, T)) = \mu((H, H, T, T)) = \mu((H, T, T, H)) = \mu((T, T, H, H)) = \mu((T, H, T, H)) = \mu((T, H, H, T)) = 0.02$. No other information is given.
   (a) Create a model frame for this situation.
   (b) Determine the beliefs associated with the results.
   (c) Determine the plausibilities of the results.
   (d) Find the probabilities of the results.
   (e) Determine $P(\{H\})$ and $P(\{T\})$.
3. An urn contains balls of three colors: blue (B), red (R), and white (W). We consider that $\mu(\{B\}) = 0.2$, $\mu(\{R\}) = 0.3$, $\mu(\{W\}) = 0.5$ for a single draw. Two balls are drawn sequentially, with remittance of the first ball in the urn.
   (a) Create a model frame for a single draw.
   (b) Create a model frame for two draws as a cartesian product of two frames.
   (c) Determine the beliefs associated with the results.
   (d) Determine the plausibilities of the results.
   (e) Determine les probabilities of the results.
   (f) Determine $P(\{R\})$, $P(\{B\})$, and $P(\{W\})$.

2.6 Contour Functions

It is possible to define the mass function using the elementary plausibilities $pl_e(\omega) = pl_e(\{\omega\})$, $\omega \in \Omega$: then, $Pl(A) = \max \{pl(\omega) : \omega \in A\}$. In the context of belief functions, pl_e is referred to as ***contour function***. Notice that at least one of the elements must receive a value equal to 1. In such a situation, we use the consonant equations to define *Pl* and, as a consequence, *Bel*. Ordering the values of pl_e in a decreasing order $pl_e(\omega_1) \geq pl_e(\omega_2) \geq \ldots \geq pl_e(\omega_n)$, such a definition corresponds to a mass function $\mu(\{\omega_1, \ldots, \omega_k\}) = \mu(\omega_k) - \mu(\omega_{k+1})$, $\mu(\Omega) = pl_e(\omega_n)$ (Denoeux 2021).

Example 2.13 Let us consider 3 possible options for a decision: O_1, O_2, O_3. Assume that an expert furnishes the estimations $pl_e(O_1) = 0.2$, $pl_e(O_2) = 0.5$, $pl_e(O_3) = 0.3$. Then, we start by normalizing the values to get a maximum equal to 1:

$$pl_e(O_2) = 1, pl_e(O_3) = 0.6, pl_e(O_1) = 0.4.$$

Then,

$$\begin{aligned} Pl(\varnothing) = 0, Pl(\{O_1\}) = 0.4, Pl(\{O_2\}) = 1, Pl(\{O_3\}) = 0.6, \\ Pl(\{O_1, O_2\}) = \max\{pl_e(O_1), pl_e(O_2)\} = 1, \\ Pl(\{O_1, O_3\}) = \max\{pl_e(O_1), pl_e(O_3)\} = 0.6, \\ Pl(\{O_2, O_3\}) = \max\{pl_e(O_2), pl_e(O_3)\} = 1, \\ Pl(\{O_1, O_2, O_3\}) = \max\{pl_e(O_1), pl_e(O_2), pl_e(O_3)\} = 1. \end{aligned}$$

We can evaluate these plausibilities using the function `evalpl_from_ple`:

```
evalpl_from_ple = function(plvec, ev){
  pln = plvec/max(plvec)
  nel = length(pln)
  nev = dim(ev)[[1]]
  resu = numeric(length = nev)
  for (i in 1:nev){
    ind = which(ev[i,] > 0)
    if (length(ind) == 0) { resu[[i]] = 0 }
    else{resu[[i]] = max(pln[ind])}
  }
  return(resu)
}
```

(continued)

Example 2.13 (continued)

For instance, with the data above:

```
source("dfram.R")

pl1 = c(0.2, 0.5, 0.3)
v1names = c("O1", "O2", "O3")

fr = dfram$new()
evc = fr$complete_frame(length(pl1))

p = evalpl_from_ple(pl1,evc)
fr$print_events_values(evc,p,v1names, "Pl")

##   O1 O2 O3  Pl
## 1  0  0  0 0.0
## 2  1  0  0 0.4
## 3  0  1  0 1.0
## 4  0  0  1 0.6
## 5  1  1  0 1.0
## 6  1  0  1 0.6
## 7  0  1  1 1.0
## 8  1  1  1 1.0
```

Since $Bel(A) = 1 - Pl(\Omega - A)$, we have

$$
\begin{aligned}
Bel(\emptyset) &= 1 - Pl(\{O_1, O_2, O_3\}) = 0,\\
Bel(\{O_1\}) &= 1 - Pl(\{O_2, O_3\}) = 0,\\
Bel(\{O_2\}) &= 1 - Pl(\{O_1, O_3\}) = 0.4,\\
Bel(\{O_3\}) &= 1 - Pl(\{O_1, O_2\}) = 0,\\
Bel(\{O_1, O_2\}) &= 1 - Pl(\{O_3\}) = 0.4,\\
Bel(\{O_1, O_3\}) &= 1 - Pl(\{O_2\}) = 0,\\
Bel(\{O_2, O_3\}) &= 1 - Pl(\{O_1\}) = 0.6,\\
Bel(\{O_1, O_2, O_3\}) &= 1
\end{aligned}
$$

(continued)

Example 2.13 (continued)

These values can be evaluated by the function `evalbel_from_ple`:

```
evalbel_from_ple = function(plvec, ev){
  pln = plvec/max(plvec)
  nel = length(pln)
  nev = dim(ev)[[1]]
  resu = numeric(length = nev)
  for (i in 1:nev){
    evaux = c(rep(1, nel)) - ev[i,]
    ind = which(evaux > 0)
    if (length(ind) == 0) { resu[[i]] = 1 }
    else{resu[[i]] = 1 - max(pln[ind])}
  }
  return(resu)
}
```

For instance, with the data above:

```
b = evalbel_from_ple(pl1,evc)
fr$print_events_values(evc,b,v1names, "Bel")

##    O1 O2 O3 Bel
## 1  0  0  0 0.0
## 2  1  0  0 0.0
## 3  0  1  0 0.4
## 4  0  0  1 0.0
## 5  1  1  0 0.4
## 6  1  0  1 0.0
## 7  0  1  1 0.6
## 8  1  1  1 1.0
```

The mass function associated with pl_e is given by

$$\mu(\{O_1, O_2, O_3\}) = 0.4$$

$$\mu(\{O_2, O_3\}) = 0.2$$

$$\mu(\{O_2\}) = 0.4$$

(continued)

Example 2.13 (continued)

We can evaluate this mass function using the function ple2mu below:

```
ple2mu = function(plvec){
  pln = plvec/max(plvec)
  nel = length(pln)
  resu = sort(pln,index.return=TRUE, decreasing = TRUE)
  aux = c(rep(0,nel))
  ev = matrix(0, nrow = nel, ncol = nel)
  mu = numeric(length = nel)
  for (i in 1: (nel-1)){
    aux[[resu$ix[[i]]]] = 1
    ev[i,] = aux
    mu[[i]] = resu$x[[i]] - resu$x[[i+1]]
  }
  mu[[nel]] = resu$x[[nel]]
  ev[nel,] =  rep(1,nel)
  result = list(ev,mu)
  names(result) = c("ev","m")
  return (result)
}
```

For instance, with the data above:

```
r = ple2mu(pl1)
fr$print_events_values(r$ev,r$m,v1names, "mass")

##    O1 O2 O3 mass
## 1  0  1  0  0.4
## 2  0  1  1  0.2
## 3  0  0  0  0.0
```

Once the mass function is determined, we can create a frame and determine *Bel* and *Pl*, as in Examples 2.3 and 2.4. For instance,

```
f1 = dfram$new(r$ev,r$m)

library(dst)
f1b = bca(tt=r$ev,m=r$m)
```

(continued)

Example 2.13 (continued)

We obtain:

```
f1$print_events_values(f1$ev,f1$m,f1$names, "mass")

##   var1 var2 var3 mass
## 1    0    1    0  0.4
## 2    0    1    1  0.2
## 3    1    1    1  0.4

f1b$tt

##             col1 col2 col3
## col2           0    1    0
## col2 + col3    0    1    1
## frame          1    1    1

f1b$spec

##      specnb mass
## [1,]      1  0.4
## [2,]      2  0.2
## [3,]      3  0.4
```

Then, we can use the frame to evaluate the plausibilities as in Example 2.3:

```
evc = fr$complete_frame(length(pl1))
p = f1$eval_at_events(fr$evalpl,evc, f1)
f1$print_events_values(evc,p,v1names, "Pl")

##   O1 O2 O3  Pl
## 1  0  0  0 0.0
## 2  1  0  0 0.4
## 3  0  1  0 1.0
## 4  0  0  1 0.6
## 5  1  1  0 1.0
## 6  1  0  1 0.6
## 7  0  1  1 1.0
## 8  1  1  1 1.0
```

The values of the beliefs can be generated analogously, using the function `evalbel` (see Example 2.3).

Exercises

1. A six-faced dice is rolled twice. Consider the contour function $pl_e(6) = 0.12$ and $pl_e(1) = 0.24$, while $pl_e(i) = 0.16,\ 2 \leq i \leq 5$.

   (a) Determine the plausibilities of the results of two rolls.
   (b) Find the probabilities of the results of two rolls.

2. A coin is tossed four times. A gambler considers that the mass function verifies $pl_e(H) = 0.55$ and $pl_e(T) = 0.45$.

   (a) Determine the plausibilities of the results.
   (b) Find the probabilities of the results.

3. An urn contains balls of three colors: blue (B), red (R), and white (W). Two balls are drawn sequentially, with remittance of the first ball in the urn. We consider that $pl_e(B) = 0.2$, $pl_e(R) = 0.3$, $pl_e(W) = 0.5$.

   (a) Determine the plausibilities of the results.
   (b) Determine les probabilities of the results.

2.7 Combination of Beliefs

Belief and plausibility functions have found applications in data fusion – see, for instance, among the large existing literature (Smets 2000; He 2013; Martin 2019; Ni et al. 2020; Zhang and Deng 2020; Laghmara et al. 2020). They had also been applied to decision making – again, many works use such an approach – see, for instance, Strat (1990), Jaffray and Wakker (1993), Fox and See (2006), Taillandier and Therond (2011), Lian (2017), Denoeux (2019). They can be connected to rough sets -see, for instance, Yao and Lingras (1998), Klopotek and Wierzchon (1998), Wu and Mi (2008), Campagner et al. (2022) – and set valued maps (Wasserman 1990; Denœux et al. 2010; Martin et al. 2010; Benavoli 2014). There are also interpretations in terms of random sets (Nguyen 1977, 1978; Dubois and Prade 1986; Nguyen and Wang 1997; Miranda et al., 2005; Couso et al., 2014; Fetz and Oberguggenberger 2016).

In the context of data fusion, it is important to know how to manage data from different sources, sometimes providing contradictory information: the theory of belief functions develops methods to combine different mass functions and to manage conflicting information. The literature considers basically three methods of combination:

- ***Disjunctive***:

$$\mu_D(A) = \sum_{A_1 \cup A_2 = A} \mu_1(A_1)\mu_2(A_2)$$

- ***Conjunctive***:

$$\mu_C(A) = \sum_{A_1 \cap A_2 = A} \mu_1(A_1)\mu_2(A_2)$$

- ***Normalized***:

$$\mu(A) = \frac{\mu_C(A)}{1 - \mu_c(\emptyset)}, A \neq \emptyset; \mu(A) = \emptyset$$

Normalized combinations can be defined only when $\mu_c(\emptyset) < 1$. The value $d_c(\mu_1, \mu_2) = \mu_C(\emptyset)$ measures the ***degree of conflict*** between μ_1 and μ_2. It is interesting to notice that the commonality function associated with the conjunctive combination is the product of the commonality functions:

$$Q_{c,\mu}(A) = Q_{c,\mu_1}(A) Q_{c,\mu_2}(A), \forall A \subset \Omega. \tag{2.2}$$

The combination of mass functions may involve a ***discounting operation***, destined to reduce the masses associated with one or both the mass functions to some elements of $\mathcal{P}(\Omega)$. The reader will find in the literature a number of texts devoted to the combination of different mass functions – it is a core subject in data fusion. See, for instance Strat (1987), Guan and Bell (1993), Mercier et al. (2005), Liu (2006), Zhou et al. (2018), Martin (2019), Denoeux (2021).

Example 2.14 Let us consider 3 possible options for a decision: O_1, O_2, O_3. Assume that the expert no. 1 furnishes the masses

$$\mu_1(\{O_1, O_2\}) = 0.2, \mu_1(\{O_1\}) = 0.4, \mu_1(\{O_1, O_3\}) = 0.3, \mu_1(\Omega) = 0.1.$$

A second expert furnishes the masses

$$\mu_2(\{O_1, O_2\}) = 0.4, \mu_2(\{O_2\}) = 0.3, \mu_2(\{O_2, O_3\}) = 0.3.$$

(continued)

Example 2.14 (continued)

Let us create the corresponding frames in R:

```
source("dfram.R")
ev1 = matrix(nrow = 4, ncol = 3)
ev1[1,] = c(1, 1, 0)
ev1[2,] = c(1, 0, 0)
ev1[3,] = c(1, 0, 1)
ev1[4,] = c(1, 1, 1)
mu1 = c(0.2,0.4,0.3, 0.1)
f1 = dfram$new(ev1,mu1)

ev2 = matrix(nrow = 3, ncol = 3)
ev2[1,] = c(0, 1, 0)
ev2[2,] = c(1, 1, 0)
ev2[3,] = c(0, 1, 1)
mu2 = c(0.3,0.4,0.3)
f2 = dfram$new(ev2,mu2)
```

Let us determine the combinations of these two mass functions. For the determination of $\mu_D(A)$, we must consider the reunions such that $A_1 \cup A_2 = A$. The possible subsets are given in Table 2.4.

Table 2.4 Disjunctive combination of the masses

A	A_1	A_2	μ_D
$\varnothing$	$\varnothing$	$\varnothing$	0
$\{O_1\}$	$\varnothing$	$\varnothing$	0
$\{O_2\}$	$\varnothing$	$\varnothing$	0
$\{O_3\}$	$\varnothing$	$\varnothing$	0
$\{O_1, O_2\}$	$\{O_1, O_2\}$ $\{O_1\}$ $\{O_1\}$ $\{O_1, O_2\}$	$\{O_1, O_2\}$ $\{O_2\}$ $\{O_1, O_2\}$ $\{O_2\}$	0.42
$\{O_1, O_3\}$	$\varnothing$	$\varnothing$	0
$\{O_2, O_3\}$	$\varnothing$	$\varnothing$	0
$\{O_1, O_2, O_3\}$	$\{O_1, O_2\}$ $\{O_1\}$ $\{O_1, O_3\}$ $\{O_1, O_3\}$ $\{O_1, O_3\}$ Ω Ω Ω	$\{O_2, O_3\}$ $\{O_2, O_3\}$ $\{O_1, O_2\}$ $\{O_2\}$ $\{O_2, O_3\}$ $\{O_1, O_2\}$ $\{O_2\}$ $\{O_2, O_3\}$	0.58

(continued)

Example 2.14 (continued)

The disjunctive combination can be determined by the function disjunct:

```
disjunct = function(f1, f2){
  nel = f1$nvar
  nc = 2^nel
  evn = matrix(nrow = nc, ncol=nel)
  k = 1
  evn[k,]=c(rep(0,nel))
  for (i in (1:nel)){
    c = combn(1:3,i)
    for ( j in 1:dim(c)[[2]]){
      ind = c[,j]
      ev = c(rep(0, nel))
      ev[ind] = 1
      k = k + 1
      evn[k,] = ev
    }
  }
  mud = numeric(length = nc)
  n1 = length(f1$m)
  n2 = length(f2$m)
  for (k in 2:nc){
    u = evn[k,]
    s = 0
    for (i in 1: n1){
      vec1 = f1$ev[i,]
      for (j in 1: n2){
        v = vec1 + f2$ev[j,]
        ind = which(v > 1)
        if (length(ind) > 0){
          v[ind] = 1
        }
        aux = v - u
        n_aux = sum(abs(aux))
        if (n_aux == 0) {
          s = s + f1$m[[i]]*f2$m[[j]]
        }
      }
    }
    mud[[k]] = s
  }
  source("dfram.R")
  fdisj = dfram$new(evn,mud)
  return(fdisj)
}
```

(continued)

Example 2.14 (continued)

Using the data furnished by the experts, we obtain

```
fd = disjunct(f1,f2)
fd$print_events_values(fd$ev,fd$m,fd$names, "mass")

##    var1 var2 var3 mass
## 1     0    0    0 0.00
## 2     1    0    0 0.00
## 3     0    1    0 0.00
## 4     0    0    1 0.00
## 5     1    1    0 0.42
## 6     1    0    1 0.00
## 7     0    1    1 0.00
## 8     1    1    1 0.58
```

For the determination of $\mu_C(A)$, we must consider the intersections such that $A_1 \cap A_2 = A$. The possible subsets are given in Table 2.5.

Table 2.5 Conjunctive combination of the masses

A	A_1	A_2	μ_D
$\varnothing$	$\{O_1\}$ $\{O_1\}$ $\{O_1, O_3\}$	$\{O_2\}$ $\{O_2, O_3\}$ $\{O_2\}$	0.33
$\{O_1\}$	$\{O_1\}$ $\{O_1, O_3\}$	$\{O_1, O_2\}$ $\{O_1, O_2\}$	0.28
$\{O_2\}$	$\{O_1, O_2\}$ $\{O_1, O_2\}$ Ω	$\{O_2\}$ $\{O_2, O_3\}$ $\{O_2\}$	0.15
$\{O_3\}$	$\{O_1, O_3\}$	$\{O_2, O_3\}$	0.09
$\{O_1, O_2\}$	$\{O_1, O_2\}$ Ω	$\{O_1, O_2\}$ $\{O_1, O_2\}$	0.12
$\{O_1, O_3\}$	$\varnothing$	$\varnothing$	0
$\{O_2, O_3\}$	Ω	$\{O_2, O_3\}$	0.03
$\{O_1, O_2, O_3\}$	$\varnothing$	$\varnothing$	0

(continued)

Example 2.14 (continued)

The conjunctive combination can be determined by the function conjunct:

```
conjunct = function(f1, f2){
  nel = f1$nvar
  nc = 2^nel
  evn = matrix(nrow = nc, ncol=nel)
  k = 1
  evn[k,]=c(rep(0,nel))
  for (i in (1:nel)){
    c = combn(1:3,i)
    for ( j in 1:dim(c)[[2]]){
      ind = c[,j]
      ev = c(rep(0, nel))
      ev[ind] = 1
      k = k + 1
      evn[k,] = ev
    }
  }
  muc = numeric(length = nc)
  n1 = length(f1$m)
  n2 = length(f2$m)
  for (k in 1:nc){
    u = evn[k,]
    s = 0
    for (i in 1: n1){
      vec1 = f1$ev[i,]
      for (j in 1: n2){
        v = vec1*f2$ev[j,]
        aux = v - u
        n_aux = sum(abs(aux))
        if (n_aux == 0) {
          s = s + f1$m[[i]]*f2$m[[j]]
        }
      }
    }
    muc[[k]] = s
  }
  source("dfram.R")
  fconj = dfram$new(evn,muc)
  return(fconj)
}
```

(continued)

Example 2.14 (continued)

For instance, using the evaluations of the experts:

```
fc = conjunct(f1,f2)
fc$print_events_values(fc$ev,fc$m,fc$names, "mass")

##   var1 var2 var3 mass
## 1    0    0    0 0.33
## 2    1    0    0 0.28
## 3    0    1    0 0.15
## 4    0    0    1 0.09
## 5    1    1    0 0.12
## 6    1    0    1 0.00
## 7    0    1    1 0.03
## 8    1    1    1 0.00
```

The package dst proposes a function dsrwon to evaluate the conjunctive combination. For instance, using the data furnished by the experts (notice that, on the one hand, the order is not the same as in Table 2.5 and, on the other hand, only the non-null elements appear):

```
library(dst)
f1b = bca(tt=ev1,m=mu1)
f2b = bca(tt=ev2,m=mu2)

fcb = dsrwon(f1b,f2b)
fcb$tt

##             col1 col2 col3
## ø              0    0    0
## col2           0    1    0
## col1           1    0    0
## col3           0    0    1
## col1 + col2    1    1    0
## col2 + col3    0    1    1
t(fcb$spec)

##        [,1] [,2] [,3] [,4] [,5] [,6]
## specnb 1.00 2.00 3.00 4.00 5.00 6.00
## mass   0.33 0.15 0.28 0.09 0.12 0.03
```

(continued)

Example 2.14 (continued)

Now, let us use the product of commonalities to determine the conjunctive combination:

```
evc = fc$complete_frame(3)

Q1 = fc$eval_at_events(fc$evalcom,evc, f1)
Q2 = fc$eval_at_events(fc$evalcom,evc, f2)
Q = Q1*Q2

vfnames = c("O1", "O2", "O3")

ff <-  dfram$new(evc, ,vfnames)

ff$Q = Q
muc = fc$eval_at_events(fc$com2mu,evc, ff)
fc$print_events_values(evc,muc,vfnames, "mu_c")

##   O1 O2 O3 mu_c
## 1  0  0  0 0.33
## 2  1  0  0 0.28
## 3  0  1  0 0.15
## 4  0  0  1 0.09
## 5  1  1  0 0.12
## 6  1  0  1 0.00
## 7  0  1  1 0.03
## 8  1  1  1 0.00
```

The package dst proposes a function dsrwon to evaluate the conjunctive combination. For instance, using the data furnished by the experts (notice that, on the one hand, the order is not the same as in Table 2.5 and, on the other hand, only the non-null elements appear):

```
fcb = dsrwon(f1,f2)
fcb$tt

##             col1 col2 col3
## ø              0    0    0
## col2           0    1    0
## col1           1    0    0
## col3           0    0    1
## col1 + col2    1    1    0
## col2 + col3    0    1    1

t(fcb$spec)

##        [,1] [,2] [,3] [,4] [,5] [,6]
## specnb 1.00 2.00 3.00 4.00 5.00 6.00
## mass   0.33 0.15 0.28 0.09 0.12 0.03
```

(continued)

Example 2.14 (continued)

The normalized combination is determined by `normalconj`:

```
normalconj = function(fc){
  muc = fc$m
  aux = muc[[1]]
  if (aux < 1){
    mun = muc/(1-aux)
    mun[[1]] = 0
  }
  source("dfram.R")
  fcn = dfram$new(fc$ev,mun,fc$names)
  return(fcn)
}
```

We obtain the results below:

```
fcn = normalconj(fc)
fcn$print_events_values(fcn$ev,fcn$m,fcn$names, "mass")

##   var1 var2 var3       mass
## 1    0    0    0 0.00000000
## 2    1    0    0 0.41791045
## 3    0    1    0 0.22388060
## 4    0    0    1 0.13432836
## 5    1    1    0 0.17910448
## 6    1    0    1 0.00000000
## 7    0    1    1 0.04477612
## 8    1    1    1 0.00000000
```

Of course, we can use the commonalities to determine the normalized combination: we determine `muc` as previously shown and we normalize the result:

(continued)

Example 2.14 (continued)

```
aux = muc[[1]]
if (aux < 1){
  mun = muc/(1-aux)
  mun[[1]] = 0
}
fc$print_events_values(evc,mun,vfnames, "mu")

##    O1 O2 O3         mu
## 1  0  0  0 0.00000000
## 2  1  0  0 0.41791045
## 3  0  1  0 0.22388060
## 4  0  0  1 0.13432836
## 5  1  1  0 0.17910448
## 6  1  0  1 0.00000000
## 7  0  1  1 0.04477612
## 8  1  1  1 0.00000000
```

The package `dst` offers the function `nzdsr` to normalize the mass function determined by `dsrwon` (again, notice that the order changes when compared with `normalconj`. In addition, only non-null elements appear):

```
fcbn = nzdsr(fcb)
fcbn$tt

##             col1 col2 col3
## col2           0    1    0
## col1           1    0    0
## col3           0    0    1
## col1 + col2    1    1    0
## col2 + col3    0    1    1

t(fcbn$spec)

##             [,1]      [,2]      [,3]      [,4]        [,
5]
## specnb 1.0000000 2.0000000 3.0000000 4.0000000 5.000000
00
## mass   0.2238806 0.4179104 0.1343284 0.1791045 0.044776
12
```

Exercises

1. Consider a six-faced dice. For a first gambler, $\mu_1(6) = 0.12$ and $\mu_1(1) = 0.24$, while $\mu_1(i) = 0.16$, $2 \leq i \leq 5$. For a second one, $\mu_2(i) = 1/6$, $1 \leq i \leq 6$.

   (a) Determine the conjunctive combination of the masses.
   (b) Find the probabilities of the results furnished by the combination.

2. A coin is tossed once. A gambler considers that the contour function verifies $pl_{e,\ 1}(H) = 0.55$ and $pl_{e,\ 1}(T) = 0.45$. A second gambler considers that $pl_{e,\ 2}(H) = 0.5$ and $pl_{e,\ 2}(T) = 0.5$.

   (a) Determine the masses μ_i associated with gambler i, $1 \leq i \leq 2$.
   (b) Determine the conjunctive combination of the masses.
   (c) Find the probabilities of the results furnished by the combination.

3. An urn contains balls of three colors: blue (B), red (R), and white (W). A ball is drawn from the urn. A first observer considers that $pl_{e,\ 1}(B) = 0.2$, $pl_{e,\ 1}(R) = 0.3$, $pl_{e,\ 1}(W) = 0.5$. A second observer considers that $pl_{e,\ 2}(B) = 0.3$, $pl_{e,\ 2}(R) = 0.25$, $pl_{e,\ 2}(W) = 0.45$.

   (a) Determine the masses μ_i associated with observer i, $1 \leq i \leq 2$.
   (b) Determine the conjunctive combination of the masses.
   (c) Find the probabilities of the results furnished by the combination.

4. A murder was committed at Pratt's manor. There are four suspects: Mr. Green, Mr. Plum, Ms. White, and Ms. Scarlett. A witness says to be sure at the level of 75 % that the murderer was a woman. Mr. Green has an alibi which was verified at 90 %.

   (a) Taking the numbers as values of a belief mass function, determine the plausibility of the following sets of suspects: {Scarlett, White}, {Green}, {Green, Scarlett, White}, {Plummer}, and {Plummer, Green}.
   (b) Assume that the masses of {Scarlett} and {White} are equal. Determine the beliefs and the plausibility of each suspect. Transform the masses into probabilities.

5. Ashes were found in the crime scene at Pratt's manor (Exercise 4). Only Mr. Plum and Scarlett are smokers, but the confidence of this index is 50 %. Generate a new mass function using this new information and the preceding ones. Then find the beliefs and the plausibility of each suspect, by combining this information with the information given in Exercise 4.

2.8 Conditional Beliefs, Plausibilities, and Frames

In the theory of Beliefs, the equivalent of conditional probabilities is furnished by conditional lausibilities:

$$Pl(B|A) = \frac{Pl(A \cap B)}{Pl(A)}. \tag{2.3}$$

As shown above, we can determine the corresponding masses and beliefs from the conditional plausibilities. For instance, we have

$$Bel(B|A) = 1 - Pl(\Omega - B|A) = \frac{Bel(B \cup (\Omega - A)) - Bel(\Omega - A)}{1 - Bel(\Omega - A)}. \tag{2.4}$$

$$\mu(B|A) = \frac{1}{Pl(A)} \sum_{S \cap A = B} \mu(S). \tag{2.5}$$

Example 2.15 Let us consider the situation presented in the preceding example: the evaluation of the experts is given as first evaluation:

$$\mu_e(\{O_2\},\{P_1\}) = 0.45, \mu_e(\{O_1\},\{P_1\}) = \mu_e(\{O_3\},\{P_1\}) = 0.025$$

$$\mu_e(\{O_3\},\{P_2\}) = 0.4, \mu_e(\{O_2\},\{P_2\}) = 0.1, \mu_e(\{O_1\},\{P_2\}) = 0.$$

Let us enter this data in R:

```
ev12 = matrix(nrow = 6, ncol = 5)
ev12[1,] = c(1,0,1,0,0) # P1 with O1
ev12[2,] = c(1,0,0,1,0) # P1 with O2
ev12[3,] = c(1,0,0,0,1) # P1 with O3
ev12[4,] = c(0,1,0,1,0) # P2 with O2
ev12[5,] = c(0,1,1,0,0) # P2 with O1
ev12[6,] = c(0,1,0,0,1) # P2 with O3
nams12 <- c("P1", "P2", "O1", "O2", "O3")
mue = c(0.025, 0.45, 0.025, 0.1, 0, 0.1)
```

(continued)

Example 2.15 (continued)

We create a product frame using these events: since the evaluations do not sum to the unity, we must normalize them to get a total sum equal to 1, so that

```
mu12 = mue/sum(mue)
nvars12 = c(2,3)
events = list(ev12,mu12,nams12,nvars12)
names(events) = c("ev","m","names","nvars")

source("pdf.R")
pf1 = pdf$new(events,'events')
```

Thus, we obtain the plausibilities:

```
f = pf1$frc
f$pl

## [1] 0.7142857 0.8571429 0.8571429 0.9285714 0.3214286 0
.3214286
```

We can use `dfram` or the package `dst`: the result is the same:

```
source("dfram.R")
f12 = dfram$new(ev12,mu12,nams12)
f12$pl

## [1] 0.7142857 0.8571429 0.8571429 0.9285714 0.3214286 0
.3214286

library(dst)
f12b = bca(tt=ev12,m=mu12,cnames=nams12)
belplau(f12b)

##              Belief Plausibility Plty Ratio
## P1 + O1 0.03571429    0.7142857  0.7407407
## P1 + O2 0.64285714    0.8571429  2.4000000
## P1 + O3 0.03571429    0.8571429  0.8888889
## P2 + O2 0.14285714    0.9285714  1.0833333
## P2 + O1 0.00000000    0.3214286  0.3214286
## P2 + O3 0.14285714    0.3214286  0.3750000
```

(continued)

Example 2.15 (continued)

Let us consider:

$$A_i = \text{the option chosen } is\ O_i, B_j = \text{the option is chosen by } P_j.$$

We have

$$A_i = \{O_i, P_1\} \cup \{O_i, P_2\}, \quad B_j = \{O_1, P_j\} \cup \{O_2, P_j\} \cup \{O_3, P_j\}.$$

Thus, $Pl(A_1) = Pl(A_2) = Pl(A_3) = 1$. Indeed:

```
A1 = as.numeric(ev12[1, ] | ev12[5, ])
pf1$evalpl(A1)
## [1] 1
A1b = c(1,1,1,0,0)
pf1$evalpl(A1b)
## [1] 1
A2 = as.numeric(ev12[2, ] | ev12[4, ])
pf1$evalpl(A2)
## [1] 1
A2b = c(1,1,0,1,0)
pf1$evalpl(A2b)
## [1] 1
A3 = as.numeric(ev12[3, ] | ev12[6, ])
pf1$evalpl(A3)
## [1] 1
A3b = c(1,1,0,0,1)
pf1$evalpl(A3b)
## [1] 1
```

Since $A_i \cap B_j = \{O_i, P_j\}$, we have $Pl(B_j|A_i) = Pl(\{O_i, P_j\})/Pl(A_i)$. Consequently, $Pl(B_j|A_i) = Pl(\{O_i, P_j\})$ and we have

$$Pl(B_1|A_1) \approx 0.7142857, Pl(B_1|A_2) \approx 0.3214286, Pl(B_1|A_3) \approx 0.8571429$$

$$Pl(B_2|A_1) \approx 0.3214286, Pl(B_2|A_2) \approx 0.9285714, Pl(B_2|A_3) \approx 0.3214286.$$

We also have $B_1 = \Omega - B_2$, so that

$$Bel(B_1|A_1) = 1, Bel(B_1|A_2) = 0.9, Bel(B_1|A_3) = 0.6$$

(continued)

Example 2.15 (continued)

$$Bel(B_2|A_1) = 0.975, Bel(B_2|A_2) = 0.55, Bel(B_2|A_3) = 0.975.$$

In addition,

$$\mu(B_1|A_1) = 1, \mu(B_1|A_2) = 0.9, \mu(B_1|A_3) = 0.6$$
$$\mu(B_2|A_1) = 0.975, \mu(B_2|A_2) = 0.55, \mu(B_2|A_3) = 0.975.$$

and

$$\mu(\{P_1\}|\{O_1\}) = 0.9, \mu(\{P_1\}|\{O_2\}) = \mu(\{P_1\}|\{O_3\}) = 0.05$$
$$\mu(\{P_2\}|\{O_1\}) = 0.8 \ \ \mu(\{P_2\}|\{O_2\}) = 0.2, \mu(\{P_2\}|\{O_3\}) = 0.$$

Person P_1 prefers O_2 at the level 90 % and is indifferent to the other options. Person P_2 prefers O_3 at the level 80 % and rejects completely O_1.

The reader can find other examples in Boivin (2022a, b, c).

Exercises

1. Consider two six-faced dice. We have $\mu((6, i)) = 0.03$ and $\mu((1, i)) = 0.02$, $1 \leq i \leq 6$.
   (a) Create a model frame for this situation.
   (b) Determine the plausibilities of the results.
   (c) Find the marginal plausibilities of each dice.
   (d) Determine the marginal masses of each dice.
   (e) Determine the conditional plausibilities of the dices.
2. Four coins are tossed together. We have $\mu((H, H, H, H)) = 0.05$ and $\mu((T, T, T, T)) = 0.05$, $\mu((H, T, H, T)) = \mu((H, H, T, T)) = \mu((H, T, T, H)) = \mu((T, T, H, H)) = \mu((T, H, T, H)) = \mu((T, H, H, T)) = 0.02$.
   (a) Create a model frame for this situation.
   (b) Determine the plausibilities of the results.
   (c) Find the marginal plausibilities of each coin.
   (d) Determine the marginal masses of each coin.
   (e) Determine the conditional plausibilities of the subset of two first coins with respect to the two last ones.

(continued)

3. An urn contains balls of three colors: blue (B), red (R), and white (W). Two balls are drawn sequentially, with remittance of the first ball in the urn.

   We consider that Bel is null, except for $Bel(\{B\}) = 0.2$, $Bel(\{R\}) = 0.3$, $Bel(\{W\}) = 0.6$, $Bel(\{B, R\}) = 0.5$, $Bel(\{B, W\}) = 0.8$, $Bel(\{R, W\}) = 0.9$, $Bel(\Omega) = 1$.

   (a) Create a model frame for this situation.
   (b) Determine the mass function associated with your model.
   (c) Determine the plausibilities of your model.
   (d) Find the marginal plausibilities of each color.
   (e) Determine the marginal masses of each color.
   (f) Determine the conditional plausibilities of the first result with respect to the results of the two last ones.

4. An expert evaluates that 1% of the parts manufactured in a factory are defective. The quality control evaluates that it accepts 1% of non-compliant parts and rejects 2% of compliant parts. What is the plausibility that an accepted part is compliant?
5. Two boxes contain the same number of pins, divided between two colors: red and green. In box 1, 25 % are red and 75 % are green; in box 2, 50 % are red pins and 50 % are green. The boxes are equiprobable.

   (a) We take a red pin: what is the probability that it came from box 1?
   (b) We take a green pin: what is the probability that it came from box 2?
   (c) How are these values modified if box 1 is much larger than box 2, so that its probability is $\frac{3}{4}$?

6. An urn contains 25 numbered balls of three colors: 1 to 10 in red, 11 to 15 in green, and 16 to 20 in blue. We draw a ball at random.

   (a) The number is odd: what is the probability that the color is green?
   (b) The number has two digits: what is the probability that the color is blue?
   (c) The number is prime: what is the probability that the color is red?

7. Three urns contain 10 numbered balls in red and blue each, but the number of red balls is different in each one: 6 in urn 1, 3 in urn 2, and 4 in urn 3. The urns are equiprobable. A red ball is drawn.

   (a) What is the probability that it came from urn 2?
   (b) What is the probability that it did not come from urn 1?
   (c) What is the probability that it came from urn 3?

2.9 Connections Between Probabilities and Beliefs

Belief and Plausibility functions can be connected to the probabilities, namely, by considering random variables affected by imprecision and uncertainty.

For instance, let us consider a numerical variable X, defined on (Ω, P) and taking its values on $\mathbb{R}$. The imprecision or uncertainty on X yields that $X(\omega)$ is ill-known and has several possible values: $X(\omega) \in T(\omega) \subset \mathbb{R}$. Conversely, a given $x \in \mathbb{R}$ can be the image of several subsets from $\Omega : X^{-1}(x) \in S(x) \subset \mathcal{P}(\Omega)$. Then, we can model both X and X^{-1}as multivalued maps: $\mathcal{X} : \Omega \longrightarrow \mathcal{P}(\mathbb{R})$ and $\mathcal{X}^{-1} : \mathbb{R} \longrightarrow \mathcal{P}(\Omega)$ such that $\mathcal{X}(\omega) = T(\omega)$ and $\mathcal{X}^{-1}(x) = S(x)$.

In such a situation, $P(X = x)$ is uncertain: since all the elements of $S(x)$ are possible values of $X^{-1}(x)$, we have $P(X = x) \in \{P(A) : A \in S(x)\}$. In (Dempster 1967), such a situation was analyzed, introducing the notions of lower and upper probabilities, analogous to the notions of inner and outer measures (see, for instance, (Halmos 1978)). The basic idea is to consider, for a given subset $A \subset \mathbb{R}$:

- On the one hand, A_* , the non-empty subsets $\varnothing \neq \Gamma \subset \Omega$ such that $\mathcal{X}(\gamma) \subset A$, $\forall \gamma \in \Gamma$, i.e., $A_* = \{ \omega \in \Omega : \mathcal{X}(\omega) \subset A\}$;
- On the other hand, A^*, formed by all the elements $\gamma \in \Omega$ such that $\mathcal{X}(\gamma)$ has at least one value in A, i.e., $A^* = \{ \omega \in \Omega : \mathcal{X}(\omega) \cap A \neq \varnothing\}$.

The lower probability is $P_*(A) = P(A_*)$ and the upper probability is $P^*(A) = P(A^*)$. Noticing that

$$(\Omega - \mathrm{A})_* = \{ \omega \in \Omega : \mathcal{X}(\omega) \cap \mathrm{A} = \varnothing\} = \Omega - A^*,$$

we have

$$P_*(\Omega - A) = 1 - P^*(A) \Leftrightarrow P^*(A) = 1 - P_*(\Omega - \mathrm{A}). \tag{2.6}$$

We define the lower CDF F_* and the upper CDF F^*as:

$$F_*(x) = P_*(X \leq x) \quad ; \quad F^*(x) = P^*(X \leq x).$$

We can define a Belief function for the possible values of X (i.e., on $\mathbb{R}$) by taking (Shafer 1990)

$$Bel(A) = P_*(A).$$

In this case, from Eq. (2.94):

$$Pl(A) = 1 - P_*(\Omega - A).$$

In addition,

$$A_* \subset A \subset A^* \Longrightarrow P_*(A) \leq P(A) \leq P^*(A).$$

Thus, we have

$$Bel(A) \leq P(A) \leq Pl(A). \tag{2.7}$$

In the general case, such a condition is called compatibility condition:

Definition 2.3 Let Ω be a frame of discernment and $Bel : \mathcal{P}(\Omega) \longrightarrow \mathbb{R}$ be a belief function on Ω. Let P be a probability on Ω. We say that P is ***compatible*** with Bel if and only if P verifies (2.95). The set of all the probabilities compatible with Bel is

$$\mathfrak{C}(Bel) = \{P : probability\ on\ \Omega, compatible\ with\ Bel\}.$$

When Bel is generated by the mass function μ, $\mathfrak{C}(\mu) = \mathfrak{C}(Bel)$ is referred to as the ***credal set*** of μ. ∎

We have

$$Bel(A) = \min\{\ P(A) : P \in \mathfrak{C}(Bel)\}, Pl(A) = \max\{\ P(A) : P \in \mathfrak{C}(Bel)\}.$$

Thus, belief functions can be interpreted as compatible lower probability (Denoeux 2019).

We shall achieve here our exploration of belief functions. The reader will find in the literature a wide range of works on this topic, namely, books and research articles.

In the sequel, we focus on Bayesian methods applied in the context of probability.

Exercises

1. Consider a six-faced dice. For a first gambler, $P_1(6) = 0.12$ and $P_1(1) = 0.24$, while $P_1(i) = 0.16$, $2 \leq i \leq 5$. For a second one, $P_2(i) = 1/6$, $1 \leq i \leq 6$. For a third gambler, $P_3(6) = 0.2$ and $P_3(1) = 0.2$, while $P_1(i) = 0.15$, $2 \leq i \leq 5$.

   (a) Determine $P_*(i)$ and $P^*(i)$, $1 \leq i \leq 6$.
   (b) Characterize the credal set of the dice.

2. A coin is tossed once. A gambler considers that the contour function verifies $P(H) = 0.55$ and $P(T) = 0.45$. A second gambler considers that $P(H) = 0.5$ and $P(T) = 0.5$. A third gambler thinks that $P(H) = 0.48$ and $P(T) = 0.52$.

(continued)

(a) Determine $P_*(H)$, $P_*(T)$, and $P^*(H)$, $P^*(T)$.
(b) Characterize the credal set of the coin.

3. An urn contains balls of three colors: blue (B), red (R), and white (W). A ball is drawn from the urn. A first observer considers that $P(B) = 0.2$, $P(R) = 0.3$, $P(W) = 0.5$. A second observer considers that $P(B) = 0.3$, $P(R) = 0.25$, $P(W) = 0.45$. A third observer considers that $P(B) = 0.25$, $P(R) = 0.35$, $P(W) = 0.4$.

(a) Determine $P_*(R)$, $P_*(B)$, $P_*(W)$, and $P^*(R)$, $P^*(B)$, $P^*(W)$.
(b) Characterize the credal set of the urn.

References

Attiaoui, D., Doré, P., Martin, A., & Ben Yaghlane, B. (2012). A Distance between Continuous Belief Functions. In E. Hüllermeier, S. Link, T. Fober, & B. Seeger (Ed.), *Scalable Uncertainty Management: SUM 2012* (pp. 194–205). Marburg, Germany: Springer. doi:https://doi.org/10.1007/978-3-642-33362-0_15

Benavoli, A. (2014). Belief function and multivalued mapping robustness in statistical estimation. *International Journal of Approximate Reasoning, 55*, 311–329. doi:https://doi.org/10.1016/j.ijar.2013.04.014

Boivin, C. (2022a, December 25). https://cran.r-project.org/web/packages/dst/vignettes/Captain_Example.html. Retrieved from Captain's Example: https://cran.r-project.org/web/packages/dst/vignettes/Captain_Example.html

Boivin, C. (2022b, 12 24). *Introduction to Belief Functions.* Retrieved from https://cran.r-project.org/web/packages/dst/vignettes/: https://cran.r-project.org/web/packages/dst/vignettes/Introduction_to_Belief_Functions.html#fn2

Boivin, C. (2022c, December 25). *Introduction to Belief Functions: The Monty Hall Game.* Retrieved 2022, from https://cran.microsoft.com/snapshot/2018-08-11/web/packages/dst/vignettes/Monty-hall-Example.html: https://cran.microsoft.com/snapshot/2018-08-11/web/packages/dst/vignettes/Monty-hall-Example.html

Campagner, A., Ciucci, D., & Denœux, T. (2022). Belief Functions and Rough Sets: Suvey and New Insights. *International Journal of Approximate Reasoning, 143*, 92–215. doi:https://doi.org/10.1016/j.ijar.2022.01.011

Cobb, B. R., & Shenoy, P. (2006, April). On the plausibility transformation method for translating belief function models to probability models. *Journal of Approximate Reasoning, 41*(3), pp. 314–330. doi:https://doi.org/10.1016/j.ijar.2005.06.008

Couso, I., Dubois, D., & Sanchez, L. (2014). *Random Sets and Random Fuzzy Sets as Ill-Perceived Random Variables.* Springer. doi:https://doi.org/10.1007/978-3-319-08611-8

Dale, A. I. (1982). Bayes or Laplace? An Examination of the Origin and Early Applications of Bayes' Theorem. *Archive for History of Exact Sciences, 27*(1), pp. 23–47.

Dale, A. I. (1999). *A History of Inverse Probability—From Thomas Bayes to Karl Pearson* (2nd. ed.). New York: Springer.

De Finetti, B. (2017). *Theory of Probability—A Critical Introductory Treatment.* (A. M. Smith, Trans.) UK: John Wiley and Sons.

DeGroot, M. H. (2004). *Optimal Statistical Decisions.* New Jersey: John Wiley & Sons.

Dempster, A. (1967, April). Upper and Lower Probabilities Induced by a Multivalued Mapping. *Ann. Math. Statist., 38*(2), pp. 325–339,. doi:https://doi.org/10.1214/aoms/1177698950

Denoeux, T. (2019, June). Decision-making with belief functions: A review. *International Journal of Approximate Reasoning, 109*, 87–110. doi:https://doi.org/10.1016/j.ijar.2019.03.009

Denoeux, T. (2021). *Uncertainty Analysis using Belief Functions: Applications to Statistical Inference and Pattern Recognition.* Université de technologie de Compiègne, Department of Computer Science, Compiègne, France. Retrieved December 20, 2022, from https://www.hds.utc.fr/~tdenoeux/dokuwiki/_media/en/book_bf.pdf

Denœux, T., Younes, Z., & Abdallah, F. (2010). Representing uncertainty on set-valued variables using belief functions. *Artificial Intelligence, 174*(7–8), 479–499. doi:https://doi.org/10.1016/j.artint.2010.02.002

Dezert, J., & Smarandache, F. (2008). A new probabilistic transformation of belief mass assignment. *Fusion 2008: International Conference on Information Fusion*, (pp. 1410–1417). Retrieved December 19, 2022, from https://hal.archives-ouvertes.fr/hal-00304319/document

Dezert, J., & Smarandache, F. (2009). Transformations of belief masses into subjective probabilities. In J. Dezert, & F. Smarandache (Eds.), *Advances and Applications of DSmT for Information Fusion* (pp. 85–136). Rehoboth: American Research Press . Retrieved December 19, 2022, from https://www.researchgate.net/publication/306413841_Transformations_of_belief_masses_into_subjective_probabilities

Dezert, J., Smarandache, F., & Daniel, M. (2004). A Generalized Pignistic Transformation. In J. Dezert, & F. Smarandache (Eds.), *Advances and Applications of DSmT for Information Fusion* (pp. 143–153). Rehoboth: American Research Press. Retrieved December 19, 2022, from https://www.onera.fr/sites/default/files/297/C023-Dezert-Fusion2004Stockholm.pdf

Doré, P. E., Fiche, A., & Martin, A. (2010). Models of belief functions—Impacts for patterns recognitions. *13th International Conference on Information Fusion.* Edinburgh, UK. doi:https://doi.org/10.1109/ICIF.2010.5711936

Doré, P., Martin, A., Abi-Zeid, I., Jousselme, A., & Maupin, P. (2011a, January). Belief functions induced by multimodalprobability density functions, an application to the search and rescue problem. *RAIRO—Operations Research, 44*(4), 323–343. doi:https://doi.org/10.1051/ro/2011001

Doré, P., Osswald, C., Martin, A., Jousselme, A., & Maupin, P. (2011b). Continuous belief functions to qualify sensors performances. In W. Liu (Ed.), *Symbolic and Quantitative Approaches to Reasoning with Uncertainty. ECSQARU 2011* (pp. 350–361). Belfast, Ireland: Springer. doi:https://doi.org/10.1007/978-3-642-22152-1_30

Dubois, D., & Prade, H. (1986). A Set-Theoretic View of Belief Functions. *International Journal of General Systems, 12*(3), 193–226. doi:https://doi.org/10.1080/03081078608934937

Fetz, T., & Oberguggenberger, M. (2016, November). Imprecise random variables, random sets, and Monte Carlo simulation. *78*, 252–264. doi:https://doi.org/10.1016/j.ijar.2016.06.012

Fiche, A., Martin, A., Cexus, J.-C., & Khenchaf, A. (2010). Continuous belief functions and α-stable distributions. *13th Conference on Information Fusion.* Edinburgh. doi:https://doi.org/10.1109/ICIF.2010.5711934

Fox, C. R., & See, K. E. (2006). Belief and Preference in Decision Under Uncertainty. In D. Hardman, & L. Macchi (Eds.), *Thinking: Psychological Perspectives on Reasoning, Judgment and Decision Making.* John Wiley & Sons. doi:https://doi.org/10.1002/047001332X.ch14

Guan, J., & Bell, D. (1993). Discounting and Combination Operations in Evidential Reasoning. In D. Heckerman, & A. Mamdani (Ed.), *Ninth Conference on Uncertainty in Artificial Intelligence* (pp. 477–484). Washington, USA: Elsevier. doi:https://doi.org/10.1016/B978-1-4832-1451-1.50062-7

Halmos, P. (1978). *Measure theory.* Berlin, Heidelberg, New York: : Springer Verlag.

Halpern, J., & Fagin, R. (1992). Two views of belief: belief as generalized probability and belief as evidence. *Artificial Intelligence,, 54*, pp. 275–317.

He, Y. (2013). *Uncertainty Quantification and Data Fusion based on Dempster-Shafer Theory.* Florida State University, Mathematics. Florida State University Libraries.

Howson, C., & Urbach, P. (2006). *Scientific reasoning : the Bayesian approach* . Chicago: Open Court.

Hüllermeier, E., Kruse, R., & Hoffmann, F. (Eds.). (2010). Consonant continuous belief functions conflicts calculation. *IPMU'10: Proceedings of the Computational intelligence for knowledge-based systems design, and 13th international conference on Information processing and management of uncertainty* (pp. 706–715). Dortmund, Germany: Springer.

Jaffray, J.-Y., & Wakker, P. (1993). Decision making with belief functions: Compatibility and incompatibility with the sure-thing principle. *Journal of Risk and Uncertainty, 7*(3), 255–271. doi:https://doi.org/10.1007/BF01079626

Jaynes, E. (1989). *Clearing up Mysteries—The Original Goal.* (J. Skilling, Ed.) Dordrecht: Springer. doi:https://doi.org/10.1007/978-94-015-7860-8_1

Jaynes, E. (2003). *Probability Theory: The Logic of Science.* Cambridge: Cambridge University Press.

Jeffreys, H. (1939). *Theory of probability.* Oxford: University Press.

Klopotek, M. A., & Wierzchon, S. T. (1998). A New Qualitative Rough-Set Approach to Modeling Belief Functions. In L. Polkowski, & A. Skowron (Ed.), *Rough Sets and Current Trends in Computing, First International Conference, RSCTC'98* (pp. 346–354). Warzsaw, Poland: Springer. doi:https://doi.org/10.1007/3-540-69115-4_47

Laghmara, H., Laurain, T., Cudel, C., & Lauffenburger, J. P. (2020). Heterogeneous sensor data fusion for multiple object association using belief functions. *Information Fusion, 57,* 44–58. doi:https://doi.org/10.1016/j.inffus.2019.11.002

Laplace, P.-S. (1774). Memoire sur la probabilité des causes par les événements. *Memoires de Mathématique et de Physique, Presentés à l'Académie Royale des Sciences par divers Savans & lus dans ses Assemblées*, pp. 621–656.

Laplace, P.-S. (1891). *Oeuvres complètes* (Vol. 8). (A. d. Paris, Ed.) Paris: Gauthiers-Villars.

Laplace, P.-S. (1986). Memoir on the Probability of the Causes of Events. *Statistical Science, 1*(3), pp. 364–78. Retrieved from http://www.jstor.org/stable/2245476

Lian, C. (2017). *Information Fusion and Decision Making using Belief Functions.* Compiègne, France: Université de Technologie de Compiègne.

Liu, W. (2006, August). Analyzing the degree of conflict among belief functions. *Artificial Intelligence, 170* (11), 909–924. doi:https://doi.org/10.1016/j.artint.2006.05.002

Martin, A. (2019). Conflict management in information fusion with belief functions. In E. Bossé, & G. Rogova (Eds.), *Information quality in information fusion and decision making* (pp. 79–97). Springer. doi:https://doi.org/10.1007/978-3-030-03643-0_4

Martin, R., Zhang, J., & Liu, C. (2010, June). Dempster–Shafer Theory and Statistical Inference with Weak Beliefs. *Statistical Science, 25*(1), 72–87. doi:https://doi.org/10.1214/10-STS322

Mercier, D., Quost, B., & Denœux, T. (2005). Contextual Discounting of Belief Functions. In L. Godo (Ed.), *ECSQARU 2005: Symbolic and Quantitative Approaches to Reasoning with Uncertainty* (pp. 552–562). Barcelona, Spain: Springer. doi:https://doi.org/10.1007/11518655_47

Miranda, E., Couso, I., & Gil, P. (2005, July). Random sets as imprecise random variables. *Journal of Mathematical Analysis and Applications, 307*(1), 32–47. doi:https://doi.org/10.1016/j.jmaa.2004.10.022

Nguyen, H. T. (1977). *On Random sets and Belief Functions.* University of California, Berkeley, EECS Department. Retrieved from http://www2.eecs.berkeley.edu/Pubs/TechRpts/1977/28879.html

Nguyen, H. T. (1978). On Random Sets and Belief Functions. *Journal of Mathematical Analysis and Applications, 65*, 531–542. doi:https://doi.org/10.1016/0022-247X(78)90161-0

Nguyen, H., & Wang, T. (1997). Belief Functions and Random Sets. In J. Goutsias, R. Mahler, & H. Nguyen (Eds.), *Random Sets* (pp. 243–255). Springer. doi:https://doi.org/10.1007/978-1-4612-1942-2_11

Ni, S., Lei, Y., & Tang, Y. (2020). Improved Base Belief Function-Based Conflict Data Fusion Approach Considering Belief Entropy in the Evidence Theory. *Entropy, 22*(8). doi:https://doi.org/10.3390/e22080801

Press, S. J., & Tanur, J. M. (2001). *The Subjectivity of Scientists and the Bayesian Approach.* New York: John Wiley & Sons.

Risti, B., & Smets, P. (2006). Belief function theory on the continuous space with an application to model based classification. In B. Bouchon-Meunier, G. Coletti, & R. R. Yager (Ed.), *Modern Information Processing: From Theory to Applications. IPMU'04* (pp. 11–24). Perugia, Italy.: Elsevier. doi:https://doi.org/10.1016/B978-044452075-3/50002-9

Savchuk, V. P., & Tsokos, C. P. (2011). *Bayesian Theory and Methods with Applications.* Paris: Atlantis Press.

Shafer, G. (1976). *A Mathematical Theory of Evidence* . New Jersey: Princeton University Press.

Shafer, G. (1990). Perspectives on the Theory and Practice of Belief Functions. *International Journal of Approximate Reasoning, 4*(5–6), 323–362. doi:https://doi.org/10.1016/0888-613X(90)90012-Q

Smets, P. (1990). Constructing the Pignistic Probability Function in a Context of Uncertainty. *Machine Intelligence and Pattern Recognition, 10*, 29–39. doi:https://doi.org/10.1016/B978-0-444-88738-2.50010-5

Smets, P. (2000). Data fusion in the transferable belief model. *Proceedings of the Third International Conference on Information Fusion. 1*, pp. 21–33. Paris, France: IEEE. doi:https://doi.org/10.1109/IFIC.2000.862713.

Smets, P. (2005). Belief functions on real numbers. *International Journal of Approximate Reasoning, 40*, 181–223. doi:https://doi.org/10.1016/j.ijar.2005.04.001

Souza de Cursi, E. (2023). *Uncertainty Quantification using R.* Springer Cham.

Stigler, S. M. (1982). Thomas Bayes's Bayesian Inference. *Journal of the Royal Statistical Society. Series A (General), 145*(2), pp. .250–258.

Strat, T. (1984). Continuous belief functions for evidential reasoning. *Proceedings of the 4th National Conference on Artificial Intelligence.* Austin, Texas. Retrieved December 18, 2022, from https://www.aaai.org/Papers/AAAI/1984/AAAI84-035.pdf

Strat, T. (1987). The Generation of Explanations within Evidential Reasoning Systems. In J. P. McDermott (Ed.), *10th. International Joint Conference on Artificial Intelligence (IJCAI)*, (pp. 1097–1104). Milan, Italy. Retrieved December 20, 2022, from https://www.ijcai.org/Proceedings/87-2/Papers/104.pdf

Strat, T. (1990). Decision analysis using belief functions. *International Journal of Approximate Reasoning, 4*(5–6), 391–417. doi:https://doi.org/10.1016/0888-613X(90)90014-S

Taillandier, P., & Therond, O. (2011). Use of the Belief Theory to formalize Agent DecisionMaking Processes : Application to cropping Plan Decision Making. *European Simulation and Modelling Conference*, (pp. 138–142). Guimaraes, Portugal. Retrieved 12 17, 2022, from https://hal.archives-ouvertes.fr/hal-00688405

Wasserman, L. A. (1990, September). Belief functions and statistical inference. *Canadian Journal of Statistics, 18*(3), 183–196. doi:https://doi.org/10.2307/3315449

Wu, W.-Z., & Mi, J.-S. (2008). An Interpretation of Belief Functions on Infinite Universes in the Theory of Rough Sets. In C. Chan, J. Grzymala-Busse, & W. P. Ziarko (Ed.), *Rough Sets and Current Trends in Computing. RSCTC 2008* (pp. 71–80). Akron, OH, USA: Springer. doi: https://doi.org/10.1007/978-3-540-88425-5_8

Yao, Y. Y., & Lingras, P. J. (1998). Interpretations of Belief Functions in the Theory of Rough Sets. *Information Sciences, 104*(1–2), 81–106. doi:https://doi.org/10.1016/S0020-0255(97)00076-5

Zhang, H., & Deng, Y. (2020). Weighted belief function of sensor data fusion in engine fault diagnosis. *Soft Comput, 24*, 2329–2339. doi:https://doi.org/10.1007/s00500-019-04063-7

Zhou, K., Martin, A., & Pan, Q. (2018). A belief combination rule for a large number of sources. . *Journal of Advances in Information Fusion, 13*(2). Retrieved December 2022, 20, from https://www.researchgate.net/deref/https%3A%2F%2Fhal.archives-ouvertes.fr%2Fhal-01883239

Chapter 3
Information and Entropy

In the fields of Statistical Physics and Statistical Communications, one of the fundamental problems to be solved was the quantification of information – or, equivalently, the measurement of uncertainty. The approach introduced consisted of the construction of two functions, usually denoted h and H. The first one takes as argument <u>*an event*</u>: $h(E)$ measures the uncertainty associated with an event E having probability $p = P(E)$. The second one takes as argument a <u>*vector of probabilities*</u> $\boldsymbol{p} = (p_1, \ldots, p_n)$ and measures the average uncertainty associated with a collection of events: $H(\boldsymbol{p}) = H(p_1, \ldots, p_n)$ gives an evaluation of the mean uncertainty of events $E_1, \ldots, E_n$ such that $P(E_i) = p_i$, $1 \leq i \leq n$ – we have $H(p_1, \ldots, p_n) = p_1 h(E_1) + \ldots + p_n h(E_n)$. H and h are both usually referred to as entropy, *uncertainty function* or *uncertainty measure*.

The determination of these functions was carried out by Claude Elwood Shannon (Shannon 1948a, b), based on a previous work by Ralph Hartley (Hartley 1928). However, previous work by Max Karl Ernst Ludwig Planck (Planck 1901, 1914) introduced an analogous equation – Planck quoted that Ludwig Eduard Boltzmann introduced the connection between probability and entropy and used analogous formulae, namely in Boltzmann (1896, 1995) – see also Boltzmann (1877 traduction 2002, 1866). Boltzmann was inspired by Rudolf Clausius (Clausius 1850, 1851 republished in 2009). Josiah Willard Gibbs introduced analogous notions and equations by considering Hamiltonian systems and Liouville's theorem (Gibbs 1901 reprinted 1960).

We shall present in this chapter some elements concerning h and H. As observed, H has probabilities as arguments and h has events as arguments. However, as shown in this section, h is a function of $P(E) : h(E) = h(P(E))$. For this reason, we can also consider h as taking probabilities as arguments and use the notation $h(p)$ to refer to $h(E)$ such that $P(E) = p$.

E. Souza de Cursi, *Uncertainty Quantification with R*, International Series in Operations Research & Management Science 352,
https://doi.org/10.1007/978-3-031-48208-3_3

3.1 Entropy

In the sequel, $\boldsymbol{p} = (p_1, \ldots, p_n)$ denotes a probability vector:

Definition 3.1 – Let $\boldsymbol{p} = (p_1, \ldots, p_n) \in \mathbb{R}^n$. We say that $\boldsymbol{p}$ is a probability vector if and only if

$$\boldsymbol{p} \geq 0 \ (i.e., p_i \geq 0, 1 \leq i \leq n); \ \ \|p\|_1 = \sum_{i=1}^{n} p_i = 1.$$

We denote by ***supp*(*p*)** the *support* of **p**, i.e., the subvector of the indexes corresponding to strictly positive probabilities: ***supp*(*p*)** is formed by the indexes $\{i : 1 \leq i \leq n \text{ and } p_i > 0\}$, ordered increasingly: ***supp*(*p*)** $= (i_1, \ldots, i_m)$, $i_1 < i_2 < \ldots < i_m$ Analogously, $\boldsymbol{p}^+$denotes the subvector of the components having strictly positive probabilities: $\boldsymbol{p}^+ = \boldsymbol{p}_{\boldsymbol{supp(p)}}$.

Shannon (1948a, b) introduced three conditions to be satisfied by *H*:

- $\boldsymbol{p} = (p_1, \ldots, p_n) \rightarrow H(\boldsymbol{p})$ is continuous;
- $n \rightarrow f(n) = H(\boldsymbol{p} = (1/n, \ldots, 1/n))$ is monotonic increasing;
- If $p_2 = (1 - p_1)q_1, p_3 = (1 - p_1)q_2$, then $H(p_1, p_2, p_3) = H(p_1, 1 - p_1) + (1 - p_1) H(q_1, q_2)$.

The original formulation by Shannon is shown in Figs. 3.1 and 3.2.
Under these assumptions, Shannon established that the general form of *H* is

$$H(p_1, \ldots, p_n) = -C \sum_{i=1}^{n} p_i \log_b(p_i), C > 0, b > 1 \tag{3.1}$$

Thus, *h* is given by

$$h(E) = -C \log_b(P(E)), C > 0, b > 1. \tag{3.2}$$

h(*E*) measures the uncertainty of the event *E* or the information given by its realization. For instance, if *E* is almost sure, we have $P(E) = 1$, so that $h(E) = \log_b 1 = 0$ – there is no uncertainty about the realization of *E*, and its observation does not give any information. On the other extreme, if *E* is almost impossible, we have $P(E) = 0$, so that $h(E) = +\infty$ – there is an infinite uncertainty about the realization, and its observation gives infinite information. One unit of information in the logarithmic base *b* corresponds to the amount of information associated with an event of probability 1/*b*.

As previously observed, *h* depends upon the probability $p = P(E)$, so we can use the notation $h(p) = -C\log_b p$ to refer to $h(E) = -C\log_b(P(E) = p)$.

If there is such a measure, say $H(p_1, p_2, \ldots, p_n)$, it is perhaps reasonable to require of it the following properties:

1. H should be continuous in the p_i.
2. If all the p_i are equal, $p_i = \frac{1}{n}$, then H should be a monotonic increasing function of n. With equally likely events there is more choice, or uncertainty, when there are more possible events.
3. If a choice be broken down into two successive choices, the original H should be the weighted sum of the individual values of H. The meaning of this is illustrated in Fig. 6. At the left we have three possibilities $p_1 = \frac{1}{2}$, $p_2 = \frac{1}{3}$, $p_3 = \frac{1}{6}$. On the right we first choose between two possibilities each with probability $\frac{1}{2}$, and if the second occurs make another choice with probabilities $\frac{2}{3}$, $\frac{1}{3}$. The final results have the same probabilities as before. We require, in this special case, that

$$H(\tfrac{1}{2}, \tfrac{1}{3}, \tfrac{1}{6}) = H(\tfrac{1}{2}, \tfrac{1}{2}) + \tfrac{1}{2}H(\tfrac{1}{3}, \tfrac{2}{3})$$

The coefficient $\frac{1}{2}$ is because this second choice only occurs half the time.

Fig. 3.1 Original text by Claude Elwood Shannon. (From Shannon, A mathematical theory of communication – Nokia Bell Labs (s.d.))

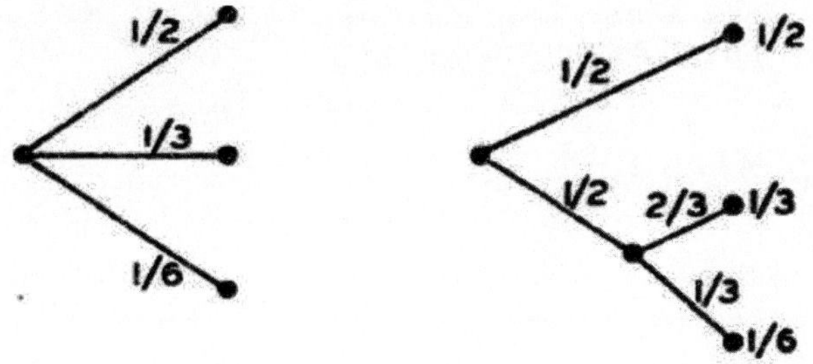

Fig. 3.2 Original Fig. 6 referred by Shannon. (From Shannon, A mathematical theory of communication – Nokia Bell Labs (s.d.))

The choice of b is, in principle, arbitrary. For instance, we can find in the literature works using the natural base $b = e$ – the corresponding unit of information is the *nat* – or the decimal base $b = 10$ – the unit of information is the *decit*, which was originally used by Hartley (1928) and also by Turing (1941), republished as Turing (2015). However, the most popular choice, made by Shannon himself, is the base $b = 2$ – the corresponding unity of information is the well-known *bit*.

The constant C is generally taken as the unity, and the corresponding entropy is usually called Shannon's entropy:

Definition 3.2 Let $\boldsymbol{p} = (p_1, \ldots, p_n)$. Shannon's entropy in logarithmic base $b > 1$ is given by

$$H(\boldsymbol{p}) = -\sum_{i \in supp(\boldsymbol{p})} p_i h(p_i), h(p) = -\log_b(p) \tag{3.3}$$

In most of the applications, the base is $b = 2$ (information measured in bits) or $b = e$ (information measured in nats).

Remark 3.1 It is usual to use, by convention that $0\log_b 0 = 0$ (what corresponds to the limit of $p\log_b p$ for $p \to 0+$). Under this convention, we can write

$$H(\boldsymbol{p}) = -\sum_{i=1}^{n} p_i h(p_i), h(p) = -\log_b(p) \tag{3.4}$$

In practice, the definition is applied to distributions formed by strictly positive values, i.e., such that $supp(\boldsymbol{p}) = (1, \ldots, n)$, i.e., $\boldsymbol{p}^+ = \boldsymbol{p}$:

$$p_i > 0, 1 \leq i \leq n. \tag{3.5}$$

In the sequel, we assume that this assumption is verified – except if the contrary is explicitly underlined.

To alleviate the notation, we do not write the base b in the sequel, except if it is necessary: we shall write log *instead* $\log_b$, *except if it is necessary to precise the value of b.*

We have:

Proposition 3.1 Let h, H be given by (3.3). Then

(i) $h(p) \geq 0$;
(ii) $p \rightarrow h(p)$ is monotonic strictly decreasing;
(iii) $A \subset B \Rightarrow h(A) \geq h(B)$;
(iv) $h(A \cup B) \leq \min \{h(A), h(B)\}$;
(v) If A and B are independent, then $h(A \cap B) = h(A) + h(B)$;
(vi) If $\boldsymbol{q} = (q_1, \ldots, q_n)$, with $|\boldsymbol{q}| = 1$, then

$$\sum_{i=1}^{n} p_i \log(p_i) \geq \sum_{i=1}^{n} p_i \log(q_i).$$

Equality holds if and only if $\boldsymbol{p} = \boldsymbol{q}$.
(vii) Let and $\boldsymbol{p} = (p_1, \ldots, p_n)$. Then $0 \leq H(\boldsymbol{p}) \leq \log n$.

Proof:

(i) Since the probabilities remain included in the interval (0,1), we have (i).
(ii) Since $b > 1$ and

$$h'(p) = -\frac{1}{p \log b} < 0$$

h is monotonically strictly decreasing.
(iii) results from (ii): $A \subset B \Rightarrow P(A) \leq P(B) \Rightarrow h(A) \geq h(B)$;
(iv) follows from (iv): $A \subset A \cup B$ yields $h(A \cup B) \leq h(A)$ and $B \subset A \cup B$ yields $h(A \cup B) \leq h(B)$.
(v) If A and B are independent, then $P(A \cap B) = P(A).\ P(B)$, so that $h(A \cap B) = -\log(P(A)) - \log(P(B)) = h(A) + h(B)$.
(vi) Let $f(x) = \log x - (x-1)$. We have $f(1) = 0$ and $f'(x) > 0$ for $0 < x < 1$, so that $f(x) < f(1) = 0$ for $0 < x < 1$. Analogously, $f'(x) < 0$ for $x > 1$, so that $f(x) < f(1) = 0$ for $x > 1$. Thus, $\log x \leq x - 1$, for $x > 0$, with a strict inequality for $x \neq 1$. Consequently,

$$\log\left(\frac{q_i}{p_i}\right) \leq \frac{q_i}{p_i} - 1 \Rightarrow \sum_{i=1}^{n} p_i \log\left(\frac{q_i}{p_i}\right) \leq 1 - 1 = 0,$$

and we have

$$\sum_{i=1}^{n} p_i \log(q_i) \leq \sum_{i=1}^{n} p_i \log(p_i).$$

In addition, if $\boldsymbol{p} \neq \boldsymbol{q}$, then $\exists i \in \{1, \ldots, n\} : p_i \neq q_i$. Thus, $q_i/p_i \neq 1 \Rightarrow \log\left(q_i/p_i\right) < q_i/p_i - 1$ and the inequality is strict.
(vii) $H(p_1, \ldots, p_n) \geq 0$ yields from (i). $H(p_1, \ldots, p_n) \leq \log_b n$ results from (iv) by taking $q_i = 1/n$, $1 \leq i \leq n$. ∎

Under R, we can use the command `Entropy` from package `DescTools` to determine the entropy of discrete distributions defined by contingency or frequency tables. For instance, let us consider two vectors:

```
x = c(1,2,3,3,3,3,2,1,3,3,2,1,1,4,4,3,1,2)
y = c(3,4,1,1,1,1,4,3,1,1,4,3,3,2,2,1,3,4)
```

The associated counts of a number of occurrences are:

```
table(x)
## x
## 1 2 3 4
## 5 4 7 2
```

```
table(y)
## y
## 1 2 3 4
## 7 2 5 4
```

We have $\boldsymbol{p_x} = (5, 4, 7, 2)/18$ and $\boldsymbol{p_y} = (7, 2, 5, 4)/18$, so that the entropy is the same for both the tables. We can evaluate them as follows:

```
library(DescTools)
Entropy(table(x),base = exp(1))
## [1] 1.301481
Entropy(table(x),base = 2)
## [1] 1.87764
Entropy(table(x),base = 10)
## [1] 0.565226
```

```
Entropy(table(y),base = exp(1))
## [1] 1.301481
Entropy(table(y),base = 2)
## [1] 1.87764
Entropy(table(y),base = 10)
## [1] 0.565226
```

A second package `entropy` also proposes tools for the evaluation of entropies:

```
library(entropy)
entropy(table(x),unit = "log")
## [1] 1.301481
entropy(table(x),unit = "log2")
## [1] 1.87764
entropy(table(x),unit="log10")
## [1] 0.565226
```

```
entropy(table(y),unit = "log")
## [1] 1.301481
entropy(table(y),unit = "log2")
## [1] 1.87764
entropy(table(y),unit="log10")
## [1] 0.565226
```

Package `infotheo` also proposes a command entropy that furnishes the result in nats.

```
library(infotheo)
entropy(x)
## [1] 1.301481
```

```
entropy(y)
## [1] 1.301481
```

Finally package `varrank` furnishes a result in bits:

```
library(varrank)
entropy.data(table(x))
## [1] 1.87764
```

```
entropy.data(table(y))
## [1] 1.87764
```

We can also create our own class to deal with entropy. For instance, class shannon below:

```
library(R6)
#
shannon=R6Class("shannon",
  private = list(
    sumlog = function(p,h){
      ind = which(p > 0)
      pp = p[ind]
      slog = 0
      for (i in (1: length(pp))){
        str = paste0(h,"(",pp[[i]],")")
        slog = slog + pp[[i]]*eval(parse(text=str))}
      return(slog)
    }
  ),
  public=list(
  freqs = NULL, data = NULL,
  probs = NULL, names = NULL,
  initialize=function(data){
   if( missing(data) == FALSE){
    if (data$type == "sample"){
      self$data = data$values
      aux = table(data$values)
      self$freqs = as.numeric(aux)
      self$names = names(aux)
      self$probs = self$freqs/sum(self$freqs)}
    else{
      if (data$type == "freqs"){
        self$freqs = data$values
        self$names = as.character(1:length(self$freqs))
        self$probs = self$freqs/sum(self$freqs)}
      else{
        if (data$type == "probs"){
          self$names = as.character(1:length(self$freqs))
          self$probs = self$freqs/sum(self$freqs)}
        else{
          if (data$type == "table"){
            self$names = names(data$values)
            self$freqs = data$values
            self$probs = self$freqs/sum(self$freqs)}}}}}
  },
  Hobj = function(sobj,h){
    if( missing(sobj) == TRUE){sobj = self}
    if ( missing(h) == TRUE ){h = "log"}
    etr = -private$sumlog(sobj$probs,h)
    return(etr)
  }
  )
)
```

We can use this class as follows:

```
source("shannon.R")
d1 = list(x,"sample")
names(d1) = c("values","type")
s1 = shannon$new(d1)
h1 = "log2"
s1$Hobj(h = h1)
## [1] 1.87764
```

```
t1 = table(x)
d2 = list(t1,"table")
names(d2) = c("values","type")
s2 = shannon$new(d2)
h2 = "log"
s2$Hobj(h = h2)
## [1] 1.301481
```

Since the works by Shannon, many other ones tend to establish an axiomatic theory of information, namely by determining what the axioms needed to define H and h – see, for instance, Brillouin (1956), Khinchin (1957), Rényi (1961), Ash (1965, reedited 2003), Nambiar et al. (1992), Chakrabarti and Indranil (2005), Kojadinovic et al. (2005), Beretta (2008), Zanchini and Beretta (2008), Csiszár (2008), and Hulse et al. (2018).

Let us highlight the work of Khinchin (1957), which proposed a different set of three axioms:

- For a given $n > 0$, $H(p_1,\ldots,p_n) \leq H(1/n,\ldots,1/n)$, $\forall\, \boldsymbol{p} = (p_1,\ldots,p_n)$, $\boldsymbol{p} \geq 0$, $|\boldsymbol{p}| = 1$;
- Let $AB = \{A_i \cap B_j : 1 \leq i \leq m, 1 \leq j \leq n\}$ be the outer product of $A = (A_1,\ldots,A_m)$ and $B = (B_1,\ldots,B_m)$. If A and B are independent, then $h(AB) = h(A) + h(B)$.

$$H(p_1,\ldots,p_n,0) = H(p_1,\ldots,p_n), \forall \boldsymbol{p} = (p_1,\ldots,p_n), \boldsymbol{p} \geq 0, |\boldsymbol{p}| = 1.$$

Under these assumptions, Khinchin (1957) established a result known as the uniqueness theorem, which shows that we must have (3.1) and (3.2).

Example 3.1 Let us consider a game with two players. In this game, there are 13 cards considered as equiprobable.

Player one picks a card of diamonds. Player 2 must find the card by asking questions to which the first player will answer with "yes" or "no." We desire to determine the minimal number of questions necessary to determine the card.

In this situation, the choice of Player 1 can be represented by a one-hot vector of dimension 13: $X = (x_1, \ldots, x_{13}) \in \{0, 1\}^{13}$, $\sum_{i=1}^{13} x_i = 1$ (a single x_i is equal to 1, the others are equal to zero). There are 13 possible choices. The initial uncertainty is $\log_2 13 \approx 3.7$ bits. Each answer corresponds to a value in $\{0,1\}$ ({"no","yes"}). Thus, it furnishes information $\log_2 2 = 1$ bit. Consequently, we shall need $3.7/1 = 3.7$ answers, i.e., four questions.

Indeed, we can split the set $E^{(0)}$, formed by the 13 diamond cards in two subsets $E_1^{(0)} = \{1, \ldots, 7\}$, $E_2^{(0)} = \{8, \ldots, 13\}$. We ask if the card is in the $E_1^{(0)}$.

(continued)

Example 3.1 (continued)

If yes, we eliminate $E_2^{(0)}$ – otherwise, we eliminate $E_2^{(0)}$. Whatever the result, we have a new set $E^{(1)}$ at most seven cards. We iterate by splitting it again into two subsets – in the worst case, $E_1^{(1)} = \{1,2,3,4\}, E_2^{(1)} = \{5,6,7\}$ and asking if the card is in $E_1^{(1)}$. As a result, we have a set $E^{(2)}$ of at most four cards, which can be split again – in the worst case, $E_1^{(2)} = \{1,2\}, E_2^{(2)} = \{3,4\}$. Asking a third question will reduce the search to one of the subsets, containing at most two cards. The last step splits this subset into two subsets containing one card each, and the fourth question will determine the card.

We can use R to determine the minimal number of questions as follows:

```
n = 13   # number of possible values
na = 2; # number of possible answers
nq = ceiling(logb(n,na))  # min number of questions
print(paste('min number of questions : ',nq))
## [1] "min number of questions :  4"
```

Example 3.2 Let us consider a set of 12 coins, all identical, except a fake one which is lighter than the others. We desire to determine the minimal number of weights necessary to identify the fake coin using a two-armed scale.

In this case, we have an initial uncertainty of $\log_2 12 \approx 3.6$ bits. For the weights, three results are possible: left is heavier, right is heavier, and both have an equal weight: one weight produces information equal to $\log_2 3 \approx 1.6$ bits (there are three possible results: heavier at left, heavier at right, equilibrated). Thus, we shall need $3.6/1.6 \approx 2.3$ weights, i.e., three weights (since two are not enough).

Indeed, we can partition the initial set $E^{(0)}$ of 12 coins in three subsets $E_1^{(0)}$, $E_2^{(0)}$, and $E_3^{(0)}$ having four coins each. We use the scale to compare $E_1^{(0)}$ and $E_2^{(0)}$. If one is lighter, we eliminate the two others. If they are equal, both are eliminated, and we keep $E_3^{(0)}$. Whatever the result, the search reduces to a subset $E^{(1)}$ of four coins. We split it into two subsets $E_1^{(1)}$ and $E_2^{(1)}$ containing two coins each, and we compare them on the scale. We select the lighter so that the final decision must be taken between two coins of a set $E^{(2)}$: the third weight determines the fake coin.

(continued)

Example 3.2 (continued)

We can use R to determine the minimal number of questions as follows:

```
n = 12  # number of possible values
na = 3; # number of possible answers
nq = ceiling(logb(n,na))  # min number of questions
print(paste('min number of questions : ',nq))
## [1] "min number of questions :  3"
```

Example 3.3 Let us consider a set of 12 coins, all identical, except a fake one which does not the same weight as the others. We desire to determine the minimal number of weights necessary to identify the fake coin using a two-armed scale.

In this case, we do not know if the fake coin is heavier or lighter than the others. Consequently, we have 24 possibilities since each coin can be a fake one (12 possibilities), and it can be heavier or lighter (two possibilities by coin). Thus, we have an initial uncertainty of $\log_2 24 \approx 4.6$ bits. Analogously to the preceding example, each weight produces an information equal to $\log_2 3 \approx 1.6$ bits. Thus, we shall need $4.6/1.6 \approx 2.9$ weights, i.e., three weights (again, two are not enough).

In this case, the solution is more complex, but it exists: we can partition the initial set $E^{(0)}$of 12 coins in three subsets $E_1^{(0)}$, $E_2^{(0)}$, and $E_3^{(0)}$ having four coins each. We use the scale to compare $E_1^{(0)}$ and $E_2^{(0)}$. There are two possible issues:

- If they have an equal weight, the fake is in $E_3^{(0)}$and the coins in $E_1^{(0)} \cup E_2^{(0)}$ are regular. We split $E_3^{(0)}$ in two subsets $E_{3,1}^{(1)}$ and $E_{3,2}^{(1)}$ of two coins. We use the scale to compare $E_{3,1}^{(1)}$ to two regular coins chosen in $E_1^{(0)} \cup E_2^{(0)}$. If $E_{3,1}^{(1)}$ does not equilibrate these two coins, the fake is in $E_{3,1}^{(1)}$, otherwise, it is in $E_{3,2}^{(1)}$. Whatever the result, the search is reduced to a subset $\{c_1, c_2\}$ of two coins. We compare c_1 to a regular one chosen in $E_1^{(0)} \cup E_2^{(0)}$: if the scale equilibrates, the fake is c_2. If not, it is c_1.
- If they do not equilibrate, the fake is either in $E_1^{(0)} = \{c_1, c_2, c_3, c_4\}$ or $E_2^{(0)} = \{c_5, c_6, c_7, c_8\}$, and the coins in $E_3^{(0)} = \{c_9, c_{10}, c_{11}, c_{12}\}$ are regular. We compare, on the scale, two subsets $E_1^{(1)}$ and $E_2^{(1)}$, such that $E_1^{(1)} = \{c_1, c_2, c_5\}$ contains two elements from $E_1^{(0)}$ and one element of

(continued)

Example 3.3 (continued)

$E_2^{(0)}$, while $E_1^{(1)} = \{c_3, c_6, c_{10}\}$ contains one element from each subset $E_1^{(0)}$, $E_2^{(0)}$, and $E_3^{(0)}$.

(i) If the scale equilibrates at the second weight, the fake is one of the coins that is not on the scale: it belongs to $\{c_4, c_7, c_8\}$. In this case, we can use the third weight to compare, on the scale, the subsets $\{c_4, c_7\}$ and $\{c_1, c_2\}$:

  (a) If the scale equilibrates, the fake is c_8;
  (b) If the heavier side is the same as in the first weight, this means that the fake coin did not change sides: it is c_4;
  (c) If the heavier side is different from the first weight, this means that the fake changed sides: it is c_7.

(ii) If the scale does not equilibrate at the second weight:

  (a) If the heavier side is the same as in the first weight, the fake did not change sides: it belongs to $\{c_1, c_2, c_6\}$ – in this case, the third weight compares $\{c_1, c_6\}$ to $\{c_{10}, c_{11}\}$. If the heavier side is always the same, the fake did not change sides: it is c_1, If the heavier side is not the same as in the preceding weights, this means that the fake changed sides: it is c_6. If the scale equilibrates, the fake is not on the scale: it is c_2.
  (b) If the heavier side is not the same as in the first weight, the fake changed sides: it belongs to $\{c_3, c_5\}$. The third weight compares c_3 and c_{10}: if the scale equilibrates, the fake is not on the scale – it is c_5. If not, the fake is on the scale: it is c_3.

We can use R to determine the minimal number of questions as follows:

```
n = 24  # number of possible valuess
na = 3; # number of possible answers
nq = ceiling(logb(n,na))  # min number of questions
print(paste('min number of questions : ',nq))
## [1] "min number of questions :  3"
```

Example 3.4 Let us consider a set of 16 balls, numbered 1 to 16. Someone chooses a ball, but not equiprobably: denoting by X the number of the ball, we have $P(X = i) = 1/2^i$, for $1 \leq i \leq 15$, $P(X = 16) = 1/2^{15}$.

We must determine the ball by asking questions where the answer is "yes" or "no." We can consider two strategies: either the strategy corresponding to the minimum number of questions to surely determine the ball;

(continued)

Example 3.4 (continued)

- *minmax*: the strategy minimizes the maximal number of questions: we look for the maximal number of questions that allows us to surely determine the ball;
- *minmean*: the strategy that minimizes the mean number of questions.

The first one is analogous to the preceding ones: we adopt a pessimistic point of view, by considering that, despite the probabilities given, any ball has the same probability, i.e., the entropy is supposed maximal: $\log_2 16 = 4$ bits. Thus, we adopt a strategy analogous to Example 3.1; the set of balls is successively divided into 2 halves containing 8, 4, 2, and 1 balls, and we ask if the ball is in the first one: the maximal number of questions necessary to retrieve the ball chosen is 4.

The second one uses the real entropy of the system:

$$\sum_{i=1}^{15} \frac{\log_2 2^i}{2^i} + \frac{\log_2 2^{15}}{2^{15}} = \frac{32767}{16384} \approx 1.99994.$$

Therefore, we can expect, in the meantime, to find the ball with two questions. In this case, the strategy consists of asking sequentially: Is it 1? Is it 2? Etc. Indeed, the maximum number of questions for this strategy is 15, but the mean number of questions necessary is

$$\sum_{i=1}^{15} \frac{i}{2^i} = \frac{65519}{32768} \approx 1.99948,$$

i.e., two questions.

We can use R to determine those results as follows:

```
n = 16  # number of possible valuess
na = 2; # number of possible answers
nq = ceiling(logb(n,na))  # min number of questions
print(paste('min number of questions : ',nq))
## [1] "min number of questions :  4"

x = 1:15
mnq = ceiling(sum(x/2^x))
print(paste('mean number of questions : ',mnq))
## [1] "mean number of questions :  2"
```

Example 3.5 Let us determine the entropy of a Bernoulli distribution $B(p)$. In this case, $\boldsymbol{p} = (p, 1 - p)$ and $H(\boldsymbol{p}) = -p \log p - (1 - p) \log (1 - p)$. Let us use R to plot the value of H for different values of p:

```
p = seq(0.01,0.99,length.out = 101)
H = numeric(length= length(p))
for (i in 1:length(p)){
  H[[i]] = -p[[i]]*log(p[[i]]) - (1-p[[i]])*log(1-p[[i]])
}
plot(p,H, type = 'n', main = "Entropy of Bernouilli",
     xlab = 'p',ylab='H')
lines(p,H,lwd=2)
```

The result appears in Fig. 3.2. The maximum is attained for $p = ½$ (Fig. 3.3).

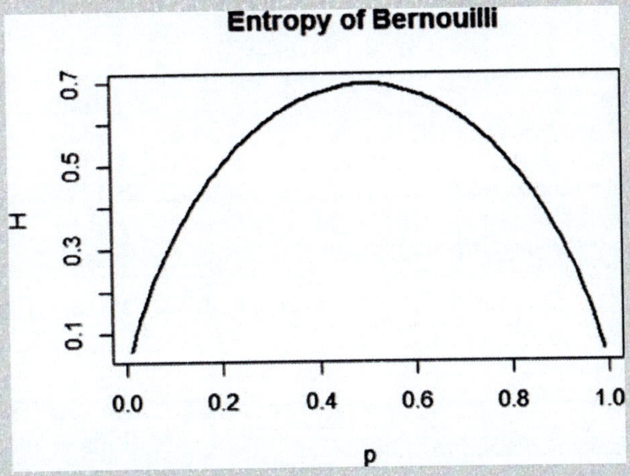

Fig. 3.3 Entropy of a Bernoulli distribution

Example 3.6 Let us use R to determine the entropy of a Binomial distribution $B(n, p)$ for different values of n with $p = 1/2$.

```
n = 100
nn = 1:n
H = numeric(length = n)
for (i in 1:n){
  s = 0
  for (j in 0:nn[[i]]){
    p = dbinom(j,nn[[i]],0.5)
    s = s + p*log(p)}
  H[[i]] = -s}
plot(nn,H, type = 'n', main = "Entropy of Binomial B(n,0.5
)",xlab = 'n',ylab='H')
lines(nn,H,lwd=2)
```

and for different values of p with n=20 :

```
p = seq(0.01,0.99,length.out = 101)
H = numeric(length= length(p))
nn = 20
for (i in 1:length(p)){
  s = 0
  for (j in 0:nn){
    pp = dbinom(j,nn,p[[i]])
    s = s + pp*log(pp)}
  H[[i]] = -s}
plot(p,H, type = 'n', main = "Entropy of Binomial B(n,p)",
     xlab = 'p',ylab='H')
lines(p,H,lwd=2)
```

The results appear in Fig. 3.4.

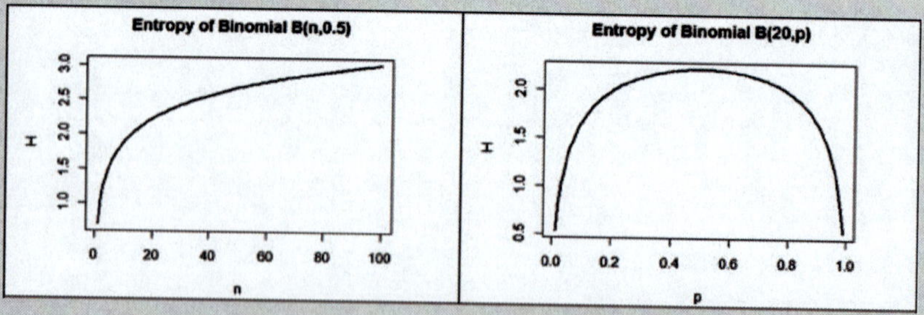

Fig. 3.4 Entropy of a binomial distribution

It is interesting to compare Shannon's entropy to the Boltzmann–Planck definition of entropy: $S_B(E) = k \log W(E)$, where $k \approx 1.381E - 23\ J/K$ is the Boltzmann constant, and W is the number of microstates that produce the macrostate E. If we consider that all the microstates of the system have the same probability (i.e., they are equiprobable and, so, uniformly distributed), then $W(E) = NP(E)$, where N is the number of all the possible microstates of the system. In the literature, you will find works showing that, for large N, we have $S_B(E) \approx kNP(E) \log P(E)$ – see, for instance, Dragulescu and Yakovenko (2000), Yakovenko and Rosser (2009), Yakovenko (2010), Sohrab (2014), and Fort (2022).

A second comparison can be made with Gibb's entropy $S_G(E) = -k\int_E p \log (p)\, dE$ – recall that Gibbs entropy is defined on the phase space of a system so that E is a subset of the phase space. Gibbs entropy is often written in discrete situations under the same form as Shannon's entropy – or sometimes as $S_G(E) = -k \sum_{A \in W(E)} P(A) \log P(A)$. Under this last form, Gibbs entropy appears as analogous to Shannon's, but some differences remain: Gibbs entropy is evaluated on a macrostate by summation on all the microstates, while Shannon's entropy applies directly to the set of all macrostates (recall that $|\boldsymbol{p}| = 1$). Of course, it is possible to consider a single macrostate as a universe and apply Shannon's definition to the microstates, considered as elementary events: in this case, the value furnished by Shannon is similar to Boltzmann's one (see example below).

In addition, a discretization of the Gibbs integral leads to $S(E) = -k\sum_i E_i p_i \log p_i$, where E_i, $1 \leq i \leq n$ is a partition of E. Such an expression is analyzed in Liboff (1974), which points out the differences with Shannon's entropy – see also Jaynes (1965).

Example 3.7 Let us toss six coins simultaneously. The coins are considered fair and independent. Let us consider the macroscopic variable $X =$ *number of Heads*. The microstates are the exact results. For instance, *HTTTTT* and *TTTTTH* are both microstates corresponding to $E =$ "$X = 1$". Let us evaluate the Boltzmann entropy for each value of X:

(continued)

Example 3.7 (continued)

```
n = 6  # number of coins
ph = 0.5  # probability of Head

# Find the number of microstates for X = i
w = numeric(length = n+1)
for (i in 0:n){
  w[[i+1]]= choose(n,i)
}
print(w)
## [1]  1  6 15 20 15  6  1

# Evaluate  Boltzmann's SB/k for X=i
SBk = log(w)
print(SBk)
## [1] 0.000000 1.791759 2.708050 2.995732 2.708050 1.7917
59 0.000000
```

Now let us determine Gibbs' entropy for each value of X:

```
# Evaluate Gibbs' SG/k for X = i
# find the probabilities of the microstates
pm = numeric(length = n+1)
for (i in 0:n){
  pm[[i+1]]= (ph^(i))*((1-ph)^(n-i))
}
print(pm)
## [1] 0.015625 0.015625 0.015625 0.015625 0.015625 0.0156
25 0.015625

#Determine the sum of Gibbs' h on the microstates
SG = -w*(pm*log(pm))
print(SG)
## [1] 0.06498255 0.38989529 0.97473822 1.29965096 0.97473
822 0.38989529 0.06498255
```

Now, let us consider "$X = i$" as a universe and apply Shannon's formula: we need to bring the global probability of "$X = i$" to 1:

(continued)

Example 3.7 (continued)

```
# Evaluate Shannon's entropy for X = i
# by considering the microstates
# find the probabilities of the macrostates
pmacro = w*pm
print(pmacro)
## [1] 0.015625 0.093750 0.234375 0.312500 0.234375 0.0937
50 0.015625
# satisfy the condition |p| = 1
pms = pm/pmacro
# evaluate Shannon's entropy on this universe
SEm = -w*(pms*log(pms))
print(SEm)
## [1] 0.000000 1.791759 2.708050 2.995732 2.708050 1.7917
59 0.000000
```

Let us compare these values to those furnished by Boltzmann: the values are equal.

```
# compare Shannon on X=i and Boltzmann
err = sqrt(sum((SEm - SBk)^2))
print(err)
## [1] 0
```

Let us evaluate Shannon's h in *nats*:

```
# Evaluate Shannon's h for X=I in nats
p = numeric(length = n+1)
for (i in 0:n){
  p[[i+1]]= dbinom(i,n,ph)
}
print(p)
## [1] 0.015625 0.093750 0.234375 0.312500 0.234375 0.0937
50 0.015625
h = -log(p)
print(h)
## [1] 4.158883 2.367124 1.450833 1.163151 1.450833 2.3671
24 4.158883
```

(continued)

Example 3.7 (continued)

Then, find Shannon's entropy for *X*:

```
# Find Shannon's entropy for X
SE = sum(p*h)
print(SE)
## [1] 1.617363
```

Compare it to the mean values of Boltzmann and Gibbs:

```
#Find the mean of Boltzmann values
SBkm = sum(pmacro*SBk)
print(SBkm)
## [1] 2.54152
#Find the mean of Gibbs values
SGm = sum(pmacro*SG)
print(SGm)

## [1] 0.9381855
```

Now, let us suppose that the coin is unfair, and the probability of a head is `ph = 0.6`. In this case:

```
print(SBk)
## [1] 0.000000 1.791759 2.708050 2.995732 2.708050 1.7917
59 0.000000

print(SG)
## [1] 0.02251876 0.18772178 0.64790519 1.18370739 1.20555
495 0.64766345 0.14299848

print(SEm)
## [1] 0.000000 1.791759 2.708050 2.995732 2.708050 1.7917
59 0.000000

print(SE)
## [1] 1.5927
print(SBkm)
## [1] 2.44537
print(SGm)
## [1] 0.9263673
```

(continued)

Example 3.7 (continued)

Boltzmann's entropy depends only on the number of microstates so that the corresponding values remain unmodified. Gibbs entropy depends on the probability of the microstates, so its values change. Analogously, Shannon's entropy has its values modified. The mean value of the Boltzmann entropies was modified because we used the probabilities generated by the microstates – in the original formulation, uniform distributions were considered.

In the sequel, we shall consider random variables. In such a situation, we can consider events connected to X, such as $E =$ "$X \leq x$", $E = X \in (a, b)$, and so on. For these events, we can determine the entropy $h(E) = h(P(E))$ – for instance, $h(X \leq x)$, $h(X \in (a, b))$. For a discrete random variable X taking the values $X_1, \ldots, X_n, \ldots$ with probabilities $P(X = X_i) = p_i$, we use Shannon's entropy as follows:

Definition 3.3 Let (Ω, P) be a probability space. Let X be a random variable defined on (Ω, P) such that $Im(X) = X(\Omega) = \{X_1, \ldots, X_n\}$, with $X_1 < X_2 < \ldots < X_n$. Let $P(X = X_i) = p_i$. The entropy of X is

$$H(X) = H(\boldsymbol{p}), \boldsymbol{p} = (p_1, \ldots, p_n). \tag{3.6}$$

Notice that the entropy of X is independent of the values X_i: it depends only on the probabilities p_i.

Remark 3.2 As indicated in Remark 1.1, we assume that all the probabilities are strictly positive, except if the contrary is explicitly indicated: Eq. (3.5) holds for all the probabilities considered in the sequel.

Let us denote

$$h_X(x) = h(X = x) = h(P(X = x)) \ . \tag{3.7}$$

We have:

Proposition 3.2 Let X be a random variable such that $Im(X) = X(\Omega) = \{X_1, \ldots, X_n\}$, with $X_1 < X_2 < \ldots < X_n$. Then

(i) $H(X) \geq 0$;
(ii) $H(X) \leq \log n$;
(iii) $H(X) = E(h(X))$.

(continued)

Proposition 3.2 (continued)

Proof:

(i) follows from proposition 3.1.(i).
(ii) results from proposition 3.1.(vii).
(iii) yields from (1.27). ∎

Entropies of discrete random variables can be managed by the class `drv`, introduced in Chap. 1, by adding a supplementary property `entropy`:

```
X=NULL,p=NULL,mean=NULL,m2=NULL,var=NULL,sd=NULL,
    cumul=NULL,med=NULL,phi=NULL,entropy = NULL,
```

which is evaluated as shown at right. This code must be added to the function `initialize` in class `drv`. It evaluates the entropy in nats.

```
pp = p[which(p>0)]
self$entropy = -sum(pp*log(pp))
```

To evaluate the entropy in a base b, we can use the method at right. It must be added to the class `drv`.

```
bentropy = function(b){
  v = self$entropy/log(b)
  return(v)}
```

We can use it as follows:

```
X = 1:4  # values of X
p = c(5,4,7,2)/18 # probabilities
source("drv.R")
va = drv$new(X,p)
```

```
va$entropy
## [1] 1.301481
va$bentropy(2)
## [1] 1.87764
va$bentropy(10)
## [1] 0.565226
```

Of course, we can also add to class `shannon` a method for the evaluation of the entropy of a discrete random variable:

```
HX = function(X,h){
  if( missing(X) == TRUE){p = self$probs}
  else{p=X$p}
  if ( missing(h) == TRUE ){h = "log"}
  etr = -private$sumlog(p,h)
  return(etr)
}
```

We can use it as follows:

```
X = 1:4  # values of X
p = c(5,4,7,2)/18 # probabilities
source("drv.R")
va = drv$new(X,p)
source("shannon.R")
s1 = shannon$new()
```

```
s1$HX(va,"log")
## [1] 1.301481
s1$HX(va,"log2")
## [1] 1.87764
s1$HX(va,"log10")
## [1] 0.565226
```

Equations (3.6) and (3.7) extend to a continuous random variable X having a density f_X as:

Definition 3.4 Let (Ω, P) be a probability space. Let X be a random variable defined on (Ω, P) having f_X as PDF. The entropy of X is

$$h_X(x) = h(f_X(x)), H(X) = E(h_X(X)) = \int_{\mathbb{R}} f_X(x) h(f_X(x)) dx \tag{3.8}$$

Remark 3.3 Contrarily to $H(X)$ for a discrete random variable, Eq. (3.8) can produce negative values (see examples below).

Indeed, Eq. (3.8) is not a continuous form of Eq. (3.3), as it can be easily seen by considering a random variable X uniformly distributed on (0,1): assume that the interval (0, 1) is discretized in N subintervals of equal length $h = 1/N$. We have $P(X \in (ih, ih + h)) = h$, $0 \le i \le N - 1$. Let us consider an approximated random variable X_N corresponding to such a discretization – for instance, $P(X_N = ih - h/2) = h$, $1 \le i \le N$. Then

$$H(X_N) = -\sum_{i=1}^{N} h \log h = -Nh \log h = -\log h = \log N.$$

Consequently, $H(X_N) \to +\infty$, for $N \to +\infty$, while $H(X) < \infty$ (see Example 3.8). Sometimes, H given by Eq. (3.8) is referred to as *differential entropy*, to underline the differences with the entropy for discrete random variables.

Analogously to Remark 1.1, Eq. (3.8) must be understood as an integration on

$$supp(f) = \{x \in \mathbb{R} : f(x) > 0\},$$

which corresponds to the convention that $0 \log 0 = 0$. In the sequel, we assume that $f_X > 0$ on $Im(X) = X(\Omega)$, except if the contrary is explicitly indicated. For instance, if $Im(X) = (a, b)$, then

(continued)

Definition 3.4 (continued)

$$f_X(x) > 0 \; on \; (a, b); H(X) = \int_a^b f_X(x) h(f_X(x)) dx \tag{3.9}$$

This situation can be managed by the class `crv`, introduced in Chap. 1.

We can introduce a new property `entropy`, which can be evaluated in nats as shown at right. Here, we use Romberg's method, but other methods are available in `ints.R`

```
g = function(x){f(x)*log(f(x))}
r = super$romb1(g,a,b,500,1e-10)
self$entropy = -r
```

Transformation to a different base is carried by the method of `bentropy` previously introduced.

Alternatively, we can add to class `shannon` the corresponding method:

```
DH1 = function(f,a,b){
  g = function(x){f(x)*log(f(x))}
  r = super$romb1(g,a,b,500,1e-10)
  return(-r)},
DHX1 = function(X){
  entr = self$DH1(X$pdf,X$lower,X$upper)
  return(entr)}
```

Finally, H can be evaluated independently from classes as follows:

```
source("ints.R")
df = function(f,param,lims){
  g = function(x){f(x,param)*log(f(x,param))}
  r = ints$public_methods$romb1(g,lims[[1]],lims[[2]],100,1e-8)
  return(-r)}
```

Example 3.8 Let us consider a variable uniformly distributed on (a, b). Then, $Im(X) = (a, b)$ and

$$f(x) = \frac{1}{b-a}, if \; x \in (a, b); f(x) = 0, otherwise.$$

(continued)

Example 3.8 (continued)

Let us evaluate the entropy of this distribution in nats:

$$h(f(x)) = -\log f(x) = \log(b-a)$$

and

$$H(X) = \int_a^b \frac{\log(b-a)}{b-a} dx = \log(b-a)$$

If $(b - a) < 1$, then $H(X) < 0$. Let us use R to evaluate such an entropy. First, we use the class `crv`: let F, f be the CDF and the PDF associated with the uniform distribution on (a, b) – they can be generated by

```
F = function(x){punif(x,a,b)}
```

```
f = function(x){dunif(x,a,b)}
```

Then, we create an instance of `crv` and evaluate its entropy:

```
source("crv.R")
va1 = crv$new(F,f,-1,1)
va1$entropy
## [1] 0.6931472
```

```
va2 = crv$new(F,f,0,0.5)
va2$entropy
## [1] -0.6931472
```

We can also use the class `shannon`:

```
s1 = shannon$new()
s1$DHX1(va1)
## [1] 0.6931472
```

```
s1$DHX1(va2)
## [1] -0.6931472
```

Now, let us make a direct evaluation using `df`:

```
f = function(x, param){dunif(x, param[[1]],param[[2]])}
```

```
ab = c(-1,1)
e1 = df(f,ab,ab)
e1
## [1] 0.6931472
```

```
ab = c(0,0.5)
e2 = df(f,ab,ab)
e2
## [1] -0.6931472
```

Example 3.9 Let us consider a Gaussian variable $N(m, \sigma)$. We have

$$f(x) = \frac{1}{\sigma\sqrt{2\pi}} \exp\left(-\frac{1}{2}\left(\frac{x-m}{\sigma}\right)^2\right).$$

Let us evaluate the entropy of this distribution in nats:

$$h(f(x)) = -\log f(x) = \frac{1}{2}\left(\frac{x-m}{\sigma}\right)^2 + \underbrace{\log\sigma + \frac{1}{2}\log(2\pi)}_{C}$$

and

$$H(X) = \frac{1}{2}\underbrace{\int_{-\infty}^{+\infty}\left(\frac{x-m}{\sigma}\right)^2 f(x)dx}_{=1} + C\underbrace{\int_{-\infty}^{+\infty} f(x)dx}_{=1},$$

So that

$$H(X) = C + \frac{1}{2} = \log\sigma + \frac{1}{2}(1 + \log(2\pi)).$$

This expression is often written as

$$H(X) = \log\left(\sigma\sqrt{2\pi e}\right).$$

If $\sigma\sqrt{2\pi e} < 1$, then $H(X) < 0$. Let us use R to evaluate such an entropy: let F, f be the CDF and the PDF associated with $N(m, \sigma)$ – they can be generated by `pnorm` and `dnorm`.

```
source("crv.R")
va1 = crv$new(F,f,-5,5)
va1$entropy
## [1] 1.41893
```

```
va2 = crv$new(F,f,-4,6)
va2$entropy
## [1] -0.1904994
```

With the class `shannon`:

```
s1 = shannon$new()
s1$DHX1(va1)
## [1] 1.41893
```

```
s1$DHX1(va2)
## [1] -0.1904994
```

Using df:

```
f = function(x, param){ dnorm(x, param[[1]],param[[2]])}
```

```
ms = c(0,1)
ab = c(-5,5)
e1 = df(f,ms,ab)
e1
## [1] 1.41893
```

```
ms = c(1,0.2)
ab = c(-4,6)
e2 = df(f,ms,ab)
e2
## [1] -0.1904994
```

The reader will find a Table giving $H(X)$ for most of the classical distributions (Lazo and Rathie 1978).

Exercises

1. Player 1 rolls n equilibrated six-faced dice simultaneously and keeps the sum of points of the faces up into a variable X. Player 2 must find that sum by asking questions that have an answer "yes" or "no." Use R to determine the minimal number of questions to surely determine the value of X for $n = 1, \ldots, 1000$.
2. Player 1 tosses a fair coin n times and keeps the number of heads into a variable X. Player 2 must find that value by asking questions that have an answer "yes" or "no." Use R to determine the minimal number of questions to surely determine the value of X for $n = 1, \ldots, 1000$.
3. Player 1 tosses an unfair coin 100 times and keeps the number of heads into a variable X. Player 2 must find that value by asking questions that have an answer "yes" or "no." Use R to determine the mean number of questions in the strategy *min mean* to determine the value of X for different values of p.
4. n six-faced dice are rolled simultaneously. Let the macrostate X be the number of "6" obtained and the microstates be the results of each face at the roll. Assume that the dices are equilibrated. Use R to determine, for $n = 1, \ldots, 10$

   (a) the probabilities of the microstates
   (b) the probabilities of the macrostates
   (c) Boltzmann's entropy of each macrostate
   (d) Gibbs' entropy of each macrostate
   (e) Shannon's entropy of each macrostate, considered as a universe
   (f) Shannon's entropy of X

5. A coin is tossed n times. Let the macrostate X be the number of heads and the microstates be the results of the tosses. Assume that the coin is fair. Use R to determine, for $n = 1, \ldots, 20$

(continued)

(a) the probabilities of the microstates
(b) the probabilities of the macrostates
(c) Boltzmann's entropy of each macrostate
(d) Gibbs' entropy of each macrostate
(e) Shannon's entropy of each macrostate, considered as a universe.
(f) Shannon's entropy of X

6. An urn contains balls of three colors: blue (B), red (R), and white (W). Three balls are drawn sequentially, with the remittance of the first ball in the urn. We consider that $P(\{B\}) = 0.2$, $P(\{R\}) = 0.3$, *and* $P(\{W\}) = 0.5$. Let the macrostate X be the number of red balls drawn and the microstates be the results of the draws.

   (a) Determine the probabilities of the microstates
   (b) Determine the probabilities of the macrostates
   (c) Determine Boltzmann's entropy of each macrostate
   (d) Determine the Gibbs' entropy of each macrostate
   (e) Determine Shannon's entropy of each macrostate, considered as a universe
   (f) Determine Shannon's entropy of X

7. Determine the entropy of a negative binomial distribution $NB(n, ½)$, for $n = 1, \ldots, 10$. $P(X = k) = \binom{k+n-1}{k} p^n (1-p)^k$.
8. Determine the differential entropy of an exponential distribution of parameter >0: $f_X(x) = \lambda e^{-\lambda x}$, *if* $x \geq 0$; $f_X(x) = 0$, *otherwise.*
9. Determine the differential entropy of a Rayleigh's distribution of parameter : $f_X(x) = x \exp\left(-x^2/2\sigma^2\right)/\sigma^2$, *if* $x \geq 0$; $f_X(x) = 0$, *otherwise.*

3.2 Joint and Conditional Entropy

Definition 3.5 Let (Ω, P) be a probability space. The joint entropy of two events $A, B \subset \Omega$, is

$$h(A, B) = -\log(P(A \cap B)). \tag{3.10}$$

The entropy of A conditionally to B is

$$h(A|B) = -\log(P(A|B)). \tag{3.11}$$

(continued)

Definition 3.5 (continued)

Let X, Y be discrete random variables defined on (Ω, P), such that $X(\Omega) = \{X_1, \ldots, X_m\}$ and $Y(\Omega) = \{Y_1, \ldots, Y_n\}$. The joint entropy of X and Y is

$$H(X, Y) = -\sum_{\substack{i=1,\ldots,m \\ j=1,\ldots,n}} p_{ij} \log p_{ij}, p_{ij} = P(X = X_i, Y = Y_j). \tag{3.12}$$

The conditional entropy of X given that $Y = Y_j$ is

$$H(X|Y = Y_j) = -\sum_{i=1,\ldots,m} P(X = X_i|Y = Y_j) \log P(X = X_i|Y = Y_j). \tag{3.13}$$

The conditional entropy of X given Y is

$$H(X|Y) = -\sum_{\substack{i=1,\ldots,m \\ j=1,\ldots,n}} p_{ij} \log P(X = X_i|Y = Y_j) \tag{3.14}$$

Remark 3.4 As indicated in the previous remarks 1.1, we assume that all the probabilities are strictly positive, except if the contrary is explicitly said: In reality, all the sums relate to the supports of the probability distributions involved.

We have:

Proposition 3.3 Let X, Y be random variable such that $Im(X) = X(\Omega) = \{X_1, \ldots, X_m\}$, with $X_1 < X_2 < \ldots < X_m$; $Im(Y) = Y(\Omega) = \{Y_1, \ldots, Y_n\}$, with $Y_1 < Y_2 < \ldots < Y_n$. Then

(i) $h(A|B) \geq 0$, $H(X, Y) \geq 0$, $H(X|Y) \geq 0$, $H(X|Y = Y_j) \geq 0$
(ii) $h(A|B) = h(A \cap B) - h(B)$
(iii) *A and B are independent* $\Leftrightarrow h(A|B) = h(A)$
(iv) $H(X, Y) \leq H(X) + H(Y)$. Equality holds if and only if X and Y are independent
(v) $H(X, Y) = H(X) + H(Y|X) = H(Y) + H(X|Y) = H(Y, X)$
(vi) $H(Y|X) \leq H(Y)$ and $H(X|Y) \leq H(X)$. Equality holds if and only if X and Y are independent

(continued)

Proposition 3.3 (continued)

Proof:

(i) yields from the fact that probabilities belong to (0, 1)
(ii) follows from $P(A \cap B) = P(B)P(A|B)$
(iii) results from the fact that $P(A \cap B) = P(B)P(B)$ if and only if A and B are independent
(iv) $H(X) = -\sum_{i=1}^{m} p_{i\blacksquare} \log p_{i\blacksquare} = -\sum_{i=1}^{m}\sum_{j=1}^{n} p_{ij} \log p_{i\blacksquare}$ (see (1.58)). Analogously, $H(Y) = -\sum_{i=1}^{m}\sum_{j=1}^{n} p_{ij} \log p_{\blacksquare j}$. Thus

$$H(X) + H(Y) = -\sum_{i=1}^{m}\sum_{j=1}^{n} p_{ij} \log q_{ij}, q_{ij} = p_{i\blacksquare} p_{\blacksquare j}$$

It follows from proposition 3.1.(vii) that $H(X, Y) \le H(X) + H(Y)$. In addition, equality holds if and only if $q_{ij} = p_{ij}$, $\forall\, i, j$, i.e., if the variables are independent (see Eq. (1.80))

We have $p_{ij} = p_{i\blacksquare} P(Y = Y_j | X = X_i)$. Thus,

$$\log p_{ij} = \log p_{i\blacksquare} + \log P(Y = Y_j | X = X_i)$$

Thus,

$$H(X, Y) = -\sum_{i=1}^{m}\sum_{j=1}^{n} p_{ij} \left(\log p_{i\blacksquare} + \log P(Y = Y_j | X = X_i)\right)$$

Since

$$\sum_{i=1}^{m}\sum_{j=1}^{n} p_{ij} \log p_{i\blacksquare} = \sum_{i=1}^{m} p_{i\blacksquare} \log p_{i\blacksquare} = -H(X),$$

We have $H(X, Y) = H(X) + H(Y|X)$.

Analogously, $p_{ij} = p_{\blacksquare j}\, P(X = X_i | Y = Y_j)$, which yields $H(X, Y) = H(X) + H(Y|X)$

(v) follows from (iv) and (v)■

The package infotheo proposes the function condentropy to evaluate conditional entropies. Otherwise, conditional entropies can be handled by the class drc, introduced in Chap. 1: we add a new property entropy, which will be a list formed by five elements, named XY, X, Y, X2Y, and Y2X, corresponding to $H(X, Y)$, $H(X)$, $H(Y)$, $H(X|Y)$, and $H(Y|X)$, respectively. The evaluation of these quantities is added to the method initialize of the class – recall that the class evaluates the conditional and marginal distributions, so that we have only to include the evaluations of Eqs. (3.12), (3.14), and the equivalent of this last equation with an exchange of the places between X and Y. Analogously, we can add to class `shannon` one or several methods for the evaluation of these quantities. We included such properties in `drc` and a method `HXY` in `shannon`.

Example 3.10 Let us consider the couple of random variables presented in Example 1.13. We have

$$H(R) = -0.1\log(0.1) - 0.9\log(0.9) \approx 0.325083$$

$$H(W) = -0.25\log(0.25) - 0.75\log(0.75) \approx 0.5623351$$

$$H(R,W) = -0.08\log(0.08) - 0.02\log(0.02) - 0.17\log(0.17) - 0.73\log(0.73) \approx 0.8112703$$

$$H(R|W) = H(R,W) - H(W) \approx 0.2489352$$

$$H(W|R) = H(R,W) - H(R) \approx 0.4861873$$

Representing "rainy" by 0, "not rainy" by 1, "windy" by 0, and "not windy" by 1, we have

```
R = c(0,1)
W = c(0,1)
p = matrix(c(0.08,0.02,0.17,0.73), nrow=length(R), ncol=length(W
),byrow=TRUE)
```

Using the class `drc` directly:

```
source("drc.R")
cva1 = drc$new(R,W,p)
cva1$entropy$XY
## [1] 0.8112703
cva1$entropy$X
## [1] 0.325083
```

```
cva1$entropy$Y
## [1] 0.5623351
cva1$entropy$X2Y
## [1] 0.2489351
cva1$entropy$Y2X
## [1] 0.4861873
```

Using the class `shannon`:

```
source("shannon.R")
s1 = shannon$new()
e1 = s1$HXY(cva1,"log")
e1$XY
## [1] 0.8112703
e1$X
## [1] 0.325083
```

```
e1$Y
## [1] 0.5623351
e1$X2Y
## [1] 0.2489351
e1$Y2X
## [1] 0.4861873
```

Example 3.11 Let us consider the couple of random variables presented in Example 1.21. We obtain

```
X = c(0,1)
Y = c(-1,0,1)
p = matrix(c(0.2,0.15,0.1,0.1,0.25,0.2), nrow=length(X), ncol=le
ngth(Y),byrow=TRUE)
```

We obtain:

```
source("drc.R")
cva1 = drc$new(X,Y,p)
cva1$entropy$XY
## [1] 1.735434
cva1$entropy$X
## [1] 0.6881388
```

```
cva1$entropy$Y
## [1] 1.0889
cva1$entropy$X2Y
## [1] 0.6465338
cva1$entropy$Y2X
## [1] 1.047295
```

and

```
source("shannon.R")
s1 = shannon$new()
e1 = s1$HXY(cva1,"log")
e1$XY
## [1] 1.735434
e1$X
## [1] 0.6881388
```

```
e1$Y
## [1] 1.0889
e1$X2Y
## [1] 0.6465338
e1$Y2X
## [1] 1.047295
```

Equations (3.12)–(3.14) extend to continuous random variables: let (X, Y) be a couple having a joint PDF $f(x, y)$: we have

$$H(X,Y) = -\int_{\mathbb{R}^2} f(x,y)\log f(x,y)\,dx\,dy\,.$$

Let the density of Y conditional to X be $f(y|x)$. We have

$$H(Y|X) = -\int_{\mathbb{R}^2} f(x,y)\log f(y|x)\,dx\,dy.$$

Analogously to the discrete case, the evaluation of these quantities can be managed by the class `crc`, introduced in Chap. 1, by adding a new property `entropy`, containing the values of $H(X, Y)$, $H(X)$, $H(Y)$, $H(X|Y)$, and $H(Y|X)$ and the corresponding evaluations in `initialize`. Alternatively, we can implement methods for their evaluation in class `shannon`. We included such properties in `crc` and a method `DHXY` in `shannon`.

Example 3.12 Let us consider the couple $(X, Y) \in (0, 1)^2$ considered in Example 1.23. Their joint PDF is

$$f(x, y) = x + y \text{ on } (0, 1)^2$$

The marginal PDFs are

$$f_X(x) = x + \frac{1}{2} \text{ on } (0, 1); \; f_Y(y) = y + \frac{1}{2} \text{ on } (0, 1).$$

The conditional PDFs are, for $0 \leq x, y \leq 1$:

$$f(x|Y = y) = 2\frac{x + y}{2y + 1}, f(y|X = x) = 2\frac{x + y}{2x + 1}.$$

Then

$$H(X, Y) = -\int_0^1 dx \int_0^1 (x + y)\log(x + y)dy = \frac{1}{6}(5 - \log 256),$$

$$H(X|Y) = -\int_0^1 dx \int_0^1 (x + y)\log\left(2\frac{x + y}{2y + 1}\right)dy = \frac{1}{3} - \frac{7}{3}\log 2 + \frac{9}{8}\log 3,$$

$$H(Y|X) = -\int_0^1 dx \int_0^1 (x + y)\log\left(2\frac{x + y}{2x + 1}\right)dy = \frac{1}{3} - \frac{7}{3}\log 2 + \frac{9}{8}\log 3,$$

$$H(X) = -\int_0^1 \left(x + \frac{1}{2}\right)\log\left(x + \frac{1}{2}\right)dx = \frac{1}{2} + \log 2 - \frac{9}{8}\log 3,$$

$$H(Y) = -\int_0^1 \left(y + \frac{1}{2}\right)\log\left(y + \frac{1}{2}\right)dx = \frac{1}{2} + \log 2 - \frac{9}{8}\log 3.$$

(continued)

Example 3.12 (continued)

Thus,

$$H(X, Y) \approx -0.09086291,$$
$$H(X|Y) = H(Y|X) \approx -0.0480713$$
$$H(X) = H(Y) \approx -0.04279164$$

Let us evaluate these quantities using the modified class `crc` – we used Romberg's method for the evaluation of the integrals:

```
F = function(x,y){0.5*(x^2*y+y^2*x)}
      f = function(x,y){x+y}
a = c(0,0) #Lower bounds for x and y
b = c(1,1) #Upper bounds for x and y
```

```
source("crc.R")
va1 = crc$new(F,f,a,b)
va1$entropy
## $XY
## [1] -0.09086295
## $X
## [1] -0.04279164
```

```
## $Y
## [1] -0.04279164
## $X2Y
## [1] -0.0480713
## $Y2X
## [1] -0.0480713
```

Now, let us use the class `shannon`:

```
source("shannon.R")
s1 = shannon$new()
s1$DHXY(va1)
## $XY
## [1] -0.09086295
## $X
## [1] -0.04279164
```

```
## $Y
## [1] -0.04279164
## $X2Y
## [1] -0.0480713
## $Y2X
## [1] -0.0480713
```

Exercises

1. Let us consider a couple of Bernoulli variables $(In, Out) \in \{0, 1\}^2$ such that

$$P(Out = 0|In = 0) = 1 - \varepsilon, P(Out = 1|In = 0) = \varepsilon,$$
$$P(Out = 0|In = 1) = \varepsilon, P(Out = 1|In = 1) = 1 - \varepsilon.$$

(continued)

Let $p = P(In = 1)$. Thus, $P(In = 0) = 1 - p$.

(a) Determine $H(In)$
(b) Determine the distribution of Out
(c) Determine $H(Out)$
(d) Determine the distribution of the couple (In, Out)
(e) Determine $H(In, Out)$
(f) Determine $H(Out| In)$
(g) Determine the conditional distribution $P(In| Out)$
(h) Determine $H(In| Out)$

2. Let us consider a couple of variables $(In, Out) \in \{0, 1, 2, 3\}^2$. The joint distribution of the couple is given in Table 3.1.

(a) Determine $H(In, Out)$
(b) Determine the marginal distributions of In and Out.
(c) Determine $H(In)$ and $H(Out)$.
(d) Determine the conditional distributions $P(In| Out)$ and $P(Out| In)$.
(e) Determine $H(Out| In)$ and $H(In| Out)$

Table 3.1 Distribution of (In, Out)

In \ Out	0	1	2	3
0	1/8	1/16	1/32	0
1	1/16	1/8	1/16	1/32
2	1/32	1/16	1/8	1/16
3	0	1/32	1/16	1/8

3. Let us consider a variable $X \in \{0, 1, 2, .., 7\}$ having the distribution below

X	0	1	2	3	4	5	6	7
$P(X = i)$	1/4	1/16	1/8	1/16	1/4	1/16	1/8	1/16

Let $Y = X \bmod 2$ (remainder of the division by 2)

(a) Determine $H(X)$
(b) Determine the distribution of Y
(c) Determine $H(Y)$
(d) Determine the conditional distribution $P(Y| X)$
(e) Determine $H(Y| X)$
(f) Determine $H(X, Y)$
(g) Determine the conditional distribution $P(X| Y)$
(h) Determine $H(X| Y)$

(continued)

4. An urn contains an equal number of red (R) and blue (B) balls. The balls are numbered: 25 % of the red balls have an odd number, while only 10 % of the blue balls have an odd number. Let X describe the color of the ball: $X = 0$, if the ball is red; $X = 1$, if the ball is blue. Let Y describe the number on the ball: $Y = 0$, if the number is a pair; $Y = 1$, if the number is odd.

   (a) Determine the conditional distribution $P(Y| X)$
   (b) Determine the distribution of the couple (X, Y)
   (c) Determine the conditional distribution $P(X| Y)$
   (d) Determine $H(X)$, $H(Y)$, $H(X, Y)$, $H(Y| X)$, $H(X| Y)$

   Let $\boldsymbol{X} = (X_1, X_2)$ be a pair of random variables having density

$$f(x_1, x_2) = \begin{cases} x_1 + x_2, \text{ if } 0 < x_1, x_2 < 1; \\ 0, \text{ otherwise}. \end{cases}.$$

   Determine $H(X)$, $H(Y)$, $H(X, Y)$, $H(Y| X)$, $H(X| Y)$.

5. Let $\boldsymbol{X} = (X_1, X_2)$ be a pair of random variables having density

$$f(x_1, x_2) = \begin{cases} \dfrac{2}{(x_2 + 1)^3}, \text{ if } 0 < x_1 < x_2; \\ 0, \text{ otherwise}. \end{cases}.$$

   Determine $H(X)$, $H(Y)$, $H(X, Y)$, $H(Y| X)$, $H(X| Y)$.

6. Let (X_1, X_2) be a pair of non-negative independent random variables such that

$$f_1(x_1) = \lambda e^{-\lambda x_1}, f_2(x_2) = \mu e^{-\mu x_2}.$$

   Let $X = X_1 - X_2$ and $Y = X_1 + X_2$.

   (a) Find the distribution of the couple (X, Y)
   (b) Determine $H(X)$, $H(Y)$, $H(X, Y)$, $H(Y| X)$, $H(X| Y)$

3.3 Mutual Information and Relative Entropy

Definition 3.6 Let (Ω, P) be a probability space. Let $\boldsymbol{p} = (p_1, \ldots, p_n)$, $\boldsymbol{q} = (q_1, \ldots, q_n)$. The relative entropy of $\boldsymbol{p}$ with respect to $\boldsymbol{q}$ is

$$D\left(\boldsymbol{p}\middle\|\boldsymbol{q}\right) = -H(\boldsymbol{p}) - \sum_{i=1}^{n} p_i \log q_i = \sum_{i=1}^{n} p_i \log\left(\frac{p_i}{q_i}\right) \tag{3.15}$$

(continued)

Definition 3.6 (continued)

Let X, Y be discrete random variables defined on (Ω, P), such that $X(\Omega) = \{X_1, \ldots, X_n\}$ and $Y(\Omega) = \{Y_1, \ldots, Y_n\}$. Let p_X be the distribution of X, and p_Y be the distribution of Y. The relative entropy of X with respect to Y is

$$D\left(X\middle\|Y\right) = D\left(p_X\middle\|p_Y\right). \tag{3.16}$$

Sometimes D is called *Kullback–Leibler divergence from* $\boldsymbol{p}$ *to* $\boldsymbol{q}$ (from X to Y, resp.) or *information gain from* $\boldsymbol{q}$ *to* $\boldsymbol{p}$ (from Y to X, resp.) – $\boldsymbol{p}$ is supposed to be the "true" distribution, which is compared to an arbitrary distribution $\boldsymbol{q}$. It was introduced by Kullback and Leibler (1951) – see also Kullback (1951 reed. 1969 reprint 1979).

Remark 3.5 The definition supposes that $supp(\boldsymbol{p}) \subset supp(\boldsymbol{q})$ and the convention that $0 \log 0 = 0$. Indeed, if $q_i = 0$ for some $i \in supp(\boldsymbol{p})$, the value of $p_i \log(q_i)$ is undefined: all the sums must be understood as related to $supp(\boldsymbol{p})$.

In addition, $D(\boldsymbol{p}\|\boldsymbol{q})$ is defined for a probability vector of the same length $n > 0$. If $\boldsymbol{p} = (p_1, \ldots, p_m)$ and $\boldsymbol{q} = (q_1, \ldots, q_n)$, with $m < n$, we can consider an extension $\tilde{\boldsymbol{p}} = (\tilde{p}_1, \ldots, \tilde{p}_n)$ such that $\tilde{p}_i = p_i, 1 \leq i \leq m$ and $\tilde{p}_i = 0, i > m$ – then, $supp(\tilde{\boldsymbol{p}}) \subset supp(\boldsymbol{q})$ and

$$\sum_{i=1}^{n} \tilde{p}_i \log q_i = \sum_{i=1}^{m} p_i \log q_i,$$

Recalling the convention that $0 \log 0 = 0$, we can define

$$D\left(\boldsymbol{p}\middle\|\boldsymbol{q}\right) = D(\tilde{\boldsymbol{p}}\,\|\boldsymbol{q}) = -H(\tilde{\boldsymbol{p}}) - \sum_{i=1}^{m} p_i \log q_i.$$

Since $H(\tilde{\boldsymbol{p}}) = H(\boldsymbol{p})$, we obtain

$$D\left(\boldsymbol{p}\middle\|\boldsymbol{q}\right) = -H(\boldsymbol{p}) - \sum_{i=1}^{m} p_i \log q_i = \sum_{i=1}^{m} p_i \log\left(\frac{p_i}{q_i}\right).$$

This approach can find an interpretation in terms of narrowing the probabilities on a subset and analyzing the resulting gain of information.

(continued)

Remark 3.5 (continued)

An analogous extension can be considered in the situation where $X(\Omega) = \{X_1, \ldots, X_m\}$ and $Y(\Omega) = \{Y_1, \ldots, Y_n\}$, with $m < n$: we can consider $D\left(X \middle\| Y\right) = D\left(\tilde{p_X} \middle\| p_Y\right)$.

Notice that $D\left(\boldsymbol{q} \middle\| \tilde{\boldsymbol{p}}\right)$ (and $D(Y\|X)$) are meaningless if $supp(\boldsymbol{q}) \neq supp\ (\boldsymbol{p})$: if $q_i > 0$ for some $i \notin supp\ (\boldsymbol{p})$, the value of $q_i \log(\tilde{p}_i)$ is undefined.

Definition 3.7 Let (Ω, P) be a probability space. Let X, Y be discrete random variables defined on (Ω, P), such that $X(\Omega) = \{X_1, \ldots, X_m\}$ and $Y(\Omega) = \{Y_1, \ldots, Y_n\}$. Let $p_{k\ell} = P(X = X_k, Y = Y_\ell)$ be the joint distribution of the couple (X, Y), $p_{k\blacksquare} = P(X = X_k)$ be the marginal distribution of X and $p_{\blacksquare\ell} = P(Y = Y_\ell)$ be the marginal distribution of Y. The mutual information between X and Y is

$$H(X, Y) = H(X) + H(Y) - H(X, Y). \tag{3.17}$$

We have:

Proposition 3.4 Let X be a random variable such that $Im(X) = X(\Omega) = \{X_1, \ldots, X_n\}$, with $X_1 < X_2 < \ldots < X_n$. Then

(i) $D(\boldsymbol{p}\|\boldsymbol{q}) \geq 0$, with equality if and only if $\boldsymbol{p} = \boldsymbol{q}$

(ii) $H(\boldsymbol{q}) - H(\boldsymbol{p}) = D(\boldsymbol{p} \mid |\boldsymbol{q}) + \sum_i (p_i - q_i) \log q_i$

(iii) Let $\boldsymbol{q}$ be the uniform distribution. Then, $H(\boldsymbol{q}) - H(\boldsymbol{p}) = D(\boldsymbol{p}\|\boldsymbol{q})$

(iv) $D(X\|Y) \geq 0$, with equality if and only if X and Y have the same distribution

(v) $I(X, Y)$ verifies

$$I(X, Y) = \sum_{\substack{k = 1, \ldots, m \\ \ell = 1, \ldots, n}} p_{k\ell}(\log(p_{k\ell}) - \log(p_{k\blacksquare}) - \log(p_{\blacksquare\ell}))$$

(continued)

Proposition 3.4 (continued)

and

$$I(X,Y) = \sum_{\substack{k=1,\ldots,m \\ \ell=1,\ldots,n}} p_{k\ell} \log\left(\frac{p_{k\ell}}{p_{k\bullet}p_{\bullet\ell}}\right) \tag{3.18}$$

(vi) $I(X,Y) \geq 0$, with equality if and only if X and Y are independent
(vii) $I(X,Y) = I(Y,X)$
(viii) $I(X,Y) = H(X) - H(X|Y) = H(Y) - H(Y|X)$

Proof:

(i) follows from proposition 3.1.(vi)
(ii) We have

$$H(\boldsymbol{q}) - H(\boldsymbol{p}) = \sum_i (p_i \log p_i - q_i \log q_i) \ .$$

Since

$$p_i \log p_i - q_i \log q_i = p_i \log\left(\frac{p_i}{q_i}\right) + (p_i - q_i) \log q_i,$$

We have

$$H(\boldsymbol{q}) - H(\boldsymbol{p}) = D(\boldsymbol{p} \mid |\boldsymbol{q}) + \sum_i (p_i - q_i) \log q_i$$

(iii) If $q_i = 1/n$, $\forall\, i$, then $\sum_i (p_i - q_i) \log q_i = 0$ and we have the result stated in the proposition

(continued)

Proposition 3.4 (continued)

(iv) results from (i)

(v) is a consequence of the equalities

$$\sum_{\substack{k=1,\ldots,m \\ \ell=1,\ldots,n}} p_{k\ell} \log(p_{k\blacksquare}) = \sum_{k=1,\ldots,m} p_{k\blacksquare} \log(p_{k\blacksquare}) = H(X),$$

$$\sum_{\substack{k=1,\ldots,m \\ \ell=1,\ldots,n}} p_{k\ell} \log(p_{\blacksquare\ell}) = \sum_{\ell=1,\ldots,n} p_{\blacksquare\ell} \log(p_{\blacksquare\ell}) = H(Y),$$

$$\sum_{\substack{k=1,\ldots,m \\ \ell=1,\ldots,n}} p_{k\ell} \log(p_{k\ell}) = H(X, Y).$$

(vi) yields from proposition 3.1.(vi) (or from (i)) and (v): on the one hand, $I(X, Y) \geq 0$; on the other hand, $I(X, Y) = 0$ if and only if $p_{k\ell} = p_{k\blacksquare} p_{\blacksquare\ell}, \forall\, k, \ell$

(vii) is immediate

(viii) follows from Proposition 3.3.(v)∎

Remark 3.6 This definition supposes the convention that $0 \log 0 = 0$. Indeed, if $q_i = 0$ for some $i \in$ *supp* ($\boldsymbol{p}$), the value of $p_i \log (q_i)$ is undefined: all the sums must be understood as related to *supp*($\boldsymbol{p}$).

Analogously to the preceding, the evaluation of the relative entropy and the mutual information can be included in the class `drc` or in the class `shannon`. We included properties `mut` and `kld` in `drc` and methods having the same names in `shannon`.

Example 3.13 Suppose that a fair coin is tossed. We define $X = face\ up$, $Y = face\ down$ .Their joint PDF is

$$P(X=H, Y=H) = P(X=T, Y=T) = 0,$$

$$P(X=T, Y=H) = P(X=H, Y=T) = \frac{1}{2}$$

The marginal PDFs are

$$P(X=H) = P(Y=H) = P(X=T) = P(Y=T) = \frac{1}{2}.$$

Thus:

$$D(X\,||Y) = D(Y\,||X) = \frac{1}{2}\log\frac{1/2}{1/2} + \frac{1}{2}\log\frac{1/2}{1/2} = 0.$$

$$I(X, Y) = \frac{1}{2}\log_2\frac{1/2}{(1/2).(1/2)} + \frac{1}{2}\log_2\frac{1/2}{(1/2).(1/2)} = 1bit.$$

Notice that $H(X) = H(Y) = 1$ *bit* and $H(X|Y) = H(Y|X) = 0$: we have $H(X, Y) = H(X) + H(Y|X) = 1$ *bit*, and $H(X) + H(Y) - H(X, Y) = 1$ *bit*.

Let us use R to make these calculations: we start by using the class `drc`:

```
X = c(0,1)
Y = c(0,1)
p = matrix(c(0,0.5,0.5,0), nrow=length(X), ncol=length(Y),byrow=
TRUE)
source("drc.R")
cva1 = drc$new(X,Y,p)
list(unlist(cva1$entropy)/log(2))
## [[1]]
##  XY   X   Y X2Y Y2X
##   1   1   1   0   0
cva1$mut/log(2)
## [1] 1
list(unlist(cva1$kld)/log(2))
## [[1]]
## X2Y Y2X
##   0   0
```

(continued)

Example 3.13 (continued)

Now, let us use the class shannon:

```
source("shannon.R")
s1 = shannon$new()
list(unlist(s1$HXY(cva1,"log2")))
## [[1]]
##  XY   X   Y X2Y Y2X
##   1   1   1   0   0
s1$mut(cva1,"log2")
## [1] 1
list(unlist(s1$kld(cva1,"log2")))
## [[1]]
## X2Y Y2X
##   0   0
```

Example 3.14 Suppose that an equilibrated six-faced dice is rolled. We define $X = \textit{face up}$, $Y = \textit{face down}$ .Their joint PDF is

$$P(X=1, Y=6) = P(X=6, Y=1) = \frac{1}{6},$$

$$P(X=2, Y=5) = P(X=5, Y=2) = \frac{1}{6},$$

$$P(X=3, Y=4) = P(X=4, Y=3) = \frac{1}{6}.$$

The other results have a probability equal to zero. Thus,

$$P(X=i) = P(Y=j) = \frac{1}{6}., 1 \leq i, j \leq 6.$$

Consequently:

$$D(X\,||Y) = D(Y\,||X) = 6\log_2\frac{1/6}{1/6} = 0.$$

$$I(X, Y) = \log_2\frac{1/6}{(1/6).(1/6)} = \log_2 6 \approx 2.58\ \textit{bits}.$$

Notice that $H(X) = H(Y) = \log_2 6$ *bits* and $H(X|Y) = H(X|Y) = 0$: we have $H(X, Y) = H(X) + H(Y|X) = \log_2 6$ *bits*, and $H(X) + H(Y) - H(X, Y) = \log_2 6$ *bits*. Using R:

(continued)

Example 3.14 (continued)

```
X = 1:6
Y = 1:6
p = matrix(0, nrow=length(X), ncol=length(Y),byrow=TRUE)
p[[1,6]] = 1/6
p[[6,1]] = 1/6
p[[2,5]] = 1/6
p[[5,2]] = 1/6
p[[3,4]] = 1/6
p[[4,3]] = 1/6

source("drc.R")
cva1 = drc$new(X,Y,p)
list(unlist(cva1$entropy)/log(2))
## [[1]]
##       XY        X        Y      X2Y      Y2X
## 2.584963 2.584963 2.584963 0.000000 0.000000
cva1$mut/log(2)
## [1] 2.584963
list(unlist(cva1$kld)/log(2))
## [[1]]
## X2Y Y2X
##   0   0

source("shannon.R")
s1 = shannon$new()
list(unlist(s1$HXY(cva1,"log2")))
## [[1]]
##       XY        X        Y      X2Y      Y2X
## 2.584963 2.584963 2.584963 0.000000 0.000000
s1$mut(cva1,"log2")
## [1] 2.584963
list(unlist(s1$kld(cva1,"log2")))
## [[1]]
## X2Y Y2X
##   0   0
```

Example 3.15 Let us consider the couple (R,W) of random variables presented in Example 1.13 and reconsidered in Example 3.10. We have

$$P(0,0)=0.08, P(0,1)=0.02, P(1,0)=0.17, P(1,1)=0.73$$

$$P(R=0)=0.1, P(R=1)=0.9, P(W=0)=0.25, P(W=1)=0.75.$$

(continued)

Example 3.15 (continued)

Thus,

$$D(X\mid|Y) = 0.1\log_2\left(\frac{0.1}{0.25}\right) + 0.9\log_2\left(\frac{0.9}{0.75}\right) \approx 0.105\ bits$$

$$D(Y\mid|X) = 0.25\log_2\left(\frac{0.25}{0.1}\right) + 0.75\log_2\left(\frac{0.75}{0.9}\right) \approx 0.133\ bits$$

$$I(X,\mathrm{Y}) = 0.08\log_2\left(\frac{0.08}{0.025}\right) + 0.02\log_2\left(\frac{0.02}{0.075}\right) + 0.17\log_2\left(\frac{0.17}{0.225}\right)$$
$$+ 0.73\log_2\left(\frac{0.73}{0.675}\right) \approx 0.110\ bits$$

Using R:

```
X = c(0,1)
Y = c(0,1)
p = matrix(c(0.08,0.02,0.17,0.73), nrow=length(X), ncol=length(Y
),byrow=TRUE)

source("drc.R")
cva1 = drc$new(X,Y,p)
list(unlist(cva1$entropy)/log(2))
## [[1]]
##        XY         X         Y       X2Y       Y2X
## 1.1704156 0.4689956 0.8112781 0.3591375 0.7014200
cva1$mut/log(2)
## [1] 0.1098581
list(unlist(cva1$kld)/log(2))
## [[1]]
##       X2Y       Y2X
## 0.1045382 0.1332062

source("shannon.R")
s1 = shannon$new()
list(unlist(s1$HXY(cva1,"log2")))
## [[1]]
##        XY         X         Y       X2Y       Y2X
## 1.1704156 0.4689956 0.8112781 0.3591375 0.7014200
s1$mut(cva1,"log2")
## [1] 0.1098581
list(unlist(s1$kld(cva1,"log2")))
## [[1]]
##       X2Y       Y2X
## 0.1045382 0.1332062
```

Relative entropy and mutual information extend to continuous variables as follows: for two PDFs f and g, we have

$$D(f\,||g) = \int_{\mathbb{R}} f(x) \log f(x)\, dx - \int_{\mathbb{R}} f(x) \log g(x)\, dx,$$

i.e.,

$$D(f\,||g) = \int_{\mathbb{R}} f(x) \log\left(\frac{f(x)}{g(x)}\right) dx.$$

For two random variables X and Y having joint PDF $f_{XY}(x, y)$ and marginal PDFs f_X, f_Y, we have

$$I(X, Y) = \int_{\mathbb{R}^2} f_{XY}(x, y) \log\left(\frac{f_{XY}(x, y)}{f_X(x) f_Y(y)}\right) dxdy,$$

and

$$D(X\,|\,|Y) = D(f_X\,|\,|f_Y).$$

We have

$$D(X\,||Y) = -H(X) - \int_{\mathbb{R}} f_X(x) \log f_Y(x)\, dx = \int_{\mathbb{R}} f_X(x) \log\left(\frac{f_X(x)}{f_Y(x)}\right) dx.$$

Remark 3.7 Analogously to the discrete case, the definition of $D(f||g)$ supposes that $supp(f) \subset supp\,(g)$. Indeed, if $x \in supp\,(f)$ and $x \notin supp\,(g)$, then $f(x) \log g(x)$ is undefined. Consequently, all the integrals must be understood as integrations on the support of f, under the assumption that $supp(f) \subset supp\,(g)$. This condition implies, for instance that $X(\Omega) \subset Y(\Omega)$.

If $supp(f_X) = (a, b) \subset supp\,(f_Y)$, then

$$D(X\,||Y) = \int_a^b f_X(x) \log\left(\frac{f_X(x)}{f_Y(x)}\right) dx.$$

Again, the evaluation of these quantities can be made by adding properties to the class `crc` or methods to the class `shannon`. We included properties `mut` and `kld` in `crc` and methods `mutXY` and `kldXY` in `shannon`.

Example 3.16 Let us consider the couple (X, Y) formed by two independent uniform variables: X on (a_X, b_X) and Y on (a_Y, b_Y). Since the variables are independent, the joint PDF is $f_{XY}(x, y) = f_X(x)f_Y(y)$, so that $I(X, Y) = 0$.

If $a_Y < a_X < b_X < b_Y$, we have $X(\Omega) \subset Y(\Omega)$, so that we can evaluate $D(X||Y)$: as previously explained, we can consider extending f_X to (a_Y, b_Y) by $f_X(x) = 0$ outside (a_X, b_X). Then (recall that we use the convention that $0 \log 0 = 0$),

$$D(X\,||Y) = \int_{a_X}^{b_X} f_X(x) \log\left(\frac{f_X(x)}{f_Y(x)}\right) dx = \log\left(\frac{b_Y - a_Y}{b_X - a_Y}\right).$$

In this situation, $D(Y\,|\,|X)$ is meaningless: for $a_Y < x < a_X, f_X(x) = 0$, while $f_Y(x) = 1/(b_Y - a_Y) > 0$ and $f_Y(x) \log\left(f_Y(x)/f_X(x)\right)$ cannot be evaluated.

Example 3.17 Let us consider the couple (X, Y) formed by two independent normally distributed variables: $X \sim N(m_X, \sigma_X)$ and $\sim N(m_Y, \sigma_Y)$. Both the variables take their values on R, so that we can evaluate

$$D(X\,||Y) = \int_{-\infty}^{+\infty} f_X(x) \log\left(\frac{f_X(x)}{f_Y(x)}\right) dx.$$

We have

$$\frac{f_X(x)}{f_Y(x)} = \frac{\sigma_Y}{\sigma_X} \exp\left(-\frac{1}{2}\left(\left(\frac{x - m_X}{\sigma_X}\right)^2 - \left(\frac{x - m_Y}{\sigma_Y}\right)^2\right)\right).$$

Thus,

$$D(X\,||Y) = -\frac{1}{2} \int_{-\infty}^{+\infty} f_X(x)\left(\left(\frac{x - m_X}{\sigma_X}\right)^2 - \left(\frac{x - m_Y}{\sigma_Y}\right)^2\right) dx + \log\left(\frac{\sigma_Y}{\sigma_X}\right),$$

and

$$D(X\,|\,|Y) = \frac{1}{2} \int_{-\infty}^{+\infty} \left(\frac{x - m_Y}{\sigma_Y}\right)^2 f_X(x)\, dx - \frac{1}{2} + \log\left(\frac{\sigma_Y}{\sigma_X}\right),$$

(continued)

Example 3.17 (continued)

i.e.,

$$D(X \mid |Y) = \frac{\sigma_X^2 + (m_X - m_Y)^2}{2\sigma_Y^2} - \frac{1}{2} + \log\left(\frac{\sigma_Y}{\sigma_X}\right).$$

Example 3.18 Let us consider the couple $(X, Y) \in (0, 1)^2$ considered in Example 1.23 and reconsidered in Example 3.12. Their joint PDF is

$$f(x, y) = x + y \ on \ (0, 1)^2$$

The marginal PDFs are

$$f_X(x) = x + \frac{1}{2} \ on \ (0, 1); \ f_Y(y) = y + \frac{1}{2} \ on \ (0, 1).$$

Then

$$I(X, Y) = \int_0^1 dx \int_0^1 (x + y) \log\left(\frac{x + y}{\left(x + \frac{1}{2}\right)\left(y + \frac{1}{2}\right)}\right) dy,$$

Thus,

$$I(X, Y) = \frac{1}{6} + \frac{10}{3} \log 2 - \frac{9}{4} \log 3 \approx 0.0053 \ nats$$

Since both X and Y take their values on (0,1), we can evaluate $D(X \mid |Y)$ and $D(Y \mid |X)$, but these values are both equal to zero. Indeed, $f_X(x) = f_Y(x)$, $\forall \, x$, so that $\log(f_X(x)/f_Y(x)) = \log(f_Y(y)/f_X(y)) = 0$

Let us evaluate these quantities using the modified class `crc` – we used Romberg's method for the evaluation of the integrals:

```r
F = function(x,y){0.5*(x^2*y+y^2*x)}
f = function(x,y){x+y}
a = c(0,0) #Lower bounds for x and y
b = c(1,1) #Upper bounds for x and y
```

```r
source("crc.R")
va1 = crc$new(F,f,a,b)
va1$mut
## [1] 0.005279659
va1$kld
## $X2Y
## [1] 0
## $Y2X
## [1] 0
```

```r
source("shannon.R")
s1 = shannon$new()
s1$mutXY(va1)
## [1] 0.005279659
s1$kldXY(va1)
## $X2Y
## [1] 0
## $Y2X
## [1] 0
```

Exercises

1. A geological cavity exists at a depth between 5 and 20 meters. We look for its position in the set {5, 6, . . ., 20}, initially supposed equiprobable.

   A first measurement shows that the depths 5 to 9 and 19 to 20 have a probability $1/34$ each, while the others have a probability $3/34$ each.

   A second measurement made at another point shows that the depths 5 to 11 and 20 have a probability $1/64$ each, while the others have a probability $7/64$ each.

   (a) Determine the initial uncertaintyH_0 before any measurement
   (b) Determine the uncertainty H_1 after the first measurement. Find its relative entropy with respect to the initial state
   (c) Determine the uncertainty H_2 associated with the second measurement taken alone. Find its relative entropy with respect to the initial state
   (d) Consider the two measurements as a couple (X, Y) of independent variables. Determine the information $H(X, Y)$ associated with it
   (e) Determine the relative entropy from Y to X
   (f) Determine the information gain from a pair of uniformly districted depths to the couple of two measurements(X,Y)

(continued)

2. An equilibrated dice is rolled. Let $X = face\ up$, $Y = face\ at\ the\ right\ vertical\ side$.

   (a) Determine the joint distribution of (X, Y)
   (b) Determine the marginal distributions of X and Y
   (c) Determine the uncertainties $H(X, Y)$, $H(X)$, $H(Y)$
   (d) Determine the uncertainties $H(X| Y)$, $H(Y| X)$
   (e) Determine the mutual information between X and Y
   (f) Determine the relative entropy from Y to X and from X to Y

3. A population is divided into two groups. in group 1, 75 % always answer the truth, 20 % always lie, and 5 % always refuse to answer; in group 2, the proportions are 20 %, 75 %, and 5 %, respectively. Let us denote by p the proportion of group 1 – then, the proportion of group 2 is $q = 1 - p$. Let us denote by X the group (1 or 2) of an individual and by Y the response ("truth"=1, "lie"=2, "and refuse"=3).

   (a) Determine the entropy internal to each group
   (b) Determine the joint distribution of (X, Y) as a function of p
   (c) Determine the marginal distribution of Y as a function of p
   (d) Determine the entropy of the population as a function of p
   (e) Determine the mutual information between X and Y, as a function of p

4. Let us consider a couple of variables $(In, Out) \in \{0, 1, 2, 3\}^2$ having the distribution given in Table 3.1.

   (a) Determine the mutual information between *In* and *Out*
   (b) Determine the relative entropy from *In* to *Out*
   (c) Determine the relative entropy from *Out* to *In*

5. An urn contains an equal number of red (R) and blue (B) balls. The balls are numbered: 25 % of the red balls have an odd number, while only 10 % of the blue balls have an odd number. Let X describe the color of the ball: $X = 0$, if the ball is red; $X = 1$, if the ball is blue. Let Y describe the number on the ball: $Y = 0$, if the number is a pair; $Y = 1$, if the number is odd.

   (a) Determine the mutual information between X and Y
   (b) Determine the relative entropy from X to Y
   (c) Determine the relative entropy from Y to X

6. Let $\boldsymbol{X} = (X_1, X_2)$ be a pair of random variables having density

$$f(x_1, x_2) = \begin{cases} x_1 + x_2, \textit{if } 0 < x_1, x_2 < 1; \\ 0, \textit{otherwise}. \end{cases}$$

   (a) Determine the mutual information between X and Y
   (b) Determine the relative entropy from X to Y
   (c) Determine the relative entropy from Y to X

(continued)

7. Let $X = (X_1, X_2)$ be a pair of random variables having density

$$\begin{cases} \frac{2}{(x_2+1)^3}, \text{if } 0 < x_1 < x_2; \\ 0, \text{otherwise}. \end{cases}$$

   (a) Determine the mutual information between X and Y
   (b) Determine the relative entropy from X to Y
   (c) Determine the relative entropy from Y to X

8. Let (X_1, X_2) be a pair of non-negative independent random variables such that

$$f_1(x_1) = \lambda e^{-\lambda x_1}, \ f_2(x_2) = \mu e^{-\mu x_2}.$$

   Let $X = X_1 - X_2$ and $Y = X_1 + X_2$.

   (a) Determine the mutual information between X and Y
   (b) Determine the relative entropy from X to Y
   (c) Determine the relative entropy from Y to X

9. Two fair coins are tossed. If the result X contains a head, we roll a die. If the result X does not contain a head, we toss the die twice and we get the largest result. Let Y be the result of the die.

   (a) Determine the joint distribution of (X, Y)
   (b) Determine mutual information between X and Y
   (c) Determine the relative entropy from X to Y
   (d) Determine the relative entropy from Y to X

3.4 Surprise Quantification

One of the applications of the preceding theory is the quantification of the new information given by observed events, i.e., the surprise generated by the observation. Although there are many statistical definitions of surprise (Modirshanechi et al. 2022), surprise quantification found applications in many fields, such as for instance, data science, economy, psychology, and neurosciences, namely for the analysis of public attention in spectacles, music, and learning. The reader can find references in the literature. To cite a few among many: (Bayarria and Morales 2003; Itti and Baldi 2009; Baldi and Itti 2010; Ostwald et al. 2012; Lee et al. 2015; Kolossa et al. 2015; Aurnhammer and Frank 2019; Cheung et al. 2019; Quiroga-Martinez et al. 2020; Grassi and Bartels 2021; Saravanan and Levine 2022; Palm 2023).

The fundamental element in the quantification of the surprise is Baye's formula (1.3) or (1.18). Indeed, Bayes' formula shows that the observation of an event A modifies the probability of an event B as:

$$P(B|A) = \frac{P(A|B)}{P(B)} P(A). \tag{3.19}$$

Thus, the observation of A causes a modification in the evaluation of the probability of event B. We can quantify the surprise by comparing the entropies $h(B)$ and $h(B|A)$:

- If $h(B|A) \leq h(B)$, the observation has given not any information: it is not a surprise
- If $h(B|A) > h(B)$, the observation has given some information: it is a surprise;

Example 3.19 Let us consider the situation described in Example 1.13 and reconsidered in Example 3.10: The events R : *weather is rainy* and $W =$ *weather is windy* verify

$$P(R) = 0.1, P(W) = 0.25, P(W|R) = 0.8, P(R|W) = 0.32$$

Suppose that the weather is rainy. Then

$$h(W|R) = -\log(0.8) \approx 0.22314.$$

Without this information, we have

$$h(W) = -\log(0.25) \approx 1.38629.$$

Since $h(W|R) < h(W)$, windy weather is not a surprise by rainy one. We have also $P(W^c|R) = 0.2$. Thus

$$h(W^c|R) = -\log(0.2) \approx 1.609438.$$

Then, $h(W^c|R) > h(W^c) \approx 0.2876821$. Thus, non-windy weather is a surprise for rainy weather.

Example 3.19 suggests that the surprise can be measured by

$$S(A,B) = h(B|A) - h(B) = \log \frac{P(B)}{P(B|A)}. \tag{3.20}$$

Notice that

$$S(A,B) = \log \frac{P(A)P(B)}{P(A \cap B)} = h(A \cap B) - h(A) - h(B). \tag{3.21}$$

If $S(A, B) > 0$, the observation of A generates a surprise concerning B. Otherwise, not. For instance, in Example 3.19:

$$S(W,R) = \log \frac{0.25}{0.8} \approx -1.163151 < 0,$$

$$S(W^c,R) = \log \frac{0.75}{0.2} \approx 1.321756 > 0.$$

This approach extends to a couple (X, Y) of discrete random variables by taking ="$X = X_i$", B = "$Y = Y_j$": the observation of the variables X can generate a surprise on the variable Y. In this case,

$$S(X = X_i, Y = Y_j) = \log \frac{P(Y = Y_j)}{P(Y = Y_j | X = X_i)} = \log \frac{p_{\blacksquare j}}{p_{ij}/p_{i\blacksquare}} = \log \frac{p_{\blacksquare j} p_{i\blacksquare}}{p_{ij}}. \quad (3.22)$$

Example 3.20 Let us consider the situation described in Example 3.19. Let us consider $X = 0$, if weather is rainy, $X = 1$, if not; $Y = 0$, if weather is windy, Y=1, if not. Then, the distribution of (X, Y) is given in Table 3.2:

Table 3.2 Distribution of (X, Y)

X \ Y	0	1	$p_{i\blacksquare}$
0	0.08	0.02	0.1
1	0.17	0.73	0.9
$p_{\blacksquare j}$	0.25	0.75	1

Then

$$S(X = 0, Y = 0) = \log \left(\frac{0.1 \times 0.25}{0.08} \right) \approx -1.163151.$$

$$S(X = 0, Y = 1) = \log \left(\frac{0.1 \times 0.75}{0.02} \right) \approx 1.321756.$$

Thus, the observation $X = 0$ is a surprise for $Y = 1$ but not for $Y = 0$.

This approach compares directly the information concerning the events so that it does not evaluate the effect of the new information on the distribution but on the probability of chosen events.

To evaluate the effects on the distribution of Y, we can evaluate the information gain from the original to the conditional distribution:

$$S_{kld}(X=X_i)=D\left(P(Y|X=X_i)\,\middle\|\,P(Y)\right). \tag{3.23}$$

Alternatively, we can compare the entropies of $P(Y)$ and $P(Y|X=X_i)$:

$$S_H(X=X_i)=H(P(Y))-H(P(Y|X=X_i)\,). \tag{3.24}$$

Analogously to the preceding approach, a surprise corresponds to a positive value.

Example 3.21 Let us consider again the situation described in Example 3.20. The distributions are shown in Table 3.3:

Table 3.3 Distribution of Y

Y	0	1
$P(Y)$	0.25	0.75
$P(Y\|X=0)$	0.8	0.2

Then,

$$S_{kld}(X=0)=0.8\log\left(\frac{0.8}{0.25}\right)+0.2\log\left(\frac{0.2}{0.75}\right)\approx 0.6661695.$$

The value is positive. Therefore, we may consider that the information generates a surprise. Analogously, we have

$$H(P(Y))=-0.25\log 0.25-0.75\log 0.75\approx 0.5623351,$$
$$H(P(Y|X=0))=-0.25\log 0.25-0.2\log 0.2\approx 0.5004024.$$

Thus

$$S_H(X=0)=0.5623351-0.5004024\approx 0.0619327.$$

Again, the value is positive: we consider that the information generates a surprise.

Example 3.22 A coin is tossed three times. We do not know if the coin is fair or not, so that we start by assuming that the probability p of a head is a number in the interval $I = (0.4, 0.6)$. Let X be the number of heads observed and Y be the value of p. Since no other information is given, we assume equiprobability, so that Y is uniformly distributed on the interval $I = (0.4,0.6)$. The distribution of X is binomial:

$$P(X=i|Y=p)=\binom{3}{i}p^i(1-p)^{3-i}.$$

Our initial estimation for p is the mean of the distribution of Y: $p_0 = 1/2$ and the initial distribution is $P_0(X = i) = P(X = i| Y = p_0)$.

Assume that we observed $X_{obs} = k$ Heads. From Bayes' formula:

$$P(Y=p|X=k)=\frac{P(X=k|Y=p)P(Y=p)}{P(X=k)}.$$

Thus, the value of $P(Y = p| X = k)$ is proportional to $P(X = k| Y = p)$:

$$P(Y=p|X=k)=\frac{1}{M}p^k(1-p)^{3-k}$$

The value of M can be determined by

$$\int_{0.4}^{0.6} P(Y=p|X=k)dp=1.$$

We have

$$M=\int_{0.4}^{0.6} p^k(1-p)^{3-k}dp.$$

We can use this information to get a new estimation of the value of p. For instance, we can use the mean value of Y:

$$p_1=E(Y|X=k)\approx\frac{1}{M}\int_{0.4}^{0.6} p^{k+1}(1-p)^{3-k}dp,.$$

(continued)

Example 3.22 (continued)

Thus, we have a new distribution $P_1(X = i) = P(X = i | p = p_1)$ for X, which can be used to determine if there is a surprise or not.

As an example, assume that we observed all heads: $X_{obs} = 3$. Then, $k = 3$ and

$$M = \frac{0.6^4 - 0.4^4}{4} = 0.026, \quad p_1 = \frac{0.6^5 - 0.4^5}{5M} \approx 0.52$$

The initial and the actual distributions are shown in Table 3.4

Table 3.4 Distribution of X

X	0	1	2	3
$P_0(X = i)$	0.125	0.375	0.375	0.125
$P_1(X = i)$	0.111	0.359	0.389	0.141

We observe that

$$D\left(P_1 \,\middle\|\, P_0\right) \approx 0.0023$$

and

$$H(P_0) - H(P_1) \approx 0.001.$$

Under R, the calculation is made as follows:

```
a = 0.4
b = 0.6
n = 3
m = (b^(n+1) - a^(n+1))/(n+1)
p = (b^(n+2) - a^(n+2))/((n+2)*m)

p0 = c(dbinom(0:n,n,0.5))
p1 = c(dbinom(0:n,n,p))

sum(p1*log(p1/p0))
## [1] 0.002255145
sum(p1*log(p1)) - sum(p0*log(p0))
## [1] 0.00101669
```

(continued)

Example 3.22 (continued)

Finally, we can consider the events $B : p \in (\alpha, \beta) \subset I$ and $A : X = X_{obs}$. Initially, the probability of B is

$$P(B) = \int_{\alpha}^{\beta} \frac{dp}{0.6 - 0.4} = 5(\beta - \alpha).$$

Considering the observation, our evaluation of the probability $P(B)$ becomes

$$P(B|A) = \frac{1}{M} \int_{\alpha}^{\beta} p^k (1-p)^{3-k} dp \underbrace{=}_{k=3} \frac{\beta^4 - \alpha^4}{4M}.$$

Thus,

$$S(A, B) = \log \frac{P(B)}{P(B|A)} = \log \left(\frac{20M(\beta - \alpha)}{\beta^4 - \alpha^4} \right) = \log \left(\frac{20M}{(\beta + \alpha)(\beta^2 + \alpha^2)} \right).$$

Let us consider $\alpha_\varepsilon = 0.5 - \varepsilon$, $\beta_\varepsilon = 0.5 + \varepsilon$, and $B_\varepsilon : p \in (\alpha_\varepsilon, \beta_\varepsilon)$ Then,

$$S(A, B_\varepsilon) = \log \left(\frac{20M}{0.5 + 2\varepsilon^2} \right).$$

The surprise is minimal for $\varepsilon = 0.1$ and maximal for $\varepsilon \to 0+$, which corresponds to $B_{0+} : p = ½$. Under R, we evaluate it as shown at right: the maximal surprise is 0.039.

```
c = 4/(b-a)
emax = 0.5*(b-a)
log(c*m/(0.5 + 2*emax^2))
## [1] 0
log(c*m/0.5)
## [1] 0.03922071
```

Although the result seems surprising, it is not: for all the criteria considered, the differences are too small to conclude that the coin is not fair. However, such a result depends on our initial estimation of the distribution of p. For instance, assume that we suppose that p can take any value on $(0, 1)$. In this case, the results are the following:

(continued)

Example 3.22 (continued)

```
a = 0
b = 1
n = 3
m = (b^(n+1) - a^(n+1))/(n+1)
p = (b^(n+2) - a^(n+2))/((n+2)*m)

p0 = c(dbinom(0:n,n,0.5))
p1 = c(dbinom(0:n,n,p))

sum(p1*log(p1/p0))
## [1] 0.5782343
sum(p1*log(p1)) - sum(p0*log(p0))
## [1] 0.281609
c = 4/(b-a)
emax = 0.5*(b-a)
log(c*m/(0.5 + 2*emax^2))
## [1] 0
log(c*m/0.5)
## [1] 0.6931472
```

Thus,

$$D\left(P_1 \,\middle\|\, P_0\right) \approx 0.58, H(P_0) - H(P_1) \approx 0.28$$

For $B : p = ½.$, we have $S(A, B) \approx 0.69$. In this case, the difference is significant for all the criteria, and we conclude that it will be surprising that the coin is fair.

Example 3.23 An useful family of surprises is the exponential family, introduced in section (sec17 (see Eq. (eq137b)).

Exercises

1. A coin is tossed $n = 5$ times and we observe the number of heads X. Assume that the probability of a head is a variable Y uniformly distributed on $(0.3, 0.7)$. The initial estimation is $p_0 = 1/2$.

   (a) Determine the probability $P(B_\varepsilon)$ of the event $B_\varepsilon : p \in \left(\frac{1}{2} - \varepsilon, \frac{1}{2} + \varepsilon\right)$.
   (b) Consider the event A^k corresponding to the observation $X_{obs} = k$. Determine, for $k = 0, \ldots, n$, a new estimation p_1^k of the probability of a Head.
   (c) Use this estimation to evaluate the probability $P(B_\varepsilon | A^k)$.
   (d) Determine $S(A^k, B_{0+})$. For which values of k do we have a surprise?
   (e) Determine the distributions $P_1^k(X = i) = P\left(X = i | p = p_1^k\right), k = 0, \ldots, n$.
   (f) Determine $D\left(P_1^k \middle\| P_0\right), k = 0, \ldots, n$. For which values of k do we have a surprise?
   (g) Determine $H(P_0) - H\left(P_1^k\right), k = 0, \ldots, n$. For which values of k do we have a surprise?

2. Repeat Exercise 1 assuming that Y is uniformly distributed on (0, 1).
3. Repeat Exercise 1 assuming that $n = 10$.
4. A dice is rolled $n = 5$ times and we observe the number of 6 obtained, denoted X. Assume that the probability of a 6 is a variable Y uniformly distributed on (0, 1/3). The initial estimation is $p_0 = 1/6$.

   (a) Determine the probability $P(B)$ of the event $B : p \in \left(\frac{1}{6} - \varepsilon, \frac{1}{6} + \varepsilon\right)$.
   (b) Consider the event A^k corresponding to the observation $X_{obs} = k$. Determine, for $k = 0, \ldots, n$, a new estimation p_1^k of the probability of a 6.
   (c) Use this estimation to evaluate the probability $P(B | A^k)$.
   (d) Determine $S(A^k, B)$. For which values of k do we have a surprise?
   (e) Determine the distributions $P_1^k(X = i) = P\left(X = i | p = p_1^k\right), k = 0, \ldots, n$.
   (f) Determine $D\left(P_1^k \middle\| P_0\right), k = 0, \ldots, n$. For which values of k do we have a surprise?
   (g) Determine $H(P_0) - H\left(P_1^k\right), k = 0, \ldots, n$. For which values of k do we have a surprise?

5. Repeat Exercise 4 assuming that Y is uniformly distributed on (0, 1).
6. Repeat Exercise 4 assuming that $n = 10$.

References

Ash, R. (1965, reedited 2003). *Information Theory*. New York, NY, USA: Dover.

Attiaoui, D., Doré, P., Martin, A., & Ben Yaghlane, B. (2012). A Distance between Continuous Belief Functions. In E. Hüllermeier, S. Link, T. Fober, & B. Seeger (Ed.), *Scalable Uncertainty Management: SUM 2012* (pp. 194–205). Marburg, Germany: Springer. doi:https://doi.org/10.1007/978-3-642-33362-0_15

Aurnhammer, C., & Frank, S. L. (2019). Evaluating information-theoretic measures of word prediction in naturalistic sentence reading. *Neuropsychologia, 134*. doi:https://doi.org/10.1016/j.neuropsychologia.2019.107198

Baldi, P., & Itti, L. (2010). Of bits and wows: A Bayesian theory of surprise with applications to attention. *Neural Networks, 23*(5), 649–666. doi:https://doi.org/10.1016/j.neunet.2009.12.007

Bayarria, M. J., & Morales, J. (2003). Bayesian measures of surprise for outlier detection. *Journal of Statistical Planning and Inference, 111*(1–2), 3–22. doi:https://doi.org/10.1016/S0378-3758(02)00282-3

Benavoli, A. (2014). Belief function and multivalued mapping robustness in statistical estimation. *International Journal of Approximate Reasoning, 55*, 311–329. doi:https://doi.org/10.1016/j.ijar.2013.04.014

Beretta, G. P. (2008). Axiomatic Definition of Entropy for Nonequilibrium States. *International Journal of Thermodynamics, 11*(2), 39–48. doi:https://doi.org/10.5541/ijot.211

Boivin, C. (2022a, December 25). https://cran.r-project.org/web/packages/dst/vignettes/Captain_Example.html. Retrieved from Captain's Example: https://cran.r-project.org/web/packages/dst/vignettes/Captain_Example.html

Boivin, C. (2022b, 12 24). *Introduction to Belief Functions*. Retrieved from https://cran.r-project.org/web/packages/dst/vignettes/: https://cran.r-project.org/web/packages/dst/vignettes/Introduction_to_Belief_Functions.html#fn2

Boivin, C. (2022c, December 25). *Introduction to Belief Functions: The Monty Hall Game*. Retrieved 2022, from https://cran.microsoft.com/snapshot/2018-08-11/web/packages/dst/vignettes/Monty-hall-Example.html: https://cran.microsoft.com/snapshot/2018-08-11/web/packages/dst/vignettes/Monty-hall-Example.html

Boltzmann, L. E. (1866). Über die mechanische Bedeutung des zweiten Hauptsatzes der Wärmetheorie. *Wiener Berichte, 53*, 195–220. Retrieved from http://opacplus.bsb-muenchen.de/title/BV020135572/ft/bsb10133426?page=3

Boltzmann, L. E. (1877 traduction 2002). On the relationship between the second main theorem of mechanical heat theory and the probability calculation with respect to the results about the heat equilibrium. *Akademie der Wissenschaften in Wien Mathematisch-naturwissenschaftliche Klasse Sitzungsberichte, 2*(76), 373–435. Retrieved February 14, 2023, from http://users.polytech.unice.fr/~leroux/boltztrad.pdf

Boltzmann, L. E. (1896). *Vorlesungen �uber Gastheorie* (Vol. 1). Leipzig: Barth.

Boltzmann, L. E. (1995). *Lectures on Gas Theory*. (S. G. Brush, Trans.) New York, NY, USA: Dover.

Boyle, G. H. (n.d.). *modelingcommons.org/file/download/6101?file_id=3384*. Retrieved from http://modelingcommons.org/file/download/6101?file_id=3384

Brillouin, L. (1956). *Science and Information Theory*. New York, NY, USA: Dover.

Campagner, A., Ciucci, D., & Denœux, T. (2022). Belief Functions and Rough Sets: Suvey and New Insights. *International Journal of Approximate Reasoning, 143*, 92–215. doi:https://doi.org/10.1016/j.ijar.2022.01.011

Chakrabarti, C. G., & Indranil, C. (2005). Shannon entropy: axiomatic characterization and application. *International Journal of Mathematics and Mathematical Sciences*. doi:https://doi.org/10.1155/IJMMS.2005.2847

Cheung, V. K., M.C., P., Meyer, L., Pearce, M. T., Haynes, J.-D., & Koelsch, S. (2019). Uncertainty and Surprise Jointly Predict Musical Pleasure and Amygdala, Hippocampus, and Auditory Cortex Activity. *Current Biology, 29*(23), 4084–4092.e4. doi:https://doi.org/10.1016/j.cub.2019.09.067

Clausius, R. J. (1850). Ueber die bewegende Kraft der Wärme und die Gesetze, welche sich daraus für die Wärmelehre selbst ableiten lassen. *Annalen der Physik*, 368–397 , 500–524.

Clausius, R. J. (1851 republished in 2009). On the Moving Force of Heat, and the Laws regarding the Nature of Heat itself which are deducible therefrom. *The London, Edinburgh, and Dublin Philosophical Magazine and Journal of Science*, 2(8), 1–21 , 102–119. doi:https://doi.org/10.1080/14786445108646819 , https://doi.org/10.1080/14786445108646840

Cobb, B. R., & Shenoy, P. (2006, April). On the plausibility transformation method for translating belief function models to probability models. *Journal of Approximate Reasoning, 41*(3), pp. 314–330. doi:https://doi.org/10.1016/j.ijar.2005.06.008

Couso, I., Dubois, D., & Sanchez, L. (2014). *Random Sets and Random Fuzzy Sets as Ill-Perceived Random Variables.* Springer. doi:https://doi.org/10.1007/978-3-319-08611-8

Csiszár, I. (2008). Axiomatic Characterizations of Information Measures. *Entropy, 10*(3), 261–273. doi:https://doi.org/10.3390/e10030261

Dale, A. I. (1982). Bayes or Laplace? An Examination of the Origin and Early Applications of Bayes' Theorem. *Archive for History of Exact Sciences, 27*(1), pp. 23–47.

Dale, A. I. (1999). *A History of Inverse Probability—From Thomas Bayes to Karl Pearson* (2nd. ed.). New York: Springer.

De Finetti, B. (2017). *Theory of Probability—A Critical Introductory Treatment.* (A. M. Smith, Trans.) UK: John Wiley and Sons.

DeGroot, M. H. (2004). *Optimal Statistical Decisions.* New Jersey: John Wiley & Sons.

Dempster, A. (1967, April). Upper and Lower Probabilities Induced by a Multivalued Mapping. *Ann. Math. Statist., 38*(2), pp. 325–339,. doi:https://doi.org/10.1214/aoms/1177698950

Denoeux, T. (2019, June). Decision-making with belief functions: A review. *International Journal of Approximate Reasoning, 109*, 87–110. doi:https://doi.org/10.1016/j.ijar.2019.03.009

Denoeux, T. (2021). *Uncertainty Analysis using Belief Functions: Applications to Statistical Inference and Pattern Recognition.* Université de technologie de Compiègne, Department of Computer Science, Compiègne, France. Retrieved December 20, 2022, from https://www.hds.utc.fr/~tdenoeux/dokuwiki/_media/en/book_bf.pdf

Denœux, T., Younes, Z., & Abdallah, F. (2010). Representing uncertainty on set-valued variables using belief functions. *Artificial Intelligence, 174*(7–8), 479–499. doi:https://doi.org/10.1016/j.artint.2010.02.002

Dezert, J., & Smarandache, F. (2008). A new probabilistic transformation of belief mass assignment. *Fusion 2008: International Conference on Information Fusion*, (pp. 1410–1417). Retrieved December 19, 2022, from https://hal.archives-ouvertes.fr/hal-00304319/document

Dezert, J., & Smarandache, F. (2009). Transformations of belief masses into subjective probabilities. In J. Dezert, & F. Smarandache (Eds.), *Advances and Applications of DSmT for Information Fusion* (pp. 85–136). Rehoboth: American Research Press . Retrieved December 19, 2022, from https://www.researchgate.net/publication/306413841_Transformations_of_belief_masses_into_subjective_probabilities

Dezert, J., Smarandache, F., & Daniel, M. (2004). A Generalized Pignistic Transformation. In J. Dezert, & F. Smarandache (Eds.), *Advances and Applications of DSmT for Information Fusion* (pp. 143–153). Rehoboth: American Research Press. Retrieved December 19, 2022, from https://www.onera.fr/sites/default/files/297/C023-Dezert-Fusion2004Stockholm.pdf

Doré, P. E., Fiche, A., & Martin, A. (2010). Models of belief functions—Impacts for patterns recognitions. *13th International Conference on Information Fusion.* Edinburgh, UK. doi:https://doi.org/10.1109/ICIF.2010.5711936

Doré, P., Martin, A., Abi-Zeid, I., Jousselme, A., & Maupin, P. (2011a, January). Belief functions induced by multimodalprobability density functions, an application to the search and rescue problem. *RAIRO—Operations Research, 44*(4), 323–343. doi:https://doi.org/10.1051/ro/2011001

Doré, P., Osswald, C., Martin, A., Jousselme, A., & Maupin, P. (2011b). Continuous belief functions to qualify sensors performances. In W. Liu (Ed.), *Symbolic and Quantitative Approaches to Reasoning with Uncertainty. ECSQARU 2011* (pp. 350–361). Belfast, Ireland: Springer. doi:https://doi.org/10.1007/978-3-642-22152-1_30

Dragulescu, A., & Yakovenko, V. (2000). Statistical mechanics of money. *Eur. Phys. J. B, 17*, 723–729. doi:https://doi.org/10.1007/s100510070114

Dubois, D., & Prade, H. (1986). A Set-Theoretic View of Belief Functions. *International Journal of General Systems, 12*(3), 193–226. doi:https://doi.org/10.1080/03081078608934937

Fetz, T., & Oberguggenberger, M. (2016, November). Imprecise random variables, random sets, and Monte Carlo simulation. *78*, 252–264. doi:https://doi.org/10.1016/j.ijar.2016.06.012

Fiche, A., Martin, A., Cexus, J.-C., & Khenchaf, A. (2010). Continuous belief functions and α-stable distributions. *13th Conference on Information Fusion.* Edinburgh. doi:https://doi.org/10.1109/ICIF.2010.5711934

Fort, H. (2022). *Forecasting with Maximum Entropy.* IOP Publishing Ltd.

Fox, C. R., & See, K. E. (2006). Belief and Preference in Decision Under Uncertainty. In D. Hardman, & L. Macchi (Eds.), *Thinking: Psychological Perspectives on Reasoning, Judgment and Decision Making.* John Wiley & Sons. doi:https://doi.org/10.1002/047001332X.ch14

Gibbs, J. W. (1901 reprinted 1960). *Elementary Principles of Statistical Mechanics.* Dover.

Grassi, P. R., & Bartels, A. (2021). Magic, Bayes and wows: A Bayesian account of magic tricks. *Neuroscience & Biobehavioral Reviews, 126*, 515–527. doi:https://doi.org/10.1016/j.neubiorev.2021.04.001

Guan, J., & Bell, D. (1993). Discounting and Combination Operations in Evidential Reasoning. In D. Heckerman, & A. Mamdani (Ed.), *Ninth Conference on Uncertainty in Artificial Intelligence* (pp. 477–484). Washington, USA: Elsevier. doi:https://doi.org/10.1016/B978-1-4832-1451-1.50062-7

Halmos, P. (1978). *Measure theory.* Berlin, Heidelberg, New York: : Springer Verlag.

Halpern, J., & Fagin, R. (1992). Two views of belief: belief as generalized probability and belief as evidence. *Artificial Intelligence,, 54*, pp. 275–317.

Hartley, R. (1928, July). Transmission of Information. *Bell System Technical Journal, 7*(3), 535–563.

He, Y. (2013). *Uncertainty Quantification and Data Fusion based on Dempster-Shafer Theory.* Florida State University, Mathematics. Florida State University Libraries.

Howson, C., & Urbach, P. (2006). *Scientific reasoning : the Bayesian approach* . Chicago: Open Court.

Hüllermeier, E., Kruse, R., & Hoffmann, F. (Eds.). (2010). Consonant continuous belief functions conflicts calculation. *IPMU'10: Proceedings of the Computational intelligence for knowledge-based systems design, and 13th international conference on Information processing and management of uncertainty* (pp. 706–715). Dortmund, Germany: Springer.

Hulse, A., Schumacher, B., & Westmoreland, M. D. (2018). Axiomatic Information Thermodynamics. *Entropy, 20*(4). doi:https://doi.org/10.3390/e20040237

Itti, L., & Baldi, P. (2009). Bayesian surprise attracts human attention. *Vision Research, 49*(10), 1295–1306. doi:https://doi.org/10.1016/j.visres.2008.09.007

Jaffray, J.-Y., & Wakker, P. (1993). Decision making with belief functions: Compatibility and incompatibility with the sure-thing principle. *Journal of Risk and Uncertainty, 7*(3), 255–271. doi:https://doi.org/10.1007/BF01079626

Jaynes, E. (1989). *Clearing up Mysteries—The Original Goal.* (J. Skilling, Ed.) Dordrecht: Springer. doi:https://doi.org/10.1007/978-94-015-7860-8_1

Jaynes, E. (2003). *Probability Theory: The Logic of Science.* Cambridge: Cambridge University Press.

Jaynes, E. T. (1965). Gibbs vs. Boltzmann Entropies. *American Journal of Physics, 33*(5), 391–398.

Jeffreys, H. (1939). *Theory of probability.* Oxford: University Press.

Khinchin, A. Y. (1957). *Mathematical Foundations of Information Theory.* New York, NY, USA: Dover.

Klopotek, M. A., & Wierzchon, S. T. (1998). A New Qualitative Rough-Set Approach to Modeling Belief Functions. In L. Polkowski, & A. Skowron (Ed.), *Rough Sets and Current Trends in Computing, First International Conference, RSCTC'98* (pp. 346–354). Warzsaw, Poland: Springer. doi:https://doi.org/10.1007/3-540-69115-4_47

Kojadinovic, I., Marichal, J.-L., & Roubens, M. (2005). An axiomatic approach to the definition of the entropy of a discrete Choquet capacity. *Information Sciences, 172*(1–2), 131–153. doi: https://doi.org/10.1016/j.ins.2004.05.011

Kolossa, A., Kopp, B., & Fingscheidt, T. (2015). A computational analysis of the neural bases of Bayesian inference. *NeuroImage, 106*, 222–237. doi:https://doi.org/10.1016/j.neuroimage.2014.11.00

Kullback, S. (1951 reed. 1969 reprint 1979). *Information Theory and Statistics*. New York: Wiley, reed. Dover.

Kullback, S., & Leibler, R. A. (1951, March). On Information and Sufficiency. *The Annals of Mathematical Statistics, 22*(1), 79–86. doi:https://doi.org/10.1214/aoms/1177729694

Laghmara, H., Laurain, T., Cudel, C., & Lauffenburger, J. P. (2020). Heterogeneous sensor data fusion for multiple object association using belief functions. *Information Fusion, 57*, 44–58. doi:https://doi.org/10.1016/j.inffus.2019.11.002

Laplace, P.-S. (1774). Memoire sur la probabilité des causes par les événements. *Memoires de Mathématique et de Physique, Presentés à l'Académie Royale des Sciences par divers Savans & lus dans ses Assemblées*, pp. 621–656.

Laplace, P.-S. (1891). *Oeuvres complètes* (Vol. 8). (A. d. Paris, Ed.) Paris: Gauthiers-Villars.

Laplace, P.-S. (1986). Memoir on the Probability of the Causes of Events. *Statistical Science, 1*(3), pp. 364–78. Retrieved from http://www.jstor.org/stable/2245476

Lazo, A. V., & Rathie, P. (1978). On the entropy of continuous probability distributions (Corresp.). *IEEE Transactions on Information Theory, 24*(1), 120–122. doi:https://doi.org/10.1109/tit.1978.1055832

Lee, J., Fan, Y., & Sisson, S. (2015). Bayesian threshold selection for extremal models using measures of surprise. *Computational Statistics and Data Analysi, 85*, 84–99. doi:https://doi.org/10.1016/j.csda.2014.12.004

Lian, C. (2017). *Information Fusion and Decision Making using Belief Functions*. Compiègne, France: Université de Technologie de Compiègne.

Liboff, R. L. (1974). Gibbs vs. Shannon entropies. *J Stat Phys, 11*, 343–357. doi:https://doi.org/10.1007/BF01009793

Liu, W. (2006, August). Analyzing the degree of conflict among belief functions. *Artificial Intelligence, 170* (11), 909–924. doi:https://doi.org/10.1016/j.artint.2006.05.002

Martin, A. (2019). Conflict management in information fusion with belief functions. In E. Bossé, & G. Rogova (Eds.), *Information quality in information fusion and decision making* (pp. 79–97). Springer. doi:https://doi.org/10.1007/978-3-030-03643-0_4

Martin, R., Zhang, J., & Liu, C. (2010, June). Dempster–Shafer Theory and Statistical Inference with Weak Beliefs. *Statistical Science, 25*(1), 72–87. doi:https://doi.org/10.1214/10-STS322

Mercier, D., Quost, B., & Denœux, T. (2005). Contextual Discounting of Belief Functions. In L. Godo (Ed.), *ECSQARU 2005: Symbolic and Quantitative Approaches to Reasoning with Uncertainty* (pp. 552–562). Barcelona, Spain: Springer. doi:https://doi.org/10.1007/11518655_47

Miranda, E., Couso, I., & Gil, P. (2005, July). Random sets as imprecise random variables. *Journal of Mathematical Analysis and Applications, 307*(1), 32–47. doi:https://doi.org/10.1016/j.jmaa.2004.10.022

Modirshanechi, A., Brea, J., & Gerstner, W. (2022). A taxonomy of surprise definitions. *Journal of Mathematical Psychology, 110*. doi:https://doi.org/10.1016/j.jmp.2022.102712

Nambiar, K. K., Varma, P. K., & Saroch, V. (1992). An axiomatic definition of Shannon's entropy. *Appl. Math. Lett., 5*(4), 45–46. doi:https://doi.org/10.1016/0893-9659(92)90084-M

Nguyen, H. T. (1977). *On Random sets and Belief Functions*. University of California, Berkeley, EECS Department. Retrieved from http://www2.eecs.berkeley.edu/Pubs/TechRpts/1977/28879.html

Nguyen, H. T. (1978). On Random Sets and Belief Functions. *Journal of Mathematical Analysis and Applications, 65*, 531–542. doi:https://doi.org/10.1016/0022-247X(78)90161-0

Nguyen, H., & Wang, T. (1997). Belief Functions and Random Sets. In J. Goutsias, R. Mahler, & H. Nguyen (Eds.), *Random Sets* (pp. 243–255). Springer. doi:https://doi.org/10.1007/978-1-4612-1942-2_11

Ni, S., Lei, Y., & Tang, Y. (2020). Improved Base Belief Function-Based Conflict Data Fusion Approach Considering Belief Entropy in the Evidence Theory. *Entropy, 22*(8). doi:https://doi.org/10.3390/e22080801

Ostwald, D., Spitzer, B., Guggenmos, M., Schmidt, T. T., Kiebel, S. J., & Blankenburg, F. (2012). Evidence for neural encoding of Bayesian surprise in human somatosensation. *NeuroImage, 62*(1), 177–188. doi:https://doi.org/10.1016/j.neuroimage.2012.04.050

Palm, G. (2023). *Novelty, Information and Surprise.* Germany: Springer-Verlag. doi:https://doi.org/10.1007/978-3-662-65875-8

Planck, M. K. (1901). Über das Gesetz der Energieverteilung im Normalspektrum. *Annalen der Physik, 309*(3), 553–563. doi:https://doi.org/10.1002/andp.19013090310

Planck, M. K. (1914). *The Theory of Heat Radiation.* (M. Masius, Trans.) Philadelphia: P. Blakiston's Son & Co.

Press, S. J., & Tanur, J. M. (2001). *The Subjectivity of Scientists and the Bayesian Approach.* New York: John Wiley & Sons.

Quiroga-Martinez, D., Hansen, N., Højlund, A., Pearce, M., Brattico, E., & Vuust, P. (2020). Decomposing neural responses to melodic surprise in musicians and non-musicians: Evidence for a hierarchy of predictions in the auditory system. *NeuroImage.* doi:https://doi.org/10.1016/j.neuroimage.2020.116816

Rényi, A. (1961). On Measures of Entropy and Information. In J. Neyman (Ed.), *4th Berkeley Symposium on Mathematics, Statistics and Probability,. 1*, pp. 547–561. University of California Press.

Risti, B., & Smets, P. (2006). Belief function theory on the continuous space with an application to model based classification. In B. Bouchon-Meunier, G. Coletti, & R. R. Yager (Ed.), *Modern Information Processing: From Theory to Applications. IPMU'04* (pp. 11–24). Perugia, Italy.: Elsevier. doi:https://doi.org/10.1016/B978-044452075-3/50002-9

Saravanan, R., & Levine, R. (2022). Surprisal analysis of diffusion processes. *Chemical Physics, 556.* doi:https://doi.org/10.1016/j.chemphys.2022.111450

Savchuk, V. P., & Tsokos, C. P. (2011). *Bayesian Theory and Methods with Applications.* Paris: Atlantis Press.

Shafer, G. (1976). *A Mathematical Theory of Evidence* . New Jersey: Princeton University Press.

Shafer, G. (1990). Perspectives on the Theory and Practice of Belief Functions. *International Journal of Approximate Reasoning, 4*(5–6), 323–362. doi:https://doi.org/10.1016/0888-613X(90)90012-Q

Shannon, C. E. (1948a). A Mathematical Theory of Communication. *The Bell System Technical Journal, 27*(3), 379–423. doi:https://doi.org/10.1002/j.1538-7305.1948.tb01338.x

Shannon, C. E. (1948b). A Mathematical Theory of Communication. *Bell System Technical Journal, 27*(4), 623–666. doi:https://doi.org/10.1002/j.1538-7305.1948.tb00917.x

Shannon, C. E. (n.d.). *A mathematical theory of comunication—Nokia Bell Labs.* Retrieved 2 13, 2023, from https://www.bell-labs.com/claude-shannon/assets/images/discoveries: https://www.bell-labs.com/claude-shannon/assets/images/discoveries/1948-04-21-a-mathematical-theory-of-communication-parts-I-and-carousel-01.pdf

Smets, P. (1990). Constructing the Pignistic Probability Function in a Context of Uncertainty. *Machine Intelligence and Pattern Recognition, 10*, 29–39. doi:https://doi.org/10.1016/B978-0-444-88738-2.50010-5

Smets, P. (2000). Data fusion in the transferable belief model. *Proceedings of the Third International Conference on Information Fusion. 1*, pp. 21–33. Paris, France: IEEE. doi:https://doi.org/10.1109/IFIC.2000.862713.

Smets, P. (2005). Belief functions on real numbers. *International Journal of Approximate Reasoning, 40*, 181–223. doi:https://doi.org/10.1016/j.ijar.2005.04.001

Sohrab, S. H. (2014). Boltzmann entropy of thermodynamics versus Shannon entropy of information theory. *International Journal of Mechanics, 8*, 73–84. Retrieved February 15, 2023, from https://www.naun.org/main/NAUN/mechanics/2014/a182003-086.pdf

Souza de Cursi, E. (2023). *Uncertainty Quantification using R.* Springer Cham.

Stigler, S. M. (1982). Thomas Bayes's Bayesian Inference. *Journal of the Royal Statistical Society. Series A (General), 145*(2), pp. .250–258.

Strat, T. (1984). Continuous belief functions for evidential reasoning. *Proceedings of the 4th National Conference on Artificial Intelligence.* Austin, Texas. Retrieved December 18, 2022, from https://www.aaai.org/Papers/AAAI/1984/AAAI84-035.pdf

Strat, T. (1987). The Generation of Explanations within Evidential Reasoning Systems. In J. P. McDermott (Ed.), *10th. International Joint Conference on Artificial Intelligence (IJCAI)*, (pp. 1097–1104). Milan, Italy. Retrieved December 20, 2022, from https://www.ijcai.org/Proceedings/87-2/Papers/104.pdf

Strat, T. (1990). Decision analysis using belief functions. *International Journal of Approximate Reasoning, 4*(5–6), 391–417. doi:https://doi.org/10.1016/0888-613X(90)90014-S

Taillandier, P., & Therond, O. (2011). Use of the Belief Theory to formalize Agent DecisionMaking Processes : Application to cropping Plan Decision Making. *European Simulation and Modelling Conference*, (pp. 138–142). Guimaraes, Portugal. Retrieved 12 17, 2022, from https://hal.archives-ouvertes.fr/hal-00688405

Turing, A. M. (1941). *The Applications of Probability to Cryptography.* Bletchley Park. Available at Archive.org and https://www.nationalarchives.gov.uk/. Retrieved from https://archive.org/details/hw-25-37

Turing, A. M. (2015, May 26). *The Applications of Probability to Cryptography.* doi:https://doi.org/10.48550/arXiv.1505.04714

Wasserman, L. A. (1990, September). Belief functions and statistical inference. *Canadian Journal of Statistics, 18*(3), 183–196. doi:https://doi.org/10.2307/3315449

Wu, W.-Z., & Mi, J.-S. (2008). An Interpretation of Belief Functions on Infinite Universes in the Theory of Rough Sets. In C. Chan, J. Grzymala-Busse, & W. P. Ziarko (Ed.), *Rough Sets and Current Trends in Computing. RSCTC 2008* (pp. 71–80). Akron, OH, USA: Springer. doi:https://doi.org/10.1007/978-3-540-88425-5_8

Yakovenko, V. M. (2010). Statistical Mechanics of Money, Debt, and Energy Comsumption. *Science and Culture, 76*(9–10), 430–436. doi:https://doi.org/10.48550/arXiv.1008.2179

Yakovenko, V. M., & Rosser, J. B. (2009, December 2). Statistical mechanics of money, wealth, and income. *Rev. Mod. Phys., 81*(4), 1703–1725. doi:https://doi.org/10.1103/RevModPhys.81.1703

Yao, Y. Y., & Lingras, P. J. (1998). Interpretations of Belief Functions in the Theory of Rough Sets. *Information Sciences, 104*(1–2), 81–106. doi:https://doi.org/10.1016/S0020-0255(97)00076-5

Zanchini, E., & Beretta, G. P. (2008). Rigorous Axiomatic Definition of Entropy Valid Also for Non-Equilibrium States. In G. P. Beretta, A. Ghoniem, & G. Hatsopoulos (Ed.), *MEETING THE ENTROPY CHALLENGE: An International Thermodynamics Symposium in Honor and Memory of Professor Joseph H. Keenan. 1033.* Cambridge, MA, USA: AIP Conference Proceedings. doi:https://doi.org/10.1063/1.2979048

Zhang, H., & Deng, Y. (2020). Weighted belief function of sensor data fusion in engine fault diagnosis. *Soft Comput, 24*, 2329–2339. doi:https://doi.org/10.1007/s00500-019-04063-7

Zhou, K., Martin, A., & Pan, Q. (2018). A belief combination rule for a large number of sources. . *Journal of Advances in Information Fusion, 13*(2). Retrieved December 2022, 20, from https://www.researchgate.net/deref/https%3A%2F%2Fhal.archives-ouvertes.fr%2Fhal-01883239

Chapter 4
Maximum Entropy

Many methods based on stochastic simulations use samples – for instance, Monte Carlo methods.

The generation of samples implies the choice of distributions from which they are generated – such a choice is a critical step that influences the results, so that it must be made carefully.

If some evidence is available, it can be used to choose a distribution – eventually by using physical, mathematical or, more generally, theoretical, or experimental arguments. If not, there is a procedure to determine a distribution, based on general properties – the principle of maximum entropy.

The reader can find in the literature presentations of the principle of maximum entropy such as, for instance, (Jaynes 1957a, b; Shore and Johnson 1980; Kapur 1989 revised 1993). In the sequel, we shall examine this principle and show how it can be used to determine distributions.

4.1 The Principle of Maximum Entropy

As introduced in Chap. 3, entropy is associated with the probability distribution. For a random variable X having a probability density $f(x)$, we have

$$H(X) = H(f) = -\int_{\mathbb{R}} f(x)\log(f(x))\,dx. \tag{4.1}$$

If f is supported by the interval (a, b), then the integration over $\mathbb{R}$ reduces to an integration over (a, b). In the continuous case, the principle of maximum entropy reads as

E. Souza de Cursi, *Uncertainty Quantification with R*, International Series in Operations Research & Management Science 352,
https://doi.org/10.1007/978-3-031-48208-3_4

Definition 4.1 Let C be the set of the admissible choices for f. Estimate the real distribution as

$$f^* = \arg\max\{H(f) : f \in C\}, \tag{4.2}$$

i.e., f^* which maximizes H on C.

C is defined by the information available on the distribution. The minimal available information is that

$$f \geq 0 \quad \text{and} \quad \int_{\mathbb{R}} f(x)dx = 1. \tag{4.3}$$

These relations define constraints to be satisfied by the solution f^*. Let us introduce Lagrange's multipliers λ_+ and λ_0 associated with these conditions. According to the standard formulation of Lagrange's multipliers, the condition $f \geq 0$ is rewritten under the equivalent form $-f \leq 0$. The Lagrangian Penalty associated with the restrictions is

$$\Pi(f; \lambda_+, \lambda_0) = \lambda_0 \left(\int_{\mathbb{R}} f(x)dx - 1 \right) - \int_{\mathbb{R}} \lambda_+(x) f(x)dx.$$

We have

$$\lambda_0 \in \mathbb{R}, \quad \lambda_+(x) \geq 0, \lambda_+(x) = 0 \text{ if } f(x) > 0.$$

In practice, we work on the support of the distribution, so that $\lambda_+ \equiv 0$ on the region of interest – in practice, λ_+ is considered null and neglected and we consider the simplified Lagrangian Penalty

$$P_0(f; \lambda_0) = \lambda_0 \left(\int_{\mathbb{R}} f(x)dx - 1 \right). \tag{4.4}$$

In addition, a situation often found in practice is the one where information is available about the support of the distribution and the moments of X.

As previously remarked, the information on the support is used to reduce the integration over $\mathbb{R}$ to the integration over the support of the distribution and to eliminate λ_+. The information about a moment $M_i(X) = m_i$ or a centered moment $CM_i(X) = cm_i$ generates a supplementary restriction to be satisfied by the elements of C. We can write the constraints about moments under the general form

$$\int_{\mathbb{R}} g_i(x) f(x) dx = c_i, \tag{4.5}$$

where

$$g_i(x) = x^i \text{ (moment) } or \ g_i(x) = (x - m_1)^i \text{ (centered moment)}.$$

The reader can find a complete analysis involving this kind of restriction (Muñoz-Cobo et al. 2017). We present below a more simplified analysis, restricted to some practical situations.

Let $\lambda_i \in \mathbb{R}$ be Lagrange's multiplier associated with this condition. The associated Lagrangian Penalty is

$$P_i(f; \lambda_i) = \lambda_i \left(\int_{\mathbb{R}} g_i(x) f(x) dx - c_i \right). \tag{4.6}$$

Let us introduce $g_0(x) = 1$, $c_0 = 1$. Then (4.4) can be put under the form (4.5). Let us consider $1 \leq i \leq k$ restrictions of the kind (4.5). Let $\boldsymbol{\Lambda} = (\lambda_0, \ldots, \lambda_k)$: the Lagrangian Penalty is

$$P(f; \boldsymbol{\Lambda}) = \sum_{i=0}^{k} P_i(f; \lambda_i) = \sum_{i=0}^{k} \int_{\mathbb{R}} \lambda_i g_i(x) f(x) dx - \sum_{i=0}^{k} \lambda_i c_i.$$

The Lagrangian associated with (4.2) is

$$\mathcal{L}(f; \boldsymbol{\Lambda}) = H(f) + P(f; \boldsymbol{\Lambda}).$$

We have

$$\mathcal{L}(f; \boldsymbol{\Lambda}) = \int_{\mathbb{R}} \ell(f(x); \boldsymbol{\Lambda}) dx - F(\boldsymbol{\Lambda}),$$

with

$$\ell(f(x);\mathbf{\Lambda}) = -f(x)\log f(x) + \sum_{i=0}^{k} \lambda_i g_i(x) f(x); F(\boldsymbol{\Lambda}) = \sum_{i=0}^{k} \lambda_i c_i.$$

Euler-Lagrange's equations associated with $\mathcal{L}$ read as

$$-1 - \log f(x) + \sum_{i=0}^{k} \lambda_i g_i(x) = 0,$$

so that

$$f(x) = \exp\left(-1 + \sum_{i=0}^{k} \lambda_i g_i(x)\right). \tag{4.7}$$

This equation can be written under the simplified form

$$f(x) = A \exp\left(\sum_{i=1}^{k} \lambda_i g_i(x)\right), \tag{4.8}$$

where

$$A = \exp(\lambda_0 - 1). \tag{4.9}$$

In practice, A is a constant of normalization, to be chosen to get $\int_{\mathbb{R}} f(x)dx = 1$.

Remark 4.1 Since

$$c_i = c_i \times 1 = c_i \times \int_{\mathbb{R}} f(x)dx = \int_{\mathbb{R}} c_i f(x)dx,$$

we can write Eq. (4.5) under the form

$$\int_{\mathbb{R}} (g_i(x) - c_i) f(x)dx = 0.$$

(continued)

Remark 4.1 (continued)

The associated Lagrangian Penalty is

$$P_i(f;\lambda_i) = \lambda_i \left(\int_{\mathbb{R}} (g_i(x) - c_i) f(x) dx \right).$$

In this formulation, we have

$$P(f;\boldsymbol{\Lambda}) = \int_{\mathbb{R}} \lambda_0 f(x) dx + \sum_{i=1}^{k} \int_{\mathbb{R}} \lambda_i (g_i(x) - c_i) f(x) dx - \lambda_0.$$

Thus

$$\mathcal{L}(f;\boldsymbol{\Lambda}) = \int_{\mathbb{R}} \ell(f(x);\boldsymbol{\Lambda}) dx - \lambda_0,$$

with

$$\ell(f(x);\boldsymbol{\Lambda}) = -f(x)\log f(x) + \lambda_0 f(x) + \sum_{i=1}^{k} \lambda_i (g_i(x) - c_i) f(x).$$

The corresponding Euler-Lagrange's equations are

$$-1 - \log f(x) + \lambda_0 + \sum_{i=1}^{k} \lambda_i (g_i(x) - c_i) = 0,$$

so that

$$f(x) = \exp\left(-1 + \lambda_0 + \sum_{i=1}^{k} \lambda_i (g_i(x) - c_i) \right).$$

The simplified form is again

$$f(x) = A \exp\left(\sum_{i=1}^{k} \lambda_i g_i(x) \right),$$

(continued)

Remark 4.1 (continued)

where

$$A = \exp\left(-1 + \lambda_0 - \sum_{i=1}^{k} \lambda_i c_i\right).$$

Again, A appears as a constant of normalization.

Example 4.1 Let us consider the maximum entropy under the information that the distribution is supported by the interval (a, b). We look for

$$f^* = \arg\max H(f), \quad H(f) = -\int_a^b f(x)\log(f(x))\, dx,$$

under the restrictions

$$f \geq 0 \quad \text{and} \quad \int_{\mathbb{R}} f(x)dx = 1.$$

Euler-Lagrange's equations read as

$$-1 - \log(f^*(x)) + \lambda_0 = 0 \quad on\ (a, b).$$

Thus, Eq. (4.7) furnishes

$$f^*(x) = \exp(-1 + \lambda_0) = A \in \mathbb{R} \quad on\ (a, b).$$

That is, f^* corresponds to the density of a uniformly distributed continuous variable on (a, b). Using the condition $\int_{\mathbb{R}} f(x)dx = 1$, we obtain $A = \frac{1}{b-a}$.

Example 4.2 Let us consider the following information:

$$M_1(X)=m_1, M_2(X)=m_2.$$

Notice that

$$m_2 - m_1^2 = V(X) = \sigma^2 \geq 0.$$

In this case, the Lagrangian density is

$$\ell = -f\log f + \lambda_0 f + \lambda_1 x f + \lambda_2 x^2 f.$$

We have

$$-1 - \log f + \lambda_0 + \lambda_1 x + \lambda_2 x^2 = 0.$$

Thus,

$$f = \exp\left(-1 + \lambda_0 + \lambda_1 x + \lambda_2 x^2\right).$$

We must determine the multipliers λ_0, λ_1, λ_2 using the conditions $M_0(X) = 1$, $M_1(X) = m_1$, $M_2(X) = m_2$. We can write the solution under the form

$$f = A\exp\left(B(x-C)^2\right),$$

with

$$A = \exp\left(-1 + \lambda_0 + \frac{\lambda_1}{4\lambda_2^2}\right), \quad B = \lambda_2, \quad C = -\frac{\lambda_1}{2\lambda_2}.$$

Let us set $B = -\frac{1}{2s^2}, C = m$. Then, $f = A\exp\left(-\frac{1}{2}\left(\frac{x-m}{s}\right)^2\right)$, so that

$$\int_{\mathbb{R}} f(x)dx = As\sqrt{2\pi} = 1,$$

$$\int_{\mathbb{R}} xf(x)dx = Ams\sqrt{2\pi} = m_1 \Longrightarrow m = m_1,$$

(continued)

Example 4.2 (continued)

$$\int_{\mathbb{R}} x^2 f(x)dx = A\left(m^2 + s^2\right)s\sqrt{2\pi} = m_2 + \Rightarrow s^2 = m_2 - m_1^2.$$

Thus, the solution is a Gaussian distribution:

$$f = \frac{1}{\sigma\sqrt{2\pi}} \exp\left(-\frac{1}{2}\left(\frac{x - m_1}{\sigma}\right)^2\right), \sigma^2 = m_2 - m_1^2.$$

Example 4.3 Let us consider the following information: the support of the distribution is (a, b) and

$$M_1(X) = m_1, M_2(X) = m_2.$$

Analogously to the preceding example, the solution reads as

$$f = A\exp\left(B(x - C)^2\right).$$

However, since the support is not $\mathbb{R}$, we must determine A, B, C by solving the nonlinear system of equations

$$\int_a^b f(x)dx = 1, \int_a^b xf(x)dx = m_1, \int_a^b x^2 f(x)dx = m_2.$$

Notice that (B, C) is the solution of

$$\int_a^b (x - m_1)\exp\left(B(x - C)^2\right)dx = \int_a^b (x^2 - m_2)\exp\left(B(x - C)^2\right)dx = 0.$$

We can use R to solve these equations and determine (B, C). Then A is determined by the equality $A\int_a^b \exp\left(B(x - C)^2\right)dx = 1$. As an example, let us consider $a = 0$, $b = 3$, $m_1 = 5/2$, $m_2 = 9$:

(continued)

Example 4.3 (continued)

```
source("ints.R")
int = ints$new()
fa = function(x,a,b,m){
  B = x[[1]]
  C = x[[2]]
  fd = function(s){s*exp(B*(s-C)^2)}
  aux1 = int$int1(fd,a,b)
  fd = function(s){s^2*exp(B*(s-C)^2)}
  aux2 = int$int1(fd,a,b)
  mm = as.matrix(c(aux1,aux2));
  dim(mm) = dim(m)
  dm = mm - m
  return(dm)}

library(pracma)
m = as.matrix(c(2.5,9))
a = 0
b = 3
f = function(x){fa(x,a,b,m)}
x0 = c(1,1)
sol = fsolve(f,x0)
B = sol$x[[1]]
C = sol$x[[2]]
fb = function(s){s*exp(B*(s-C)^2)}
A = 1/int$int1(fb,a,b)
print(paste("A = ",A))
## [1] "A =  0.215330350883315"
print(paste("B = ",B))
## [1] "B =  0.060985082207061"
print(paste("C = ",C))
## [1] "C =  1.91409344871721"
```

The PDF determined is shown in Fig. 4.1.

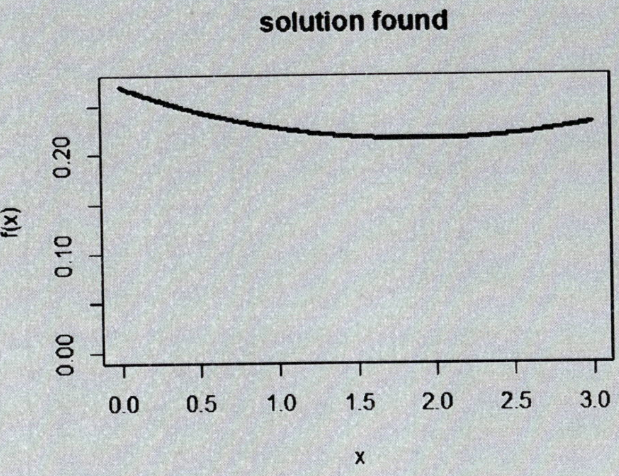

Fig. 4.1 An example of PDF determined by the principle of maximum entropy

Example 4.4 Let us consider the following information: the support of the distribution is $(0, +\infty)$ and

$$M_1(X) = m_1.$$

In this case, the Lagrangian density is

$$\ell = -f \log f + \lambda_0 f + \lambda_1 x f,$$

and we have

$$f = A \exp\left(-\frac{x}{m}\right), \quad A = \exp(-1 + \lambda_0), m = -\frac{1}{\lambda_1} \quad (x > 0).$$

Thus, the solution is an exponential distribution $f = \frac{1}{m_1} \exp\left(-\frac{x}{m_1}\right)$.

Example 4.5 Let us consider the following information: the support of the distribution is $(0, +\infty)$ and

$$M_1(X) = m_1, \quad E(\log X) = \eta.$$

In this case, the Lagrangian density is

$$\ell = -f \log f + \lambda_0 f + \lambda_1 x f + \lambda_2 \log x,$$

and we have

$$f = A\, x^{\lambda_2} \exp(\lambda_1 x), \quad A = \exp(-1 + \lambda_0) \quad (x > 0).$$

The solution is a Gamma distribution

$$f = \frac{x^{a-1} \exp\left(-\frac{x}{\theta}\right)}{\theta^a \Gamma(a)}.$$

The parameters of the distribution can be determined from the properties of the Gamma distribution:

$$a\theta = m_1, \quad \frac{\Gamma'(a)}{\Gamma(a)} + \log \theta = \eta.$$

For a discrete random variable such that $P(X = X_i) = p_i > 0$, $1 \le i \le n$, we have

$$H(X) = H(\boldsymbol{p} = (p_1, \ldots, p_n)) = -\sum_{i=1}^{n} p_i \log p_i. \tag{4.10}$$

In the discrete case, the principle of maximum entropy reads as

Definition 4.2 Let C be the set of the admissible choices for $\boldsymbol{p}$. Estimate the real distribution as

$$\boldsymbol{p}^* = \arg\max\{H(\boldsymbol{p}) : \boldsymbol{p} \in C\}, \tag{4.11}$$

i.e., $\boldsymbol{p}^*$ which maximizes H on C.

Analogously to the continuous case, the minimal information is that

$$\boldsymbol{p} \ge 0, \ |\boldsymbol{p}|_1 = 1$$

and we must consider multipliers $\boldsymbol{\lambda}_+$ and λ_0 associated with these conditions. $\boldsymbol{\lambda}_+$ is eliminated by the same way as in the continuous case: we consider only the values $p_1 > 0$ (as previously mentioned). If additional information is available, under the form

$$\sum_{j=1}^{n} p_j g_i(X_j) = c_i, \ 1 \le i \le k, \tag{4.12}$$

we must consider also the associated Lagrange's multipliers, so that the Lagrangian is

$$\mathcal{L}(\boldsymbol{p}; \boldsymbol{\Lambda}) = H(\boldsymbol{p}) + P(\boldsymbol{p}; \boldsymbol{\Lambda}),$$

where

$$P(\boldsymbol{p}; \boldsymbol{\Lambda}) = \lambda_0 \sum_{r=1}^{n} p_r + \sum_{j=1}^{k} \lambda_j \sum_{r=1}^{n} p_r g_j(X_r) - \lambda_0 - \sum_{j=1}^{k} \lambda_j c_j$$

Thus

$$-1 - \log p_i + \lambda_0 + \sum_{j=1}^{k} \lambda_i g_j(X_i) = 0, 1 \le i \le n.$$

Consequently,

$$p_i = \exp\left(-1 + \lambda_0 + \sum_{j=1}^{k} \lambda_i g_j(X_i)\right), 1 \leq i \leq n.$$

Example 4.6 Let us consider that not any supplementary information is available. Then

$$p_i = \exp(-1 + \lambda_0), 1 \leq i \leq n.$$

Thus, the solution is a uniform distribution $p_i = \frac{1}{n}, 1 \leq i \leq n.$

Example 4.7 Let us consider that the supplementary information is

$$E(X) = m_1.$$

Then, $g_1(x) = x$ and

$$p_i = A \exp(\lambda_1 X_i), A = \exp(-1 + \lambda_0), 1 \leq i \leq n\ .$$

Thus, the solution must be determined numerically, by solving the system of equations

$$A \sum_{i=1}^{k} \exp(\lambda_1 X_i) = 1, A \sum_{i=1}^{k} X_i \exp(\lambda_1 X_i) = m_1.$$

Notice that λ_1 verifies

$$\sum_{i=1}^{k} X_i \exp(\lambda_1 X_i) = m_1 \sum_{i=1}^{k} exp(\lambda_1 X_i).$$

Example 4.8 Let us consider the Jayne's dice problem: a six-faced dice is tossed a large number of times and we observe that the result X verifies

$$E(X) = m.$$

What is the probability distribution of the dice? For a six-faced dice, $X_i = i$ and $n = 6$. According to the preceding example, we have

$$\sum_{i=1}^{6} X_i \exp(i\lambda_1) = m \sum_{i=1}^{k} \exp(i\lambda_1).$$

In his original example, Jaynes used $m = 4.5$. We can determine λ_1, A and the probabilities p_i under R as follows:

```
fa = function(lbda,m){
  s1 = 0
  s2 = 0
  for (i in 1:6){
    s1 = s1 + exp(lbda*i)
    s2 = s2 + i*exp(lbda*i)}
  return(s2 - m*s1)}

library(pracma)
f = function(x){fa(x,4.5)}
sol = fzero(f,0,10)
lbda = sol$x
s = 0
for (i in 1:6){s = s + exp(lbda*i)}
A = 1/s
p = numeric(length=6)
for (i in 1:6){p[[i]] = A*exp(lbda*i)}
print(p)
## [1] 0.05435317 0.07877155 0.11415998 0.16544680 0.23977444 0.
34749407
```

In Fig. 4.2, we show the values of A and λ_1 for different values of m. Figure 4.3 shows the values of p_1 and p_6 for different values of m.

(continued)

Example 4.8 (continued)

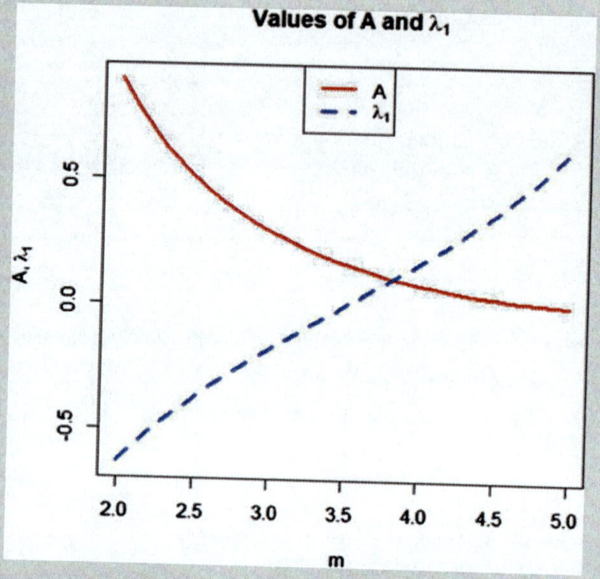

Fig. 4.2 Solution of Jayne's dice for different values of the mean m

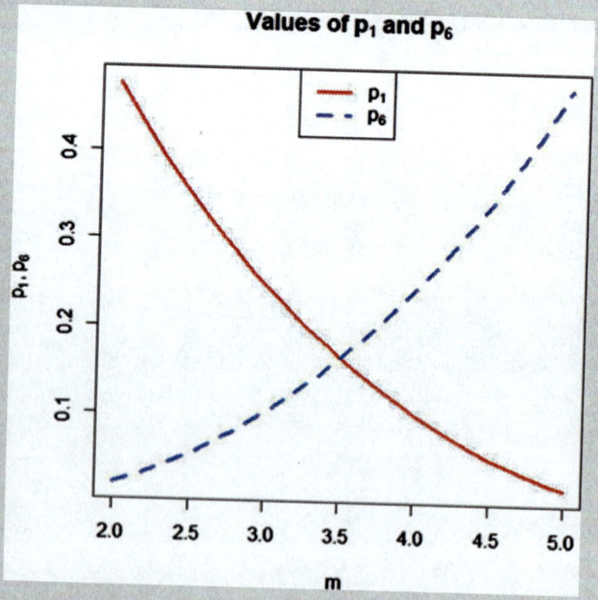

Fig. 4.3 Probabilities p_1 and p_6 of Jayne's dice for different values of the mean m

The principle of maximum entropy also applies to random vectors and random matrices. The results are analogous to those previously presented. Although matrices can be mapped to random vectors, some works consider specific analysis for matrices, namely, in higher dimensions – see, for instance, (Soize 2005, 2008; Batou and Soize 2013; Souza de Cursi and Sampaio 2015).

Example 4.9 Let us consider a vector $\boldsymbol{X} = (X_1, \ldots, X_n) \in \mathbb{R}^n$, supported by $Q = \prod_{i=1}^{n} (a_i, b_i)$. We look for

$$f^* = \arg\max H(f), \quad H(f) = -\int_Q f(\boldsymbol{x}) \log(f(\boldsymbol{x}))d\boldsymbol{x},$$

under the restrictions

$$f \geq 0 \quad \text{and} \quad \int_{\mathbb{R}} f(\boldsymbol{x})d\boldsymbol{x} = 1.$$

Euler-Lagrange's equations read as

$$-1 - \log(f^*(\boldsymbol{x})) + \lambda_0 = 0 \ \ on\ Q.$$

Thus, we have yet

$$f^*(x) = \exp(-1 + \lambda_0) = A \in \mathbb{R} \ \ on\ Q.$$

Again, f^* corresponds to the density of a uniformly distributed continuous variable on Q. Using the condition $\int_{\mathbb{R}} f(x)dx = 1$, we obtain $A = 1/\prod_{i=1}^{n} (b_i - a_i)$.

Example 4.10 Let us consider a vector $\boldsymbol{X} = (X_1, \ldots, X_n) \in \mathbb{R}^n$, supported by $Q = \prod_{i=1}^{n} (0, +\infty)$. Assume that we have the supplementary information

$$E(\boldsymbol{X}) = \boldsymbol{m} \ \ , \boldsymbol{m} = (m_1, \ldots m_n).$$

(continued)

Example 4.10 (continued)

Here, we have n restrictions, corresponding to $g_i(\boldsymbol{x}) = x_i$, $1 \leq i \leq n$. Euler-Lagrange's equations read as

$$-1 - \log(f^*(\boldsymbol{x})) + \lambda_0 + \sum_{i=1}^{n} \lambda_i x_i - \sum_{i=1}^{n} \lambda_i m_i = 0 \quad on \ Q.$$

Thus, we have

$$f^*(\boldsymbol{x}) = A \exp\left(\sum_{i=1}^{n} \lambda_i x_i\right) \quad on \ Q.$$

Thus, f^* corresponds to the density of a vector of independent exponential variables.

Example 4.11 Let us consider a vector $\boldsymbol{X} = (X_1, \ldots, X_n) \in \mathbb{R}^n$. Assume that we have the supplementary information

$$E(\boldsymbol{X}) = \boldsymbol{m} \ , \boldsymbol{m} = (m_1, \ldots m_n);$$

$$cov(\boldsymbol{X}, \boldsymbol{X}) = \boldsymbol{C}, C_{ij} = cov(X_i, X_j).$$

In this case, the principle of maximum entropy furnishes a Gaussian distribution $N(\boldsymbol{m}, \boldsymbol{C})$.

Exercises

1. Consider the following information of the distribution of X:

$$E(X) = m_1, E(|X - m_1|) = c > 0.$$

- Consider the Laplace's distribution $f(x) = \exp(-|x - m_1|/c)/(2c)$. Let g be the density of a variable Y which satisfies the same conditions.

$$E(Y) = m_1, E(|Y - m_1|) = c > 0.$$

(continued)

(a) Show that

- $\int g(x) \log f(x)\, dx = -(1 + \log(2c))$.

(b) Verify that

- $\int f(x) \log f(x)\, dx = -(1 + \log(2c))$

(c) Show that $D(g||f) = H(X) - H(Y)$.

(d) Conclude that f is the Maximum Entropy distribution associated with the given information.

2. Let $X \in \mathbb{N}$ be an enumerable random variable taking the values $X_0, X_1, \ldots$, with $X_i \in \mathbb{N}$, $P(X = X_i) = p_i > 0$, $i \in \mathbb{N}$. Assume that we have the information

- $\sum_{i=0}^{+\infty} p_i X_i = m > 0$.
- Let $\boldsymbol{p} = (p_0, p_1, \ldots) = (p_i\text{: } i \in \mathbb{N})$. Recall that $\sum_{i=0}^{+\infty} p_i = 1$.

(a) Introduce Lagrange's multipliers $\boldsymbol{\Lambda} = (\lambda_0, \lambda_1)$ and write the Lagrangian $\mathcal{L}(\boldsymbol{p}; \boldsymbol{\Lambda})$.

(b) Show that $p_i = A\, B^{X_i}$, with $A, B > 0$.

(c) Conclude that we must have $\sum_{i=0}^{+\infty} B^{X_i} < \infty$.

(d) In the sequel, assume that $X_i = i$. Show that $0 < B < 1$.

(e) Show that $A = 1 - B$.

(f) Show that $B = \frac{m}{1+m}$.

(g) Determine p_i as a function of m.

3. Let $X \in \mathbb{N}$ be an enumerable random variable taking the values $1, 2, \ldots$, with $P(X = i) = p_i > 0$, $i \in \mathbb{N}^*$. Let $0 < p < 1$. Assume that we have the information

- $\sum_{i=1}^{+\infty} i p_i \leq \frac{1}{p}$.
- Let $\boldsymbol{p} = (p_1, p_2, \ldots) = (p_i\text{: } i \in \mathbb{N}^*)$ be such that $p_i = pq^{i-1}$, $q = 1 - p$.

(a) Show that $\sum_{i=1}^{+\infty} i p_i = 1/p$.

(b) Show that $H(\boldsymbol{p}) = -\frac{q}{p} \log q - \log p$.

(c) Let $\boldsymbol{r} = (r_1, r_2, \ldots) = (r_i\text{: } i \in \mathbb{N}^*)$ be other distribution such that $\sum_{i=1}^{+\infty} i r_i \leq \frac{1}{p}$. Verify that $-r_i \log r_i = r_i \log\left(\frac{q^{i-1}}{r_i}\right) - r_i \log(q^{i-1})$.

(continued)

(d) Show that $H(\boldsymbol{r}) = \sum_{i=1}^{+\infty} r_i \log\left(\frac{q^{i-1}}{r_i}\right) - \sum_{i=1}^{+\infty} r_i \log(q^{i-1})$.

(e) Show that $H(\boldsymbol{r}) \le \log \sum_{i=1}^{+\infty} q^{i-1} - \sum_{i=0}^{+\infty} r_i \log(q^{i-1})$.

(f) Verify that $H(\boldsymbol{r}) \le H(\boldsymbol{p})$.

(g) Conclude that $\boldsymbol{p}$ is the Maximum Entropy distribution associated with the given information.

4. A 16-faced dice is rolled a large number of times and we observe that the result X verifies $E(X) = m$. Use R to determine the values of the probabilities $p_1 = P(X = 1)$ and $p_{16} = P(X = 16)$ for $m \in (7, 10)$.
5. Let X be a random variable supported by (0,1). Assume that
   - $E(\log X) = c_1,\ E(\log(1 - X)) = c_2$.
   - Show that the Maximum Entropy distribution associated is a Beta distribution.

4.2 Karhunen-Loève Expansions for the Numerical Generation of Stochastic Processes

The principle of maximum entropy can be used to determine distributions from available information but does not furnish a method for the generation of random numbers. Or Monte Carlo methods request the generation of large samples formed by variates from a random variable, vector, or process. The generation of such large samples may result expensive in terms of computational cost and introduce numerical bias. To avoid such a difficulty, Karhunen-Loève expansions can be used for the generation of stochastic processes.

Karhunen-Loève expansions represent a stochastic process as an infinite sum of time functions multiplied by convenient random vectors. The infinite sums can be truncated to generate a finite approximation of the process, which can be used to generate variates.

Let us consider a stochastic process X_t, indexed by time $t \in (0, T)$. We denote by m_X its mean and by $C(s, t)$ its covariance function (Cf. Sect. sec113). Let us introduce the integral operator $K : L^2(0, T) \to L^2(0, T)$ given by

$$K(\psi)(t) = \int C(t, s)\psi(s)ds. \tag{4.13}$$

We have

Proposition 4.1 Let $\int_0^T E(X_t^2)dt < \infty$. Then C is a Hilbert-Schmidt kernel, so that K is positive, self-adjoint, and compact. Consequently, $L^2(0,T)$ has a Hilbert basis $\{\psi_i : i \in \mathbb{N}^*\}$ formed by eigenfunctions of K. In addition, all the eigenvalues of K are non-negative real numbers and have zero as unique possible cluster point. Moreover, the eigenspaces associated with different eigenvalues are mutually orthogonal, and have finite dimension if the eigenvalues are strictly positive. Finally, if λ_i is the eigenvalue associated with ψ_i, we can represent

$$C(s,t) = \sum_{i\in\mathbb{N}^*} \lambda_i \psi_i(s)\psi_i(t).$$

Proof:

(i) See (Yosida 1995): C is a Hilbert-Schmidt kernel, since (from Schwarz's inequality):

$$|E(X_tX_s)| \le \sqrt{E(X_t^2)}\ \sqrt{E(X_s^2)},$$

so that

$$\int_0^T\int_0^T |C(t,s)|^2 ds\, dt = \int_0^T\int_0^T |E(X_tX_s)|^2 ds\, dt \le \left(\int_0^T E(X_t^2)dt\right)^2 < \infty.$$

(ii) K is positive:

$$\int_0^T K(\psi)(t)\psi(t)dt = \int_0^T \psi(t)dt \int_0^T C(t,s)\psi(s)ds.$$

Thus,

$$\int_0^T K(\psi)(t)\psi(t)dt = \int_0^T \psi(t)dt \int_0^T E(X_tX_s)\psi(s)ds,$$

and

(continued)

Proposition 4.1 (continued)

$$\int_0^T K(\psi)(t)\psi(t)dt = E\left(\int_0^T \psi(s)X_s ds \times \int_0^T \psi(t)X_t dt\right),$$

and

$$\int_0^T K(\psi)(t)\psi(t)dt = E\left(\left(\int_0^T \psi(t)X_t dt\right)^2\right) \geq 0.$$

(iii) K is self-adjoint: since $C(t, s) = C(s, t)$.
(iv) Since C is a Hilbert-Schmidt Kernel, K is compact.
(v) The other properties result from the general properties of compact operators (see, for instance, (Dieudonné 1969)) and the representation yields from Mercer's theorem (Mercer 1909; Ash 1965, reedited 2003).▪

Let us consider the eigenvalue problem

Problem 4.1 Determine $(\lambda, \psi) \in \mathbb{R} \times L^2(0, \mathrm{T})$ such that

$$K(\psi)(t) = \lambda\psi(t) \ , \|\psi\|_{L^2(0,T)} = 1. \tag{4.14}$$

Notice that ψ is a unitary eigenvalue of K. From the proposition above, the eigenvalues are always positive $\lambda_i \geq 0, \forall i \in N^*$, and the problem has a countable set of solutions $\{(\lambda_i, \psi_i) : i \in N^*\}$, which are positive and may be ordered decreasingly: $0 \leq \lambda_{i+1} \leq \lambda_i, \forall i \in N^*$. Karhunen-Loève expansions are defined as follows:

Definition 4.3 Let $\{Z_i : i \in N^*\}$ be a sequence of random variables given by

$$Z_i = \frac{1}{\sqrt{\lambda_i}} \int_0^T (X_t - m_X(t))\psi_i(t)dt \quad \forall i \in \mathbb{N}. \tag{4.15}$$

The Karhunen-Loève expansion of X_t is

$$X_t = m_X(t) + \sum_{i=1}^{\infty} \sqrt{\lambda_i}\ \psi_i(t)\ Z_i. \tag{4.16}$$

Notice that the variables $\{Z_i : i \in N^*\}$ are orthonormal (see, for instance, (Wang 2008; Giambartolomei 2015)):

$$E(Z_i) = 0 \quad \text{and} \quad cov(Z_i, Z_j) = E(Z_i Z_j) = \delta_{ij} \quad \forall i, j \in \mathbb{N}^*. \tag{4.17}$$

The Karhunen-Loève expansion furnishes a representation of the stochastic process X_t as a sum of orthonormal processes, each one defined by the product of a time function $\psi_i(t)$ and a random variable Z_i. The expansion furnishes a practical way for the generation of trajectories of the process, by generating variates of the random variables. In practice, the infinite sum is truncated at a finite number $k > 0$:

$$X_t \approx P_k X_t = m_X(t) + \sum_{i=1}^{k} \sqrt{\lambda_i} \psi_i(t) Z_i. \tag{4.18}$$

Recalling that the eigenvalues λ_i are ranged in decreasing order and that both ψ_i and Z_i are orthonormalized, the truncation error can be estimated as

$$\|X_t - P_k X_t\|^2 = \int_0^T E\left((X_t - P_k X_t)^2\right) dt = \sum_{i=k+1}^{+\infty} \lambda_i^2. \tag{4.19}$$

Example 4.12 (see also (Wang 2008)). Let us consider the Wiener process W_t, $t \in (0, T)$. We have

$$C(s,t) = \min(s,t).$$

Equation (4.14) yields that

$$\int_0^T \min(s,t) \psi(s) ds = \lambda \psi(t).$$

Notice that $\lambda \psi(0) = 0$, so that, for $\lambda > 0$, $\psi(0) = 0$.
Since

$$\frac{d}{dt} \min(s,t) = 1_{t<s}(s) = \begin{cases} 0, & t > s \\ 1, & t < s \end{cases}.$$

We have

(continued)

Example 4.12 (continued)

$$\lambda\psi'(t) = \int_0^T 1_{t<s}(s)\psi(s)ds = \int_t^T \psi(s)ds.$$

Thus, we have

$$\psi''(t) + \omega^2\psi(t) = 0, \quad \omega^2 = \frac{1}{\lambda}.$$

Considering that $\psi(0) = 0$, we have

$$\psi(t) = B_\omega \sin(\omega t).$$

Thus

$$\lambda\omega B_\omega \cos(\omega t) = B_\omega \int_t^T \sin(\omega s)ds = \frac{B_\omega}{\omega}(\cos(\omega t) - \cos(\omega T)),$$

and

$$\cos(\omega T) = 0 \Longrightarrow \omega_i = (2i-1)\frac{\pi}{2T} \Longrightarrow \lambda_i = \frac{4T^2}{(2i-1)^2\pi^2} \quad (i \in \mathbb{N}^*).$$

Let $B_i = B_{\omega_i}$. We have also

$$B_i^2 \int_0^T \sin^2(\omega_i s)ds = 1 \Longrightarrow B_i^2 \frac{T}{2} = 1 \Longrightarrow B_i = \sqrt{\frac{2}{T}}.$$

Consequently,

$$\psi_i(t) = \sqrt{\frac{2}{T}} \sin\left((2i-1)\frac{\pi}{2}\frac{t}{T}\right)$$

the Karhunen-Loève expansion of the Brownian motion on (0,T) is

$$W_t = \sqrt{8T} \sum_{i=1}^{+\infty} \frac{Z_i}{(2i-1)} \quad \sin\left((2i-1)\frac{\pi}{2}\frac{t}{T}\right),$$

with $Z_i \sim N(0, 1)$.

Example 4.13 (see also (Wang 2008)). Let us consider the Ornstein-Uhlenbeck process X_t, $t \in (0, T)$ such that

$$dZ_t = -aZ_t dt + bdW_t, a > 0$$

where W_t is the Wiener process. We have

$$Z_t = Z_0 e^{-at} + b\int_0^t e^{-a(t-s)} dW_t$$

and

$$C(s,t) = c\ \exp(-a|s-t|), \quad c = \frac{b^2}{2a} > 0$$

Equation (4.14) yields that

$$c\int_0^T \exp(-a|s-t|)\psi(s)ds = \lambda\psi(t).$$

Thus, we have

$$ac\int_0^T \text{sign}(s-t)\exp(-a|s-t|)\psi(s)ds = \lambda\psi'(t).$$

Notice that

$$\lambda\psi'(0) = \lambda a\psi(0), \lambda\psi'(T) = -\lambda a\psi(T).$$

Thus, for $\lambda \neq 0$:

$$\psi'(0) - a\psi(0) = \psi'(T) + a\psi(T) = 0.$$

The derivative of *sign* is a multiple of Dirac's delta: *sign'* = 2δ. Hence

(continued)

Example 4.13 (continued)

$$-2ac\int_0^T \delta(s-t)\exp(-a|s-t|)\psi(s)ds + a^2c\lambda\psi(t) = \lambda\psi''(t)$$

and

$$-2ac\psi(t) + a^2c\lambda\psi(t) = \lambda\psi''(t),$$

i.e.,

$$\psi''(t) + \omega^2\psi(t) = 0, \quad \omega^2 = \frac{2ac - a^2c\lambda}{\lambda}.$$

The general solution is

$$\psi(t) = A_\omega\cos(\omega t) + B_\omega\sin(\omega t) = (\cos(\omega t) \quad \sin(\omega t))\begin{pmatrix} A_\omega \\ B_\omega \end{pmatrix},$$

and

$$\psi'(t) = -\omega A_\omega\sin(\omega t) + \omega B_\omega\cos(\omega t) = \omega(-\sin(\omega t) \quad \cos(\omega t))\begin{pmatrix} A_\omega \\ B_\omega \end{pmatrix}.$$

Noticing that

$$\psi(t) = P(t)\begin{pmatrix} A_\omega \\ B_\omega \end{pmatrix}, \quad \psi'(t) = Q(t)\begin{pmatrix} A_\omega \\ B_\omega \end{pmatrix},$$

with

$$P(t) = (\cos(\omega t) \quad \sin(\omega t)), \quad Q(t) = \omega(-\sin(\omega t) \quad \cos(\omega t)),$$

we can write the boundary conditions as

$$\psi'(0) - a\psi(0) = (Q(0) - aP(0))\begin{pmatrix} A_\omega \\ B_\omega \end{pmatrix} = 0;$$

(continued)

Example 4.13 (continued)

$$\psi'(T) + a\psi(T) = (Q(T) + aP(T))\begin{pmatrix} A_\omega \\ B_\omega \end{pmatrix} = 0.$$

Hence, the boundary conditions form a linear system:

$$M\begin{pmatrix} A_\omega \\ B_\omega \end{pmatrix} = \begin{pmatrix} 0 \\ 0 \end{pmatrix},$$

with

$$M = \begin{pmatrix} Q_1(0) - aP_1(0) & Q_2(0) - aP_2(0) \\ Q_1(T) + aP_1(T) & Q_2(T) + aP_2(T) \end{pmatrix},$$

i.e.,

$$M = \begin{pmatrix} -a & \omega \\ -\omega\sin(\omega T) + a\cos(\omega T) & a\sin(\omega T) + \omega\cos(\omega T) \end{pmatrix}.$$

To get a non-null solution, we must have $\det(M) = 0$, so that

$$-(a\sin(\omega T) + \omega\cos(\omega T))a - (a\cos(\omega T) - \omega\sin(\omega T))\omega = 0,$$

i.e.,

$$(\omega^2 - a^2)\sin(\omega T) - 2\omega a\cos(\omega T) = 0.$$

The solutions must verify

$$\cot(\omega_i T) = \frac{\omega_i^2 - a^2}{2a\omega_i}.$$

We can determine the constants $A_i = A_{\omega_i}$, $B_i = B_{\omega_i}$: on the one hand, $\omega_i B_i = aA_i$, so that

$$\psi_i(t) = B_i\left(\frac{\omega_i\cos(\omega_i t)}{a} + \sin(\omega_i t)\right);$$

on the other hand,

(continued)

Example 4.13 (continued)

$$\int_0^T \psi_i^2(t)ds = 1.$$

For each ω_i:

$$A_i = \sqrt{\frac{2\omega_i^2}{2a + T(\omega_i^2 + a^2)}}, \quad B_i = \sqrt{\frac{2a^2}{2a + T(\omega_i^2 + a^2)}}, \quad \lambda_i = \frac{2ac}{\omega_i^2 + a^2 c}.$$

Exercises

1. Consider the Brownian bridge on $(0, T)$:

$$X_t = W_t - \left(\frac{t}{T}\right) W_T.$$

   (a) Show that $E(X_t) = 0$.
   (b) Verify that $E(X_t X_s) = \min(s, t) - \frac{st}{T}$.
   (c) Conclude that $C(t, s) = \min(s, t) - \frac{st}{T}$.
   (d) Let $\lambda\psi(t) = \int_0^T C(t, s)\psi(s)ds$. Verify that $\lambda\psi(0) = 0$.
   (e) Show that $\lambda\psi'(t) = \int_t^T \psi(s)ds - \int_0^T \frac{s\psi(s)}{T} ds$.
   (f) Verify that $\lambda\psi''(t) = -\psi(t)$.
   (g) Show that $\psi(t) = A \sin(\omega t)$, with $\omega = 1/\sqrt{\lambda}$.
   (h) Use (e) to show that $\sin(\omega T) = 0$.
   (i) Conclude that $\psi_i(t) = A_i \sin(\omega_i t), \omega_i = \frac{i\pi}{T}, \ A_i = \sqrt{\frac{2}{T}}, \ \lambda_i = \frac{T^2}{i^2\pi^2}$.
   (j) Verify that $X_t = \sqrt{2T} \sum_{i=1}^{+\infty} \frac{Z_i}{i\pi} \sin\left(\frac{i\pi t}{T}\right), Z_i \sim N(0, 1)$.

2. Consider the covariance function

$$C(t, s) = \cos(a(t - s)) \ (a > 0, 0 \le s, t \le T).$$

   - Let $\lambda\psi(t) = \int_0^T C(t, s)\psi(s)ds$.

   (a) Show that $\lambda\psi(t) = A(\psi) \cos(at) + B(\psi) \sin(at), A(\psi), B(\psi) \in \mathbb{R}$.
   (b) Verify that $\psi''(t) = -\omega^2\psi(t), \omega^2 = \frac{a^2}{\lambda}$.
   (c) Conclude that $\psi(t) = A_\omega \cos(\omega t) + B_\omega \sin(\omega t)$.

(continued)

(d) Show that $\lambda = 1$, $\omega = a$.
(e) Conclude that $X_t = m(t) + AZ_1 \cos(at) + B Z_2 \sin(at)$.

3. Consider the covariance function

$$C(t,s) = -a|t-s| \ (a > 0).$$

- Let $\lambda\psi(t) = \int_0^T C(t,s)\psi(s)ds$.

(a) Show that $\lambda\psi'(t) = -a\int_0^T \mathrm{sign}(t-s)\ \psi(s)ds$.
(b) Verify that $\lambda\psi'(0) = a\int_0^T \psi(s)ds$ and $\lambda\psi'(T) = -a\int_0^T \psi(s)ds$.
(c) Verify that $\lambda\psi''(t) = -a\psi(t)$.
(d) Conclude that $\psi(t) = A\cos(\omega t) + B\sin(\omega t), \omega^2 = \frac{a}{\lambda}$.
(e) Use (b) to show that non-null solutions ψ exist if and only if .$\sin(\omega T) = 0$. Conclude that $\omega_i = \frac{i\pi}{T}, i \in \mathbb{N}^*$.
(f) Use (b) and (e) to show that $B = 0$.
(g) Conclude that $\psi_i(t) = A_i \cos(\omega_i t), \omega_i = \frac{i\pi}{T}, A_i = \sqrt{\frac{2}{T}}, \lambda_i = \frac{T^2}{i^2\pi^2}$.

4.3 Numerical Construction of Karhunen-Loève Expansions

In most of the practical situations, the analytical determination of the eigenvalues λ_i, the eigenfunctions ψ_i cannot be made and the use of numerical methods is requested. Consequently, the determination of Karhunen-Loève expansions is generally made by numerical procedures. Equation (4.13) defines a Fredholm operator and (4.14) is the associated eigenvalue problem: the reader will find in the literature works about this topic. In the framework of Karhunen-Loeve expansions, (Wang 2008) presents a summary of the most popular approaches.

If the analytical expression of the covariance function C is known, we can consider numerical approximations of the operator K defined by (4.13) and use them to determine the solutions of (4.14). Once they are determined, we can generate approximations given by (4.18).

The first approach consists in using a discretization of the integral: assume that the interval $(0, T)$ is discretized in $N > 0$ subintervals (t_{i-1}, t_i), $i = 1, \ldots, N$, $t_i = i\Delta t$, $0 \le i \le N$, $\Delta t = T/nt$. Then, we can approximate the integral using a trapezoidal rule:

$$\int_0^T C(t,s)\psi(s)ds \approx \Delta t\, \boldsymbol{C}(t)\boldsymbol{\Psi}, \tag{4.20}$$

where

$$\boldsymbol{C}(t) = \left(\frac{1}{2}C(t,t_0), C(t,t_1), \ldots, C(t,t_{nt-1}), \frac{1}{2}C(t,t_N)\right),$$
$$\boldsymbol{\Psi} = (\psi_0, \ldots, \psi_{nt})^t; \quad \psi_i = \psi(t_i)\ 0 \le i \le N.$$

Let us write Eq. (4.14) for $t = t_i$, $0 \le i \le N$, using the approximation (4.20). Then,

$$\Delta t\, \mathbb{C}\boldsymbol{\Psi} = \lambda\boldsymbol{\Psi}, \tag{4.21}$$

with

$$\mathbb{C} = (\boldsymbol{C}(t_0), \boldsymbol{C}(t_1), \ldots, \boldsymbol{C}(t_N))^t \ \left(i.e., \ \mathbb{C}_{ij} = C(t_i, t_j)\right). \tag{4.22}$$

Thus, we determine approximations of the eigenvalues and eigenvectors using $\Delta t\, \mathbb{C}$.

Example 4.14 Let us consider $T = 1$ and

$$C(t,s) = \min(t,s).$$

We can generate the matrix $\mathbb{C}$ as follows:

```
T = 1
fC = function(t,s){min(c(t,s))}
N = 1000
t = seq(0,T,length.out = N+1)
dt = T/N
matC = matrix(0, nrow = N+1, ncol = N+1)
for (i in 1: (N+1)){
  ti = t[[i]]
  for (j in 1: (N+1)){
    tj = t[[j]]
    if (j == 0){matC[[i,j]] = 0.5*fC(ti,tj)}
    else{
      if (j == N){matC[[i,j]] = 0.5*fC(ti,tj)}
      else{matC[[i,j]] = fC(ti,tj)}}}}
```

(continued)

Example 4.14 (continued)

Then, the eigenvalues are evaluated as

```
sol1 = eigen(dt*matC,symmetric=TRUE)
vp1 = sol1$values
```

Figures 4.4 and 4.5 show the values of λ_i furnished by the approximation (4.21), with $N = 100$, 1000. The first 50 eigenvalues are compared to the exact ones (Fig. 4.6).

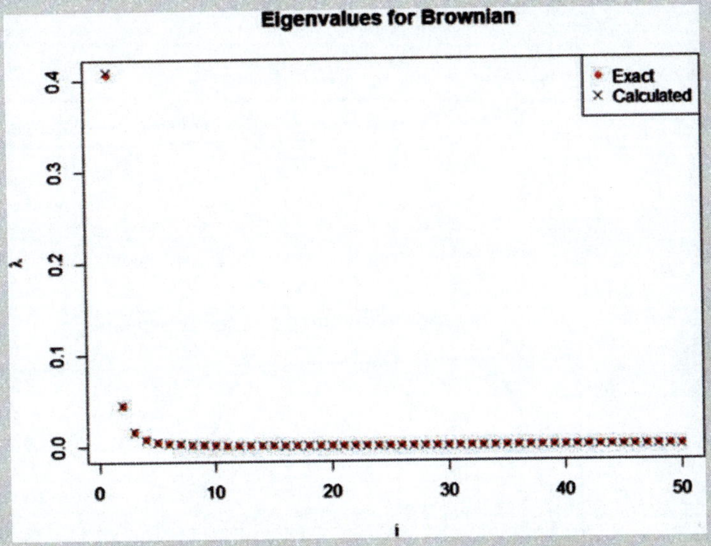

Fig. 4.4 Values of λ generated by the approximation (4.21) (N =**100**)

(continued)

Example 4.14 (continued)

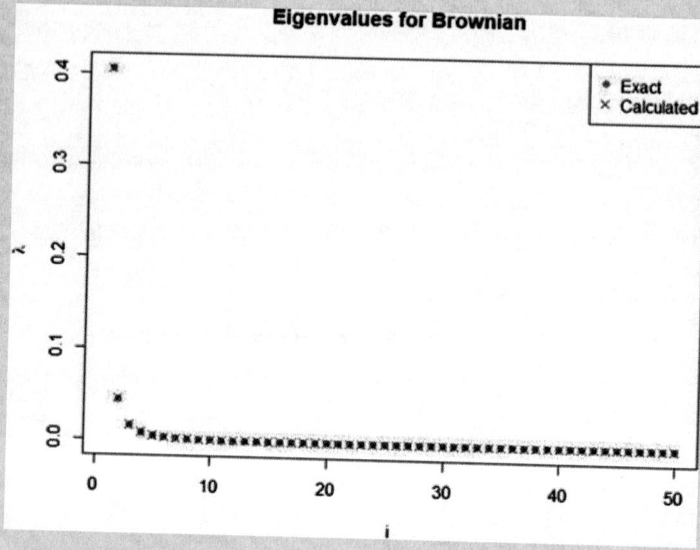

Fig. 4.5 Values of λ generated by the approximation (4.21) ($N = \mathbf{1000}$)

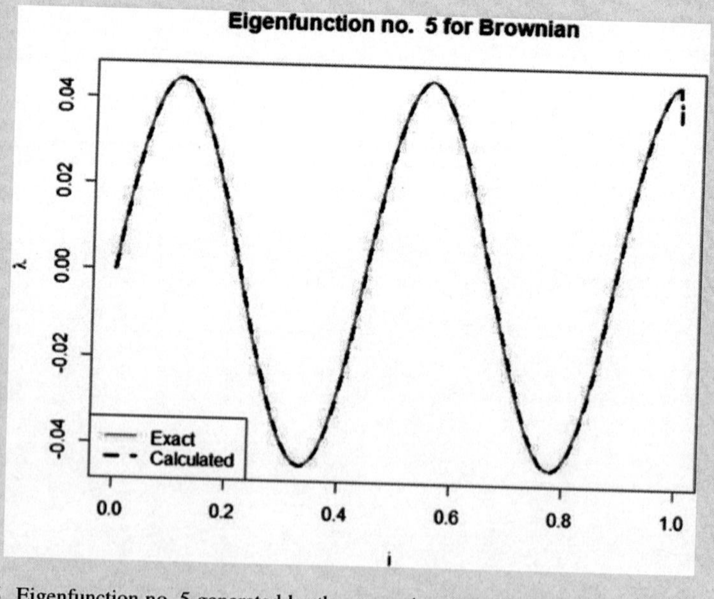

Fig. 4.6 Eigenfunction no. 5 generated by the approximation (4.21) ($N = \mathbf{1000}$)

Let now us consider the situation where the analytical expression of $C(t, s)$ is unknown. In such a situation, we must use the data furnished by samples of X_t–a sample consists in a finite number ns of trajectories $X_t^{(n)}, n = 1, \ldots, ns$. In practice, the trajectories are determined in a finite number of time moments $0 = t_0, \ldots, t_{nt} = T$–for instance, $t_i = i\Delta t$, $i = 0, \ldots, nt$, $\Delta t = T/nt$. In such a situation, each realization consists in a vector $X^{(n)} = \left(X_0^{(n)}, \ldots, X_{nt}^{(n)}\right)$, where $X_i^{(n)} = X^{(n)}(t_i)$. Then, we can estimate the mean and the covariance as follows:

$$m_X(t_i) \approx m_i = \frac{1}{ns} \sum_{n=1}^{ns} X_i^{(n)}, 0 \leq n \leq nt, \tag{4.23}$$

$$C(t_i, t_j) \approx \frac{1}{ns} \sum_{n=1}^{ns} \left(X_i^{(n)} - m_i\right)\left(X_j^{(n)} - m_j\right), 0 \leq n \leq nt. \tag{4.24}$$

Then, we can use these approximations to generate the matrix $\mathbb{C}$ and determine the approximated eigenvalues.

Example 4.15 Let us consider $T = 1$ and the Wiener process W_t, $t \in (0, T)$.

We can generate a sample of trajectories using the package Sim.DifProc as follows (we use a time step $1E - 3$):

```
library(Sim.DiffProc)
xx0 = 0
ns = 1E3
N = 1000
T = 1
dt = T/N
a <- expression(sig)
b <- expression(-teta*x)
X = BM(N = N, M = ns, x0 = xx0, t0 = 0, T = 1, Dt = dt)
```

Then, we determine the matrix $\mathbb{C}$ using the approximation (4.23) and (4.24):

```
matC = matrix(nrow = N+1, ncol = N+1)
for (i in 1:(N+1)){
  for (j in (1:(N+1))){
    matC[[i,j]] = mean( (X[i,] - m[[i]]) *(X[j,] - m[[j]]))
    if ( j == 1){matC[[i,j]]=0.5*matC[[i,j]]}
    if ( j == N){matC[[i,j]]=0.5*matC[[i,j]]}}}
```

(continued)

Example 4.15 (continued)

Figures 4.7 and 4.8 show the values of λ_i furnished by this method with $ns = 1E3$, $1E5$. The first 50 eigenvalues are compared to the exact ones. Figure 4.9 shows the eigenfunction no. 5 (corresponding to λ_5).

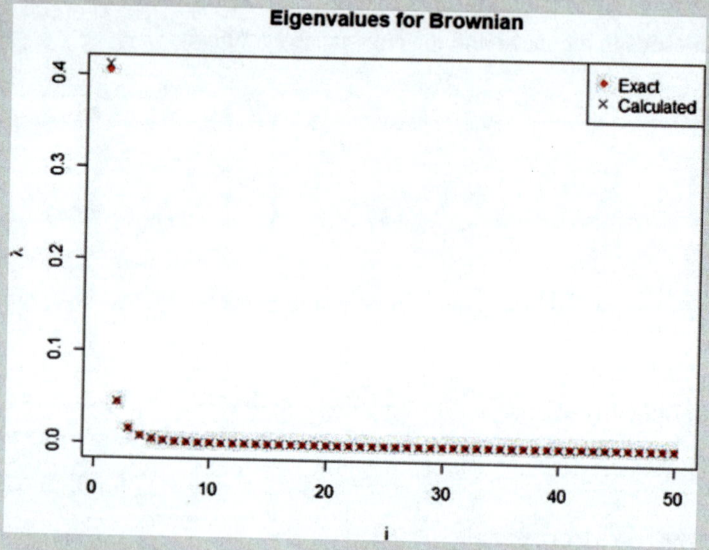

Fig. 4.7 Values of λ generated by the approximation (4.21) ($ns = 1E3$)

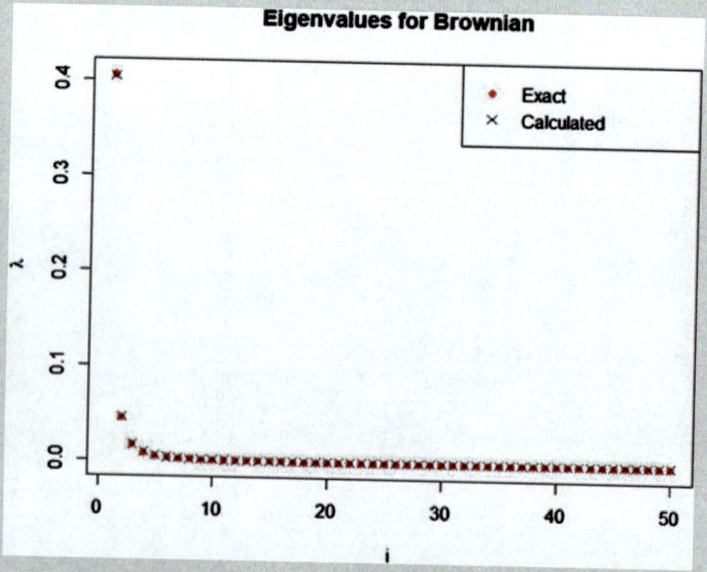

Fig. 4.8 Values of λ generated by the approximation (4.21) ($ns = 1E5$)

(continued)

Example 4.15 (continued)

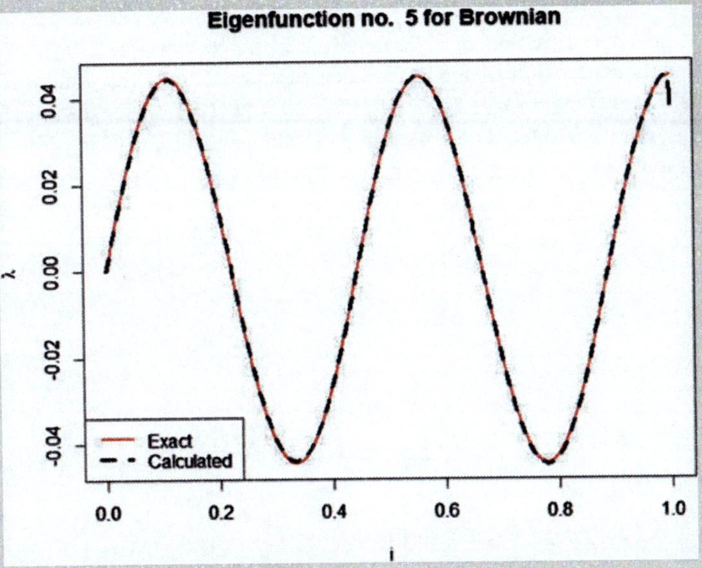

Fig. 4.9 Eigenfunction no. 5 generated by the approximation (4.21) (***ns*** = **1*E*5**)

An alternative approach consists in considering a Hilbert basis or total family $\mathcal{F} = \{\varphi_i : i \in \mathbb{N}\} \subset V = L^2(0,T)$ and the associated sequence of finite-dimensional subspaces

$$V_N = [\{\varphi_i : 0 \leq i \leq N\}] = \left\{ u = \sum_{i=0}^{N} u_i \varphi_i, (u_0, \ldots, u_N) \in \mathbb{R}^{N+1} \right\}$$

We shall assume that uniqueness of the projections onto V_N-such a property is verified by Hilbert basis and many total families, such as wavelets and finite elements. Then, the matrix $\boldsymbol{A} = \{A_{ij} : 0 \leq i,j \leq N\}$, where

$$A_{r\ell} = A_{\ell r} = \int_0^T \varphi_r(s)\varphi_\ell(s)ds$$

has an inverse $\boldsymbol{A}^{-1}$, $\forall\ N \in \mathcal{N}^*$ and the coefficients $(u_0, \ldots, u_N)$ are uniquely determined, $\forall u \in V$. Since $\psi \in L^2(0,T)$, we have

$$\psi = \sum_{r\in\mathbb{N}} \psi_r \varphi_r \approx P_N \psi = \sum_{r=0}^{N} \psi_r \varphi_r, \tag{4.25}$$

so that an approximation of ψ is determined by $\boldsymbol{\Psi} = (\psi_1, \ldots, \psi_N)^t$. We can also consider the expansion of the covariance function C in a convenient Hilbert basis (or total family). $\mathcal{H} = \{\eta_{ij} : i, j \in \mathbb{N}^*\} \subset L^2(0, T; L^2(0, T))$–for instance, we can and take $\eta_{ij}(t, s) = \varphi_i(t)\varphi_j(s)$. Then, we can generate an approximation of the kernel C by using a matrix $\boldsymbol{C} = (C_{\ell j} : 0 \le \ell, j \le N)$. Indeed:

$$C(t,s) = \sum_{j,\ell\in\mathbb{N}} C_{j\ell}\varphi_j(t)\varphi_\ell(s) \approx P_M C(t,s) = \sum_{j,\ell=0}^{N} C_{j\ell}\varphi_j(t)\varphi_\ell(s). \tag{4.26}$$

Then,

$$K(\varphi_r) \approx \sum_{j=0}^{N} \varphi_j(t) \sum_{\ell=0}^{N} C_{j\ell} \int_0^T \varphi_\ell(s)\varphi_r(s)ds = \sum_{j=0}^{N} \mathbb{T}_{jr}\varphi_j(t),$$

where $\mathbb{T} = \{\mathbb{T}_{jr} : 0 \le j, r \le N\}$ verifies

$$\mathbb{T}_{jr} = \sum_{\ell=0}^{N} C_{j\ell} A_{\ell r} \; (\; i.e., \mathbb{T} = \boldsymbol{CA}). \tag{4.27}$$

Then,

$$K(\psi) \approx \sum_{r=0}^{N} \psi_r K(\varphi_r) = \sum_{j=0}^{N} \varphi_j \sum_{r=0}^{N} \mathbb{T}_{jr}\psi_r,$$

Thus, the equality $K(\psi) = \lambda\psi$ can be approximated as

$$\sum_{j=0}^{N} \varphi_j \sum_{r=0}^{N} \mathbb{T}_{jr}\psi_r = \lambda \sum_{j=0}^{N} \psi_j \varphi_j(t).$$

The uniqueness of the coefficients shows that

$$\sum_{r=0}^{N} \mathbb{T}_{jr}\psi_r = \lambda\psi_j, 0 \le j \le N,$$

i.e.,

$$\mathbb{T}\boldsymbol{\Psi} = \lambda\boldsymbol{\Psi}. \tag{4.28}$$

And we can look for the eigenvalues and eigenvectors of $\mathbb{T} = \boldsymbol{CA}$.

Example 4.16 Let us consider $T = 1$ and

$$C(t, s) = \min(t, s).$$

Let us consider the approximation of C by finite elements. We consider the $P1$-family generated by the nodes $t_i = ih$, $h = T/N$, $i = 0, \ldots, N$:

$$\varphi_i(t) = 1 - \frac{|t - t_i|}{h}, \textit{if } |t - t_i| \le h; \varphi_i(t) = 0, \text{otherwise}.$$

It is usual to introduce the functions $N_1(\xi) = 1 - \xi$, $N_2(\xi) = \xi$, and write

$$\varphi_i(t) = \begin{cases} N_2(1 + \xi), \textit{if } -1 \le \xi \le 0 \\ N_1(\xi), \textit{if } 0 \le \xi \le 1 \\ 0, \text{otherwise} \end{cases}, \quad \xi = \frac{t - t_i}{h}.$$

On the element (t_i, t_{i+1}), the value of $u(t)$ is approximated by

$$u(t) \approx u(t_i)N_1(\xi) + u(t_{i+1})N_2(\xi), \ \xi = \frac{t - t_i}{h}.$$

In this case, $\boldsymbol{A}$ is a tridiagonal matrix, corresponding to the Mass Matrix of the finite elements. We can generate the approximation using the class below:

(continued)

Example 4.16 (continued)

```
library(R6)
#
feP1=R6Class("feP1",
 public=list(
  lb = 0,ub = 1,nel = 10,h=NA, nodes=NA, nn = NA,
  N1=NA,N2 = NA, shaperef=NA, shape = NA, element = NA,
  initialize=function(lb1,ub1,nel1)
   {if (missing(lb1)==FALSE) {self$lb = lb1}
    if (missing(ub1)==FALSE) {self$ub = ub1}
    if (missing(nel1)==FALSE) {self$nel = nel1}
    self$h = (self$ub -self$lb)/self$nel
    self$nn = self$nel + 1
    self$nodes = seq(self$lb,self$ub,length.out = self$nn)
    self$element = vector(length=self$nel,'list')
    for (i in 1:self$nel){
      self$element[[i]] = c(i, i+1)}
    self$N1 = function(ksi){1 - ksi}
    self$N2 = function(ksi){ksi}
    self$shaperef = function(ksi){
      v = 0
      if ((ksi < 0) && ksi > -1){v = self$N2(1+ksi)}
      if ((ksi > 0) && ksi < 1){v = self$N1(ksi)}
      if (ksi == 0){v = 1}
      return(v)}
    self$shape = function(i,x){
      ksi = (x - self$nodes[[i]])/self$h
      v = self$shaperef(ksi)
      return(v) }
  },
  value_at = function(x,coef){
   val = 0
   for (i in 1:self$nn){
     val = val + coef[[i]]*self$value(i,x)}
   return(val)},
  value = function(i, x){
   val = 0
   cval = FALSE
   if (i == 1) {
    if (x < self$nodes[[i+1]]) {cval = TRUE}}
   else{
    if (i == (self$nn)){
     if (x > self$nodes[[i-1]]) {cval = TRUE}}
     else
      if ( x > self$nodes[[i-1]]){
       if ( x < self$nodes[[i+1]]){cval = TRUE}}}
```

(continued)

Example 4.16 (continued)

```
    if (cval) {val = 1 - abs(x - self$nodes[[i]])/self$h}
    return(val)},
   massmat = function(k){
    if (missing(k)==TRUE){k = self$nn}
    mel = matrix( c(2, 1,1,2),nrow=2,byrow=TRUE)
    M = matrix(0,nrow=k,ncol=k)
    for (i in 1: (k-1)){
     ind = matrix(c(i,i+1),nrow = 1)
     M[ind,ind] = M[ind,ind] + mel}
    M = M*self$h/6
    return(M)}
 )
)
```

Then, we can consider the basis generated by $\eta_{ij}(t,s) = \varphi_i(t)\varphi_j(s)$, which can be defined by the class

```
library(R6)
source("feP1.R")
#
feP1P1=R6Class("feP1P1",
 public=list(
  lb = NA,ub = NA,h=NA, nodes=NA, nel = NA,
  nn = NA, P1x = NA, P1y = NA, indbasis = NA,
  initialize=function(P1x,P1y)
  {if (missing(P1x)==TRUE) {self$P1x = fep1$new(0,1,10)}
    else{self$P1x=P1x}
  if (missing(P1y)==TRUE) {self$P1y = fep1$new(0,1,10)}
    else{self$P1y=P1y}
  self$h = c(self$P1x$h,self$P1y$h)
  self$lb = c(self$P1x$lb,self$P1y$lb)
  self$ub = c(self$P1x$ub,self$P1y$ub)
  self$nel = self$P1x$nel*self$P1y$nel
  self$nn = self$P1x$nn*self$P1y$nn
  self$nodes = vector('list', length = self$nn)
  for (i in 1:self$P1x$nn){
   for (j in 1:self$P1y$nn){
    k = self$number_node(i,j)
    self$nodes[[k]] = c(self$P1x$nodes[[i]],self$P1y$nodes[[j]])
}}
```

(continued)

Example 4.16 (continued)

```
    self$indbasis = self$create_indexes(self$P1x$nn,self$P1x$nn)
    },
  number_node = function(i,j){
    k = (i-1)*self$P1x$nn + j
    return(k)},
  create_indexes=function(k1, k2){
    v = list(rep(NA,k1*k2))
    k = 0
    for (i in 1 : k1){
      for (j in 1 : k2){
        k = k + 1
        v[[k]] = c(i,j)}}
    return(v)},
  value_at = function(x,coef){
    val = 0
    for (k in 1:self$nn){
      val = val + coef[[k]]*self$value(k,x)}
    return(val)},
  value= function(k, x){
     aux = self$indbasis[[k]]
     i = aux[[1]]
     j = aux[[2]]
     aux2 = self$value1d(i, x[[1]],self$P1x) * self$value1d(j, x
[[2]],self$P1y)
      return(aux2)},
    value1d = function(i, x,p){
     val = 0
     cval = FALSE
     if (i == 1) {
      if (x < p$nodes[[i+1]]) {cval = TRUE}}
     else{
      if (i == (p$nn)){
       if (x > p$nodes[[i-1]]) {cval = TRUE}}
       else
        if ( x > p$nodes[[i-1]]){
         if ( x < p$nodes[[i+1]]){cval = TRUE}}}
    if (cval) {val = 1 - abs(x - p$nodes[[i]])/p$h}
    return(val)}
 )
)
```

(continued)

Example 4.16 (continued)

In this case, $C_{j\ell} = C(t_j, t_\ell)$. For instance, we can generate an approximation of C with $N = 20$ as follows:

```
source("feP1.R")
p1x = feP1$new(0,1,20)
p1y = feP1$new(0,1,20)

source("feP1P1.R")
p12d = feP1P1$new(p1x,p1y)

vf = function (f){
  vf = function(x,y){
    v = matrix(NA,nrow = length(x),ncol = length(y))
    for (i in (1:length(x)) ){
      for (j in (1: length(y)) ){
        aux = f(x[[i]],y[[j]])
        v[[i,j]] = f(x[[i]],y[[j]]) }}
    return(v)}
  return(vf)}

kern = function(t,s){min(c(t,s))}
x = p1x$nodes
y = p1y$nodes
vkern = vf(kern)
uu = vkern(x,y)
Pkern = function(x,y){ p12d$value_at(c(x,y),uu)}
```

Figure 4.10 compares the approximation and the exact values on a uniform grid 100 × 100. The RMS error is 2E-4.

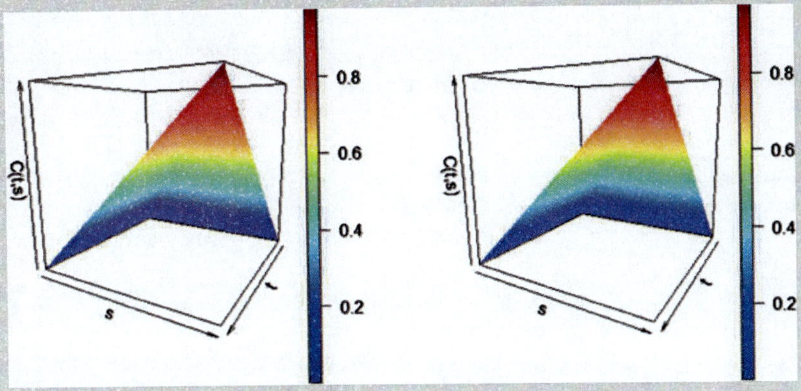

Fig. 4.10 Approximation of C using a family of 1D-finite elements ($N = 20$)

(continued)

Example 4.16 (continued)

We can use (4.28) to determine the eigenvalues. Notice that the class `feP1P1` is not requested for the determination of the eigenvalues – an example of code is given below. Figure 4.11 shows the values of λ_i determined by these commands. Figure 4.12 exhibits an eigenfunction.

```
A1 = p1x$massmat()
matT1 = uu %*% A1
sol1 = eigen(matT1,symmetric = TRUE)
vp1 = sol1$values
```

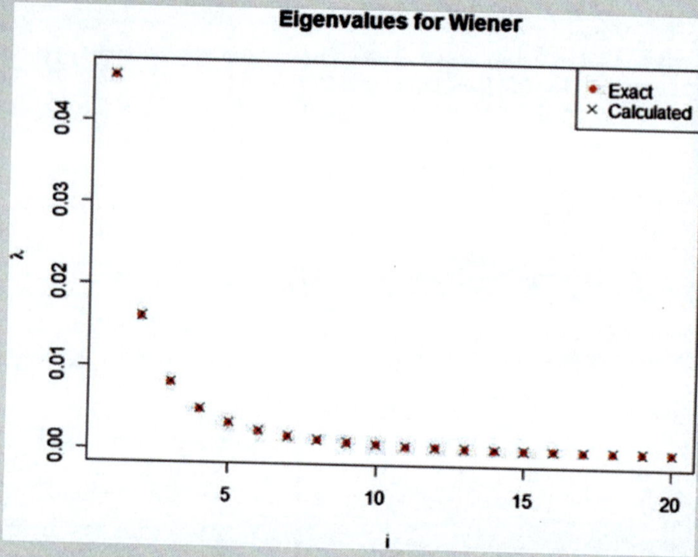

Fig. 4.11 Values of λ generated by the family of 1D-finite elements ($N = 20$)

(continued)

Example 4.16 (continued)

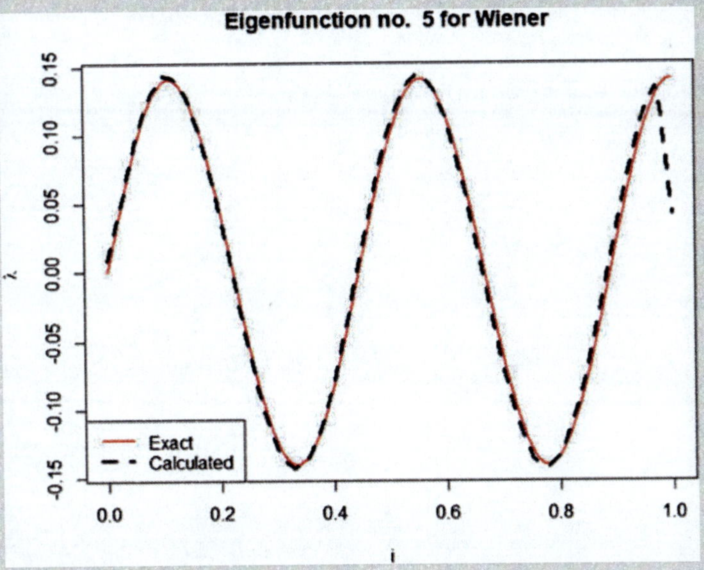

Fig. 4.12 Eigenfunction no. 5 generated by1D-finite elements (*N* = **20**)

Of course, you can prefer to use a Hilbert basis (or total family). From $\mathcal{H} = \{\eta_{ij} : i, j \in \mathbb{N}^*\} \subset L^2(0, T; L^2(0, T))$, such as, for instance, 2D finite elements. In this case,

$$C(t,s) = \sum_{\ell \in \mathbb{N}} C_\ell \eta_\ell(t,s) \approx P_M C(t,s) = \sum_{\ell=0}^{M} C_\ell \eta_\ell(t,s). \tag{4.29}$$

Then,

$$K(\varphi_r) \approx \sum_{\ell=0}^{M} C_\ell H_{\ell r}(t), H_{\ell r}(t) = \int_0^T \eta_\ell(t,s)\varphi_r(s)ds, \tag{4.30}$$

and

$$K(\psi) \approx \sum_{r=0}^{N} \psi_r K(\varphi_r) = \sum_{r=0}^{N} \psi_r \sum_{\ell=0}^{M} C_\ell H_{\ell r}(t),$$

Thus, the equality $K(\psi) = \lambda\psi$ can be approximated as

$$\sum_{r=0}^{N} \psi_r \sum_{\ell=0}^{M} C_\ell H_{\ell r}(t) = \lambda \sum_{r=0}^{N} \psi_r \varphi_r(t).$$

Taking the scalar product by φ_j, we obtain

$$\sum_{r=0}^{N} \mathbb{M}_{jr} \psi_r = \lambda \sum_{r=0}^{N} \psi_r A_{jr} \ , 0 \leq j \leq N,$$

where

$$\mathbb{M}_{jr} = \sum_{\ell=0}^{M} C_\ell \int_0^T H_{\ell r}(t) \varphi_j(t) dt, 0 \leq j, r \leq N, \tag{4.31}$$

i.e.,

$$\mathbb{M}_{jr} = \sum_{\ell=0}^{M} C_\ell D_{\ell jr}, 0 \leq j, r \leq N, \tag{4.32}$$

with

$$D_{\ell jr} = \int_0^T \int_0^T \eta_\ell(t, s) \varphi_r(s) \varphi_j(t) \, ds \, dt. \tag{4.33}$$

Again, we can determine approximations of the eigenvalues and the eigenvectors by solving a finite-dimensional eigenvalue problem: here, we have

$$\mathbb{M}\boldsymbol{\psi} = \lambda \mathbf{A}\boldsymbol{\psi} \Leftrightarrow (\mathbf{A}^{-1}\mathbb{M})\boldsymbol{\psi} = \lambda\boldsymbol{\psi}. \tag{4.34}$$

Example 4.17 Let us consider again $T = 1$ and

$$C(t, s) = \min(t, s).$$

Let us consider the approximation of C by $Q1$-finite elements. In this case, the region $(0, T) \times (0, T)$ is discretized using the nodes $P_\ell = (x_\ell, y_\ell)$ distributed on a grid on the region. We can generate the approximation using the class feQ1 below, which uses a mesh $n_x \times n_y$, based on the coordinates $(i - 1)h_x$, $(j - 1)h_y$, $h_x = T/n_x$, $h_y = T/n_y$.

(continued)

Example 4.17 (continued)

```
library(R6)
#
feQ1=R6Class("feQ1",
 public=list(
  lb = c(0,0),ub = c(1,1),nint = c(10,10),
  h=NA, nodes=NA, nn = NA, nel = NA, xs = NA,
  ys = NA, nnx = NA, nny = NA, element = NA,
  N1 = NA, N2 = NA, N3 = NA, N4 = NA, x=NA,y=NA,
  nelnode = NA, shaperef = NA, shape = NA,
  elnode=NA,elij=NA,
  initialize=function(lb1,ub1,nint1){
   if (missing(lb1)==FALSE) {self$lb = lb1}
    if (missing(ub1)==FALSE) {self$ub = ub1}
    if (missing(nint1)==FALSE) {self$nint = nint1}
    self$h = (self$ub -self$lb)/self$nint
    self$nnx = self$nint[[1]] + 1
    self$nny = self$nint[[2]] + 1
    self$nn = self$nnx*self$nny
    self$nel = self$nint[[1]]*self$nint[[2]]
    self$x = seq(self$lb[[1]],self$ub[[1]],length.out = self$nnx
)
    self$y = seq(self$lb[[2]],self$ub[[2]],length.out = self$nny
)
    self$xs = numeric(length = self$nn)
    self$ys = numeric(length = self$nn)
    for (i in 1:self$nnx){
      for (j in 1:self$nny){
        k = self$number_node(i,j)
        self$xs[[k]] = self$x[[i]]
        self$ys[[k]] = self$y[[j]]}}
    self$element = vector('list', length = self$nel)
    self$nelnode = numeric( length = self$nn)
    self$elnode = matrix(list(), nrow = self$nn,ncol=1)
    self$elij = matrix(list(), nrow = self$nnx, ncol = self$nny)
    for (i in 1:self$nint[[1]]){
      for (j in 1:self$nint[[2]]){
        k1 = self$number_node(i,j)
        k2 = self$number_node(i+1,j)
        k3 = self$number_node(i+1,j+1)
        k4 = self$number_node(i,j+1)
        k = self$number_element(i,j)
        self$element[[k]] = c(k1,k2,k3,k4)
        self$nelnode[[k1]] = self$nelnode[[k1]] + 1
        self$nelnode[[k2]] = self$nelnode[[k2]] + 1
        self$nelnode[[k3]] = self$nelnode[[k3]] + 1
        self$nelnode[[k4]] = self$nelnode[[k4]] + 1
        self$elnode[[k1]] = c(self$elnode[[k1]],k)
        self$elnode[[k2]] = c(self$elnode[[k2]],k)
```

(continued)

Example 4.17 (continued)

```
        self$elnode[[k3]] = c(self$elnode[[k3]],k)
        self$elnode[[k4]] = c(self$elnode[[k4]],k)
        self$elij[[i,j]] = c(self$elij[[i,j]],k)
        self$elij[[i+1,j]] = c(self$elij[[i+1,j]],k)
        self$elij[[i+1,j+1]] = c(self$elij[[i+1,j+1]],k)
        self$elij[[i,j+1]] = c(self$elij[[i,j+1]],k)}}
    self$nodes = vector('list', length = self$nn)
    for (i in 1:self$nnx){
      for (j in 1:self$nny){
        k = self$number_node(i,j)
        self$nodes[[k]] = c(self$xs[[k]],self$ys[[k]])}}
    self$N1 = function(ksi,eta){(1-ksi)*(1-eta)}
    self$N2 = function(ksi,eta){ksi*(1-eta)}
    self$N3 = function(ksi,eta){ksi*eta}
    self$N4 = function(ksi,eta){(1-ksi)*eta}
    self$shaperef = function(ksi,eta){
      v = 0
      if ((ksi < 0) && (ksi > -1)){
        if ((eta <= 0) && (eta > -1)){v = self$N3(1+ksi,1+eta)}
        if ((eta >= 0) && (eta < 1)){v = self$N2(1+ksi,eta)}
        if (eta == 0){v = self$N3(1+ksi,1)} }
      if ((ksi > 0) && (ksi < 1)){
        if ((eta <= 0) && (eta > -1)){v = self$N4(ksi,1+eta)}
        if ((eta >= 0) && (eta < 1)){v = self$N1(ksi,eta)}
        if (eta == 0){v = self$N4(ksi,1)} }
      if ((ksi == 0)){
        if ((eta <= 0) && (eta > -1)){v = self$N3(1,1+eta)}
        if ((eta >= 0) && (eta < 1)){v = self$N1(0,eta)}
        if (eta == 0){v = 1}}
      return(v)}
    self$shape = function(i,x){
      aux = self$nodes[[i]]
      ksi = (x[[1]] - aux[[1]])/self$h[[1]]
      eta = (x[[2]] - aux[[2]])/self$h[[2]]
      v = self$shaperef(ksi,eta)
      return(v)}},
  number_node = function(i,j){
    k = (j-1)*self$nnx + i
    return(k)},
  number_element = function(i,j){
    k = (j-1)*self$nint[[1]] + i
    return(k)},
  invnumber_node = function(k){
    i = k %% self$nnx
    if (i == 0) {i = self$nnx}
    j = 1 + (k - i)/self$nnx
    return(c(i,j))},
  value_at = function(x,coef){
    val = 0
```

(continued)

Example 4.17 (continued)

```
    for (k in 1:self$nel){
      ind = self$element[[k]]
      cof = coef[ind]
      w = self$nelnode[ind]
      vel = self$valelem(ind,x)
      val = val + sum(cof*vel/w)}
    return(val)},
  valelem = function(ind,x){
    resu = rep(0,1,length(ind))
    for (i in 1:length(ind)){
     xp = self$xs[[ ind[[i]] ]]
     yp = self$ys[[ ind[[i]] ]]
     ksi = (x[[1]] - xp)/self$h[[1]]
     eta = (x[[2]] - yp)/self$h[[2]]
     resu[[i]] = self$shape(ind[[i]],x)
    }
    return(resu)
  },
  intrombel = function(i,f,integ1){
    ind = self$element[[i]]
    i1 = ind[[1]]
    i2 = ind[[2]]
    i3 = ind[[3]]
    i4 = ind[[4]]
    P1 = self$nodes[[i1]]
    P2 = self$nodes[[i2]]
    P3 = self$nodes[[i3]]
    P4 = self$nodes[[i4]]
    ax = P1[[1]]
    bx = P2[[1]]
    ay = P2[[2]]
    by = P3[[2]]
    a = c(ax,ay)
    b = c(bx,by)
    s = integ1$romb2(f,a,b,40,1e-10)
    return(s)},
  matMg = function(pg,P1,coef){
    source("ints.R")
    int1 = ints$new()
      M = matrix(0,nrow = P1$nn, ncol = P1$nn)
      for (n in 1: self$nel){
        indel = self$element[[n]]
        ind1 = self$invnumber_node( indel[[1]])
        ii = ind1[[1]]
        jj = ind1[[2]]
        indsel = list(c(ii,jj),c(ii+1,jj),c(ii+1,jj+1),c(ii,jj+1
))
```

(continued)

Example 4.17 (continued)

```
        for (i in 1: length(indel)){
          ele = indel[[i]]
          for (k in 1: length(indsel)){
            aux = indsel[[k]]
            j = aux[[1]]
            r = aux[[2]]
            fonc = function(x,y){
              v1 = self$shape(ele,c(x,y))
              v2 = P1$shape(j,x)
              v3 = P1$shape(r,y)
              return(v1*v2*v3)}
            M[[j,r]] = M[[j,r]] + coef[[ele]]*self$intrombel(n,f
onc,int1)}}}
      return(M)}
  )
)
```

In this case, $C_\ell = C(x_\ell, y_\ell)$. We can generate an approximation of C with $n_1 = n_2 = 20$ using

```
source("feQ1.R")
fe1 = feQ1$new(c(0,0),c(1,1),c(20,20))
x = fe1$x
y = fe1$y
vkern = vf(kern)
uu = vkern(x,y)
Pkern = function(x,y){ fe1$value_at(c(x,y),uu)}
```

Figure 4.13 compares the approximation and the exact values on a uniform grid 100 × 100. The RMS error is 2E-4 yet. Figure 4.14 exhibits the eigenvalues determined by this approach. Figure 4.15 exhibits an eigenfunction.

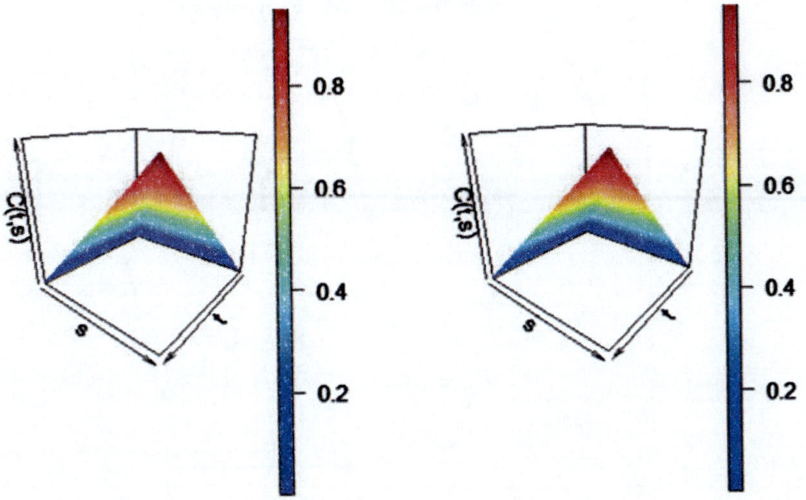

Fig. 4.13 Approximation of C using a family of Q1-finite elements (*mesh* **40 × 40**)

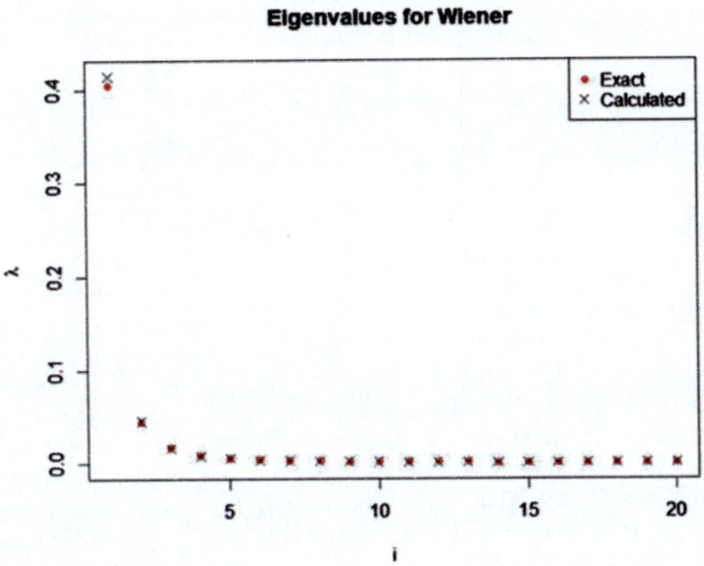

Fig. 4.14 Values of λ generated by tQ1-finite elements (*mesh* **40 × 40**)

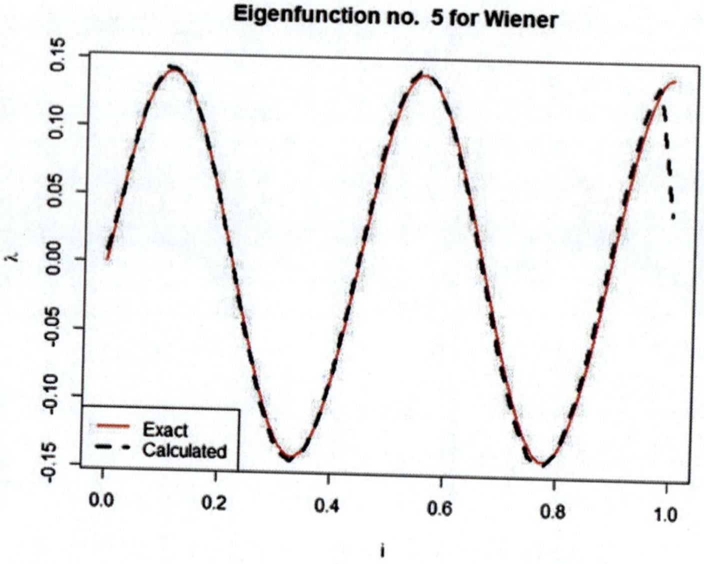

Fig. 4.15 Eigenfunction no. 5 generated byQ1-finite elements (*mesh* **40 × 40**)

Exercises

1. Consider the Brownian bridge on (0, 1):

$$X_t = W_t - tW_1, \quad C(s,t) = \min(s,t) - st.$$

(a) Determine the 20 first eigenvalues for the Karhunen-Loève expansion, using the method shown in Example 4.14. Plot the eigenfunctions 2, 3, and 4.
(b) Determine the 20 first eigenvalues for the Karhunen-Loève expansion, using the method shown in Example 4.15. Plot the eigenfunctions 2, 3, and 4.
(c) Determine the 20 first eigenvalues for the Karhunen-Loève expansion, using the method shown in Example 4.16. Plot the eigenfunctions 2, 3, and 4.
(d) Determine the 20 first eigenvalues for the Karhunen-Loève expansion, using the method shown in Example 4.17. Plot the eigenfunctions 2, 3, and 4.

(continued)

2. Consider the covariance function

$$C(t,s) = \exp(-|t-s|)\ (\ 0 \le s,t \le 1).$$

   (a) Determine the 20 first eigenvalues for the Karhunen-Loève expansion, using the method shown in Example 4.14. Plot the eigenfunctions 2, 3, and 4.
   (b) Determine the 20 first eigenvalues for the Karhunen-Loève expansion, using the method shown in Example 4.16. Plot the eigenfunctions 2, 3, and 4.
   (c) Determine the 20 first eigenvalues for the Karhunen-Loève expansion, using the method shown in Example 4.17. Plot the eigenfunctions 2, 3, and 4.

3. Consider the covariance function

$$C(t,s) = -|t-s|\ \ (0 \le t,s \le 1.$$

   (a) Determine the 20 first eigenvalues for the Karhunen-Loève expansion, using the method shown in Example 4.14. Plot the eigenfunctions 2, 3, and 4.
   (b) Determine the 20 first eigenvalues for the Karhunen-Loève expansion, using the method shown in Example 4.16. Plot the eigenfunctions 2, 3, and 4.
   (c) Determine the 20 first eigenvalues for the Karhunen-Loève expansion, using the method shown in Example 4.17. Plot the eigenfunctions 2, 3, and 4.

References

Ash, R. (1965, reedited 2003). *Information Theory*. New York, NY, USA: Dover.

Attiaoui, D., Doré, P., Martin, A., & Ben Yaghlane, B. (2012). A Distance between Continuous Belief Functions. In E. Hüllermeier, S. Link, T. Fober, & B. Seeger (Ed.), *Scalable Uncertainty Management: SUM 2012* (pp. 194–205). Marburg, Germany: Springer. doi:https://doi.org/10.1007/978-3-642-33362-0_15

Aurnhammer, C., & Frank, S. L. (2019). Evaluating information-theoretic measures of word prediction in naturalistic sentence reading. *Neuropsychologia, 134*. doi:https://doi.org/10.1016/j.neuropsychologia.2019.107198

Baldi, P., & Itti, L. (2010). Of bits and wows: A Bayesian theory of surprise with applications to attention. *Neural Networks, 23*(5), 649–666. doi:https://doi.org/10.1016/j.neunet.2009.12.007

Batou, A., & Soize, C. (2013). Calculation of Lagrange Multipliers in the Construction of Maximum Entropy Distributions in High Stochastic Dimension. *SIAM/ASA Journal on Uncertainty Quantification, 1*(1). doi:https://doi.org/10.1137/120901386

Bayarria, M. J., & Morales, J. (2003). Bayesian measures of surprise for outlier detection. *Journal of Statistical Planning and Inference, 111*(1–2), 3–22. doi:https://doi.org/10.1016/S0378-3758(02)00282-3

Benavoli, A. (2014). Belief function and multivalued mapping robustness in statistical estimation. *International Journal of Approximate Reasoning, 55*, 311–329. doi:https://doi.org/10.1016/j.ijar.2013.04.014

Beretta, G. P. (2008). Axiomatic Definition of Entropy for Nonequilibrium States. *International Journal of Thermodynamics, 11*(2), 39–48. doi:https://doi.org/10.5541/ijot.211

Boivin, C. (2022a, December 25). https://cran.r-project.org/web/packages/dst/vignettes/Captain_Example.html. Retrieved from Captain's Example: https://cran.r-project.org/web/packages/dst/vignettes/Captain_Example.html

Boivin, C. (2022b, 12 24). *Introduction to Belief Functions.* Retrieved from https://cran.r-project.org/web/packages/dst/vignettes/: https://cran.r-project.org/web/packages/dst/vignettes/Introduction_to_Belief_Functions.html#fn2

Boivin, C. (2022c, December 25). *Introduction to Belief Functions: The Monty Hall Game.* Retrieved 2022, from https://cran.microsoft.com/snapshot/2018-08-11/web/packages/dst/vignettes/Monty-hall-Example.html: https://cran.microsoft.com/snapshot/2018-08-11/web/packages/dst/vignettes/Monty-hall-Example.html

Boltzmann, L. E. (1866). Über die mechanische Bedeutung des zweiten Hauptsatzes der Wärmetheorie. *Wiener Berichte, 53*, 195–220. Retrieved from http://opacplus.bsb-muenchen.de/title/BV020135572/ft/bsb10133426?page=3

Boltzmann, L. E. (1877 traduction 2002). On the relationship between the second main theorem of mechanical heat theory and the probability calculation with respect to the results about the heat equilibrium. *Akademie der Wissenschaften in Wien Mathematisch-naturwissenschaftliche Klasse Sitzungsberichte, 2*(76), 373–435. Retrieved February 14, 2023, from http://users.polytech.unice.fr/~leroux/boltztrad.pdf

Boltzmann, L. E. (1896). *Vorlesungen �uber Gastheorie* (Vol. 1). Leipzig: Barth.

Boltzmann, L. E. (1995). *Lectures on Gas Theory.* (S. G. Brush, Trans.) New York, NY, USA: Dover.

Boyle, G. H. (n.d.). *modelingcommons.org/file/download/6101?file_id=3384.* Retrieved from http://modelingcommons.org/file/download/6101?file_id=3384

Brillouin, L. (1956). *Science and Information Theory.* New York, NY, USA: Dover.

Campagner, A., Ciucci, D., & Denœux, T. (2022). Belief Functions and Rough Sets: Suvey and New Insights. *International Journal of Approximate Reasoning, 143*, 92–215. doi:https://doi.org/10.1016/j.ijar.2022.01.011

Chakrabarti, C. G., & Indranil, C. (2005). Shannon entropy: axiomatic characterization and application. *International Journal of Mathematics and Mathematical Sciences*. doi:https://doi.org/10.1155/IJMMS.2005.2847

Cheung, V. K., M.C., P., Meyer, L., Pearce, M. T., Haynes, J.-D., & Koelsch, S. (2019). Uncertainty and Surprise Jointly Predict Musical Pleasure and Amygdala, Hippocampus, and Auditory Cortex Activity. *Current Biology, 29*(23), 4084–4092.e4. doi:https://doi.org/10.1016/j.cub.2019.09.067

Clausius, R. J. (1850). Ueber die bewegende Kraft der Wärme und die Gesetze, welche sich daraus für die Wärmelehre selbst ableiten lassen. *Annalen der Physik*, 368–397 , 500–524.

Clausius, R. J. (1851 republished in 2009). On the Moving Force of Heat, and the Laws regarding the Nature of Heat itself which are deducible therefrom. *The London, Edinburgh, and Dublin Philosophical Magazine and Journal of Science, 2*(8), 1–21 , 102–119. doi:https://doi.org/10.1080/14786445108646819 , https://doi.org/10.1080/14786445108646840

Cobb, B. R., & Shenoy, P. (2006, April). On the plausibility transformation method for translating belief function models to probability models. *Journal of Approximate Reasoning, 41*(3), pp. 314–330. doi:https://doi.org/10.1016/j.ijar.2005.06.008

Couso, I., Dubois, D., & Sanchez, L. (2014). *Random Sets and Random Fuzzy Sets as Ill-Perceived Random Variables.* Springer. doi:https://doi.org/10.1007/978-3-319-08611-8

Csiszár, I. (2008). Axiomatic Characterizations of Information Measures. *Entropy, 10*(3), 261–273. doi:https://doi.org/10.3390/e10030261

Dale, A. I. (1982). Bayes or Laplace? An Examination of the Origin and Early Applications of Bayes' Theorem. *Archive for History of Exact Sciences, 27*(1), pp. 23–47.

Dale, A. I. (1999). *A History of Inverse Probability—From Thomas Bayes to Karl Pearson* (2nd. ed.). New York: Springer.

De Finetti, B. (2017). *Theory of Probability—A Critical Introductory Treatment.* (A. M. Smith, Trans.) UK: John Wiley and Sons.

DeGroot, M. H. (2004). *Optimal Statistical Decisions.* New Jersey: John Wiley & Sons.

Dempster, A. (1967, April). Upper and Lower Probabilities Induced by a Multivalued Mapping. *Ann. Math. Statist., 38*(2), pp. 325–339,. doi:https://doi.org/10.1214/aoms/1177698950

Denoeux, T. (2019, June). Decision-making with belief functions: A review. *International Journal of Approximate Reasoning, 109*, 87–110. doi:https://doi.org/10.1016/j.ijar.2019.03.009

Denoeux, T. (2021). *Uncertainty Analysis using Belief Functions: Applications to Statistical Inference and Pattern Recognition.* Université de technologie de Compiègne, Department of Computer Science, Compiègne, France. Retrieved December 20, 2022, from https://www.hds.utc.fr/~tdenoeux/dokuwiki/_media/en/book_bf.pdf

Denœux, T., Younes, Z., & Abdallah, F. (2010). Representing uncertainty on set-valued variables using belief functions. *Artificial Intelligence, 174*(7–8), 479–499. doi:https://doi.org/10.1016/j.artint.2010.02.002

Dezert, J., & Smarandache, F. (2008). A new probabilistic transformation of belief mass assignment. *Fusion 2008: International Conference on Information Fusion,* (pp. 1410–1417). Retrieved December 19, 2022, from https://hal.archives-ouvertes.fr/hal-00304319/document

Dezert, J., & Smarandache, F. (2009). Transformations of belief masses into subjective probabilities. In J. Dezert, & F. Smarandache (Eds.), *Advances and Applications of DSmT for Information Fusion* (pp. 85–136). Rehoboth: American Research Press . Retrieved December 19, 2022, from https://www.researchgate.net/publication/306413841_Transformations_of_belief_masses_into_subjective_probabilities

Dezert, J., Smarandache, F., & Daniel, M. (2004). A Generalized Pignistic Transformation. In J. Dezert, & F. Smarandache (Eds.), *Advances and Applications of DSmT for Information Fusion* (pp. 143-153). Rehoboth: American Research Press. Retrieved December 19, 2022, from https://www.onera.fr/sites/default/files/297/C023-Dezert-Fusion2004Stockholm.pdf

Dieudonné, J. (1969). *Foundations of Modern Analysis.* London: Academic Press.

Doré, P. E., Fiche, A., & Martin, A. (2010). Models of belief functions—Impacts for patterns recognitions. *13th International Conference on Information Fusion.* Edinburgh, UK. doi:https://doi.org/10.1109/ICIF.2010.5711936

Doré, P., Martin, A., Abi-Zeid, I., Jousselme, A., & Maupin, P. (2011a, January). Belief functions induced by multimodalprobability density functions, an application to the search and rescue problem. *RAIRO—Operations Research, 44*(4), 323–343. doi:https://doi.org/10.1051/ro/2011001

Doré, P., Osswald, C., Martin, A., Jousselme, A., & Maupin, P. (2011b). Continuous belief functions to qualify sensors performances. In W. Liu (Ed.), *Symbolic and Quantitative Approaches to Reasoning with Uncertainty. ECSQARU 2011* (pp. 350–361). Belfast, Ireland: Springer. doi:https://doi.org/10.1007/978-3-642-22152-1_30

Dragulescu, A., & Yakovenko, V. (2000). Statistical mechanics of money. *Eur. Phys. J. B, 17*, 723–729. doi:https://doi.org/10.1007/s100510070114

Dubois, D., & Prade, H. (1986). A Set-Theoretic View of Belief Functions. *International Journal of General Systems, 12*(3), 193–226. doi:https://doi.org/10.1080/03081078608934937

Fetz, T., & Oberguggenberger, M. (2016, November). Imprecise random variables, random sets, and Monte Carlo simulation. *78*, 252–264. doi:https://doi.org/10.1016/j.ijar.2016.06.012

Fiche, A., Martin, A., Cexus, J.-C., & Khenchaf, A. (2010). Continuous belief functions and α-stable distributions. *13th Conference on Information Fusion.* Edinburgh. doi:https://doi.org/10.1109/ICIF.2010.5711934

Fort, H. (2022). *Forecasting with Maximum Entropy.* IOP Publishing Ltd.
Fox, C. R., & See, K. E. (2006). Belief and Preference in Decision Under Uncertainty. In D. Hardman, & L. Macchi (Eds.), *Thinking: Psychological Perspectives on Reasoning, Judgment and Decision Making.* John Wiley & Sons. doi:https://doi.org/10.1002/047001332X.ch14
Giambartolomei, G. (2015). *The Karhunen-Loeve Theorem.* Tesi di Laurea, Universita di Bologna, Bologna, Italy. Retrieved March 25, 2023, from https://amslaurea.unibo.it/10169/1/Giambartolomei_Giordano_Tesi.pdf
Gibbs, J. W. (1901 reprinted 1960). *Elementary Principles of Statistical Mechanics.* Dover.
Grassi, P. R., & Bartels, A. (2021). Magic, Bayes and wows: A Bayesian account of magic tricks. *Neuroscience & Biobehavioral Reviews, 126*, 515–527. doi:https://doi.org/10.1016/j.neubiorev.2021.04.001
Guan, J., & Bell, D. (1993). Discounting and Combination Operations in Evidential Reasoning. In D. Heckerman, & A. Mamdani (Ed.), *Ninth Conference on Uncertainty in Artificial Intelligence* (pp. 477–484). Washington, USA: Elsevier. doi:https://doi.org/10.1016/B978-1-4832-1451-1.50062-7
Halmos, P. (1978). *Measure theory.* Berlin, Heidelberg, New York: : Springer Verlag.
Halpern, J., & Fagin, R. (1992). Two views of belief: belief as generalized probability and belief as evidence. *Artificial Intelligence,, 54*, pp. 275–317.
Hartley, R. (1928, July). Transmission of Information. *Bell System Technical Journal, 7*(3), 535–563.
He, Y. (2013). *Uncertainty Quantification and Data Fusion based on Dempster-Shafer Theory.* Florida State University, Mathematics. Florida State University Libraries.
Howson, C., & Urbach, P. (2006). *Scientific reasoning : the Bayesian approach* . Chicago: Open Court.
Hüllermeier, E., Kruse, R., & Hoffmann, F. (Eds.). (2010). Consonant continuous belief functions conflicts calculation. *IPMU'10: Proceedings of the Computational intelligence for knowledge-based systems design, and 13th international conference on Information processing and management of uncertainty* (pp. 706–715). Dortmund, Germany: Springer.
Hulse, A., Schumacher, B., & Westmoreland, M. D. (2018). Axiomatic Information Thermodynamics. *Entropy, 20*(4). doi:https://doi.org/10.3390/e20040237
Itti, L., & Baldi, P. (2009). Bayesian surprise attracts human attention. *Vision Research, 49*(10), 1295–1306. doi:https://doi.org/10.1016/j.visres.2008.09.007
Jaffray, J.-Y., & Wakker, P. (1993). Decision making with belief functions: Compatibility and incompatibility with the sure-thing principle. *Journal of Risk and Uncertainty, 7*(3), 255–271. doi:https://doi.org/10.1007/BF01079626
Jaynes, E. (1989). *Clearing up Mysteries—The Original Goal.* (J. Skilling, Ed.) Dordrecht: Springer. doi:https://doi.org/10.1007/978-94-015-7860-8_1
Jaynes, E. (2003). *Probability Theory: The Logic of Science.* Cambridge: Cambridge University Press.
Jaynes, E. T. (1957a, May). Information Theory and Statistical Mechanics. *Phys. Rev., 106*(4), 620–630. doi:https://doi.org/10.1103/PhysRev.106.620
Jaynes, E. T. (1957b, October). Information Theory and Statistical Mechanics. II. *Phys. Rev., 108*(2), 171–190. doi:https://doi.org/10.1103/PhysRev.108.171
Jaynes, E. T. (1965). Gibbs vs. Boltzmann Entropies. *American Journal of Physics, 33*(5), 391–398.
Jeffreys, H. (1939). *Theory of probability.* Oxford: University Press.
Kapur, J. N. (1989 revised 1993). *Maximum Entropy Models in Science and Technology.* New Delhi, India: Wiley Eastern.
Khinchin, A. Y. (1957). *Mathematical Foundations of Information Theory.* New York, NY, USA: Dover.
Klopotek, M. A., & Wierzchon, S. T. (1998). A New Qualitative Rough-Set Approach to Modeling Belief Functions. In L. Polkowski, & A. Skowron (Ed.), *Rough Sets and Current Trends in Computing, First International Conference, RSCTC'98* (pp. 346–354). Warzsaw, Poland: Springer. doi:https://doi.org/10.1007/3-540-69115-4_47

Kojadinovic, I., Marichal, J.-L., & Roubens, M. (2005). An axiomatic approach to the definition of the entropy of a discrete Choquet capacity. *Information Sciences, 172*(1–2), 131–153. doi: https://doi.org/10.1016/j.ins.2004.05.011

Kolossa, A., Kopp, B., & Fingscheidt, T. (2015). A computational analysis of the neural bases of Bayesian inference. *NeuroImage, 106*, 222–237. doi:https://doi.org/10.1016/j.neuroimage.2014.11.00

Kullback, S. (1951 reed. 1969 reprint 1979). *Information Theory and Statistics.* New York: Wiley, reed. Dover.

Kullback, S., & Leibler, R. A. (1951, March). On Information and Sufficiency. *The Annals of Mathematical Statistics, 22*(1), 79–86. doi:https://doi.org/10.1214/aoms/1177729694

Laghmara, H., Laurain, T., Cudel, C., & Lauffenburger, J. P. (2020). Heterogeneous sensor data fusion for multiple object association using belief functions. *Information Fusion, 57*, 44–58. doi:https://doi.org/10.1016/j.inffus.2019.11.002

Laplace, P.-S. (1774). Memoire sur la probabilité des causes par les événements. *Memoires de Mathématique et de Physique, Presentés à l'Académie Royale des Sciences par divers Savans & lus dans ses Assemblées*, pp. 621–656.

Laplace, P.-S. (1891). *Oeuvres complètes* (Vol. 8). (A. d. Paris, Ed.) Paris: Gauthiers-Villars.

Laplace, P.-S. (1986). Memoir on the Probability of the Causes of Events. *Statistical Science, 1*(3), pp. 364–78. Retrieved from http://www.jstor.org/stable/2245476

Lazo, A. V., & Rathie, P. (1978). On the entropy of continuous probability distributions (Corresp.). *IEEE Transactions on Information Theory, 24*(1), 120–122. doi:https://doi.org/10.1109/tit.1978.1055832

Lee, J., Fan, Y., & Sisson, S. (2015). Bayesian threshold selection for extremal models using measures of surprise. *Computational Statistics and Data Analysi, 85*, 84–99. doi:https://doi.org/10.1016/j.csda.2014.12.004

Lian, C. (2017). *Information Fusion and Decision Making using Belief Functions.* Compiègne, France: Université de Technologie de Compiègne.

Liboff, R. L. (1974). Gibbs vs. Shannon entropies. *J Stat Phys, 11*, 343–357. doi:https://doi.org/10.1007/BF01009793

Liu, W. (2006, August). Analyzing the degree of conflict among belief functions. *Artificial Intelligence, 170* (11), 909–924. doi:https://doi.org/10.1016/j.artint.2006.05.002

Martin, A. (2019). Conflict management in information fusion with belief functions. In E. Bossé, & G. Rogova (Eds.), *Information quality in information fusion and decision making* (pp. 79–97). Springer. doi:https://doi.org/10.1007/978-3-030-03643-0_4

Martin, R., Zhang, J., & Liu, C. (2010, June). Dempster–Shafer Theory and Statistical Inference with Weak Beliefs. *Statistical Science, 25*(1), 72–87. doi:https://doi.org/10.1214/10-STS322

Mercer, J. (1909). Functions of positive and negative type, and their connection the theory of integral equations. *Philosophical Transactions of the Royal Society of London. Series A., 209*(441-458), 415–446. doi:https://doi.org/10.1098/rsta.1909.0016

Mercier, D., Quost, B., & Denœux, T. (2005). Contextual Discounting of Belief Functions. In L. Godo (Ed.), *ECSQARU 2005: Symbolic and Quantitative Approaches to Reasoning with Uncertainty* (pp. 552–562). Barcelona, Spain: Springer. doi:https://doi.org/10.1007/11518655_47

Miranda, E., Couso, I., & Gil, P. (2005, July). Random sets as imprecise random variables. *Journal of Mathematical Analysis and Applications, 307*(1), 32–47. doi:https://doi.org/10.1016/j.jmaa.2004.10.022

Modirshanechi, A., Brea, J., & Gerstner, W. (2022). A taxonomy of surprise definitions. *Journal of Mathematical Psychology, 110*. doi:https://doi.org/10.1016/j.jmp.2022.102712

Muñoz-Cobo, J.-L., Mendizábal, R., Miquel, A., Berna, C., & Escrivá, A. (2017). Use of the Principles of Maximum Entropy and Maximum Relative Entropy for the Determination of Uncertain Parameter Distributions in Engineering Applications. *Entropy, 19*(9). doi:https://doi.org/10.3390/e19090486

Nambiar, K. K., Varma, P. K., & Saroch, V. (1992). An axiomatic definition of Shannon's entropy. *Appl. Math. Lett., 5*(4), 45–46. doi:https://doi.org/10.1016/0893-9659(92)90084-M

Nguyen, H. T. (1977). *On Random sets and Belief Functions.* University of California, Berkeley, EECS Department. Retrieved from http://www2.eecs.berkeley.edu/Pubs/TechRpts/1977/28879.html

Nguyen, H. T. (1978). On Random Sets and Belief Functions. *Journal of Mathematical Analysis and Applications, 65*, 531–542. doi:https://doi.org/10.1016/0022-247X(78)90161-0

Nguyen, H., & Wang, T. (1997). Belief Functions and Random Sets. In J. Goutsias, R. Mahler, & H. Nguyen (Eds.), *Random Sets* (pp. 243–255). Springer. doi:https://doi.org/10.1007/978-1-4612-1942-2_11

Ni, S., Lei, Y., & Tang, Y. (2020). Improved Base Belief Function-Based Conflict Data Fusion Approach Considering Belief Entropy in the Evidence Theory. *Entropy, 22*(8). doi:https://doi.org/10.3390/e22080801

Ostwald, D., Spitzer, B., Guggenmos, M., Schmidt, T. T., Kiebel, S. J., & Blankenburg, F. (2012). Evidence for neural encoding of Bayesian surprise in human somatosensation. *NeuroImage, 62*(1), 177–188. doi:https://doi.org/10.1016/j.neuroimage.2012.04.050

Palm, G. (2023). *Novelty, Information and Surprise.* Germany: Springer-Verlag. doi:https://doi.org/10.1007/978-3-662-65875-8

Planck, M. K. (1901). Über das Gesetz der Energieverteilung im Normalspektrum. *Annalen der Physik, 309*(3), 553–563. doi:https://doi.org/10.1002/andp.19013090310

Planck, M. K. (1914). *The Theory of Heat Radiation.* (M. Masius, Trans.) Philadelphia: P. Blakiston's Son & Co.

Press, S. J., & Tanur, J. M. (2001). *The Subjectivity of Scientists and the Bayesian Approach.* New York: John Wiley & Sons.

Quiroga-Martinez, D., Hansen, N., Højlund, A., Pearce, M., Brattico, E., & Vuust, P. (2020). Decomposing neural responses to melodic surprise in musicians and non-musicians: Evidence for a hierarchy of predictions in the auditory system. *NeuroImage*. doi:https://doi.org/10.1016/j.neuroimage.2020.116816

Rényi, A. (1961). On Measures of Entropy and Information. In J. Neyman (Ed.), *4th Berkeley Symposium on Mathematics, Statistics and Probability,. 1*, pp. 547–561. University of California Press.

Risti, B., & Smets, P. (2006). Belief function theory on the continuous space with an application to model based classification. In B. Bouchon-Meunier, G. Coletti, & R. R. Yager (Ed.), *Modern Information Processing: From Theory to Applications. IPMU'04* (pp. 11–24). Perugia, Italy.: Elsevier. doi:https://doi.org/10.1016/B978-044452075-3/50002-9

Saravanan, R., & Levine, R. (2022). Surprisal analysis of diffusion processes. *Chemical Physics, 556*. doi:https://doi.org/10.1016/j.chemphys.2022.111450

Savchuk, V. P., & Tsokos, C. P. (2011). *Bayesian Theory and Methods with Applications.* Paris: Atlantis Press.

Shafer, G. (1976). *A Mathematical Theory of Evidence* . New Jersey: Princeton University Press.

Shafer, G. (1990). Perspectives on the Theory and Practice of Belief Functions. *International Journal of Approximate Reasoning, 4*(5–6), 323–362. doi:https://doi.org/10.1016/0888-613X(90)90012-Q

Shannon, C. E. (1948a). A Mathematical Theory of Communication. *The Bell System Technical Journal, 27*(3), 379–423. doi:https://doi.org/10.1002/j.1538-7305.1948.tb01338.x

Shannon, C. E. (1948b). A Mathematical Theory of Communication. *Bell System Technical Journal, 27*(4), 623–666. doi:https://doi.org/10.1002/j.1538-7305.1948.tb00917.x

Shannon, C. E. (n.d.). *A mathematical theory of comunication—Nokia Bell Labs*. Retrieved 2 13, 2023, from https://www.bell-labs.com/claude-shannon/assets/images/discoveries: https://www.bell-labs.com/claude-shannon/assets/images/discoveries/1948-04-21-a-mathematical-theory-of-communication-parts-I-and-carousel-01.pdf

Shore, J., & Johnson, R. (1980, January). Axiomatic derivation of the principle of maximum entropy and the principle of minimum cross-entropy. *IEEE Transactions on Information Theory, 26*(1), 26–37. doi:https://doi.org/10.1109/TIT.1980.1056144.

Smets, P. (1990). Constructing the Pignistic Probability Function in a Context of Uncertainty. *Machine Intelligence and Pattern Recognition, 10*, 29–39. doi:https://doi.org/10.1016/B978-0-444-88738-2.50010-5

Smets, P. (2000). Data fusion in the transferable belief model. *Proceedings of the Third International Conference on Information Fusion. 1*, pp. 21–33. Paris, France: IEEE. doi:https://doi.org/10.1109/IFIC.2000.862713.

Smets, P. (2005). Belief functions on real numbers. *International Journal of Approximate Reasoning, 40*, 181–223. doi:https://doi.org/10.1016/j.ijar.2005.04.001

Sohrab, S. H. (2014). Boltzmann entropy of thermodynamics versus Shannon entropy of information theory. *International Journal of Mechanics, 8*, 73–84. Retrieved February 15, 2023, from https://www.naun.org/main/NAUN/mechanics/2014/a182003-086.pdf

Soize, C. (2005, April). Random matrix theory for modeling uncertainties in computational mechanics. *Computer Methods in Applied Mechanics and Engineering, 194*(12–16), 1333–1366. doi:https://doi.org/10.1016/j.cma.2004.06.038

Soize, C. (2008). Construction of probability distributions in high dimension using the maximum entropy principle: Applications to stochastic processes, random fields and random matrices. *Int. J. Numer. Meth. Engng, 76*, 1583–1611. doi:https://doi.org/10.1002/nme.2385

Souza de Cursi, E. (2023). *Uncertainty Quantification using R.* Springer Cham.

Souza de Cursi, E., & Sampaio, R. (2015). *Uncertainty Quantification and Stochastic Modeling with Matlab.* ISTE/Elsevier. doi:https://doi.org/10.1016/C2014-0-04713-2

Stigler, S. M. (1982). Thomas Bayes's Bayesian Inference. *Journal of the Royal Statistical Society. Series A (General), 145*(2), pp. .250–258.

Strat, T. (1984). Continuous belief functions for evidential reasoning. *Proceedings of the 4th National Conference on Artificial Intelligence.* Austin, Texas. Retrieved December 18, 2022, from https://www.aaai.org/Papers/AAAI/1984/AAAI84-035.pdf

Strat, T. (1987). The Generation of Explanations within Evidential Reasoning Systems. In J. P. McDermott (Ed.), *10th. International Joint Conference on Artificial Intelligence (IJCAI)*, (pp. 1097–1104). Milan, Italy. Retrieved December 20, 2022, from https://www.ijcai.org/Proceedings/87-2/Papers/104.pdf

Strat, T. (1990). Decision analysis using belief functions. *International Journal of Approximate Reasoning, 4*(5-6), 391–417. doi:https://doi.org/10.1016/0888-613X(90)90014-S

Taillandier, P., & Therond, O. (2011). Use of the Belief Theory to formalize Agent DecisionMaking Processes : Application to cropping Plan Decision Making. *European Simulation and Modelling Conference*, (pp. 138–142). Guimaraes, Portugal. Retrieved 12 17, 2022, from https://hal.archives-ouvertes.fr/hal-00688405

Turing, A. M. (1941). *The Applications of Probability to Cryptography.* Bletchley Park. Available at Archive.org and https://www.nationalarchives.gov.uk/. Retrieved from https://archive.org/details/hw-25-37

Turing, A. M. (2015, May 26). *The Applications of Probability to Cryptography.* doi:https://doi.org/10.48550/arXiv.1505.04714

Wang, L. (2008). *K arhunen-Loeve E xpansions and their Applications.* The London School of Economics and Political Science. Ann Arbor, MI, USA: ProQuest LLC. Retrieved March 25, 2023, from http://etheses.lse.ac.uk/2950/1/U615901.pdf

Wasserman, L. A. (1990, September). Belief functions and statistical inference. *Canadian Journal of Statistics, 18*(3), 183–196. doi:https://doi.org/10.2307/3315449

Wu, W.-Z., & Mi, J.-S. (2008). An Interpretation of Belief Functions on Infinite Universes in the Theory of Rough Sets. In C. Chan, J. Grzymala-Busse, & W. P. Ziarko (Ed.), *Rough Sets and Current Trends in Computing. RSCTC 2008* (pp. 71–80). Akron, OH, USA: Springer. doi: https://doi.org/10.1007/978-3-540-88425-5_8

Yakovenko, V. M. (2010). Statistical Mechanics of Money, Debt, and Energy Comsumption. *Science and Culture, 76*(9–10), 430–436. doi:https://doi.org/10.48550/arXiv.1008.2179

Yakovenko, V. M., & Rosser, J. B. (2009, December 2). Statistical mechanics of money, wealth, and income. *Rev. Mod. Phys., 81*(4), 1703–1725. doi:https://doi.org/10.1103/RevModPhys.81.1703

Yao, Y. Y., & Lingras, P. J. (1998). Interpretations of Belief Functions in the Theory of Rough Sets. *Information Sciences, 104*(1–2), 81–106. doi:https://doi.org/10.1016/S0020-0255(97)00076-5

Yosida, K. (1995). *Functional Analysis*. Berlin Heidelberg: Springer-Verlag . doi:https://doi.org/10.1007/978-3-642-61859-8

Zanchini, E., & Beretta, G. P. (2008). Rigorous Axiomatic Definition of Entropy Valid Also for Non-Equilibrium States. In G. P. Beretta, A. Ghoniem, & G. Hatsopoulos (Ed.), *MEETING THE ENTROPY CHALLENGE: An International Thermodynamics Symposium in Honor and Memory of Professor Joseph H. Keenan. 1033*. Cambridge, MA, USA: AIP Conference Proceedings. doi:https://doi.org/10.1063/1.2979048

Zhang, H., & Deng, Y. (2020). Weighted belief function of sensor data fusion in engine fault diagnosis. *Soft Comput, 24*, 2329–2339. doi:https://doi.org/10.1007/s00500-019-04063-7

Zhou, K., Martin, A., & Pan, Q. (2018). A belief combination rule for a large number of sources. . *Journal of Advances in Information Fusion, 13*(2). Retrieved December 2022, 20, from https://www.researchgate.net/deref/https%3A%2F%2Fhal.archives-ouvertes.fr%2Fhal-01883239

Chapter 5
Bayesian Inference

Data analysis often needs to fit distributions to observed data, compare series of observation, decide about the coherence between observations and theoretical assumptions, forecast from collected data. These tasks are usually performed by statistical tools of estimation, hypothesis testing, and regression. In the companion book (Souza de Cursi 2023), we presented Fisher's point of view about these tasks. Here, we shall consider the Bayesian point of view, which is distinct from Fisher's one by the fact that it considers the parameters to be determined as random variables.

5.1 Bayesian Estimation

To understand the difference between Fisher's and Bayes' estimation, we may consider a simple situation: the determination of the probability p of a result "head" when tossing a coin.

In Fisher's approach, p is a parameter to be determined from the observations, by constructing an estimator $\widehat{p}$ – generally by using the maximum of likelihood. To determine it, we make an experiment: the coin is tossed n times and we get as result a sample $\mathcal{X} = (X_1, \ldots, X_n)$. In such a situation, the *model* for the result X_i is a Bernoulli variable $\mathcal{B}(p)$, taking the value 1 if the result is a head and 0, otherwise. The number of heads in the sample are

$$S_n = \sum_{i=1}^{n} X_i,$$

which is binomially $\mathcal{B}(n, p)$ distributed. The likelihood associated with the experiment is the probability of the results, i.e.,

E. Souza de Cursi, *Uncertainty Quantification with R*, International Series in Operations Research & Management Science 352,
https://doi.org/10.1007/978-3-031-48208-3_5

$$f(X|p) = p^{S_n}(1-p)^{n-S_n}.$$

Fisher's approach considers the maximum of likelihood estimator (MLE):

$$\widehat{p} = \frac{1}{n}S_n$$

$\widehat{p}$ is a random variable, whose distribution is known and can be used to generate confidence intervals or hypothesis testing. For instance, let the sample contain exactly n_H heads: we can estimate p by the observed value

$$\widehat{p}_{obs} = \frac{n_H}{n}$$

and generate a confidence interval $(\widehat{p}_{obs} - \Delta, \widehat{p}_{obs} + \Delta)$ using the distribution of S_n – eventually approximated by Gaussian distributions if n is large enough (see the companion book (Souza de Cursi 2023) for details).

The Bayesian approach considers that, since p is uncertain, it can be modeled by a random variable, having a distribution f_{pr} chosen by the analyst – it is the *prior distribution*, which reflects the beliefs of the analyst about the coin (for instance, we can assume that p is uniformly distributed on (a, b), $0 \leq a < b \leq 1$). In the Bayesian approach, the aim is to reduce the uncertainty using the observations. This is achieved by determining a new distribution f_{po} for p, using the data in combination with the Bayes' formula:

$$f_{po}(p) = f(p|X) = \frac{f(X|p)}{f(X)} f_{pr}(p). \tag{5.1}$$

We call this procedure the Bayesian Correction (or Bayesian Update) of the initial belief. The new distribution f_{po} is called the *posterior distribution*. The value of $f(X)$ is determined by integration:

$$f(X) = \int f(X|p) f_{pr}(p) dp$$

In practice, we can write

$$f_{po}(p) = f(p|X) = A f(X|p) f_{pr}(p) \tag{5.2}$$

and determine A by normalizing f_{po}:

$$\int f_{po}(p)dp = 1 \Rightarrow A = \left(\int f(X|p) f_{pr}(p) dp \right)^{-1}.$$

In the Bayesian approach, the estimator $\widehat{p}$ is determined by using a loss function $\ell(p,\widehat{p})$ and the distribution f_{po}: for instance, we look for the minimal value of the risk function $R(\widehat{p}) = E_{f_{po}}(\ell(p,\widehat{p}))$. The reader can find in the literature many propositions of loss functions and procedures for the determination of their optimal point – see, for instance, (Berger 1985; Zellner 1986a; Chang and Hung 2007; Masnadi-Shirazi 2011; Islam 2011; Chadli et al. 2017; Zhang 2020; Nassar et al. 2022).

For instance, we can consider the Mean Squared Error (MSE): $\ell(p,\widehat{p}) = (p-\widehat{p})^2$ – in this case, the estimation is $\widehat{p} = E_{f_{po}}(p)$. We can also consider the generalized linear-exponential (GLINEXP):

$$\ell(p,\widehat{p}) = \exp(\phi(p,\widehat{p})) - \phi(p,\widehat{p}) - 1.$$

In general,

$$\phi(p,\widehat{p}) = c\left(\frac{p}{\widehat{p}} - 1\right) \quad or \quad \phi(p,\widehat{p}) = c(p-\widehat{p})$$

Notice that both Fisher's and Bayes' approaches have as result random variables – but Fisher's approach constructs an estimator, while Bayes' approach updates a belief. Thus, these approaches are based on different ideas and may lead to different results.

Example 5.1 Let us consider $n = 5$ tosses, with $n_H = 3$. Let us apply Fisher's approach: we have

$$\widehat{p} = \frac{n_H}{n} \Longrightarrow \widehat{p}_{obs} = \frac{3}{5} = 0.6.$$

Let us apply Bayes' approach: assume as prior the uniform distribution on $(0, 1)$: $f_{pr}(p) = 1,\ on\ (0, 1)$. Since

$$f(X|p) = p^3(1-p)^2,$$

we have

$$f_{po}(p) = Ap^3(1-p)^2$$

(continued)

Example 5.1 (continued)

The normalization furnishes

$$A = \frac{\Gamma(7)}{\Gamma(3)\Gamma(4)} = \frac{6!}{2!3!} = 60.$$

We can use R for these calculations:

```
a = 0
b = 1
fpr = function(x){1/(b-a)}
nh = 3
n = 5
fa = function(x){x^(nh)*(1-x)^(n-nh)*fpr(x)}
source("ints.R")
ints1 = ints$new()
aux = ints1$romb1(fa,a,b,100,1e-10)
A = 1/aux
print(paste("A = ",A))
## [1] "A =  60"
```

Using the MSE loss function, the estimation is

$$\widehat{p} = E_{f_{po}}(p) = \frac{4}{7} \approx 0.57.$$

We can use R for this evaluation:

```
fpo = function(x){A*fa(x)}
los = function(p,pest){(p-pest)^2}
fobj = function(pest){
  ff = function(p){fpo(p)*los(p,pest)}
  resu = ints1$romb1(ff,a,b,100,1e-10)
  return(resu)
}

library(pracma)
sol = fminbnd(fobj, a, b, maxiter = 5000)
p_est = sol$xmin
print(paste("Bayes estimation 1 for p = ",p_est))
## [1] "Bayes estimation 1 for p =  0.571428571428572"
```

(continued)

Example 5.1 (continued)

Let us use the standard GLINEX with $c = 1, \phi(p, \widehat{p}) = p/\widehat{p} - 1$. We modify the loss function as follows:

```
los = function(p,pest){
  d = p/pest - 1
  resu = exp(d) - d - 1
  return(resu)}
```

The result is:

```
## [1] "Bayes estimation 2 for p =  0.644637233576548"
```

Using $c = 1, \phi(p, \widehat{p}) = p - \widehat{p}$, we obtain

```
## [1] "Bayes estimation 3 for p =  0.586551865503052"
```

Assume that you have a different belief on p: you assume that p is uniformly distributed on $(0.2, 0.8)$. In this case,

```
## [1] "A =  40.7166123491566"
```

and the results are, respectively:

```
## [1] "Bayes estimation 1 for p =  0.54760353662192"

## [1] "Bayes estimation 2 for p =  0.601599723295224"

## [1] "Bayes estimation 3 for p =  0.558209259895054"
```

Example 5.2 If the number of observations increase, Bayesian estimation and Fisher estimation tend to furnish close results. Let us consider the preceding example, but with $n = 20$ tosses, with $n_H = 12$. The results by Fisher's approach are not modified (except the distribution of $\widehat{p}$) and we have yet $\widehat{p}_{obs} = 0.6$. Bayes' results for a uniform prior on (0,1) are modified as follows:

(continued)

Example 5.2 (continued)

```
## [1] "A =  2649629.8212712"
## [1] "Bayes estimation 1 for p =  0.590909090925965"
## [1] "Bayes estimation 2 for p =  0.616663698869956"
## [1] "Bayes estimation 3 for p =  0.596136733488492"
```

For $n = 100$ tosses, with $n_H = 60$, we obtain

```
## [1] "A =  1.90147590034234e+30"
## [1] "Bayes estimation 1 for p =  0.598039214735798"
## [1] "Bayes estimation 2 for p =  0.603845338676152"
## [1] "Bayes estimation 3 for p =  0.599204697401522"
```

Example 5.3 Let us consider the situation where $X \sim N(m, \sigma)$ and $\boldsymbol{\theta} = (m, \sigma)$. Then, our model for X is

$$f_{mod}(x, \boldsymbol{\theta}) = \frac{1}{\sigma\sqrt{2\pi}} \exp\left(-\frac{1}{2}\left(\frac{x-m}{\sigma}\right)^2\right).$$

Assume that we have the data below(generated using `rnorm`, with $m = 1$, $\sigma = 0.5$):

$$X = (0.97, 1.41, 1.14, 0.82, 0.67).$$

This dataset does not give good information about the distribution, namely its variance: the dataset has an empirical variance of 0.08 (corresponding to a standard deviation of 0.27), while the exact value is 0.25 (corresponding to a standard deviation of 0.5). If we use this set of data to evaluate probabilities, we get $P(X \le 0.8) = 0.2$, $P(X \le 1) = 0.6$, $P(X \le 1.2) = 0.8$, while the real values are 0.34, 0.5, 0.66, respectively. In this example and the next ones, we shall examine the results of different approaches to the analysis of this dataset.

(continued)

Example 5.3 (continued)

Since we assume a Gaussian model for *X*, the likelihood is

$$f(X|\boldsymbol{\theta}) = \prod_{i=1}^{ns} f_{mod}(X_i, \boldsymbol{\theta}) = \frac{1}{\sigma^{ns}(2\pi)^{\frac{ns}{2}}} \exp\left(-\frac{1}{2}\sum_{i=1}^{ns}\left(\frac{X_i - m}{\sigma}\right)^2\right),$$

where *ns* is the size of the sample (here, $ns = 5$) and Fisher's estimation (MLE) produces

$$\widehat{m}_{obs} \approx 1.0, \widehat{\sigma}_{obs} \approx 0.26\ (MLE)\ or\ \widehat{\sigma}_{obs} \approx 0.29\ (Unbiaised)\ .$$

Let us use BI: we assume that $m \in (0, 2)$ and $\sigma \in (0.1, 1.1)$ and a uniformly distributed prior, so that

$$f_{pr}(m, \sigma) = 0.5\ \ on\ (0, 2) \times (0.1, 1.1).$$

We have

$$f_{po}(m, \sigma) = \frac{\mathrm{A}}{\sigma^n} \exp\left(-\frac{1}{2}\sum_{i=1}^{n}\left(\frac{X_i - m}{\sigma}\right)^2\right)\ on\ (0, 2) \times (0.1, 1.1).$$

The constant A is determined as follows:

```
am = 0
bm = 2
as = 0.1
bs = 1.1

fpr = function(x){1/((bs-as)*(bm-am))}

fa = function(m,s){
  sa = sum((X-m)^2)
  sb = exp(-0.5*sa/s^2)/(s^length(X))
  sc = sb*fpr(m,s)
  return(sc)}

source("ints.R")
ints1 = ints$new()
aux = ints1$romb2(fa,c(am,as),c(bm,bs),500,1e-10)
A = 1/aux
print(paste("A = ",A))
## [1] "A =  0.279751633050455"
```

(continued)

Example 5.3 (continued)

Then we can estimate m and σ using the loss functions previously presented. Since we have two parameters, the global loss function is the sum of the uniparametric for each one. Using `fminunc` from package `pracma`, we obtain

```
## [1] "Bayes estimation 1 for m =  1.002 , for sigma =  0.28717
5904281679"

## [1] "Bayes estimation 2 for m =  1.06036091293791 , for sigma
=  0.541885568863211"

## [1] "Bayes estimation 3 for m =  1.02203359003532 , for sigma
=  0.432006245809168"
```

Notice that the estimates furnished by GLINEXP for σ are closer to the exact ones than those furnished by the quadratic loss.

If we are interested in the evaluation of the probabilities of some events, we can integrate our model for X, using the parameters $\widehat{\boldsymbol{\theta}}$ estimated:

$$P(X \leq a) = \int_{-\infty}^{a} f_{mod}\left(x, \widehat{\boldsymbol{\theta}}\right) dx.$$

An example of result is shown in Table 5.1.

Table 5.1 Results for the evaluation of $\boldsymbol{P(X \leq a)}$, using the MLE and BI with the three loss functions introduced. Results for a sample of size 5

a	Exact	MLE	Bayes 1	Bayes 2	Bayes 3
0.8	0.3446	0.2409	0.2409	0.3154	0.2502
1	0.5000	0.4972	0.4972	0.4557	0.4115
1.2	0.6554	0.7547	0.7547	0.6017	0.5892

Analogously to the preceding examples, the results tend to come closer together as the number of measurements increase: for $n = 25$, with

$$X = (0.97, 1.41, 1.14, 0.82, 0.67, 1.04, 0.72, 1.44, 1.29, 1.61, 0.90, 0.72,$$
$$0.70, 0.80, 0.97, 0.17, 1.50, 1.03, 1.47, 0.80, 1.58, 1.33, 0.92, 1.17, 2.22),$$

This dataset carries better information about the distribution. For instance, its variance is 0.18 (corresponding to a standard deviation of 0.42) – recall that the exact value is 0.25 (corresponding to a standard deviation of 0.5). If we use this new dataset to evaluate probabilities, we get $P(X \leq 0.8) = 0.28$,

(continued)

Example 5.3 (continued)

$P(X \leq 1) = 0.48$, $P(X \leq 1.2) = 0.64$ – the real values are 0.34, 0.5, 0.66, respectively. MLE estimates are $\widehat{m}_{obs} \approx 1.0, \widehat{\sigma}_{obs} \approx 0.42$. Using the previous approaches, we obtain

```
## [1] "A =  0.00359681753823587"

## [1] "Bayes estimation 1 for m =  1.09559983310955 , for sigma
=  0.442984856660676"

## [1] "Bayes estimation 2 for m =  1.106548082653 , for sigma =
0.459581845760112"

## [1] "Bayes estimation 3 for m =  1.09961992323163 , for sigma
=  0.445393032569115"
```

Remark 5.1 Bayesian estimation tends to reduce the uncertainty about the parameters under consideration. For instance, let us consider again the situation in Example 5.1. Assume a uniform prior on (0,1) and that you desire to generate a confidence interval with risk $\alpha = 0.05$ for p. Using the prior, a convenient interval is $(0.025, 0.975)$, which has as center the mean of the prior. If you use the observed value $\widehat{p}_{obs} = 0.6$ and the prior, the interval is (0.05,1). After the Bayesian Update, the distribution is $f_{po}(p) = 60p^3(1 - p)^2$, with mean 4/7. The corresponding confidence interval are (0.243, 0.900) and $(0.264, 0.936)$, respectively. The initial length of the confidence interval is 0.95, The posterior confidence interval has a length inferior to 0.675 – the uncertainty is reduced. The intervals can be determined by R as follows:

```
pcenter = 4/7
alfa = 0.05
fz = function(d){
  r = pbeta(pcenter+d,4,3)-pbeta(pcenter-d,4,3) - (1-alfa)
  return(r)
}
library(pracma)
sol = fzero(fz,0.3)
```

(continued)

Remark 5.1 (continued)

The result is

```
## [1] "d =  0.328362395471095 : confidence interval = ( 0.24306
6175957476 ,  0.899790966899667 )"
```

For `pcenter` = `0.6`, we obtain

```
## [1] "d =  0.335996297321333 : confidence interval = ( 0.26400
3702678667 ,  0.935996297321333 )"
```

Now, let us consider the situation of Example 5.3. For the first loss function, the confidence intervals for m and σ, with $\alpha = 0.05$, are

$$m \in (0.75, 1.25); \sigma \in (0.17, 0.83).$$

GLINEXP loss functions reduce the interval for m, but not for σ.

The other side of the coin is that you need some uncertainty to improve the estimate. If you are almost sure of the values of the parameters, you cannot expect a better result. For instance, if you are sure that $m = 1.5$, then your prior for m is $\delta_{1.5}(m)$: In this case, the Bayesian Update will return your own certainty: $f_{po}(m) = \delta_{1.5}(m)$. Such a behavior can be observed whenever your prior is highly informative – for instance, in Example 5.1, if you consider as prior the uniform distribution on (0.55,0.65), you will get as Bayesian estimation $p \approx 0.600$.

Exercises

1. Let X be a variable Poisson distributed $\mathcal{P}(\lambda)$. A sample from X is given:

$$X = (6, 2, 7, 4, 8, 5).$$

- Estimate λ using MSE and GLINEXP.

2. Let X be a variable Poisson distributed $\mathcal{P}(\lambda)$. A sample from X is given:

$$X = (3, 2, 5, 9, 10, 5).$$

- Estimate λ using MSE and GLINEXP.

(continued)

3. Let X be a variable Binomially distributed $B(10, p)$. A sample from X is given:

$$X = (4, 5, 4, 5, 3, 4).$$

   - Estimate p using MSE and GLINEXP.

4. Let X be a variable Binomially distributed $B(10, p)$. A sample from X is given:

$$X = (8, 5, 6, 7, 6, 6, 7).$$

   - Estimate p using MSE and GLINEXP.

5. Let X be a Gaussian variable $N(m, \sigma)$. A sample from X is given:

$$X = (0.613, -0.777, 2.431, -1.305, 3.908, 0.203, 1.045).$$

   - Estimate m, σ using MSE and GLINEXP.

6. Let X be a Gaussian variable $N(m, \sigma)$. A sample from X is given:

$$X = (-1.487, -2.472, 0.935, -3.639, -4.646, -0.915, -3.328).$$

   - Estimate m, σ using MSE and GLINEXP.

5.2 Jeffreys' Priors

The critical point in the Bayes' approach is the choice of the prior distribution, which highly impacts the results – see, for instance, (Morita et al. 2010; Ley et al. 2017; Ghaderinezhad and Ley 2019; Ghaderinezhad and Ley 2020; Depaoli et al. 2020; Jones et al. 2022; Ghaderinezhad et al. 2022).

The cruciality of the choice of the prior makes that many works considered the development of methods for its determination – see, for instance, (Winkler 1966, 1967; Jaynes 1968a, b; Robert 1994; Kass and Wasserman 1996; Lavenda 2001; Gelman 2002; Rahman et al. 2016; Gelman et al. 2017; Zondervan-Zwijnenburg et al. 2017; Consonni et al. 2018; Banner et al. 2020; Wesner and Pomeranz 2021; Falconer et al. 2022).

The methods existing in the literature are generally classified in regularizing priors, reference priors, eliciting priors, and noninformative priors. Regularizing priors are often used in the context of data science and optimization, where regularizing methods for approximated minimization can be interpreted in Bayesian terms. Reference priors look for a desired behavior of the estimators (Berger and Bernardo 1989, 1992; Berger et al. 2009). Eliciting priors are developed to integrate information furnished by experts (Johnson et al. 2010; Garthwaite et al. 2005; Albert et al. 2002; Falconer et al. 2021). Noninformative priors correspond to Jeffrey's views on the topic (Kass and Wasserman 1996) and try to assume the less information as possible on the distributions under analysis – they are the opposite of eliciting priors, which look for the inclusion of information – thus, they are informative priors.

The main difficult considered by Jeffreys concerning priors was the effect of reparameterizations: his main concern was to determine a general method to finding priors which remain invariant under parameterizations (Jeffreys 1946). In the modern views, the ideas of Jeffrey were reinterpreted in terms of invariance for groups of transformations (Jaynes 1968a, b; Harney 2003). The *Jeffreys prior* is based on Fisher's information matrix and is defined as follows:

Definition 5.1 Let X be a random variable having a density $\varphi(x, \boldsymbol{\theta})$, where $\boldsymbol{\theta} = (\boldsymbol{\theta}_1, \ldots, \boldsymbol{\theta}_m)$ is a vector of parameters to be estimated. Let $f(\mathcal{X}|\boldsymbol{\theta})$ be the likelihood associated with a sample $\mathcal{X} = (X_1, \ldots, X_n)$. Fisher's information matrix is

$$I(\boldsymbol{\theta}) = \left(E\left(-\frac{\partial^2}{\partial \theta_i \partial \theta_j} \log f(\mathcal{X}|\boldsymbol{\theta}) \right), 1 \leq i, j \leq m \right).$$

Jeffreys' prior is defined as

$$f_{pr}(\boldsymbol{\theta}) = A|\det(I(\boldsymbol{\theta}))|^{1/2}.$$

The constant A in Jeffrey's prior is destined to normalization, to get a probability distribution such that $\int f_{pr}(\boldsymbol{\theta})d\boldsymbol{\theta} = 1$. Such a normalization generally supposes that the distribution is defined on a finite interval. On infinite intervals, we can obtain improper priors, which cannot be normalized (see examples below).

Example 5.4 Let us consider again the situation in Example 5.1. The likelihood is

$$f(\mathcal{X}|p) = p^{S_n}(1-p)^{n-S_n}.$$

Thus,

$$\log f(\mathcal{X}|p) = S_n \log p + (n - S_n) \log(1-p),$$

and

(continued)

Example 5.4 (continued)

$$-\frac{\partial^2}{\partial p^2}\log f(X|p) = \frac{S_n}{p^2} + \frac{n-S_n}{(1-p)^2}.$$

Since $E(S_n) = np$, we have

$$I(p) = \frac{n}{p} + \frac{n}{1-p} = \frac{n}{p(1-p)}.$$

Thus, Jeffreys prior is

$$f_{pr}(p) = \frac{A}{\sqrt{p(1-p)}}.$$

Let us use Jeffreys prior to estimate p in the same situation as in the example 5.1: we normalize Jeffreys prior as follows:

```
fpra = function(x){1/sqrt(x*(1-x))}
auxr = ints1$romb1(fpra,a,b,100,1e-10)
Ar = 1/auxr
fpr = function(x){Ar*fpra(x)}
```

Then, we obtain the following results on (0,1):

```
## [1] "Bayes estimation 1 for p =  0.583333884675848"

## [1] "Bayes estimation 2 for p =  0.663745995146376"

## [1] "Bayes estimation 3 for p =  0.600425328904299"
```

For $n = 20$, $n_H = 12$, we obtain

```
## [1] "Bayes estimation 1 for p =  0.595238102945786"

## [1] "Bayes estimation 2 for p =  0.621837303505783"

## [1] "Bayes estimation 3 for p =  0.600682495258656"
```

For $n = 100$, $n_H = 60$, we obtain

```
## [1] "Bayes estimation 1 for p =  0.599009899977607"

## [1] "Bayes estimation 2 for p =  0.604858315523045"

## [1] "Bayes estimation 3 for p =  0.600185841922231"
```

Example 5.5 Let us consider the situation where $X \sim N(m, \sigma)$ and $\boldsymbol{\theta} = (m, \sigma)$. The likelihood is:

$$f(X|\boldsymbol{\theta}) = \frac{1}{\sigma^n (2\pi)^{\frac{n}{2}}} \exp\left(-\frac{1}{2} \sum_{i=1}^{n} \left(\frac{X_i - m}{\sigma} \right)^2 \right).$$

Thus,

$$\log f(X|\boldsymbol{\theta}) = -\frac{1}{2\sigma^2} \sum_{i=1}^{n} (X_i - m)^2 - n \log \sigma - \frac{n}{2} \log 2\pi,$$

and

$$\frac{\partial}{\partial m} \log f(X|\boldsymbol{\theta}) = -\frac{1}{\sigma^2} \left(nm - \sum_{i=1}^{n} X_i \right),$$

$$\frac{\partial}{\partial \sigma} \log f(X|\boldsymbol{\theta}) = \frac{1}{\sigma^3} \sum_{i=1}^{n} (X_i - m)^2 - \frac{n}{\sigma},$$

so that

$$I(\boldsymbol{\theta}) = E\left(\begin{pmatrix} \frac{n}{\sigma^2} & -\frac{2}{\sigma^3}\left(nm - \sum_{i=1}^{n} X_i \right) \\ \frac{-2}{\sigma^3}\left(nm - \sum_{i=1}^{n} X_i \right) & \frac{3}{\sigma^4} \sum_{i=1}^{n} (X_i - m)^2 - \frac{n}{\sigma^2} \end{pmatrix} \right).$$

Thus

$$I(\boldsymbol{\theta}) = \begin{pmatrix} \frac{n}{\sigma^2} & 0 \\ 0 & \frac{2n}{\sigma^2} \end{pmatrix}.$$

Thus, Jeffreys prior is (the factor $n\sqrt{2}$ is integrated in the constant)

$$f_{pr}(\boldsymbol{\theta}) = \frac{A}{\sigma^2},$$

i.e., the priori s proportional to $1/\sigma^2 : f_{pr}(\theta) \; \alpha \; 1/\sigma^2$.

(continued)

Example 5.5 (continued)

Let us use Jeffreys prior to estimate m, σ in the same situation as in the example 5.3. For $n = 5$, we obtain:

```
## [1] " Bayes estimation 1 for m =  1.002 , for sigma =  0.2871
75904281679"

## [1] " Bayes estimation 2 for m =  1.03325230411367 , for sigm
a =  0.386346805116346"
```

```
## [1] " Bayes estimation 3 for m =  1.01259062657727 , for sigm
a =  0.311063503341073"
```

Table 5.2 shows the evaluation of probabilities $P(X \le a)$.

Table 5.2 Results for the evaluation of $P(X \le a)$, using the MLE and BI with the three loss functions introduced. Results for a sample of size 5

a	Exact	MLE	Bayes 1	Bayes 2	Bayes 3
0.8	0.3446	0.2409	0.2409	0.2730	0.2472
1	0.5000	0.4972	0.4972	0.4657	0.4839
1.2	0.6554	0.7547	0.7547	0.6670	0.7266

Analogously to the preceding examples, the results tend to become close together if the size of the sample increases. For $n = 25$, we obtain

```
## [1] " Bayes estimation 1 for m =  1.09560026712422 , for sigm
a =  0.42372542006978"

## [1] " Bayes estimation 2 for m =  1.10560034812416 , for sigm
a =  0.438157727275677"

## [1] " Bayes estimation 3 for m =  1.09927017141381 , for sigm
a =  0.425730066302123"
```

Example 5.6 Let us consider the situation where $X \sim N(m,\sigma)$. Assume that we desire to estimate m for a given σ. Then, $\theta = m$ and

$$f(X|m) = \frac{1}{\sigma^n (2\pi)^{\frac{n}{2}}} \exp\left(-\frac{1}{2} \sum_{i=1}^{n} \left(\frac{X_i - m}{\sigma}\right)^2 \right).$$

Thus,

$$I(m) = \frac{n}{\sigma^2}$$

and Jeffreys prior is (we integrate the factor $\sqrt{n}$ in the constant)

$$f_{pr}(m) = \frac{A}{\sigma},$$

which corresponds to a uniform distribution (since it is independent from m). Notice that it cannot be normalized if m is taken on an infinite interval.

Now, assume that we desire to estimate σ for a given m. Then $\theta = \sigma$ and

$$I(\sigma) = \frac{2n}{\sigma^2}$$

and Jeffreys prior is (again, the factor $\sqrt{2n}$ in the constant)

$$f_{pr}(\sigma) = \frac{A}{\sigma} \ .$$

In both the cases, the prior is proportional to $1/\sigma : f_{pr}(\theta)\ \alpha\ 1/\sigma$. This fact can be exploited for hierarchical estimations: estimate $m \rightarrow$ estimate σ and iterate if necessary.

Let us use this hierarchical procedure to estimate m, σ in the same situation as in the Example 5.3. We use the iterations described above and the three loss functions previously introduced.

For $n = 5$, we obtain :

```
## [1] "Bayes estimation 1 for m =  1.00199999972895 , for sigma
=  0.303487146185095"

## [1] "Bayes estimation 2 for m =  1.04684037000472 , for sigma
=  0.391695154153477"

## [1] "Bayes estimation 3 for m =  1.01169638965244 , for sigma
=  0.311390319423956"
```

(continued)

Example 5.6 (continued)

Table 5.3 shows the evaluation of probabilities $P(X \leq a)$.

Table 5.3 Results for the evaluation of $P(X \leq a)$, using the MLE and BI with the three loss functions introduced. Results for a sample of size 5

a	Exact	MLE	Bayes 1	Bayes 2	Bayes 3
0.8	0.3446	0.2409	0.2528	0.2643	0.2483
1	0.5000	0.4972	0.4974	0.4524	0.4850
1.2	0.6554	0.7547	0.7429	0.6521	0.7273

For $n = 25$, we obtain close results for all the loss functions:

```
## [1] "Bayes estimation 1 for m =  1.09559998752209 , for sigma
=  0.423725148695016"
```

```
## [1] "Bayes estimation 2 for m =  1.10606526520233 , for sigma
=  0.438299869980701"
```

```
## [1] "Bayes estimation 3 for m =  1.09922512022266 , for sigma
=  0.425746710865184"
```

Remark 5.2 Bayesian Conjugates The update equation

$$f_{po}(\theta) = f(\theta|X) = A\, f(X|\theta) f_{pr}(\theta)$$

shows that the posterior distribution f_{po} is determined by the likelihood $f(X|\theta)$ and the prior f_{pr}. Thus, there exists a relation between the couple (likelihood, prior) and the posterior distribution, which is called conjugacy.

Moreover, the likelihood is determined from the model chosen for the random variable X under analysis: if its model density is $\varphi(x, \theta)$, then $f(X|\theta) = \prod_{i=1}^{n} \varphi(X_i, \theta)$. Consequently, there is also a connection between the couple (model, prior) and the posterior distribution – the notion of conjugacy extends to this situation.

A particular situation which can be interesting in practice is the one where all the three distributions – the likelihood, the prior, and the posterior – belong to the same family, i.e., the family is self-conjugated. This is the case of the exponential family. Indeed, as shown in Eq. (eq137b), the exponential family satisfies

(continued)

Remark 5.2 (continued)

$$f(X|\boldsymbol{\theta}) = \exp\left(\boldsymbol{\eta}^t \sum \boldsymbol{T}_m(X_i) - nA_m(\boldsymbol{\eta})\right) \prod h_m(X_i), \boldsymbol{\eta} = \boldsymbol{\psi}_m(\boldsymbol{\theta}).$$

If

$$f_{pr}(\boldsymbol{\theta}) = h_{pr}(\boldsymbol{\theta}) \exp\left(\boldsymbol{\kappa}^t \boldsymbol{T}_{pr}(\boldsymbol{\theta}) - A_{pr}(\boldsymbol{\kappa})\right).$$

Then

$$f_{po}(\boldsymbol{\theta}) = g(\boldsymbol{\theta}) \exp\left(\boldsymbol{\xi}^t \boldsymbol{P}(\boldsymbol{\theta}) - B(\boldsymbol{P}(\boldsymbol{\theta})\right),$$

with $\boldsymbol{\xi} = (\boldsymbol{\eta}; \boldsymbol{\kappa})$,

$$\boldsymbol{P}(\boldsymbol{\theta}) = \left(\sum \boldsymbol{T}_m(X_i); \boldsymbol{T}_{pr}(\boldsymbol{\theta})\right), B(\boldsymbol{P}(\boldsymbol{\theta})) = nA_m(\boldsymbol{\psi}(\boldsymbol{\theta})) + A_{pr}(\boldsymbol{\theta}),$$

$$g(\boldsymbol{\theta}) = Ah_{pr}(\boldsymbol{\theta}) \prod h_m(X_i),$$

where A is the normalizing constant. Unfortunately these expressions can be difficulty to use in practice, except in some particular situations.

As an example, let us consider a binomial likelihood:

$$f(X|p) = p^k(1-p)^{n-k},$$

and a prior uniformly distributed on (0, 1):

$$f_{pr}(p) = 1, \textit{if } 0 < p < 1; f_{pr}(p) = 0, \textit{otherwise}.$$

Then,

$$f_{po}(p) = Ap^k(1-p)^{n-k}, \textit{if } 0 < p < 1; f_{po}(p) = 0, \textit{otherwise}.$$

The normalization furnishes

$$A = \frac{\Gamma(n+k+2)}{\Gamma(k+1)\Gamma(n-k+1)}.$$

Thus, f_{po} corresponds to a Beta distribution $B(k+1, n-k+1)$: the couple (binomial likelihood, uniform prior on (0, 1)) has as conjugate a beta posterior.

(continued)

Remark 5.2 (continued)

Analogously, the couple (Bernoulli model, uniform prior on $(0, 1)$) has a beta posterior.

As a second example, let us consider an exponential likelihood:

$$f(X|p) = \lambda^n e^{-\lambda S_n}$$

and a prior uniformly distributed on $(0, \lambda_{max})$:

$$f_{pr}(\lambda) = 1/\lambda_{max}, \textit{ if } 0 < \lambda < \lambda_{max}; f_{pr}(p) = 0, \textit{ otherwise}.$$

Then,

$$f_{po}(\lambda) = A\lambda^n e^{-\lambda S_n}, \textit{ if } 0 < \lambda < \lambda_{max}; f_{po}(\lambda) = 0, \textit{ otherwise}.$$

Thus, f_{po} corresponds to a truncated gamma distribution $\Gamma(n + 1, S_n)$: the couple (exponential likelihood, uniform prior) has as conjugate a truncated gamma posterior. Analogously, the couple (exponential model, uniform prior) has as conjugate a truncated gamma posterior.

A particular case concerns the estimation of the mean under known variance in a Gaussian distribution $N(m, \sigma)$, with a normal prior $N(m_{pr}, \sigma_{pr})$. Indeed, the likelihood is

$$f(X|m) = \frac{1}{\sigma^n \sqrt{(2\pi)^n}} \exp\left(-\frac{1}{2\sigma^2} \sum_{i=1}^{n} (X_i - m)^2 \right),$$

so that

$$f_{po}(m) = \frac{A}{\sigma^n \sigma_{pr}} \exp\left(-\frac{1}{2\sigma_{pr}^2} (m - m_0)^2 - \frac{1}{2\sigma^2} \sum_{i=1}^{n} (X_i - m)^2 \right).$$

This posterior is Gaussian $N(m_{po}, \sigma_{po})$, with

$$\sigma_{po}^2 = \frac{\sigma^2 \sigma_{pr}^2}{n\sigma_{pr}^2 + \sigma^2}, m_{po} = \sigma_{po}^2 \left(\frac{m_{pr}}{\sigma_{pr}^2} + \frac{1}{\sigma^2} \sum_{i=1}^{n} X_i \right).$$

The reader will find in the literature examples of Bayesian Conjugates – see, for instance, (Diaconis and Ylvisaker 1979; Fink 1997; Murphy 2007).

Exercises

1. Let X be a variable Poisson distributed $\mathcal{P}(\lambda)$. A sample from X is given:

$$\mathcal{X} = (6, 2, 7, 4, 8, 5).$$

- Estimate λ using Jeffreys prior.

2. Let X be a variable Poisson distributed $\mathcal{P}(\lambda)$. A sample from X is given:

$$\mathcal{X} = (3, 2, 5, 9, 10, 5).$$

- Estimate λ using Jeffreys prior.

3. Let X be a variable Binomially distributed $B(10, p)$. A sample from X is given:

$$\mathcal{X} = (4, 5, 4, 5, 3, 4).$$

- Estimate p using Jeffreys prior.

4. Let X be a variable Binomially distributed $B(10, p)$. A sample from X is given:

$$\mathcal{X} = (8, 5, 6, 7, 6, 6, 7).$$

- Estimate p using Jeffreys prior.

5. Let X be a Gaussian variable $N(m, \sigma)$. A sample from X is given:

$$\mathcal{X} = (0.613, -0.777, 2.431, -1.305, 3.908, 0.203, 1.045).$$

- Estimate m, σ using Jeffreys prior.

6. Let X be a Gaussian variable $N(m, \sigma)$. A sample from X is given:

$$\mathcal{X} = (-1.487, -2.472, 0.935, -3.639, -4.646, -0.915, -3.328).$$

- Estimate m, σ using Jeffreys prior.

5.3 Using UQ in Bayesian Estimation

As shown in Chap. 1 and in the companion book (Souza de Cursi 2023), uncertainty quantification furnishes tools for the determination of distributions for limited available data. These tools can be combined with the Bayesian approach: we can use UQ to determine priors or to generate models and use them in the Bayesian update (5.1).

Example 5.7 Let us consider the situation presented in Example 5.3. We can use a bootstrap to generate a sample from the couple (m, σ) of the random variable X. For instance, we can generate a sample of 250 variates from (m, σ) as follows:

```
nboot = 250
library(boot)
stat3 = function(ddata, index){
  dd = ddata[index]
  resu = c(mean(dd), sd(dd))
  return(resu)}
b3 = boot(data = X, statistic = stat3, R = nboot)
aux = sort(b3$t[,1],index.return=TRUE)
mb1 = aux$x
sb1 = b3$t[aux$ix,2]
```

Once the sample is generated, we can use UQ methods to generate a pdf for m and σ: we start be determining a representation of m as a function of an artificial variable U. For instance, using a Gaussian $N(0, 1)$ and the representation by a polynomial of degree 5:

```
Ub = sort(rnorm(length(mb1)))
ab = min(Ub)
bb = max(Ub)
ndb = 5
source("polynom.R")
phib = polynom$new(ab,bb,1e-3)
source("expansion1D.R")
expb = expansion1D$new(phib)
cmb1 = expb$colloc(nd,0,Ub,mb1)
csb1 = expb$colloc(nd,0,Ub,sb1)
```

Then, we use the representation to generate a large sample from m and determine its pdf:

```
nsg = 1e5
Ug = rnorm(nsg)
Pm1g = expb$proj(cmb1,Ug)
Ps1g = expb$proj(csb1,Ug)
source("sampleX.R")
saux = sampleX$new()
nam = 50
s1 = saux$create(Pm1g)
sm1 = saux$empcdfpdf(s1,1,nam)
ss1 = saux$create(Ps1g)
sb1 = saux$empcdfpdf(ss1,2,nam)
```

Alternatively, we can generate a representation for X itself, and use this representation to generate a sample from m. Then, we can determine a PDF for m.

(continued)

Example 5.7 (continued)

```
a = min(X)
b = max(X)
Xs = sort(X)
Us = sort(rnorm(length(X)))
nd = 3
source("polynom.R")
phiX = polynom$new(a,b,1e-3)
source("expansion1D.R")
expX = expansion1D$new(phiX)
cX = expX$colloc(nd,0,Us,Xs)
ns = 50
mba = numeric(length = nboot)
sba = numeric(length = nboot)
for (i in 1: nboot){
  U = rnorm(ns)
  ind = which((U >= a) & (U <= b))
  U = U[ind]
  aux = expX$proj(cX,U)
  mba[[i]] = mean(aux)
  sba[[i]] = sd(aux)}
mb2 = sort(mba)
sb2 = sort(sba)
cmb2 = expb$colloc(nd,0,Ub,mb2)
csb2 = expb$colloc(nd,0,Ub,sb2)
Pm2g = expb$proj(cmb2,Ug)
Ps2g = expb$proj(csb2,Ug)
s1 = saux$create(Pm1g)
sm1 = saux$empcdfpdf(s1,1,nam)
s2 = saux$create(Pm2g)
sm2 = saux$empcdfpdf(s2,1,nam)
ss2 = saux$create(Ps2g)
sb2 = saux$empcdfpdf(ss2,1,nam)
```

An example of result is shown in Fig. 5.1.

(continued)

Example 5.7 (continued)

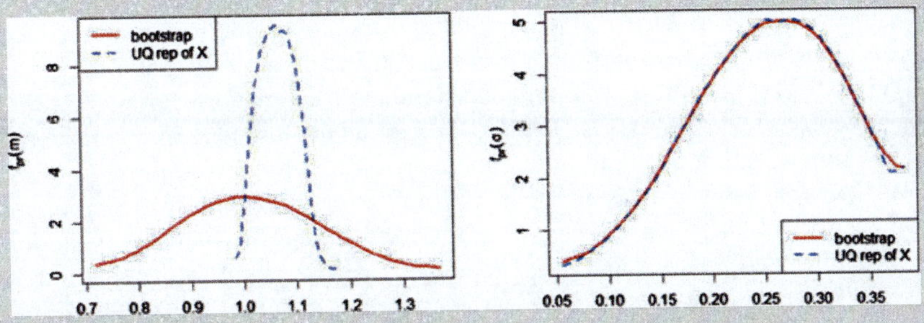

Fig. 5.1. Priors generated by the UQ approach, with $U_{m,\,\sigma} \sim N(0, 1)$

We can use a non-Gaussian distribution for U. For instance, using U uniformly distributed on $(-1, 1)$, we obtain the results shown in Fig. 5.2.

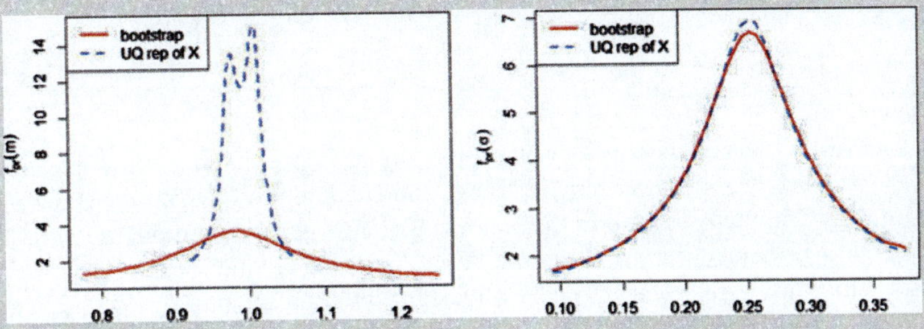

Fig. 5.2 Priors generated by the UQ approach, with $U_{m,\,\sigma} \sim Unif(-1, 1)$

We can use different variables U_m and U_σ for m and σ. An example is shown in Fig. 5.3, where the prior for m was generated using $U_m \sim N(0, 1)$ and the prior for σ was generated using $U_\sigma \sim \chi^2(4)$.

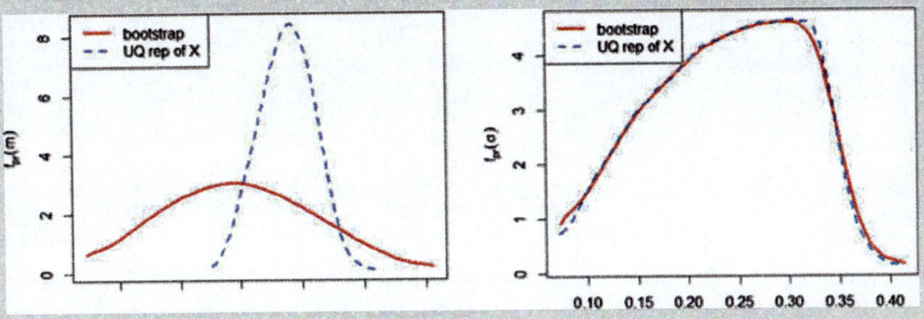

Fig. 5.3 Priors generated by the UQ approach, with $U_m \sim N(0, 1)$ and $U_\sigma \sim \chi^2(4)$

(continued)

Example 5.7 (continued)

These priors can be used in the Bayesian Update: for instance, we can take as prior for the couple (m, σ) the product of the priors previously determined. Using the same Gaussian model introduced in Example 5.3, and the priors generated by bootstrap, we obtain close punctual estimations for all the priors above: for $n = 5$, we got $\widehat{m} \approx 1.0$, $\widehat{\sigma} \approx 0.28$ for all the loss functions considered. The minimization of the loss functions was carried by `fminunc` from package `pracma`, with a starting point chosen as corresponding to the minimum on a grid 21×21 on the support of the priors. The final values of the loss functions were inferior to $1E - 2$ in all the cases.

When using the dataset of 25 variates, we got $\widehat{m} \approx 1.0$, $\widehat{\sigma} \approx 0.42$ for all the approaches (in this case, we used $U_\sigma \sim \chi^2(24)$). The final values of the loss functions were inferior to $1E - 8$ in all the cases.

Notice that the priors generated for m by using the UQ representation of X are more informative than those generated by using the bootstrap approach. The evaluation of the probabilities in the Tables 5.1, 5.2, and 5.3 furnishes results similar to the MLE estimations.

Example 5.8 Let us consider again the situation presented in Example 5.3. Let us use UQ to generate a model for X – and a likelihood. Assume that we are interested in the estimation of the mean: we must generate a model $f(x, m)$. In such a situation, we need to determine a representation of X under the constraint of a given mean, i.e., we look for the representation

$$X \approx P_n X(U) = \sum_{j=0}^{n} x_j \varphi_j(U), \quad E(X) = m \tag{5.3}$$

Using the sample $\boldsymbol{X} = (X_1, \ldots, X_{ns})'$, and the values $\mathcal{U} = (U_1, \ldots, U_{ns})$, we can approximate (5.3) by using Constrained Least Squares: let us denote $\boldsymbol{A} = (A_{ij} : 1 \le i \le ns, 0 \le j \le n)$, $\boldsymbol{B} = (B_0, \ldots, B_n)'$, given by

$$A_{ij} = \varphi_j(U_i) \ , B_j = \sum_{i=1}^{ns} \varphi_j(U_i). \tag{5.4}$$

(continued)

Example 5.8 (continued)

Introducing the Lagrange Multiplier λ associated with the constraint on the mean, we look for $\boldsymbol{x} = (x_0, \ldots, x_n)^t$ corresponding to the minimum of

$$\frac{1}{2}\|\boldsymbol{Ax} - \boldsymbol{X}\|^2 + \lambda(\boldsymbol{B}^t\boldsymbol{x} - m \times ns). \tag{5.5}$$

The solution corresponds to a linear system

$$\boldsymbol{A}^t\boldsymbol{Ax} - \boldsymbol{A}^t\boldsymbol{X} + \lambda\boldsymbol{B} = 0, \quad \boldsymbol{B}^t\boldsymbol{x} = m \times ns. \tag{5.6}$$

It can be solved under R as follows:

```
collocmean = function(k,eps,u,x,m,phi){
  vec1 = matrix(1,nrow = length(x), ncol = 1)
  Aaux = matrix(nrow = length(u), ncol = k+1)
  A = matrix(nrow = k+2, ncol = k+2)
  B = matrix(nrow = k+2, ncol = 1)
  for (i in 1:length(u)){
    uu = u[[i]]
    for (j in 1:(k+1)){
      Aaux[[i,j]] = phi$value(j-1,uu)}}
  A[1:(k+1), 1:(k+1)] = t(Aaux) %*% Aaux
  B[1:(k+1)] = t(Aaux) %*% x
  A[k+2, 1: (k+1)] = t(vec1) %*% Aaux
  A[1:(k+1) ,k+2] = t(Aaux) %*% vec1
  A[[k+2,k+2]] = 0
  B[[k+2]] = sum(vec1)*m
  aux1 = A + eps*diag(k+2)
  cc = solve(aux1,B)
  sol = list(cc[1:(k+1)],cc[k+2])
  names(sol) = c('coeffs','lambda')
  return(sol)}
```

(continued)

Example 5.8 (continued)

As an example, let us consider the representation of the data by a polynomial of degree 3:

```
X = c(0.97,1.41,1.14,0.82,0.67)
Xs = sort(X)
Us = sort(rnorm(length(X)))
a = min(Us)
b = max(Us)
nd = 3
source("polynom.R")
phiX = polynom$new(a,b,1e-3)
nsgX = 1e5
UgX = rnorm(nsgX)
ind = which((UgX >= a) & (UgX <= b))
UgX = UgX[ind]
proj = function(cc,phi,u) {
  ns = length(u) # points
  kp1 = length(cc) # coefficients
  px = numeric(length = ns)
  for (i in 1:ns){
    s = 0
    for (j in 1:kp1){
      s = s + cc[[j]] * phiX$value(j - 1, u[[i]])}
    px[[i]] = s}
  return(px)}
m1 = 1
sol1 = collocmean(3,0,Us,Xs,m1,phiX)
c1 = sol1$coeffs
px1 = proj(c1,phiX,Us)
mc1 = mean(px1)
print(paste('mean: desired = ',m1,', generated = ',mc1))
```

We obtain results as:

```
## [1] "mean: desired =  1 , generated =  1"
## [1] "mean: desired =  2 , generated =  2"
## [1] "mean: desired =  3 , generated =  3"
```

(continued)

Example 5.8 (continued)

Then, we can generate a model distribution $f(x|m)$ as follows:

```
source("sampleX.R")
saux = sampleX$new()
source("ints.R")
int1 = ints$new()
nintX = 50
nsmooth = 3
calc_fmod = function(m,phi,Us,Xs,Ug,nsmooth,nintX){
  sol = collocmean(nd,0,Us,Xs,m,phi)
  coef = sol$coeffs
  PXg = proj(coef, phi,Ug)
  sX = saux$create(PXg)
  sX1 = saux$empcdfpdf(sX,nsmooth,nintX)
  fXaux = approxfun(sX1$pdf$x,sX1$pdf$f,rule = 2)
  xmin = min(sX1$pdf$x)
  xmax = max(sX1$pdf$x)
  auxnorm = int1$romb1(fXaux,xmin,xmax,100,1e-8)
  fmodX = function(x){
    if ((x>= xmin) & (x <= xmax)){
      fXaux(x)/auxnorm}
    else{0}
  }
  resu = list(fmodX,coef,xmin,xmax)
  names(resu) = c('fmod','coeffs','xmin','xmax')
  return(resu)}
```

Examples of results are shown in Fig. 5.4.

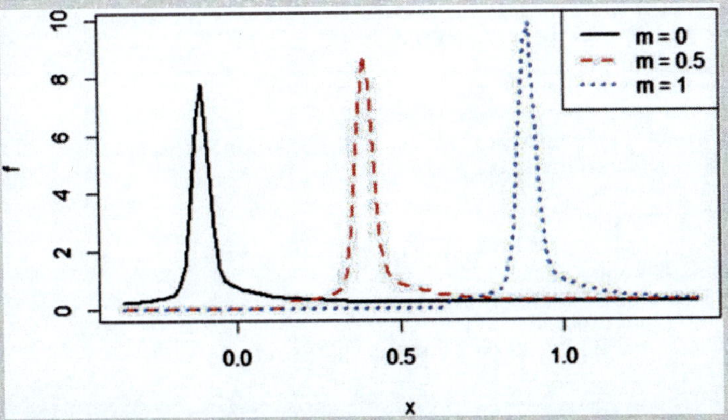

Fig. 5.4 Examples of models $f(x|m)$ generated by UQ a polynomial basis, degree 3, dataset of five elements

(continued)

Example 5.8 (continued)

The likelihood is generated as follows:

```
genlikelihood = function(m,phi,Us,Xs,Ug,nsmooth,nintX){
  rr = calc_fmod(m,phi,Us,Xs,Ug,nsmooth,nintX)
  fmod = rr$fmod
  s = 1
  for (j in 1: length(Xs)){
    s = s* fmod(Xs[[j]])}
  return(s)}
nsg = 1e5
Ug = rnorm(nsg)
ind = which((Ug >= a) & (Ug <= b))
Ug = Ug[ind]
likelihood = function(m){genlikelihood(m,phiX,Us,Xs,Ug,nsmooth,nintX)}
```

To avoid zeros in the likelihood, we set

```
xmin = min(c(sX1$pdf$x,Xs[[1]]))
xmax = max(c(sX1$pdf$x,Xs[[ length(Xs) ]]))
```

and the likelihood was normalized by using

```
mest1 = mean(X)
sd1 = sd(X)
alfa = 0.01
tlim = qt(1-alfa/2,ndf)
am = mest1 - tlim*sd1/sqrt(nd)
bm = mest1 + tlim*sd1/sqrt(nd)
nint = 250
mc = seq(am,bm,length.out = nint+1)
lkl = numeric(length=length(mc))
for (i in (1:length(mc))){lkl[[i]]= likelihood(mc[[i]]) }
lkf1 = approxfun(mc,lkl,rule = 2)
mmin1 = min(mc)
mmax1 = max(mc)
auxn1 = int1$romb1(lkf1,mmin1,mmax1,100,1e-8)
lkl1 = lkl/auxn1
lkfct1 = function(m){lkf1(m)/auxnorm1}
lkfct = function(m){
  if ((m >= mmin1) & (m <= mmax1)){lkfct1(m)}
  else{0}}
```

(continued)

Example 5.8 (continued)

An example of result is shown in Fig. 5.5.

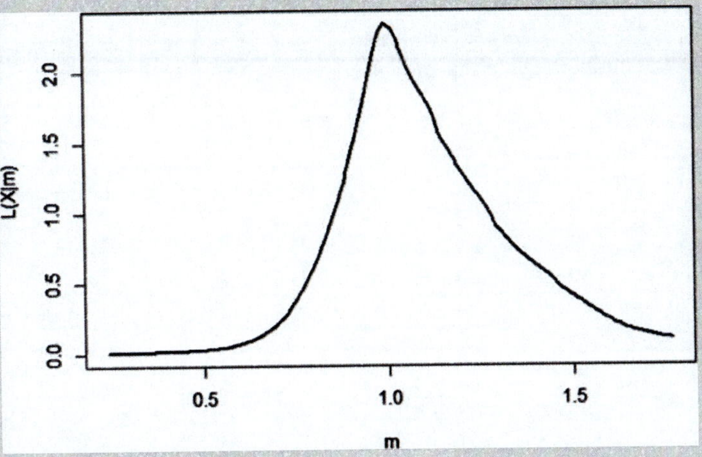

Fig. 5.5 Example of likelihood $L(X|m)$ generated by UQ, with a polynomial basis, degree 3, $U \sim N(0,1)$, dataset of five elements

The likelihood generated can be used to generate an MLE for m. For instance, we can determine the value $\widehat{m}$ corresponding to the maximum of $L(X|m)$. For the results in Fig. 5.5, we have $\widehat{m} \approx 1.0$.

We can also use the likelihood generated to evaluate the posterior distribution, using the Bayesian Update (5.1). For this, we can either use a prior chosen by the user or use the prior generated by the UQ approach introduced in the Example 5.7. Using these last priors, we obtain posterior distributions as shown in Fig. 5.6.

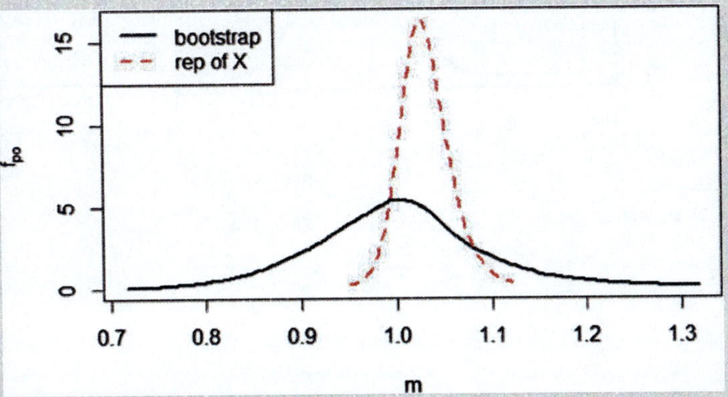

Fig. 5.6 Example of posterior distributions generated by UQ, with a polynomial basis, degree 3, $U \sim N(0,1)$, dataset of five elements

(continued)

Example 5.8 (continued)

These distributions can be used to make estimations $\widehat{m}$ of the mean m, with the loss functions previously introduced. If we are interested in the evaluation of probabilities of some events, we must use the equality $P(X \leq a) = \int_{-\infty}^{a} f(x|\widehat{m})dx$. In practice, $-\infty$ is replaced by a large enough negative number. Table 5.4 shows examples of results.

Table 5.4 Examples of results for the evaluation of $P(X \leq a)$, using the MLE generated and BI with the three loss functions introduced. Results for a sample of size 5, with a polynomial basis of degree 3. The integrals were evaluated by Romberg's method using package PRACMA

			Bayes 1		Bayes 2		Bayes 3	
a	Exact	MLE generated	boot	P_nX	boot	P_nX	boot	P_nX
0.8	0.3446	0.2152	0.1978	0.1855	0.1872	0.1854	0.1977	0.1854
1	0.5000	0.6196	0.5983	0.5803	0.5830	0.5803	0.5983	0.5803
1.2	0.6554	0.8661	0.8535	0.8428	0.8444	0.8428	0.8536	0.8428
$\widehat{m}$	1	1.00	1.02	1.03	1.03	1.03	1.02	1.03

When the dataset of 25 elements is considered, the results improve: Fig. 5.7 shows the likelihood, Fig. 5.8 shows the priors and posterior distributions by using the bootstrap and the representation of X. Table 5.5 exhibits the results for the probabilities and the mean.

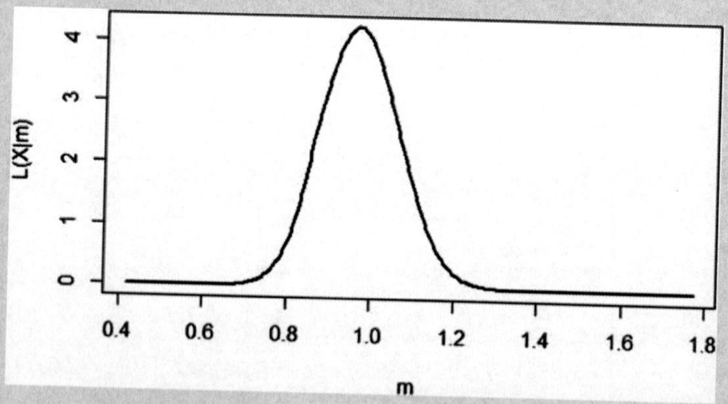

Fig. 5.7 Example of likelihood generated by UQ, with a polynomial basis, degree 3, $U \sim N(0, 1)$, dataset of 25 elements

(continued)

Example 5.8 (continued)

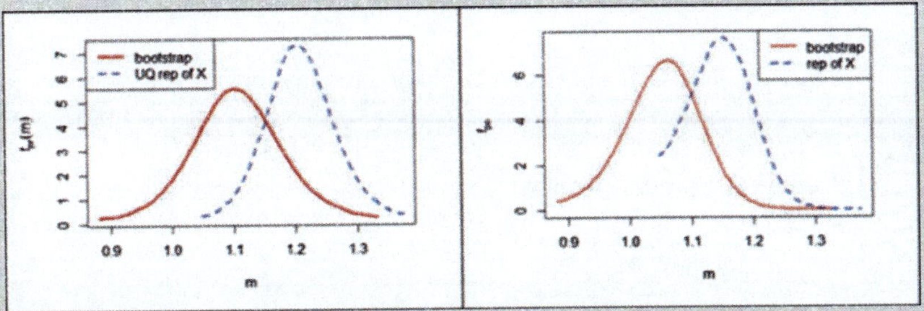

Fig. 5.8 Examples of prior and posterior generated by UQ, with a polynomial basis, degree 3, $U \sim N(\mathbf{0}, \mathbf{1})$, dataset of 25 elements. Prior is shown at left, posterior appears at right

Table 5.5 Examples of results for the evaluation of $P(X \le a)$, using the MLE generated and BI with the three loss functions introduced. Results for a sample of size 25, with a polynomial basis of degree 3. The integrals were evaluated by Romberg's method using package PRACMA

			Bayes 1		Bayes 2		Bayes 3	
a	Exact	MLE generated	boot	P_nX	boot	P_nX	boot	P_nX
0.8	0.3446	0.2818	0.2438	0.2117	0.2417	0.2108	0.2438	0.2113
1	0.5000	0.4483	0.3927	0.3371	0.3983	0.3354	0.3927	0.3365
1.2	0.6554	0.6262	0.5685	0.5006	0.5646	0.4982	0.5685	0.4997
$\widehat{m}$	1	0.97	1.05	1.14	1.06	1.15	1.05	1.15

Example 5.9 Let us consider again the situation presented in Example 5.3, but assume that we are interested in $E(X^2) = M_2$. Again, let us use UQ to generate a model $f(x, m_2)$ for X – and a likelihood. Here, we look for

$$X \approx P_nX(U) = \sum_{j=0}^{n} x_j\varphi_j(U), \quad E(X^2) = M_2. \tag{5.7}$$

Using the sample $\boldsymbol{X} = (X_1, \ldots, X_{ns})^t$, and the values $\mathcal{U} = (U_1, \ldots, U_{ns})$, we can approximate the constraint $E(X^2) = M_2$ (5.7) as

$$E\Big((P_nX(U))^2\Big) = M_2,$$

(continued)

Example 5.9 (continued)

i.e.,

$$\sum_{i,j=0}^{n} x_i x_j E\big(\varphi_i(U)\varphi_j(U)\big) = M_2.$$

This equation is approximated by

$$\sum_{i,j=0}^{n} x_i x_j \sum_{k=1}^{ns} \varphi_i(U_k)\varphi_j(U_k) = ns \times M_2,$$

i.e.,

$$\boldsymbol{x}^t \boldsymbol{A}^t \boldsymbol{A} \boldsymbol{x} = M_2 \times ns, \tag{5.8}$$

where $\boldsymbol{A}$ was defined in Example 5.8. Thus we look for the solution of

$$\boldsymbol{x}^* = \arg\min\left\{\frac{1}{2}\|\boldsymbol{A}\boldsymbol{x} - \boldsymbol{X}\|^2 : \boldsymbol{x} \text{ satisfies } (5.8)\right\}. \tag{5.9}$$

This problem can be solved by Constrained Nonlinear Optimization methods, such as SQP. For instance, we can use `fmincon` from package PRACMA. Alternatively, we can introduce the Lagrange Multiplier λ associated with the restriction. Then, the solution corresponds to a nonlinear algebraic system

$$\boldsymbol{A}^t\boldsymbol{A}\boldsymbol{x} - \boldsymbol{A}^t\boldsymbol{X} + 2\lambda \boldsymbol{A}^t\boldsymbol{A}\boldsymbol{x} = 0, \quad \boldsymbol{x}^t\boldsymbol{A}^t\boldsymbol{A}\boldsymbol{x} - M_2 \times ns = 0. \tag{5.10}$$

Consequently,

$$(1 + 2\lambda)\boldsymbol{A}^t\boldsymbol{A}\boldsymbol{x} = \boldsymbol{A}^t\boldsymbol{X}, \quad \boldsymbol{x}^t(1 + 2\lambda)\boldsymbol{A}^t\boldsymbol{A}\boldsymbol{x} = (1 + 2\lambda)M_2 \times ns. \tag{5.11}$$

Thus,

$$(1 + 2\lambda)\boldsymbol{x} = (\boldsymbol{A}^t\boldsymbol{A})^{-1}\boldsymbol{A}^t\boldsymbol{X}, \quad \boldsymbol{x}^t\boldsymbol{A}^t\boldsymbol{X} = (1 + 2\lambda)M_2 \times ns \tag{5.12}$$

and

$$\boldsymbol{X}^t\boldsymbol{A}\left(\left(\boldsymbol{A}^t\boldsymbol{A}\right)^{-1}\right)^t\boldsymbol{A}^t\boldsymbol{X} = (1 + 2\lambda)\boldsymbol{x}^t\boldsymbol{A}^t\boldsymbol{X} = (1 + 2\lambda)^2 M_2 \times ns,$$

(continued)

Example 5.9 (continued)

i.e.,

$$(1 + 2\lambda)^2 M_2 \times ns = \boldsymbol{X}^t \boldsymbol{A}(\boldsymbol{A}^t \boldsymbol{A})^{-1} \boldsymbol{A}^t \boldsymbol{X}. \tag{5.13}$$

This equation can be used to determine λ and, then, $\boldsymbol{x}$ by solving the linear system $(1 + 2\lambda)\boldsymbol{A}^t\boldsymbol{A}\boldsymbol{x} = \boldsymbol{A}^t\boldsymbol{X}$.

This approach can be implemented under R as follows:

```
collocm2 = function(k,eps,u,x,m2,phi){
  vec1 = matrix(1,nrow = length(x), ncol = 1)
  A = matrix(nrow = length(u), ncol = k+1)
  B = matrix(nrow = length(X), ncol = 1)
  for (i in 1:length(u)){
    uu = u[[i]]
    B[[i]] = x[[i]]
    for (j in 1:(k+1)){
      A[[i,j]] = phi$value(j-1,uu)}}
  ATA = t(A) %*% A + eps*diag(k+1)
  mat1 = solve(ATA)
  mat2 = t(A) %*% B
  aux3 = t(mat2) %*% mat1 %*% mat2 /(m2*length(x))
  aux4 = as.numeric(sqrt(aux3))
  mat5 = mat2/aux4
  cc = solve(ATA,mat5)
  sol = list(cc,0.5*(aux4 - 1))
  names(sol) = c('coeffs','lambda')
  return(sol)}
```

As an example, let us consider the representation of the data by a polynomial of degree 3, as in Example 5.8:

```
m2_1 = 1
r1 = collocm2(nd,0,Us,Xs,m2_1,phiX)
c1 = r1$coeffs
y1 = proj(c1,phiX,Us)
m2_1b = mean(y1^2)
print(paste('M2 desired = ',m2_1,', obtained = ',m2_1b))
```

We obtain results as:

```
## [1] "M2 desired =  1 , obtained =  1.0000000000001"
## [1] "M2 desired =  2 , obtained =  2.0000000000002"
## [1] "M2 desired =  3 , obtained =  3.0000000000003"
```

(continued)

Example 5.9 (continued)

Then, we can generate a model distribution $f(x|M_2)$ as follows:

```
calc_fmod = function(m2,phi,Us,Xs,Ug,nsmooth,nintX){
  sol = collocm2(nd,0,Us,Xs,m2,phi)
  coef = sol$coeffs
  PXg = proj(coef, phi,Ug)
  sX = saux$create(PXg)
  sX1 = saux$empcdfpdf(sX,nsmooth,nintX)
  fXaux = approxfun(sX1$pdf$x,sX1$pdf$f,rule = 2)
  # xmin = min(c(sX1$pdf$x,Xs[[1]]))
  # xmax = max(c(sX1$pdf$x,Xs[[ length(Xs) ]]))
  xmin = min(sX1$pdf$x)
  xmax = max(sX1$pdf$x)
  auxnorm = int1$romb1(fXaux,xmin,xmax,100,1e-8)
  fmodX = function(x){
    if ((x>= xmin) & (x <= xmax)){
      fXaux(x)/auxnorm}
    else{0}
  }
  resu = list(fmodX,coef,xmin,xmax)
  names(resu) = c('fmod','coeffs','xmin','xmax')
  return(resu)}
```

Examples of results are shown in Fig. 5.9.

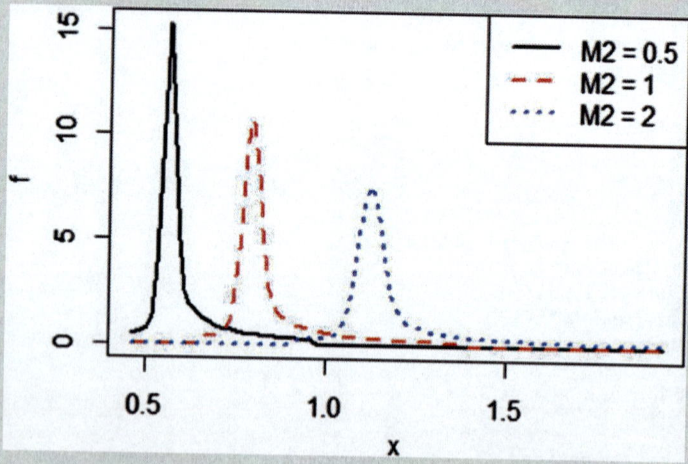

Fig. 5.9 Examples of models $f(x|M_2)$ generated by UQ a polynomial basis, degree 3

(continued)

Example 5.9 (continued)

Notice that (5.12) can also be solved by a fixed point strategy:

$$\left(1+2\lambda^{(k)}\right)\boldsymbol{A}^t\boldsymbol{A}\boldsymbol{x}^{(k)}=\boldsymbol{A}^t\boldsymbol{X},\quad \lambda^{(k+1)}=\frac{1}{2}\left(\frac{\boldsymbol{x}^{(k)^t}\boldsymbol{A}^t\boldsymbol{X}}{M_2\times ns}-1\right). \tag{5.14}$$

These iterations can be implemented under R as follows:

```
itersolm2 = function(k,eps,u,x,m2,phi,nitmax,precmin,om){
  vec1 = matrix(1,nrow = length(x), ncol = 1)
  A = matrix(nrow = length(u), ncol = k+1)
  B = matrix(nrow = length(X), ncol = 1)
  for (i in 1:length(u)){
    uu = u[[i]]
    B[[i]] = x[[i]]
    for (j in 1:(k+1)){
      A[[i,j]] = phi$value(j-1,uu)}}
  ATA = t(A) %*% A + eps*diag(k+1)
  mat2 = t(A) %*% B
  lbdac = runif(1,0,2)
  cac = solve(ATA,mat2/(1+2*lbdac))
  nit = 0
  prec = precmin+1
  while ((nit < nitmax) & (prec >precmin)){
    aux3 = t(cac) %*% mat2 /(m2*length(x))
    aux4 = as.numeric(aux3)
    lbdanew = 0.5*(aux4 - 1)
    cnew = solve(ATA,mat2/(1+2*lbdanew))
    prec = sqrt(sum((cnew-cac)^2)/length(cnew)) + abs(lbdanew -
lbdac)
    nit = nit+1
    lbdac = om*lbdanew + (1-om)*lbdac
    cac = om*cnew + (1-om)*cac
  }
  sol = list(cac,lbdac,prec,nit)
  names(sol) = c('coeffs','lambda','prec','nit')
  return(sol)}
```

Considering again the representation of the data by a polynomial of degree 3, we obtain results as:

```
## [1] "M2 desired =  1 , obtained =  1"
## [1] "M2 desired =  2 , obtained =  2"
## [1] "M2 desired =  3 , obtained =  3"
```

(continued)

Example 5.9 (continued)

The likelihood is generated as in Example 5.8. An example of result is shown in Fig. 5.10.

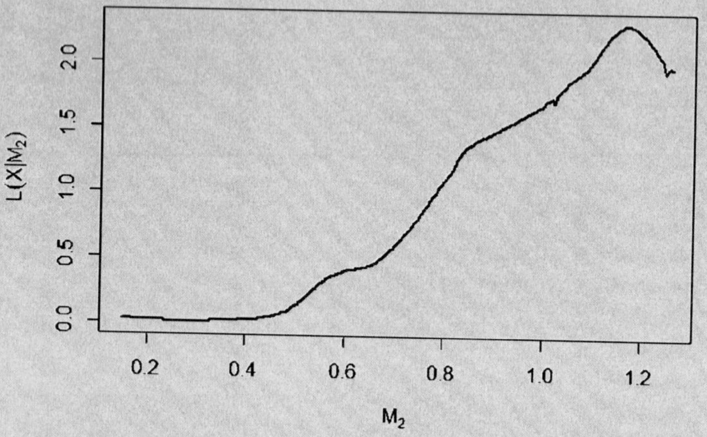

Fig. 5.10 Example of likelihood $L(X|M_2)$ generated by UQ, with a polynomial basis, degree 3, $U \sim N(0,1)$, dataset of five elements

The likelihood generated can be used to generate an MLE for M_2. For instance, we can determine the value $\widehat{M_2}$ corresponding to the maximum of $L(X|M_2)$. For the results in Fig. 5.10 , we have $\widehat{M_2} \approx 1.1$.

We can also use the likelihood generated to evaluate the posterior distribution, using the Bayesian Update (5.1). For this, we can either use a prior chosen by the user or use the prior generated by the UQ approaches introduced in the Example 5.7. An example of results generated using these last priors is shown in Figs. 5.11 and 5.12.

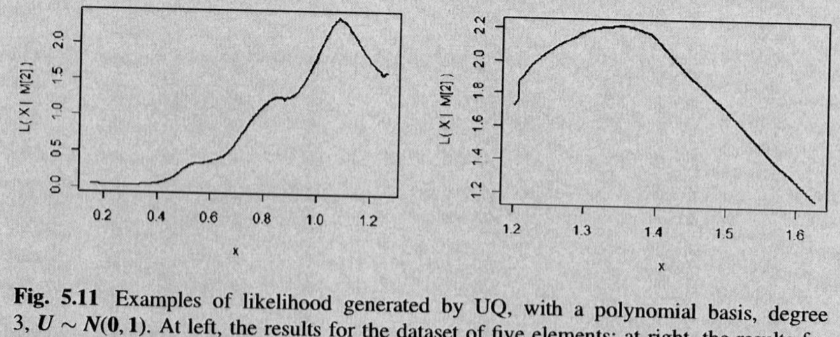

Fig. 5.11 Examples of likelihood generated by UQ, with a polynomial basis, degree 3, $U \sim N(0,1)$. At left, the results for the dataset of five elements; at right, the results for the dataset of 25 elements

(continued)

Example 5.9 (continued)

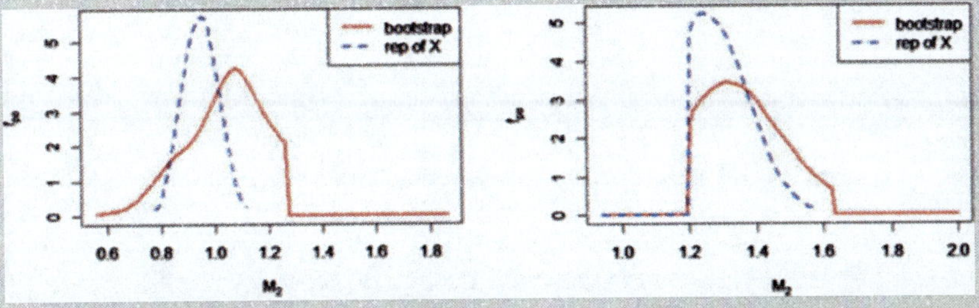

Fig. 5.12 Examples of posterior distributions generated by UQ, with a polynomial basis, degree 3, $\boldsymbol{U} \sim \boldsymbol{N}(\boldsymbol{0}, \boldsymbol{1})$. At left, the results for the dataset of five elements; at right, the results for the dataset of 25 elements

These distributions can be used to make estimations $\widehat{M_2}$ of the moment M_2, with the loss functions previously introduced. If we are interested in the evaluation of probabilities of some events, we must use the equality $P(X \leq a) = \int_{-\infty}^{a} f\left(x|\widehat{M_2}\right) dx$. Table 5.6 shows examples of results.

Table 5.6 Examples of results for the evaluation of $\boldsymbol{P}(\boldsymbol{X} \leq \boldsymbol{a})$, using the MLE generated and BI with the three loss functions introduced. Results for a sample of size 5, with a polynomial basis of degree 5. The integrals were evaluated by Romberg's method using package PRACMA

			Bayes 1		Bayes 2		Bayes 3	
a	Exact	MLE generated	boot	P_nX	boot	P_nX	boot	P_nX
0.8	0.3446	0.2762	0.2493	0.2841	0.2493	0.2766	0.2493	0.2695
1	0.5000	0.4667	0.4273	0.4778	0.4273	0.4673	0.4273	0.4572
1.2	0.6554	0.6553	0.6141	0.6663	0.614	0.6559	0.614	0.6456
$\widehat{M_2}$	1.25	1.35	1.47	1.32	1.47	1.34	1.47	1.38

Example 5.10 Let us consider again the situation presented in Example 5.3, but assume that we are interested in $E(X) = m$ and $\sigma(X) = \sigma$. This last equation corresponds to $E(X^2) = m^2 + \sigma^2$. Let us use UQ to generate a model $f(x, \boldsymbol{\theta})$ for X and a likelihood, – with $\boldsymbol{\theta} = (m, \sigma)$. Now, we look for

$$X \approx P_nX(U) = \sum_{j=0}^{n} x_j \varphi_j(U), \quad E(X) = m, \quad E\left(X^2\right) = m^2 + \sigma^2.\# \tag{5.15}$$

(continued)

Example 5.10 (continued)

Using the sample $\boldsymbol{X} = (X_1, \ldots, X_{ns})^t$, and the values $\mathcal{U} = (U_1, \ldots, U_{ns})$, we can approximate the constraints above as

$$E(P_n X(U)) = m, \quad E\left((P_n X(U))^2\right) = M_2 = m^2 + \sigma^2,$$

i.e.,

$$\boldsymbol{B}^t \boldsymbol{x} = m, \quad \boldsymbol{x}^t \boldsymbol{A}^t \boldsymbol{A} \boldsymbol{x} = M_2 \times ns. \tag{5.16}$$

Thus we look for the solution of

$$\boldsymbol{x}^* = \arg\min\left\{\frac{1}{2}\|\boldsymbol{A}\boldsymbol{x} - \boldsymbol{X}\|^2 : \boldsymbol{x} \text{ satisfies } (5.16)\right\}. \tag{5.17}$$

Let us introduce the Lagrange Multipliers λ_1, λ_2 associated with the restrictions. Then, the solution corresponds to a nonlinear algebraic system

$$(1 + 2\lambda_2)\boldsymbol{A}^t\boldsymbol{A}\boldsymbol{x} + \lambda_1\boldsymbol{B} = \boldsymbol{A}^t\boldsymbol{X}, \quad \boldsymbol{B}^t\boldsymbol{x} = m, \quad \boldsymbol{x}^t\boldsymbol{A}^t\boldsymbol{A}\boldsymbol{x} = M_2 \times ns. \tag{5.18}$$

Then,

$$(1 + 2\lambda_2)\boldsymbol{x}^t\boldsymbol{A}^t\boldsymbol{A}\boldsymbol{x} + \lambda_1\boldsymbol{x}^t\boldsymbol{B} = \boldsymbol{x}^t\boldsymbol{A}^t\boldsymbol{X}$$

and

$$(1 + 2\lambda_2)M_2 \times ns + \lambda_1 m = \boldsymbol{x}^t\boldsymbol{A}^t\boldsymbol{X}. \tag{5.19}$$

Again, we can use a fixed point strategy:

$$\left(1 + 2\lambda_2^{(k)}\right)\boldsymbol{A}^t\boldsymbol{A}\boldsymbol{x}^{(k)} + \lambda_1^{(k)}\boldsymbol{B} = \boldsymbol{A}^t\boldsymbol{X}, \quad \boldsymbol{B}^t\boldsymbol{x}^{(k)} = m, \tag{5.20}$$

$$\lambda_2^{(k+1)} = \frac{1}{2}\left(\frac{\boldsymbol{x}^{(k)^t}\boldsymbol{A}^t\boldsymbol{X} - \lambda_1^{(k)}m}{M_2 \times ns} - 1\right). \tag{5.21}$$

(continued)

Example 5.10 (continued)

This approach can be implemented under R as follows:

```
collocm1lbda = function(k,eps,u,x,m1,phi,lbda2,Aaux,vec1){
  A = matrix(nrow = k+2, ncol = k+2)
  B = matrix(nrow = k+2, ncol = 1)
  A[1:(k+1), 1:(k+1)] = (1+2*lbda2)*t(Aaux) %*% Aaux
  B[1:(k+1)] = t(Aaux) %*% x
  A[k+2, 1: (k+1)] = t(vec1) %*% Aaux
  A[1:(k+1) ,k+2] = t(Aaux) %*% vec1
  A[[k+2,k+2]] = 0
  B[[k+2]] = sum(vec1)*m1
  aux1 = A + eps*diag(k+2)
  cc = solve(aux1,B)
  sol = list(cc[1:(k+1)],cc[k+2])
  names(sol) = c('coeffs','lambda')
  return(sol)}
itersolm1sig = function(k,eps,u,x,m1,sig,phi,nitmax,precmin,om){
  m2 = m1^2 + sig^2
  vec1 = matrix(1,nrow = length(x), ncol = 1)
  Aaux = matrix(nrow = length(u), ncol = k+1)
  for (i in 1:length(u)){
    uu = u[[i]]
    for (j in 1:(k+1)){
      Aaux[[i,j]] = phi$value(j-1,uu)}}
  mat2 = t(Aaux) %*% x
  lbd2ac = runif(1,0,2)
  sol1ac = collocm1lbda(k,eps,u,x,m1,phi,lbd2ac,Aaux,vec1)
  cac = sol1ac$coeffs
  lbd1ac = sol1ac$lambda
  nit = 0
  prec = precmin+1
  while ((nit < nitmax) & (prec >precmin)){
    aux3 = (t(cac) %*% mat2 - lbd1ac*m1*length(x))
/(m2*length(x))
    aux4 = as.numeric(aux3)
    lbd2new = 0.5*(aux4 - 1)
    sol1new = collocm1lbda(k,eps,u,x,m1,phi,lbd2new,Aaux,vec1)
    cnew = sol1new$coeffs
    lbd1new = sol1new$lambda
    prec = sqrt(sum((cnew-cac)^2)/length(cnew)) + abs(lbd1new -
lbd1ac) + abs(lbd2new -lbd2ac)
    # print(prec)
    nit = nit+1
    lbd1ac = om*lbd1new + (1-om)*lbd1ac
    cac = om*cnew + (1-om)*cac
    lbd2ac = om*lbd2new + (1-om)*lbd2ac}
  sol = list(cac,lbd1ac,lbd2ac,prec,nit)
  names(sol) = c('coeffs','lambda1','lambda2','prec','nit')
  return(sol)}
## [1] "M2 desired =  3 , obtained =  3"
```

(continued)

Example 5.10 (continued)

An example of likelihood generated by this method is shown in Fig. 5.13.

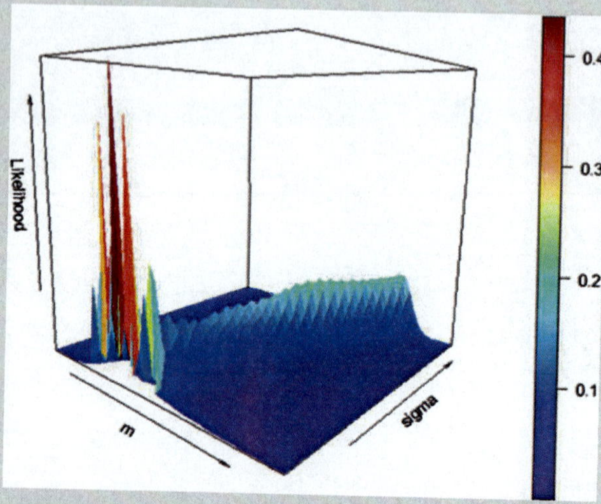

Fig. 5.13 Example of likelihood generated by UQ, with a polynomial basis, degree 3, $\boldsymbol{U} \sim \boldsymbol{N}(\mathbf{0}, \mathbf{1})$, dataset of five elements

The likelihood generated can be used to generate an MLE for (m, σ) . For instance, we can determine the value $\widehat{m} \approx 1.1, \widehat{\sigma} \approx 0.15$ corresponding to the maximum of likelihood. For instance, we can generate a sample from (m, σ) using bootstrap and use these results to fit a Hilbert representation of these variables, which is used to determine their distribution.

We can also use the data to generate a prior distribution. For instance, we can generate samples from m and σ by bootstrap. These samples can be used to fit priors from each variable, as in the preceding examples. An example of result is shown in Fig. 5.14.

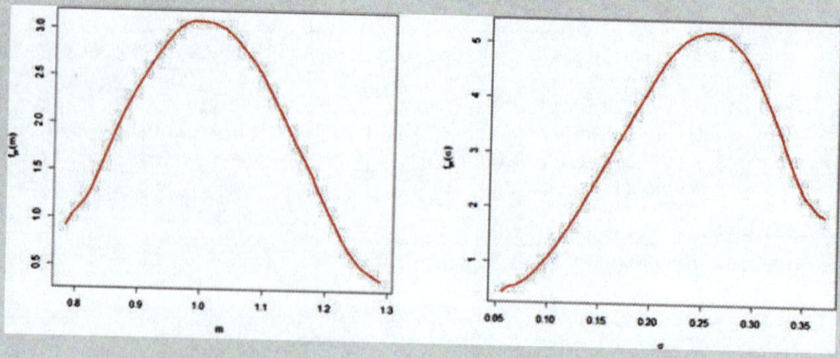

Fig. 5.14 Examples of priors generated by bootstrap, with a polynomial basis, degree 3, $\boldsymbol{U} \sim \boldsymbol{N}(\mathbf{0}, \mathbf{1})$, and a dataset of five elements

(continued)

Example 5.10 (continued)

Bayesian updating can be applied as usual, using as prior for (m, σ) the product of the priors for each variable. An example of results generated using these last priors is shown in Fig. 5.15.

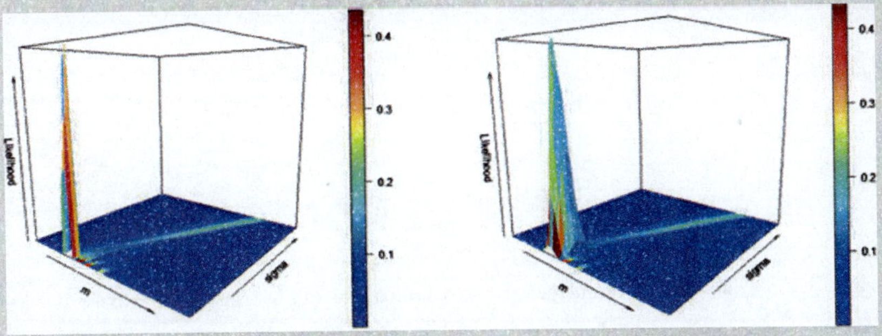

Fig. 5.15 Examples of posterior distributions generated with a polynomial basis, degree 3, $\boldsymbol{U} \sim N(\mathbf{0}, \mathbf{1})$, and a dataset of five elements. At left, the results obtained using the likelihood generated; at right, the results obtained using a Gaussian likelihood

These distributions can be used to make estimations or (m, σ). With the distribution generated using the Gaussian likelihood, we obtain $\widehat{m} \approx 1.0, \widehat{\sigma} \approx 0.28$.

For the dataset of 25 elements and polynomial approximations of degree 5, we obtain the results in Figs. 5.16, 5.17, and 5.18. In this case, the estimations from the MLE generated are value $\widehat{m} \approx 1.1, \widehat{\sigma} \approx 0.48$. The posterior estimations using the quadratic loss function and the Gaussian likelihood are $\widehat{m} \approx 1.1, \widehat{\sigma} \approx 0.45$.

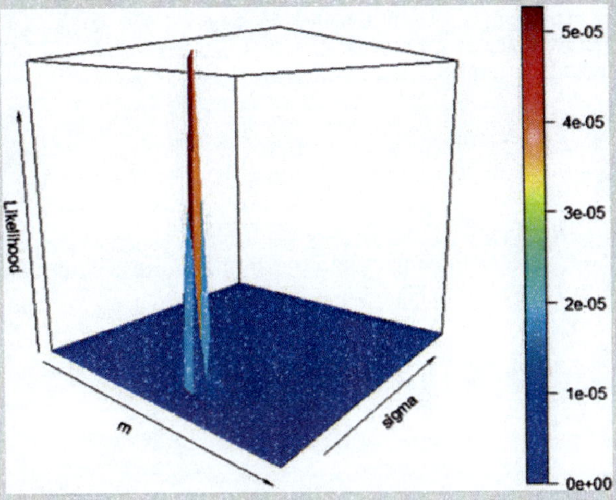

Fig. 5.16 Example of likelihood generated by UQ, with a polynomial basis, degree 5, $\boldsymbol{U} \sim N(\mathbf{0}, \mathbf{1})$, dataset of 25 elements

(continued)

Example 5.10 (continued)

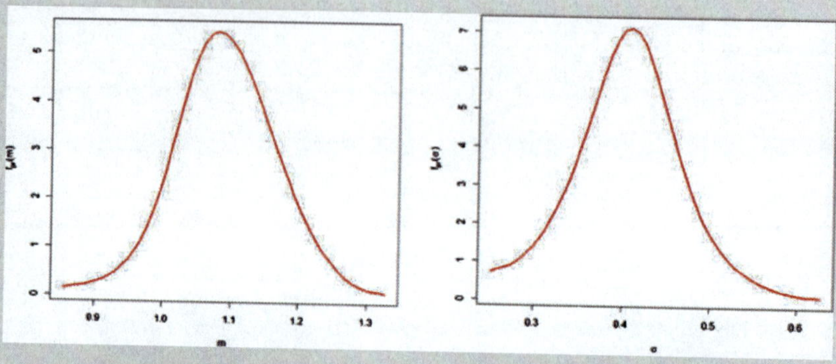

Fig. 5.17 Examples of priors generated by bootstrap, with a polynomial basis, degree 5, $U \sim N(0, 1)$, and a dataset of 25 elements

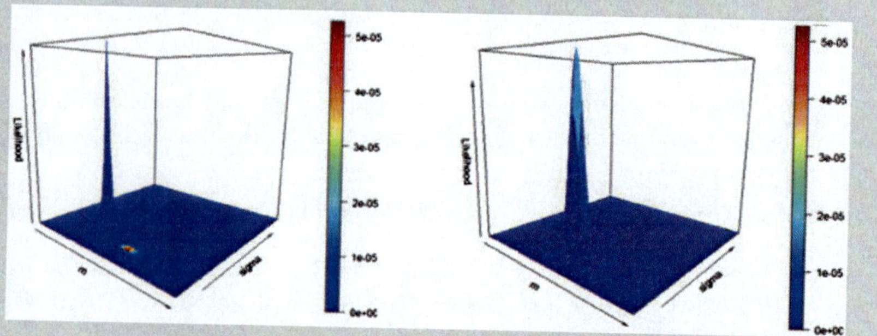

Fig. 5.18 Examples of posterior distributions generated with a polynomial basis, degree 5, $U \sim N(0, 1)$, and a dataset of 25 elements. At left, the results obtained using the likelihood generated; at right, the results obtained using a Gaussian likelihood

Example 5.11 Let us consider again the situation presented in Example 5.3, but assume that we are interested in $E(X) = m$, $\sigma(X) = \sigma$. We can consider

$$X \approx P_n X(U) = \sum_{j=0}^{n} x_j \varphi_j(U), \quad E(X) = m, \quad \sigma(X) = \sigma. \tag{5.22}$$

(continued)

Example 5.11 (continued)

Using the sample $\boldsymbol{X} = (X_1, \ldots, X_{ns})^t$, and the values $\mathcal{U} = (U_1, \ldots, U_{ns})$, we can approximate the constraints above as

$$E(P_n X(U)) = m, \quad \sigma(P_n X(U)) = \sigma. \tag{5.23}$$

Thus we look for the solution of

$$\boldsymbol{x}^* = \arg\min \left\{ \frac{1}{2} \|\boldsymbol{A}\boldsymbol{x} - \boldsymbol{X}\|^2 : \boldsymbol{x} \text{ such that Eq.(5.23) is satisfied} \right\}. \tag{5.24}$$

Equation (5.24) can be solved by numerical methods. For instance, we can use `fmincon` from the package `PRACMA`.

An example of likelihood generated by this method is shown in Fig. 5.19.

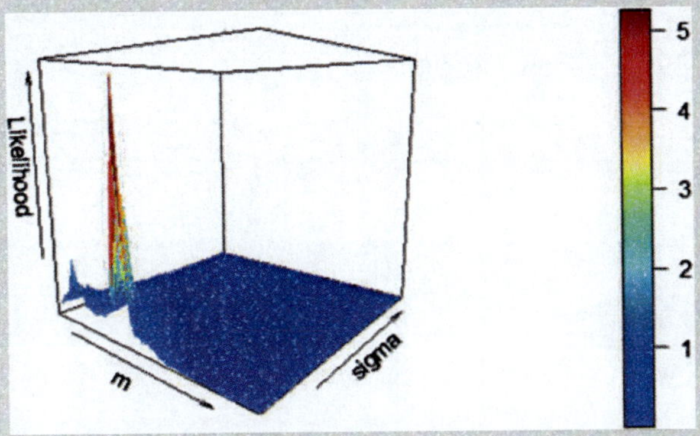

Fig. 5.19 Example of likelihood generated by UQ, with a polynomial basis, degree 3, $\boldsymbol{U} \sim N(\boldsymbol{0}, \boldsymbol{1})$, dataset of five elements

The likelihood generated can be used to generate an MLE for (m, σ) . For instance, we can determine the value $\widehat{m} \approx 1.1, \widehat{\sigma} \approx 0.21$ corresponding to the maximum of likelihood. For instance, we can generate a sample from (m, σ) using bootstrap and use these results to fit a Hilbert representation of these variables, which is used to determine their distribution.

We can also use the data to generate a prior distribution. For instance, we can generate samples from m and σ by bootstrap. These samples can be used to fit priors from each variable, as in the preceding examples. An example of result is shown in Fig. 5.20.

(continued)

Example 5.11 (continued)

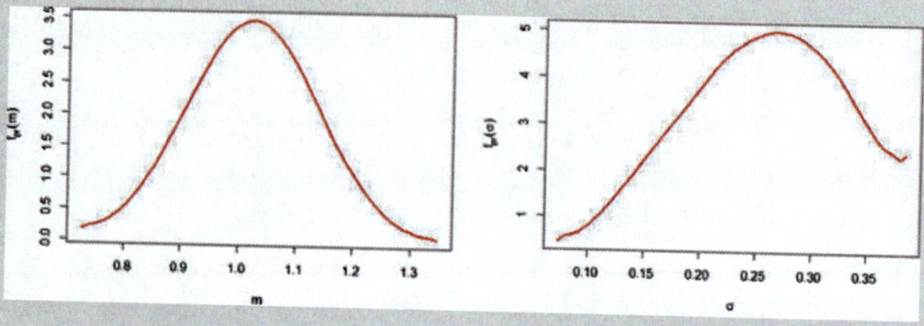

Fig. 5.20 Examples of priors generated by bootstrap, with a polynomial basis, degree 3, $\boldsymbol{U} \sim \boldsymbol{N}(\mathbf{0}, \mathbf{1})$, and a dataset of five elements

Bayesian updating can be applied as usual, using as prior for (m, σ) the product of the priors for each variable. An example of results generated using these last priors is shown in Fig. 5.21.

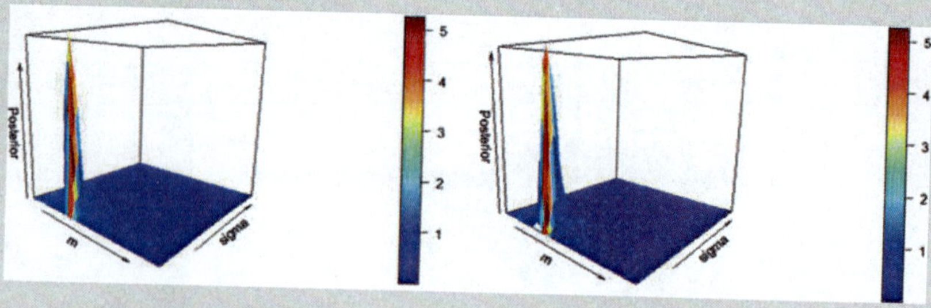

Fig. 5.21 Examples of posterior distributions generated with a polynomial basis, degree 3, $\boldsymbol{U} \sim \boldsymbol{N}(\mathbf{0}, \mathbf{1})$, and a dataset of five elements. At left, the results obtained using the likelihood generated; at right, the results obtained using a Gaussian likelihood

These distributions can be used to make estimations or (m, σ). With the distribution generated using the Gaussian likelihood, we obtain $\widehat{m} \approx 1.0, \widehat{\sigma} \approx 0.28$.

For the dataset of 25 elements and polynomial approximations of degree 5, we obtain the results in Figs. 5.22, 5.23, and 5.24. In this case, the estimations from the MLE generated are value $\widehat{m} \approx 1.0, \widehat{\sigma} \approx 0.56$. The posterior estimations using the quadratic loss function and the Gaussian likelihood are $\widehat{m} \approx 1.1, \widehat{\sigma} \approx 0.47$.

(continued)

Example 5.11 (continued)

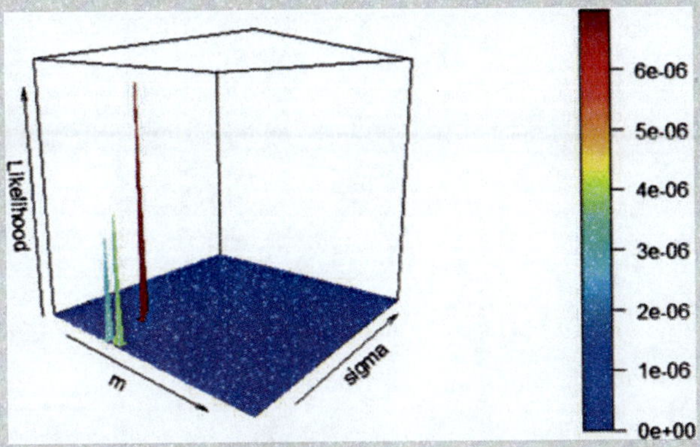

Fig. 5.22 Example of likelihood generated by UQ, with a polynomial basis, degree 5, $U \sim N(0, 1)$, dataset of 25 elements

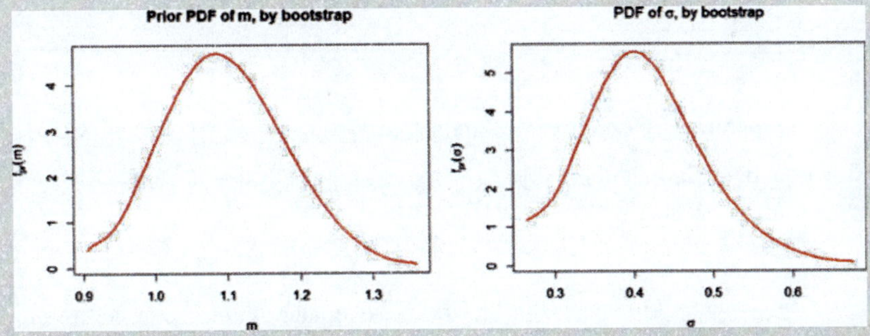

Fig. 5.23 Examples of priors generated by bootstrap, with a polynomial basis, degree 5, $U \sim N(0, 1)$, and a dataset of 25 elements

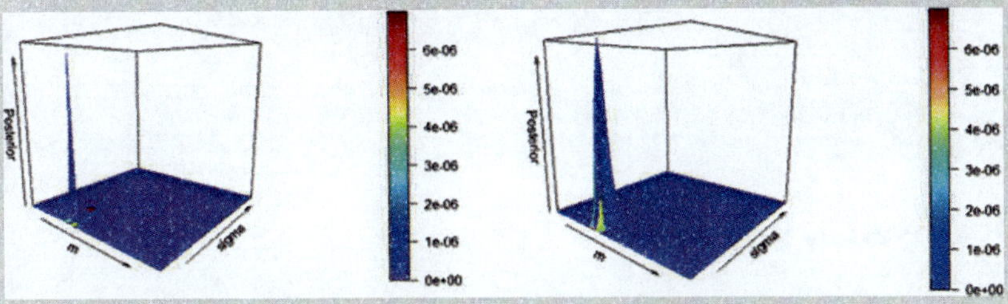

Fig. 5.24 Examples of posterior distributions generated with a polynomial basis, degree 5, $U \sim N(0, 1)$, and a dataset of 25 elements. At left, the results obtained using the likelihood generated; at right, the results obtained using a Gaussian likelihood

Exercises

1. Let X be a variable Poisson distributed $\mathcal{P}(\lambda)$. A sample from X is given:

$$\mathcal{X} = (6, 2, 7, 4, 8, 5).$$

   - Generate an UQ prior for λ and estimate λ using MSE and GLINEXP.

2. Let X be a variable Poisson distributed $\mathcal{P}(\lambda)$. A sample from X is given:

$$\mathcal{X} = (3, 2, 5, 9, 10, 5).$$

   - Generate an UQ prior for λ and estimate λ using MSE and GLINEXP.

3. Let X be a variable Binomially distributed $B(10, p)$. A sample from X is given:

$$\mathcal{X} = (4, 5, 4, 5, 3, 4).$$

   - Generate an UQ prior for p and estimate p using MSE and GLINEXP.

4. Let X be a variable Binomially distributed $B(10, p)$. A sample from X is given:

$$\mathcal{X} = (8, 5, 6, 7, 6, 6, 7).$$

   - Generate an UQ prior for p and estimate p using MSE and GLINEXP.

5. Let X be a Gaussian variable $N(m, \sigma)$. A sample from X is given:

$$\mathcal{X} = (0.613, -0.777, 2.431, -1.305, 3.908, 0.203, 1.045).$$

   - Generate UQ priors for m, σ and estimate them using MSE and GLINEXP.

6. Let X be a Gaussian variable $N(m, \sigma)$. A sample from X is given:

$$\mathcal{X} = (-1.487, -2.472, 0.935, -3.639, -4.646, -0.915, -3.328).$$

   - Generate UQ priors for m, σ and estimate them using MSE and GLINEXP.

5.4 Bayesian Regression

Let us consider observations $(\boldsymbol{x}_i, y_i)$, $i = 1, \ldots, ns$ from a couple of variables. In regression, we look for the coefficients $\boldsymbol{\theta} = (\theta_0, \ldots, \theta_k)^t$ such that

$$y_i = \varepsilon_i + \sum_{j=0}^{k} \theta_j \varphi_j(\boldsymbol{x}_i), \quad i = 1, \ldots, ns.$$

Here, $\{\varphi_j : 0 \leq j \leq k\}$ is a convenient basis, chosen by the user; ε is a random variable. We can consider a model ψ for ε–for instance, $\varepsilon \sim N(0, \sigma)$. Then, we can generate a likelihood using $\boldsymbol{\mathcal{E}} = (\varepsilon_1, \ldots, \varepsilon_n)^t$ as sample and $\boldsymbol{\theta} = (\theta_0, \ldots, \theta_n)^t$ as parameters:

$$f(\mathcal{E}|\boldsymbol{\theta}) = \prod_{i=1}^{ns} \psi(\varepsilon_i) - \prod_{i=1}^{ns} \psi\left(y_i - \sum_{j=0}^{k} \theta_j \varphi_j(\boldsymbol{x}_i)\right).$$

Then a prior distribution for $\boldsymbol{\theta}$ can be updated by the Bayesian Update procedure (5.1) to estimate the parameters by minimizing a loss function. Let us introduce

$$\boldsymbol{\phi}(\boldsymbol{x}) = (\varphi_0(x). \ldots, \varphi_k(\boldsymbol{x})).$$

Then

$$f(\boldsymbol{\mathcal{E}}|\boldsymbol{\theta}) = \prod_{i=1}^{ns} \psi(y_i - \boldsymbol{\phi}(\boldsymbol{x}_i)\boldsymbol{\theta}). \tag{5.25}$$

For linear regression, $k = 1$, $\varphi_0 = 1$, $\varphi_1 = x$. In multilinear regression, $\boldsymbol{x} \in \mathbb{R}^n$ is a vector and $\boldsymbol{x}_i = \left(x_i^{(1)}, .., x_i^{(n)}\right)$, $k = n$, $\varphi_0 = 1$; $\varphi_j = x^{(j)}$, $1 \leq j \leq k$. Nonlinear regression can be realized by introducing basis involving nonlinear functions.

To generalize (5.25), let us set $\boldsymbol{\mathcal{X}} = (\boldsymbol{x}_1, \ldots, \boldsymbol{x}_k)^t$ and

$$\boldsymbol{\Phi}(\boldsymbol{\mathcal{X}}) = \left(\Phi_{ij}(\boldsymbol{\mathcal{X}}), 0 \leq j \leq k, 0 \leq i \leq ns\right), \quad \Phi_{ij}(\boldsymbol{\mathcal{X}}) = \varphi_j(x_i).$$

Then,

$$\boldsymbol{\mathcal{E}} = \boldsymbol{\mathcal{Y}} - \boldsymbol{\Phi}(\boldsymbol{\mathcal{X}})\boldsymbol{\theta}, \quad \boldsymbol{\mathcal{Y}} = (y_1, \ldots, y_{ns})^t.$$

Assuming the PDF of $\boldsymbol{\mathcal{E}}$ is f_ε, the likelihood is

$$f(\boldsymbol{\mathcal{Y}}|\boldsymbol{\theta}) = f_\varepsilon(\boldsymbol{\mathcal{Y}} - \boldsymbol{\Phi}(\boldsymbol{\mathcal{X}})\boldsymbol{\theta}).$$

For instance, assume that $y \sim N\left(\sum_{j=0}^{k} \theta_j \varphi_j(\boldsymbol{x}), \sigma^2\right)$, so that $\boldsymbol{\mathcal{Y}} \sim N(\boldsymbol{\Phi}(\boldsymbol{\mathcal{X}})\boldsymbol{\theta}, \sigma^2 \boldsymbol{Id})$. Then, the likelihood is defined by the density associated with the multivariate Gaussian distribution $N(\boldsymbol{m}(\boldsymbol{\theta}), \boldsymbol{\Sigma})$, with $\boldsymbol{m}(\boldsymbol{\theta}) = \boldsymbol{\Phi}(\boldsymbol{\mathcal{X}})\boldsymbol{\theta}, \boldsymbol{\Sigma} = \sigma^2 \boldsymbol{Id}$:

$$f_\varepsilon(\mathcal{E}) = \frac{1}{(2\pi)^{ns/2}\,|\det(\boldsymbol{\Sigma})|^{1/2}} \exp\left(-\frac{1}{2}\mathcal{E}^t\boldsymbol{\Sigma}^{-1}\mathcal{E}\right).$$

Notice that the likelihood is a function of $\mathcal{E}$. Indeed,

$$f(\mathcal{Y}|\boldsymbol{\theta}) = f_\varepsilon(\mathcal{Y} - \boldsymbol{\Phi}(x)\boldsymbol{\theta}) = f_\varepsilon(\mathcal{E}).$$

To achieve the Bayesian Updating, we must also introduce a prior distribution for the parameters. For instance, we can assume a given distribution, such as Gaussian or uniform. Nevertheless, we can observe that

$$\mathcal{E}^t\mathcal{E} = \boldsymbol{\theta}^t\boldsymbol{\Phi}^t(\mathcal{X})\boldsymbol{\Phi}(\mathcal{X})\boldsymbol{\theta} - 2\mathcal{Y}^t\boldsymbol{\Phi}(x)\boldsymbol{\theta} + \mathcal{Y}^t\mathcal{Y}.$$

This expression suggests the use of a prior for $\boldsymbol{\theta}$ involving the covariance matrix $\sigma^2(\boldsymbol{\Phi}^t(\mathcal{X})\boldsymbol{\Phi}(\mathcal{X}))^{-1}$. Indeed, we can find in the literature about Bayesian Regression the analogous of Jeffreys prior: the ***g-prior***, introduced by (Zellner 1986b), (Agliari and Parisetti 1986). It corresponds to a Gaussian prior with covariance matrix $\boldsymbol{C} = g\sigma^2(\boldsymbol{\Phi}^t(\mathcal{X})\boldsymbol{\Phi}(\mathcal{X}))^{-1}$:

$$f_{pr}(\boldsymbol{\theta}) = \frac{1}{(2\pi)^{ns/2}\,|\det(\boldsymbol{C})|^{1/2}} \exp\left(-\frac{1}{2}\mathcal{E}^t\boldsymbol{C}^{-1}\mathcal{E}\right).$$

It is g is a scalar parameter to be determined. In practice, σ may also be considered as a parameter to be estimated, so that we must have priors for σ and $\boldsymbol{\theta}$ – for instance, we can use Jeffrey's priors for σ.

Once the prior $f_{pr}(\boldsymbol{\theta})$ and the likelihood $f(\mathcal{Y}|\boldsymbol{\theta})$ are chosen, we can generate a posterior distribution $f_{po}(\boldsymbol{\theta}|\mathcal{E})$, which can be used to determine the estimator $\widehat{\boldsymbol{\theta}}$, by minimizing the mean of a loss function, as in the previous examples.

Example 5.12 Let us consider the data $y_i = \sin(\pi x_i) + \varepsilon_i$, $x_i = -1 + 0.1(i - 1)$, $1 \le i \le 21$. Assume that $\varepsilon \sim N(0, 0.5)$. Let us generate an approximation corresponding to a polynomial of degree 3 using the Bayesian Regression. The values of $\mathcal{Y}$ are the following:

$$\begin{aligned}\mathcal{Y} = (&0.06921225, -0.68162941, -0.77869179, -0.09715108,\\ &-1.03287726, -1.70973858, -1.56817775, -0.38221393,\\ &-0.55459491, 0.57335248 - 0.08165291, 0.15512639,\\ &0.93126625, 0.38045524, 0.30685379, 1.19225918,\\ &0.86594269, 0.10178639, 0.69654522, 0.54656532,\\ &0.19013219).\end{aligned}$$

(continued)

Example 5.12 (continued)

Let us consider a prior for the distribution of the coefficients: let us consider with a mean equal to the coefficients $\boldsymbol{\theta}_{col}$ furnished by collocation and a coefficient of variation $CV = 0.25$. For the likelihood, let us consider that the errors $\boldsymbol{\mathcal{E}}$ are independent and Gaussian: $\boldsymbol{\mathcal{E}} \sim N(\mathbf{0}, \sigma \boldsymbol{Id})$, $\sigma = 0.5$. Package `mvtnorm` deals with Multivariate Gaussian Distributions and its instruction `dmvnorm` furnishes the corresponding Gaussian density. We can consider a quadratic loss function and estimate $\boldsymbol{\theta} = (\theta_0, \ldots, \theta_3)$ using `fmincon` of `pracma`. The admissible region is taken as a product of intervals $(\theta_{col,\ i} - k_\theta \sigma_i, \theta_{col,\ i} + k_\theta \sigma_i)$, $\sigma_i = CV * |\theta_{col,\ i}|$. The starting point for the minimization of the loss function is the vector of collocation coefficients $\boldsymbol{\theta}_{col}$. The implementation in R is shown below: we start by introducing the data:

```
X = seq(-1,1,by=0.1)
ns = length(X)
Yex = sin(pi*X) sig = 0.5
Y = c(0.06921225, -0.68162941, -0.77869179, -0.09715108, -1.0328
7726,
      -1.70973858, -1.56817775,-0.38221393, -0.55459491, 0.57335
248,
      -0.08165291, 0.15512639, 0.93126625, 0.38045524, 0.3068537
9, 1.19225918,
      0.86594269, 0.10178639, 0.69654522, 0.54656532, 0.19013219
)
```

Then, we determine the collocation coefficients:

```
a = min(X)
b = max(X)
phiX = polynom$new(a,b,1e-3)
expX = expansion1D$new(phiX)
nd = 3
eps = 0
coef0 = matrix(expX$colloc(nd,eps,X,Y), ncol = 1)
```

(continued)

Example 5.12 (continued)

In the next step, we define the prior, the likelihood and we evaluate the posterior distribution of the coefficients:

```
# Prior for the coefficients

mteta = coef0
coefvar = 0.25
sigteta = coefvar*abs(coef0)
Cteta = diag(as.numeric(sigteta^2))
mmteta = matrix(mteta, nrow = 1)
fpr = function(teta){
  tteta = matrix(teta, nrow = 1)
  resu = dmvnorm(tteta,mmteta, Cteta)
  return(resu)}

# Likelihood

lklh = function(err){
  rr = dmvnorm(err,rep(0,ns),sig^2*diag(ns))
  return(rr)}

# Posterior

fpo = function(teta){
  err = Y - expX$proj(teta,X)
  resu = lklh(err)*fpr(teta)
  return(resu)}
```

Finally, we introduce the loss function, the objective to be minimized – which is the mean of the loss function – and we call `fmincon` to determine the minimum

(continued)

Example 5.12 (continued)

```
# Loss function

los1 = function(tetach,teta){sum((teta - tetach)^2)}

# Search Intervals

kt = 5
tetamin = matrix(mteta - kt*sigteta, ncol = 1)
tetamax = matrix(mteta + kt*sigteta, ncol = 1)

# Objective function

fobj5 = function(tetaest){
  ff = function(t1,t2,t3,t4){
    teta = c(t1,t2,t3,t4)
    rrr = fpo(teta)*los1(tetaest,teta)
    if (is.nan(rrr)){rrr = 1e10}
    if (is.na(rrr)){rrr = 1e10}
    return(rrr)}
  resu = int1$trapz4(ff,tetamin,tetamax,100,1e-5)
  if (is.nan(resu)){resu = 1e10}
  if (is.na(resu)){resu = 1e10}
  return(resu)
}

hi1 = function(tetach){
  tetac = matrix(tetach, ncol = 1)
  aux1 = tetac - tetamax
  aux2 = tetamin - tetac
  rres = matrix(c(aux1,aux2),ncol = 1)
  return(rres)
}

sol5c = fmincon(x0=coef0, fn=fobj5,hin = hi1)
coef5c = sol5c$par
```

The results appear in Fig. 5.25. The results for $k_\theta = 5$, $k_\theta = 10$ are almost identical. The final values of the loss function were 1E-12, 2E-27, respectively.

(continued)

Example 5.12 (continued)

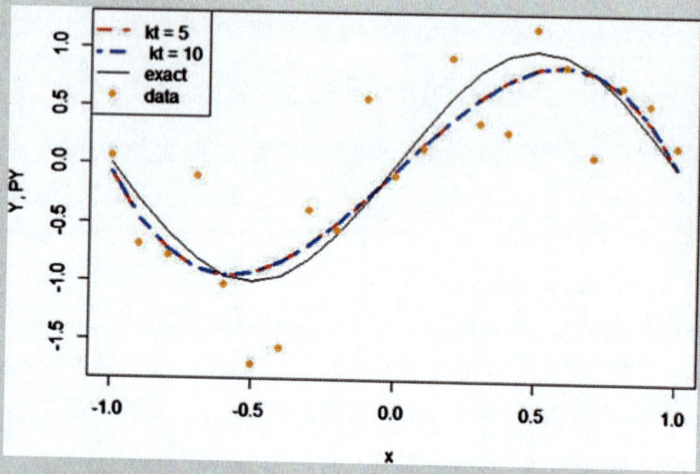

Fig. 5.25 Example of Bayesian regression by a polynomial of degree 3, using a Gaussian prior. Minimization over a product of intervals $((1 - k_t CV)\theta_{col,\,i}, (1 + k_t CV)\theta_{col,\,i})$ around the collocation coefficients by fmincon furnishes close results for $k_t = 5, 10$

Now, let us use the g-prior – we must modify the code as follows:

```
# Prior for the coefficients

PHI = expX$matA(nd,X)
PTP = t(PHI) %*% PHI
g = 1
Cteta = solve(PTP)
fgpr = function(teta){
  tteta = matrix(teta, nrow = 1)
  resu = dmvnorm(tteta,rep(0,nd+1), g*sig^2*Cteta)
  return(resu)
}
```

(continued)

Example 5.12 (continued)

The results appear in Fig. 5.26: they are close to the preceding ones. The final values of the loss function were 1E-12, 2E-27, respectively.

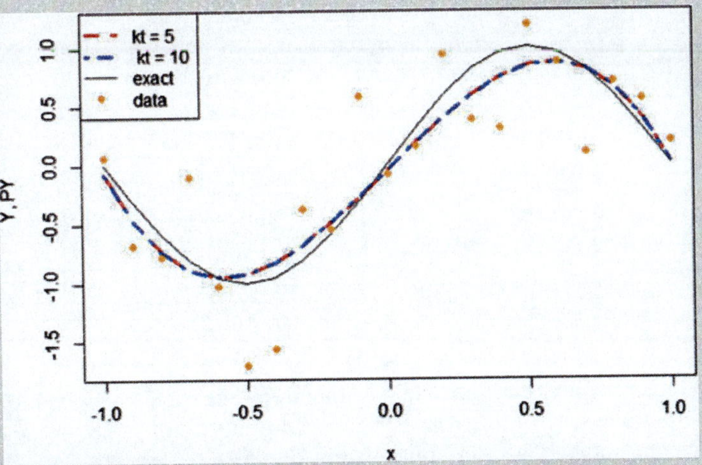

Fig. 5.26 Example of Bayesian regression by a polynomial of degree 3, using the g-prior. Minimization over a product of intervals $((1 - k_t CV)\theta_{col,\ i}, (1 + k_t CV)\theta_{col,\ i})$ around the collocation coefficients by fmincon furnishes close results for $k_t = 5, 10$

Using different values of g led to analogous results (see Fig. 5.27).

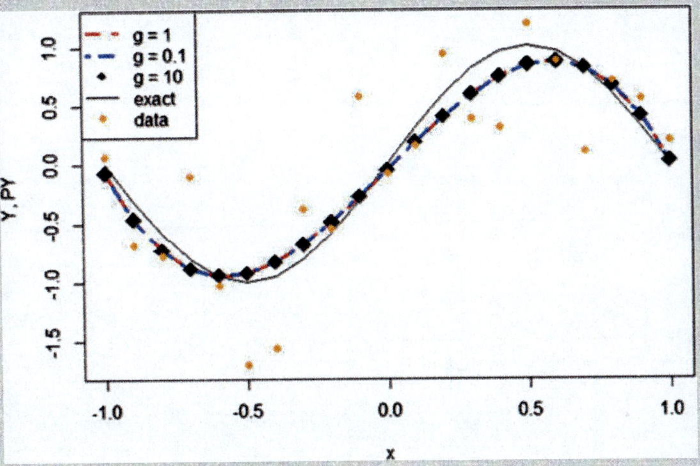

Fig. 5.27 Example of Bayesian regression by a polynomial of degree 3, using the g-prior, with different values of g, with $k_t = 10$

Example 5.13 In practice, we often must deal with multiplicative errors. Let us apply the Bayesian regression to this situation – which does not correspond to the framework introduced. As an example, let us consider the data $y_i = (1 + \eta_i)\sin(\pi x_i) + \varepsilon_i$, $x_i = -1 + 0.1(i - 1)$, $1 \le i \le 21$. Assume that $\eta \sim N(0, \sigma_\eta)$ – σ_η appears as a level of percentual error in the observations, while $\varepsilon \sim N(0, \sigma_\varepsilon)$ can be interpreted as a systematic deviation. Taking $\sigma_\eta = 10\%$, $\sigma_\varepsilon = 0.1$, we can generate the following data:

$$\mathcal{Y}=(-0.14344129,-0.32361108,-0.31770055,-0.91672618,\\ -1.04634799,-1.17487797,-0.97156951,-0.72279109,\\ -0.73226451,-0.40574030,0.02692649,0.36166421,\\ 0.72053927,0.75649930,1.01058074,0.67863824,\\ 0.88237085,0.91680675,0.68404702,0.35779675,\\ 0.08994831)$$

We can apply the procedures shown in the preceding example, by reinterpreting $\eta_i \sin(\pi x_i) + \varepsilon_i$ as being a variate from $\mathcal{E}$. For instance, the results generated by g-prior are shown in Fig. 5.28.

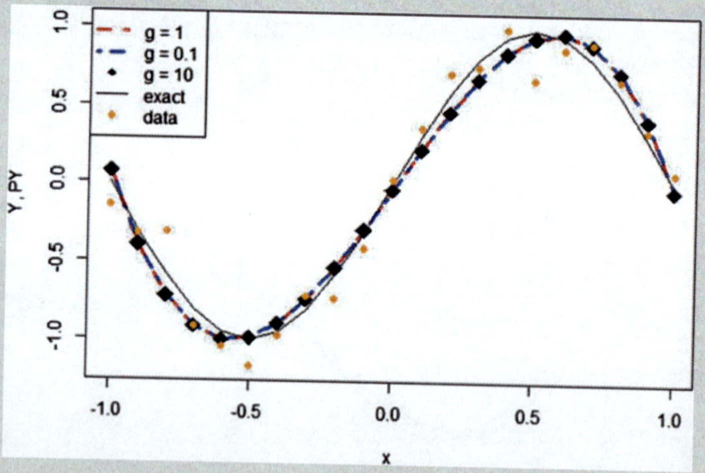

Fig. 5.28 Example of Bayesian regression by a polynomial of degree 3, using the g-prior, with different values of g, with $k_\theta = 10$

(continued)

Example 5.13 (continued)

Exercises

1. Load the data in `data1.Rmd` and generate a regression for Y as a polynomial of degree 3 in X.
2. Load the data in `data2.Rmd` and generate a regression for Y as a polynomial of degree 3 in X.
3. Load the data in `data3.Rmd` and generate a regression for Y as a polynomial of degree 3 in X.
4. Load the data in `data4.Rmd` and generate a regression for Y as a polynomial of degree 3 in X.

5.5 Model Selection and Hypothesis Testing

One of the classical applications of Statistical Inference is the decision between hypothesis or models for a situation under analysis.

Model selection and hypothesis testing can be summarized as the determination of a measurement of the quality of the models when they are applied to the available data: we shall adopt the model and the decision corresponding to the best values of an indicator, generated by the results of the measurements.

For instance, we can find in the literature the chi-squared test (Pearson 1900; Pearson 1904) and the procedures based on significance levels (Neyman and Pearson 1933; Fisher 1955), and approaches based on the likelihood (Wilks 1938; Wald 1939, 1943; Rao 1948).

The Bayesian approach is based on the computation of the posterior probabilities of each model or hypothesis and their comparison. As an example, let us assume that two models M_1 and M_2 are in competition to explain the data $\mathcal{Y}$: we can compare the values of $P(M_1|\mathcal{Y})$ and $P(M_2|\mathcal{Y})$ to take a decision. The basic Bayesian approach uses as decision variable the *Bayes Factor* (see, for instance, (Jeffreys 1935, 1939)), given by

$$K = \frac{P(M_1|\mathcal{Y})}{P(M_2|\mathcal{Y})}. \tag{5.26}$$

Since

$$P(M_i|\mathcal{Y}) = \frac{P(\mathcal{Y}|M_i)P(M_i)}{P(\mathcal{Y})},$$

we have

$$K = \frac{P(\mathcal{Y}|M_1)\,P(M_1)}{P(\mathcal{Y}|M_2)\,P(M_2)}. \tag{5.27}$$

$P(M_i)$ is our prior evaluation of the probability of the model M_i. In most of the practical applications, $P(M_1) = P(M_2)$, so that

$$K = \frac{P(\mathcal{Y}|M_1)}{P(\mathcal{Y}|M_2)}.$$

Notice that the evaluation of the likelihood $P(\mathcal{Y}|M_i)$ requests a model $f(\mathcal{Y}|M_i)$ to be furnished by the user. In practice, such a model is often defined by a vector $\boldsymbol{\theta}_i$ of parameters: $f(\mathcal{Y}|M_i) = f(\mathcal{Y}|\boldsymbol{\theta}_i)$. If the model involves randomness on $\boldsymbol{\theta}_i$, a model $\mu_i(\boldsymbol{\theta}_i)$ for such a randomness must be given – for instance, if we assume that $\mathcal{Y}$ is Gaussian $N(m, \sigma)$ and that m is random, we must also furnish a prior distribution μ for m. Then, $P(\mathcal{Y}|M_i) = \int f(\mathcal{Y}|\boldsymbol{\theta}_i)\mu_i(\boldsymbol{\theta}_i)d\boldsymbol{\theta}_i$.

$K > 1$ means that the data support better M_1than M_2; while $K < 1$ means the inverse. In the appendix of his book (Jeffreys 1939), Jeffreys proposed a scale for the interpretation of K in the classical test of a hypothesis H_0, which must be accepted or rejected against an alternative H_1: in such a situation $M_1 = H_0$, $M_2 = H_1$. In Table 5.7, you can find the scale proposed by Jeffreys: it is based on the value of $J = -2\log_{10}K$ and synthetizes the evidence ***against*** H_0. The original Table proposed by (Jeffreys 1939) appears in Fig. 5.29.

Kass and Raftery (1995) proposed a modification of the scale introduced be Jeffreys, based on $J_{KR} = -2 \log K$. Table 5.8 gives the scale. Figure 5.30 shows the original table.

Table 5.7 Jeffreys scale for the decision on accepting or rejecting a hypothesis. The scale is based on the value of $\boldsymbol{J = -2\, log_{10}K}$ and gives the evidence ***against*** $\boldsymbol{H_0}$

J	$(-\infty, 0)$	$(0, 1)$	$(1, 2)$	$(2, 3)$	$(3, 4)$	$(4, +\infty)$
Evidence	Negligible	Bare mention	Substantial	Strong	Very Strong	Decisive

Grade 0. $K > 1$. **Null hypothesis supported.**
Grade 1. $1 > K > 10^{-1/2}$. **Evidence against q, but not worth more than a bare mention.**
Grade 2. $10^{-1/2} > K > 10^{-1}$. **Evidence against q substantial.**
Grade 3. $10^{-1} > K > 10^{-3/2}$. **Evidence against q strong.**
Grade 4. $10^{-3/2} > K > 10^{-2}$. **Evidence against q very strong.**
Grade 5. $10^{-2} > K$. **Evidence against q decisive.**

Fig. 5.29 The original table from (Jeffreys 1939). "Null hypothesis" or "q" refers to $M_1 = H_0$

Table 5.8 Kass and Raftery scale for the decision on accepting or rejecting a hypothesis. The scale is based on the value of $\boldsymbol{J_{KR} = -2\, log\, K}$ and gives the evidence ***against*** $\boldsymbol{H_0}$

J_{KR}	$(-\infty, 0)$	$(0, 2)$	$(2, 6)$	$(6, 10)$	$(10, +\infty)$
Evidence	Negligible	Bare mention	Positive	Strong	Very Strong

$2\log_e(B_{10})$	(B_{10})	Evidence against H_0
0 to 2	1 to 3	Not worth more than a bare mention
2 to 6	3 to 20	Positive
6 to 10	20 to 150	Strong
>10	>150	Very strong

Fig. 5.30 The original table from (Kass and Raftery 1995). B_{10} refers to $1/K$

Alternative indicators for the quality of a model can be found in the literature. For instance, we can consider the AIC (Akaike Information Criterion), introduced in (Akaike 1998) and the BIC (Bayesian Information Criterion), introduced in (Schwarz 1978). These indicators are defined as:

$$AIC = 2k - 2\log L,$$

$$BIC = k\log n - 2\log L.$$

Here, k is the number of parameters to be estimated, n is the number of observations, and L is the maximized likelihood. The model to be selected corresponds to the minimal value of the indicator used (AIC or BIC). The reader can find in the literature number of variations of these indicators, namely corrections for small samples. For instance, (Hurwich and Tsai 1989) proposed the following correction to AIC:

$$AICc = AIC + \frac{2k(k+1)}{n-p+1}.$$

Example 5.14 Let us consider again the data in Example 5.1. Assume that we desire to test the hypothesis $H_0 : p = 0.5$ against $H_1 : p$ *uniformly distributed on* $(0, 1)$. In this case, the distribution of p corresponds to the densities:

$$\mu_0(p) = \delta\left(p - \frac{1}{2}\right), \quad \mu_1(p) = 1.$$

Since we observed three heads, the likelihood corresponding to a given p is

$$f(X|p) = Ap^3(1-p)^2, \quad A = \binom{5}{2}.$$

(continued)

Example 5.14 (continued)

Then

$$f(x|H_0) = \frac{A}{2^5} = \frac{A}{32}, f(x|H_1) = A\int_0^1 p^3(1-p)^2\mu_1(p)dp = \frac{A}{60}.$$

Thus,

$$K = \frac{f(x|H_0)}{f(x|H_1)} = \frac{60}{32} > 1$$

and we accept H_0. We have $J = -0.5460025$, $J_{KR} = -1.257217$. We also have

$$AIC_0 = -2\log L \approx 6.9, AIC_1 = 1 - 2\log L \approx 7.7,$$

so that the model corresponding to H_0 is preferred. Analogously,

$$BIC_0 = -2\log L \approx 6.9, BIC_1 = \log 5 - 2\log L \approx 8.3,$$

and BIC furnishes the same result as the other approaches.

Now, let us use the classical procedure based on the significance level: we assume that H_0 is true and we determine the probability of the event observed: it is $10/32 \approx 0.3125$, what means that H_0 is accepted at very high significance levels – under the standard threshold of 5%, it is clearly accepted. Notice that H_1 corresponds to a probability of the event observed of $10/60 \approx 0.1667$. Then, H_1 is accepted at high significance levels, but about a half of the level of H_0.

Finally, we can use the χ^2 to measure the quality of H_0: we have $\chi^2_{obs} = 0.2$, which corresponds to a probability 0.3543, so that H_0 is accepted at very high significance levels – again, under the standard threshold of 5%, it is evidently accepted.

We can use R for these evaluations:

(continued)

Example 5.14 (continued)

```
nh = 3
ns = 5

p0 = 0.5
k0 = (p0^nh)*(1-p0)^(ns-nh)
nest0 = 0

k1 = beta(nh+1,ns-nh+1)
nest1 = 1
pest1 = nh/ns

K = k0/k1
print(paste('Bayes Factor = ', K))

## [1] "Bayes Factor =  1.875"

J = -2*log10(K)
print(paste('J by Jeffreys = ', J))

## [1] "J by Jeffreys =  -0.546002544127475"

Jkr = -2*log(K)
print(paste('J by Kass & Raftery = ', Jkr))

## [1] "J by Kass & Raftery =  -1.25721731884475"

L0 = (p0^nh)*(1-p0)^(ns-nh)
AIC0 = nest0 - 2*log(L0)
L1 = (pest1^nh)*(1-pest1)^(ns-nh)
AIC1 = nest1 - 2*log(L1)
print(paste('AIC: for H0 = ', AIC0,', for H1 = ',AIC1))

## [1] "AIC: for H0 =  6.93147180559945 , for H1 =  7.7301166700
9256"

BIC0 = nest0*log(ns) - 2*log(L0)
BIC1 = nest1*log(ns) - 2*log(L1)
print(paste('BIC: for H0 = ', BIC0,', for H1 = ',BIC1))

## [1] "BIC: for H0 =  6.93147180559945 , for H1 =  8.3395545825
2666"
```

(continued)

Example 5.14 (continued)

```
binom.test(nh, ns, p = 0.5, alternative = "two.sided")

##
##  Exact binomial test
##
## data:  nh and ns
## number of successes = 3, number of trials = 5, p-value = 1
## alternative hypothesis: true probability of success is not eq
ual to 0.5
## 95 percent confidence interval:
##  0.1466328 0.9472550
## sample estimates:
## probability of success
##                    0.6

chi2obs = (nh - ns/2)^2/(ns/2) + (ns-nh - ns/2)^2/(ns/2)
pchisq(chi2obs, 1)

## [1] 0.3452792

chisq.test(c(nh,ns-nh), p = c(0.5,0.5))

##
##  Chi-squared test for given probabilities
##
## data:  c(nh, ns - nh)
## X-squared = 0.2, df = 1, p-value = 0.6547
```

Example 5.15 Let us consider again the data in Example 5.2. Assume that we desire to test the hypothesis $H_0 : p = 0.5$ against H_1 : *p uniformly distributed on* $(0, 1)$. Our prior for H_0 is that $P(H_0) = P(H_1) = 1/2$. We observed 12 heads on 20 tosses. Then (setting $A = 1$):

$$f(X|H_0) = \frac{1}{2^{20}} \approx 9.5E-7, f(X|H_1) = \int_0^1 p^{12}(1-p)^8 dp \approx 3.8E-7.$$

Then

$$K \approx 2.5 > 1$$

and we accept H_0. The result is analogous for the other approaches used in the Example 5.14. Using R:

(continued)

Example 5.15 (continued)

```
nh = 12
ns = 20

p0 = 0.5
k0 = (p0^nh)*(1-p0)^(ns-nh)
nest0 = 0

k1 = beta(nh+1,ns-nh+1)
nest1 = 1
pest1 = nh/ns

K = k0/k1
print(paste('Bayes Factor = ', K))

## [1] "Bayes Factor =  2.5228214263916"

J = -2*log10(K)
print(paste('J by Jeffreys = ', J))

## [1] "J by Jeffreys =  -0.803773021623795"

Jkr = -2*log(K)
print(paste('J by Kass & Raftery = ', Jkr))

## [1] "J by Kass & Raftery =  -1.85075577774173"

L0 = (p0^nh)*(1-p0)^(ns-nh)
AIC0 = nest0 - 2*log(L0)
L1 = (pest1^nh)*(1-pest1)^(ns-nh)
AIC1 = nest1 - 2*log(L1)
print(paste('AIC: for H0 = ', AIC0,', for H1 = ',AIC1))

## [1] "AIC: for H0 =  27.7258872223978 , for H1 =  27.920466680
3703"

BIC0 = nest0*log(ns) - 2*log(L0)
BIC1 = nest1*log(ns) - 2*log(L1)
print(paste('BIC: for H0 = ', BIC0,', for H1 = ',BIC1))

## [1] "BIC: for H0 =  27.7258872223978 , for H1 =  29.916198953
9242"

binom.test(nh, ns, p = 0.5, alternative = "two.sided")
```

(continued)

Example 5.15 (continued)

```
##
##  Exact binomial test
##
## data:  nh and ns
## number of successes = 12, number of trials = 20, p-value = 0.
5034
## alternative hypothesis: true probability of success is not eq
ual to 0.5
## 95 percent confidence interval:
##  0.3605426 0.8088099
## sample estimates:
## probability of success
##                    0.6

chi2obs = (nh - ns/2)^2/(ns/2) + (ns-nh - ns/2)^2/(ns/2)
pchisq(chi2obs, 1)

## [1] 0.6289066

chisq.test(c(nh,ns-nh), p = c(0.5,0.5))

##
##  Chi-squared test for given probabilities
##
## data:  c(nh, ns - nh)
## X-squared = 0.8, df = 1, p-value = 0.3711
```

Example 5.16 Let us consider again the data in Example 5.3. Assume that $\sigma = 0.5$ and we desire to test the hypothesis $H_0 : m = m_0$ against $H_0 : m$ *is uniformly distributed on* $(m_0 - 0.5, m_0 + 0.5)$. Let us test this hypothesis with $m_0 = 1.2$.

Since σ is known, we can use the classical significance test based on $Z = (\bar{X}_n - m_0)/(\sigma/\sqrt{n})$. For the data under consideration, $Z \approx -0.89$, so that $|Z| < 1.96$ and the hypothesis H_0 is accepted at the significance level $\alpha = 5\%$ – it will be accepted up to levels of $\alpha = 35\%$. Even if σ is known, the hypothesis will be accepted up to significance levels of about 20% – for an unknown σ, we use the classical T-test involving the parameter $T = (\bar{X}_n - 1)/(s_n/\sqrt{n})$. For the sample under consideration, we have $T \approx -1.54$, while the critical value for the rejection is 2.78.

Now, let us use the Bayesian approach: The likelihood is Gaussian:

(continued)

Example 5.16 (continued)

$$f(\mathcal{Y}|m) = \frac{1}{\sigma^{ns}(2\pi)^{\frac{ns}{2}}} \exp\left(-\frac{1}{2}\sum_{i=1}^{ns}\left(\frac{X_i - m}{\sigma}\right)^2\right).$$

We have $K \approx 1.3$, so that

```
source("ints.R")
int1 = ints$new()

X = c(0.97,1.41,1.14,0.82,0.67)
ns = length(X)

mest1 = mean(X)
sd1 = sd(X)
sigex = 0.5
m0 = 1.2
dm0 = 0.5

alfa = 0.05

zval = (mest1 - m0)/(sigex/sqrt(ns))
zcr = qnorm(1-0.5*alfa)
print(paste("| Z-val | = ",abs(zval),', Z critical = ', zcr, exp
ression(alpha),' = ',alfa))

## [1] "| Z-val | =  0.885482919089916 , Z critical =  1.9599639
8454005 alpha  =  0.05"

tval = (mest1 - m0)/(sd1/sqrt(ns))
tcr = qt(1-0.5*alfa,length(X) - 1)
print(paste("| T-val | = ",abs(tval),', T critical = ', tcr, exp
ression(alpha),' = ',alfa))

## [1] "| T-val | =  1.54170824551732 , T critical =  2.77644510
519779 alpha  =  0.05"

lklh = function(m){
  s = 1
  for (i in (1:ns)){
    aux = dnorm(X[[i]],m,sigex)
    s = s*aux}
  return(s)}

k0 = lklh(m0)
k1 = int1$trapz1(lklh,m0-dm0, m0+dm0,100,1e-6)

K = k0/k1
print(paste('Bayes Factor = ', K))

## [1] "Bayes Factor =  1.32372001982999"
```

(continued)

Example 5.16 (continued)

```
J = -2*log10(K)
print(paste('J by Jeffreys = ', J))

## [1] "J by Jeffreys =  -0.243592274284195"

Jkr = -2*log(K)
print(paste('J by Kass & Raftery = ', Jkr))

## [1] "J by Kass & Raftery =  -0.560891939535304"

L0 = k0
nest0 = 0
AIC0 = nest0 - 2*log(L0)
L1 = k1
nest1 = 0
AIC1 = nest1 - 2*log(L1)
print(paste('AIC: for H0 = ', AIC0,', for H1 = ',AIC1))

## [1] "AIC: for H0 =  4.36151352644727 , for H1 =  4.9224054659
8258"

BIC0 = nest0*log(ns) - 2*log(L0)
BIC1 = nest1*log(ns) - 2*log(L1)
print(paste('BIC: for H0 = ', BIC0,', for H1 = ',BIC1))

## [1] "BIC: for H0 =  4.36151352644727 , for H1 =  4.9224054659
8258"
```

If we consider the sample with 25 variates, we get

```
## [1] "| Z-val | =  1.044 , Z critical =  1.95996398454005 alph
a  =  0.05"

## [1] "| T-val | =  1.24482638237959 , T critical =  2.06389856
162803 alpha  =  0.05"

## [1] "Bayes Factor =  1.99931103170551"

## [1] "J by Jeffreys =  -0.601760724650186"

## [1] "J by Kass & Raftery =  -1.38560527412881"

## [1] "AIC: for H0 =  29.2603676322364 , for H1 =  30.645972906
3652"

## [1] "BIC: for H0 =  29.2603676322364 , for H1 =  30.645972906
3652"
```

Again, H_0 is accepted.

These approaches can also be used for variable selection, by considering models involving different subsets of variables: let $(x_1, \ldots, x_n)$ be the set of the possible variables for a model. We set $I = \{1, \ldots, n\}$ and we look for the set of indexes $I_\ell \in 2^I$ that better explain the observations $\mathcal{Y}$. To attain this goal, we can consider $\boldsymbol{x}_{I_\ell} = (x_{i_1}, \ldots, x_{i_\ell}), I_\ell = (i_1, \ldots, i_\ell)$ and look for a regression

$$Y \approx PY = \sum_{j=0}^{k} \theta_j \varphi_j(\boldsymbol{x}_{I_\ell}).$$

Once the regression coefficients are determined, we can determine an indicator of the quality of the model: for instance, we can determine the error $\boldsymbol{\mathcal{E}} = \boldsymbol{\mathcal{Y}} - \boldsymbol{\Phi}(\boldsymbol{x})\boldsymbol{\theta}$ and use its norm to evaluate the quality of the approximation. But we can also use other indicators such as the variance explained:

$$R^2 = \frac{\sum_{i=1}^{ns} (PY_i - \overline{Y})^2}{S^2(Y)} = 1 - \frac{\sum_{i=1}^{ns} \mathcal{E}_i^2}{S^2(Y)} = 1 - \frac{\|\mathcal{E}\|^2}{S^2(Y)}, S^2(Y) = \sum_{i=1}^{ns} (Y_i - \overline{Y})^2.$$

A value of $R^2 \approx 1$ means a good model, while a value $R^2 \approx 0$ means a bad model. We can also look for I_ℓ such that $P(I_\ell|\mathcal{Y})$ is maximum. Two models can be compared by using the Bayes factor or the criteria AIC and BIC.

Example 5.17 Let us consider data Y which can be represented as a function of two variables $X = (X_1, X_2)$. We can consider two regression models: $M_i : Y \approx PY(X_i)$, $i = 1, 2$ and select a variable by selecting a model. As an example, let us consider $Y = X_1 + X_2/100$, $-1 \leq X_1, X_2 \leq 1$ – it is expected that variable X_1 has the most important rule. We consider the data corresponding to $(X_{1,\,i}, X_{2,\,j})$, $X_{k,\ell} = -1 + \frac{\ell}{4}, 0 \leq \ell \leq 8$ and we look for a polynomial regression of order 3, with $\varphi_i(x) = (x + 1)/2$. We start by defining the parameters

```
nd = 3
epsi = 0
int1 = ints$new()

Ybar = mean(Y)
S2Y = sum((Y - Ybar)^2)
```

Then we determine a regression involving variable x_1 . We use the g-prior, with . $g = 0.1$.

(continued)

Example 5.17 (continued)

```
sig = 1
coefvar = 0.25
kt = 10
gj = 0.1

# take Y as function of X1
a1 = min(X[ ,1])
b1 = max(X[ ,1])
phi1 = polynom$new(a1,b1,1e-3)
exp1 = expansion1D$new(phi1)
c1 = exp1$colloc(nd,epsi,X[ ,1],Y)
PY1 = exp1$proj(c1,X[ ,1])
ve1 = sum((PY1 - Y)^2)
er1 = sqrt(ve1)
ev1 = 1 - ve1/S2Y

# Prior for the coefficients

PHI1 = exp1$matA(nd,X[ ,1])
PTP1 = t(PHI1) %*% PHI1
Cteta1 = solve(PTP1)
fpr1 = function(teta){
  tteta = matrix(teta, nrow = 1)
  resu = dmvnorm(tteta,rep(0,nd+1), gj*sig^2*Cteta1)
  return(resu)}

mteta1 = c1
sigteta1 = coefvar*abs(c1)

tetamin1 = matrix(mteta1 - kt*sigteta1, ncol = 1)
tetamax1 = matrix(mteta1 + kt*sigteta1, ncol = 1)

# Likelihood

lklh = function(err){
  rr = dmvnorm(matrix(err, nrow = 1),rep(0,ns),sig^2*diag(ns))
  return(rr)}

# Posterior

fpo1 = function(teta){
  err = Y - exp1$proj(teta,X[,1])
  resu = lklh(err)*fpr1(teta)
  return(resu)}
```

(continued)

Example 5.17 (continued)

```
vlk1 = lklh(Y - PY1)
vpr1 = fpr1(c1)
vpo1 = fpo1(c1)

los1 = function(tetach,teta){sum((teta - tetach)^2)}
fobj1 = function(tetaest){
  ff = function(t1,t2,t3,t4){
    teta = c(t1,t2,t3,t4)
    rrr = fpo1(teta)*los1(tetaest,teta)
    if (is.nan(rrr)){rrr = 1e10}
    if (is.na(rrr)){rrr = 1e10}
    return(rrr)}
  resu = int1$trapz4(ff,tetamin1,tetamax1,100,1e-5)
  if (is.nan(resu)){resu = 1e10}
  if (is.na(resu)){resu = 1e10}
  return(resu)}
#
hi1 = function(tetach){
  tetac = matrix(tetach, ncol = 1)
  aux1 = tetac - tetamax1
  aux2 = tetamin1 - tetac
  rres = matrix(c(aux1,aux2),ncol = 1)
  return(rres)}
#
sol1 = fmincon(x0=c1, fn=fobj1,hin = hi1)
coef1 = sol1$par
PY1a = exp1$proj(coef1,X[ ,1])
ve1a = sum((PY1a - Y)^2)
er1a = sqrt(ve1a)
ev1a = 1 - ve1a/S2Y
```

We obtain:

```
## [1] "coef1 =  -1"                    "coef1 =  2.000000000000
07"
## [3] "coef1 =  -1.75716067923089e-13" "coef1 =  1.136443061698
98e-13"
## [1] "error =  0.0580947501931112"
## [1] "explained variance =  0.999900009999"
```

(continued)

Example 5.17 (continued)

The determination of a regression for variable x_2 is analogous. We obtain:

```
## [1] "coef2 =  -0.00999999999999658"  "coef2 =  0.019999999999
9358"
## [3] "coef2 =  1.72231898396821e-13"  "coef2 =  -1.15067341004
205e-13"
## [1] "error =  5.80947501931113"
## [1] "explained variance =  9.99900009998056e-05"
```

Notice that the results correspond to $PY(X_1) \approx X_1$ and $PY(X_2) \approx X_2/100$. In addition, the error corresponding to $PY(X_2)$ is larger – as expected. The R^2 and the explained variance show that variable x_1 is more influent than variable x_2.

Let us evaluate the Bayes factor, AIC and BIC using a Gaussian model $N(0, \boldsymbol{Id})$ for the errors $Y - PY$. The number of parameters estimated are the number of coefficients: 4. We obtain the results below: the model M_1 using x_1 is preferred, so that x_1 appears as more influent than variable x_2.

```
## [1] "Bayes Factor =  21280729.4390269"

## [1] "J by Jeffreys =  -14.6559730203583"

## [1] "J by Kass & Raftery =  -33.746625"

## [1] "AIC : for M1 =  152.871417379157 , for M2 =  186.6180423
79157"

## [1] "BIC : for M1 =  166.449213997847 , for M2 =  200.1958389
97847"
```

To alleviate the computational cost, we can dispense with the step of minimizing the loss function and use the collocation coefficients. The results obtained are very close to the preceding.

The problem of variable selection can be considered as a problem of Discrete Optimization: let $(x_1, \ldots, x_n)$ be the set of the possible variables. We set $I = \{1, \ldots, n\}$ and we look for the set of indexes $I_\ell \in 2^I$ such that $AIC(I_\ell)$ (or $BIC(I_\ell)$ is minimum). The solution of such an optimization problem can be carried by adequate methods. The reader will find in the literature many variations about this basic approach – see, for instance, (O'Hara and Sillanpää 2009; Lu and Lou 2022). Exhaustive search can be used only for a small number of variables (see examples below). Exhaustive search can be performed if the number of variables n is small. To reduce the number of models to be examined, a first selection using $||\mathcal{E}||$ or R^2 can be used.

Example 5.18 Let us consider the situation where the set of possible variables is $(x_1, \ldots, x_{10})$, and the data correspond to the values $x_i \in \{-1, 0, 1\}$, $1 \le i \le 10$; $Y = 2x_1 + 4x_5 + x_{10}/100$. The set of all the possible selections of the variables contains $2^{10} = 1024$ elements. We can look for an affine approximation PY and perform an exhaustive search among all the possible choices of variables. The optimal values found are shown in Table 5.9 and Fig. 5.31. Observe that the optima found correspond to the selection of variables 1, 5, and 10. As expected, variable 10 appears as less influent than x_1, x_5.

Table 5.9 Examples of choices that minimize $\|\mathcal{E}\|^2$ and maximize R^2

ℓ	1	2	3	4	5	6
R^2	0.999496	0.9999975	1	1	1	1
$\|\mathcal{E}\|^2$	396.8223	1.984087	1.07E-12	1.07E-12	1.07E-12	1.07E-12
AIC	2.7*E*5	1.1*E*5	1.1*E*5	1.1*E*5	1.1*E*5	1.1*E*5
BIC	2.7*E*5	1.1*E*5	1.1*E*5	1.1*E*5	1.1*E*5	1.1*E*5
I_ℓ	5	1,5	1,5,10	1,2,5,10	1,2,3,5,10	1,2,3,4,5,10

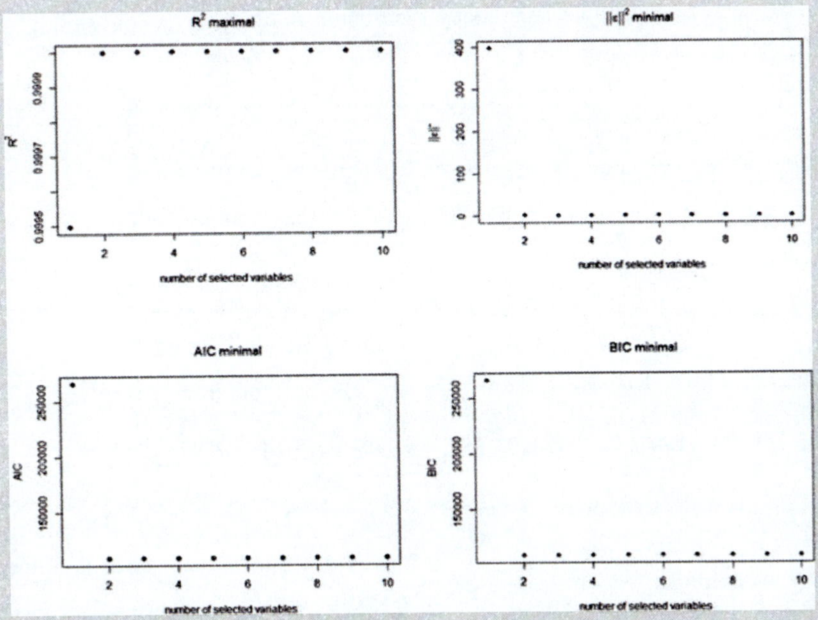

Fig. 5.31 Values of R^2, $\|\mathcal{E}\|^2$, AIC, and BIC for each number of variables

Exercises

1. Let X be a variable Poisson distributed $\mathcal{P}(\lambda)$. A sample from X is given:

$$\mathcal{X} = (6, 2, 7, 4, 8, 5).$$

   - Test the hypothesis $H_0 : \lambda = 5$ against $H_1 : \lambda \sim U(4, 6)$.

2. Let X be a variable Poisson distributed $\mathcal{P}(\lambda)$. A sample from X is given:

$$\mathcal{X} = (3, 2, 5, 9, 10, 5).$$

   - Test the hypothesis $H_0 : \lambda = 6$ against $H_1 : \lambda \sim U(5, 7)$.

3. Let X be a variable Binomially distributed $B(10, p)$. A sample from X is given:

$$\mathcal{X} = (4, 5, 4, 5, 3, 4).$$

   - Test the hypothesis $H_0 : p = 5$ against $H_1 : \lambda \sim U(4, 6)$.

4. Let X be a variable Binomially distributed $B(10, p)$. A sample from X is given:

$$\mathcal{X} = (8, 5, 6, 7, 6, 6, 7).$$

   - Test the hypothesis $H_0 : p = 6$ against $H_1 : \lambda \sim U(5, 7)$.

5. Let X be a Gaussian variable $N(m, \sigma = 2)$. A sample from X is given:

$$\mathcal{X} = (0.613, -0.777, 2.431, -1.305, 3.908, 0.203, 1.045).$$

   - Test the hypothesis $H_0 : m = 1$ against $H_1 : \lambda \sim U(0.5, 1.5)$.

6. Let X be a Gaussian variable $N(m, \sigma = 2)$. A sample from X is given:

$$\mathcal{X} = (-1.487, -2.472, 0.935, -3.639, -4.646, -0.915, -3.328).$$

   - Test the hypothesis $H_0 : m = -2$ against $H_1 : \lambda \sim U(-3, -1)$.

7. Load the file `data5.RData`. Assuming a linear model, select the three most influent variables among $(x_1, \ldots, x_{10})$.
8. Load the file `data6.RData`. Assuming a linear model, select the three most influent variables among $(x_1, \ldots, x_{10})$.
9. Load the file `data7.RData`. Select the most influent variable between x_1 and x_2.

5.6 The Expectation-Maximization Algorithm

When the model involves unobserved variables – also said hidden or latent variables, the procedures previously presented must be adapted. Indeed, assume that our model contains variables $\boldsymbol{X}$, which are observed, and $\boldsymbol{U}$, which are hidden and, so, unobserved. In this case, our complete set of variables is $\boldsymbol{V} = (\boldsymbol{X}, \boldsymbol{U})$. Thus, the sample $\mathcal{X}$ corresponds to unknown values of H. The complete sample is $\mathcal{V} = \{V_1 = (X_1, U_1), \ldots, V_n = (X_n, U_n)\}$, but only the values of $\mathcal{X} = \{X_1, \ldots, X_n\}$ are known. In such a situation, the values of the hidden variables must be estimated too. Since

$$f((\boldsymbol{X}, \boldsymbol{U})|\boldsymbol{\theta}) = f(\boldsymbol{U}|(\boldsymbol{X}, \boldsymbol{\theta}))f(\boldsymbol{X}|\boldsymbol{\theta}),$$

we have

$$\log f((\boldsymbol{X}, \boldsymbol{U})|\boldsymbol{\theta}) = \log f(\boldsymbol{U}|(\boldsymbol{X}, \boldsymbol{\theta})) + \log f(\boldsymbol{X}|\boldsymbol{\theta}).$$

Let $f_\theta(\boldsymbol{U}) = f(\boldsymbol{U}|(\boldsymbol{X}, \boldsymbol{\theta}))$ be the distribution of $\boldsymbol{U}$, conditional to the couple $(\boldsymbol{X}, \boldsymbol{\theta})$. Then,

$$f_\theta(\boldsymbol{U}) \log f((\boldsymbol{X}, \boldsymbol{U})|\boldsymbol{\theta}) = f_\theta(\boldsymbol{U}) \log \bar{f}(\boldsymbol{U}) + f_\theta(\boldsymbol{U}) \log f(\boldsymbol{X}|\boldsymbol{\theta}).$$

Integrating with respect to $\boldsymbol{U}$, we have

$$E_{f_\theta}(\log f((\boldsymbol{X}, \boldsymbol{U})|\boldsymbol{\theta}) = E_{f_\theta}(f_\theta) + E_{f_\theta}(\log f(\boldsymbol{X}|\boldsymbol{\theta})),$$

i.e.,

$$E_{f_\theta}(\log f((\boldsymbol{X}, \boldsymbol{U})|\boldsymbol{\theta}) = -H(f_\theta) + \log f(\boldsymbol{X}|\boldsymbol{\theta}),$$

where $H(f_\theta)$ is the entropy associated with f_θ. Thus, the MLE estimator $\widehat{\boldsymbol{\theta}}$ of $\boldsymbol{\theta}$ verifies

$$\widehat{\boldsymbol{\theta}} = \arg\max\{\mathcal{F}(\boldsymbol{\theta}) = E_{f_\theta}(\log f((\boldsymbol{X}, \boldsymbol{U})|\boldsymbol{\theta})) + H(f_\theta)\}.$$

We can rewrite this equality as

$$\widehat{\boldsymbol{\theta}} = \arg\max\{\overline{\mathcal{F}}(\boldsymbol{\theta}, \bar{f}) = E_{\bar{f}}(\log f((\boldsymbol{X}, \boldsymbol{U})|\boldsymbol{\theta})) + H(\bar{f}), \bar{f} = f_\theta\}.$$

Under this last formulation, the equality $\bar{f} = f_\theta$ appears as a constraint of optimization, leading to classical methods of constrained optimization, such as penalty methods, augmented Lagrangian approaches or Duality methods. Nevertheless, it can also be solved iteratively, by generating a sequence $\boldsymbol{\theta}^{(k)}$, starting from an initial guess $\boldsymbol{\theta}^{(0)}$. At step $k > 0$, $\boldsymbol{\theta}^{(k)}$ is known: we start by determining $\overline{f}^{(k)}$ such that

$$\overline{f}^{(k)}(\boldsymbol{U}) = f\left(\boldsymbol{U}|\left(\boldsymbol{X}, \boldsymbol{\theta}^{(k)}\right)\right).$$

Then, we determine

$$\boldsymbol{\theta}^{(k+1)} = \arg\max\left\{\mathcal{F}^{(k)}(\boldsymbol{\theta}) = \mathcal{F}\left(\boldsymbol{\theta}, \overline{f}^{(k)}\right)\right\}.$$

The first step (determination of $\overline{f}^{(k)}$) is called the *expectation step* or *E-step*. The second step (determination of $\boldsymbol{\theta}^{(k+1)}$) involves a maximization procedure, and is called *maximization step* or *M-step*. Notice that we can neglect the term $H\left(\overline{f}^{(k)}\right)$ in the M-step, since it is a constant: in the M-step, we can limit ourselves to maximize

$$\mathbb{F}^{(k)}(\boldsymbol{\theta}) = E_{\overline{f}^{(k)}}(\log f((\boldsymbol{X}, \boldsymbol{U})|\boldsymbol{\theta})) .$$

The community of the users of R proposes implementations of this algorithm. For instance, the sites (Zhu 2016; Hansen, 2022; Ruggieri, 2020; Le Roux 2021); the packages `em` (see Wu 2023), `sommer` (see Covarrubias-Pazaran 2023), `mixtools` (see Benaglia et al. 2009), GMM (see Thrun 2020),.... EM is often use to solve mixture distributions, namely Gaussian Mixture Problems. EM can be combined with UQ representation (see example).

Example 5.19 Let us consider

$$X = UX_1 + (1 - U)X_2,$$

where $X_1 \sim N(2, 1)$, $X_2 \sim N(-1, 2)$, and $U \sim \mathcal{B}(0.4)$. The data are formed by a sample of 100 variates of X, available in file `mixdata1.Rdata`. We start by loading the data:

```
load("mixdata1.Rdata")
```

Let us follow the implementation furnished in (Zhu 2016) : we determine starting values by using the clustering algorithm k-means:

(continued)

Example 5.19 (continued)

```
cl1 <- kmeans(xs,2)$cluster

teta_ini = vector(length = 2, 'list')
m_ini_1 <- mean(xs[cl1 == 1])
m_ini_2 <- mean(xs[cl1 == 2])
sig_ini_1 <- sd(xs[cl1 == 1])
sig_ini_2 <- sd(xs[cl1 == 2])
teta_ini[[1]] = c(m_ini_1, sig_ini_1)
teta_ini[[2]] = c(m_ini_2, sig_ini_2)

p_ini = numeric(length = 2)
p_ini[[1]] = sum(cl1 == 1)/length(cl1)
p_ini[[2]] = sum(cl1 == 2)/length(cl1)
```

Then, we can use a R package – for instance, using `mixtools`:

```
library(mixtools)
gm = normalmixEM(xs,k=2,lambda=p_ini,mu=c(m_ini_1,m_ini_2),sigm
a = c(sig_ini_1, sig_ini_2))
```

An example of result is the following:

```
p initial =  0.36  final =  0.586597580107295
q initial =  0.64  final =  0.413402419892705
m1 initial =  -2.48527899332303  final =  -1.17159407449779
m2 initial =  1.62161362664083  final =  2.00864653121052
s1 initial =  1.47456300527891  final =  2.12531359039875
s2 initial =  1.24299989634402  final =  1.11958527448805
```

Let us use the implementation furnished by (Zhu 2016). We start by defining the distributions under consideration:

```
distrib = vector(length = 2, 'list')
for (i in 1: length(distrib)){
  distrib[[i]] = function(x,teta){
    # teta[[1]] : mean
    # teta[[2]] : sigma
    resu = dnorm(x,teta[[1]],teta[[2]])
    return(resu)}}
```

We also need the procedure to update the parameters:

```
update_teta = function(x,p){
  mu = sum(finite(p * x)) / sum(finite(p))
  sigma <- sqrt(sum(finite(p * (x-mu)^2)) / sum(finite(p)))
  teta_new = c(mu, sigma)
  return(teta_new)}
```

(continued)

Example 5.19 (continued)

The EM iterations can be realized as follows:

```
finite <- function(x) {
  x[is.finite(x)]}
emsolve = function(x,p_ini, teta_ini, distrib, precmin,
nitmax){
  pi_ac = p_ini
  teta_ac = teta_ini
  Q = 0
  # starting value of expected value of the log likelihood
  s = 0
  for (i in 1: length(x)){
    saux = 0
    for (j in 1 : length(pi_ac)){
      fj = distrib[[j]]
      val = fj(x[i, ], teta_ac[[j]])
      saux = saux + log(val) + log(pi_ac[[j]])
    }
    s = s + saux
  }
  Q[[2]] = s
  nit = 2

  while ((abs(Q[nit]-Q[nit-1]) >= precmin) & (nit < nitmax)) {
    # E step
    scomp = numeric(length = ndata)
    for (i in 1 : ndata){
      scomp[[i]] =  0}
    comp = vector(length = length(pi_ac), 'list')
    for (j in 1: length(pi_ac)){
      aux = numeric(length = ndata)
      fj = distrib[[j]]
      for (i in 1 : ndata){
        auxxx = fj(x[i, ], teta_ac[[j]])
        aux[[i]] =  pi_ac[[j]]*auxxx}
      comp[[j]] = aux
      scomp = scomp + aux}
    p = vector(length = length(pi_ac),'list')
    for (j in 1: length(pi_ac)){
      p[[j]] = comp[[j]]/scomp}
    # M step
    for (j in 1: length(pi_ac)){
      pi_ac[[j]] = sum(finite(p[[j]])) / length(x)
      teta_ac[[j]] = update_teta(x,p[[j]])}
    p = pi_ac

    nit = nit + 1
    Q[[nit]] = sum(log(scomp))}
  sol = list(p,teta_ac, Q)
  names(sol) = c("prob","teta","Q")
  return(sol)}
```

(continued)

Example 5.19 (continued)

We call this procedure as follows:

```
precmin = 1e-6
nitmax = 1000
ndata = length(xs)
xsm = matrix(xs, nrow = ndata, ncol = 1)
sol1 = emsolve(xsm,p_ini, teta_ini,distrib,precmin,nitmax)
```

An example of result is given below. The PDF generated is shown in Fig. 5.32.

```
p initial =  0.36  final =  0.585233113502252
q initial =  0.64  final =  0.414766886497748
m1 initial =  -2.48527899332303  final =  -1.17698393656418
m2 initial =  1.62161362664083  final =  2.123814757629
s1 initial =  1.47456300527891  final =  2.00580683818873
s2 initial =  1.24299989634402  final =  1.12159052464896
```

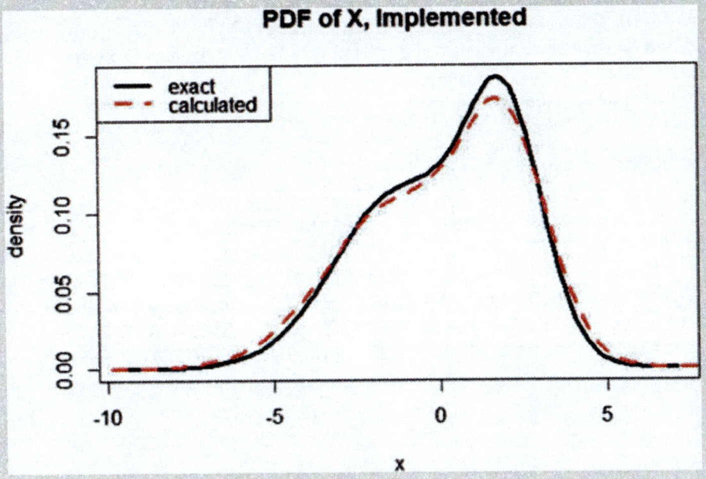

Fig. 5.32 PDF furnished by the implementation of (Zhu 2016)

Example 5.20 Let us consider

$$X = UX_1 + (1 - U)X_2,$$

where $X_1 \sim Laplace(1, 1)$, $X_2 \sim Laplace(5, 1)$, and $U \sim \mathcal{B}(0.4)$. The data are formed by a sample of 100 variates of X, available in file `mixdata2.Rdata`. We use the implementation of (Zhu 2016). In this case, the distributions are

```
distrib = vector(length = 2, 'list')
for (i in 1: length(distrib)){
  distrib[[i]] = function(x,teta){
    # teta[[1]] : mean
    # teta[[2]] : sigma
    resu = dlaplace(x,teta[[1]],teta[[2]]/sqrt(2))
    return(resu)}}
```

and the update of the parameters is made as follows:

```
update_teta = function(x,p){
  mu = sum(finite(p * x)) / sum(finite(p))
  sigma = sqrt(sum(finite(p * (x-mu)^2)) / sum(finite(p)))
  teta_new = c(mu, sigma)
  return(teta_new)}
```

We obtain

```
p initial =  0.39  final =  0.401160100988685
q initial =  0.61  final =  0.598839899011315
m1 initial =  1.16744398288578  final =  1.34620479787549
m2 initial =  5.14009750077873  final =  1.29587376845539
s1 initial =  1.07224794644653  final =  5.09438162214982
s2 initial =  1.04327156092553  final =  1.15281659565614
```

(continued)

Example 5.20 (continued)

The PDF generated is shown in Fig. 5.33.

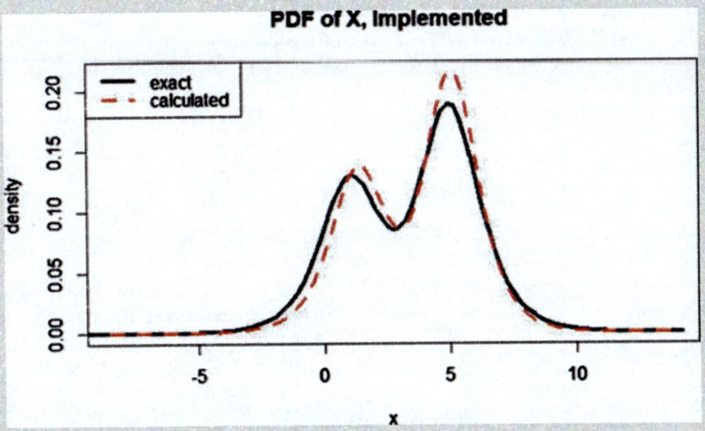

Fig. 5.33 PDF furnished by the implementation of (Zhu 2016)

Example 5.21 Let us consider

$$X = UX_1 + (1 - U)X_2,$$

where $X_1 \sim Laplace(1, 1)$, $X_2 \sim N(-5, 1)$, and $U \sim \mathcal{B}(0.3)$. The data are formed by a sample of 100 variates of X, available in file `mixdata3.Rdata`. We use again the implementation of (Zhu 2016). In this case, the distributions are

```
distrib = vector(length = 2, 'list')
distrib[[1]] = function(x,teta){
    # teta[[1]] : mean
    # teta[[2]] : sigma
    resu = dlaplace(x,teta[[1]],teta[[2]]/sqrt(2))
    return(resu)}
distrib[[2]] = function(x,teta){
  # teta[[1]] : mean
  # teta[[2]] : sigma
  resu = dnorm(x,teta[[1]],teta[[2]])
  return(resu)}
```

(continued)

Example 5.21 (continued)

and the update of the parameters is made as follows:

```
update_teta = function(x,p){
  mu = sum(finite(p * x)) / sum(finite(p))
  sigma = sqrt(sum(finite(p * (x-mu)^2)) / sum(finite(p)))
  teta_new = c(mu, sigma)
  return(teta_new)}
```

We obtain

```
p initial =  0.68  final =  0.684427672173836
q initial =  0.32  final =  0.315572327826164
m1 initial =  -5.1634829044797  final =  -5.13820853573293
m2 initial =  1.11175698448645  final =  1.02872453786613
s1 initial =  0.989117445247671  final =  1.14498622149128
s2 initial =  1.27370316406965  final =  1.22824323856616
```

The PDF generated is shown in Fig. 5.34.

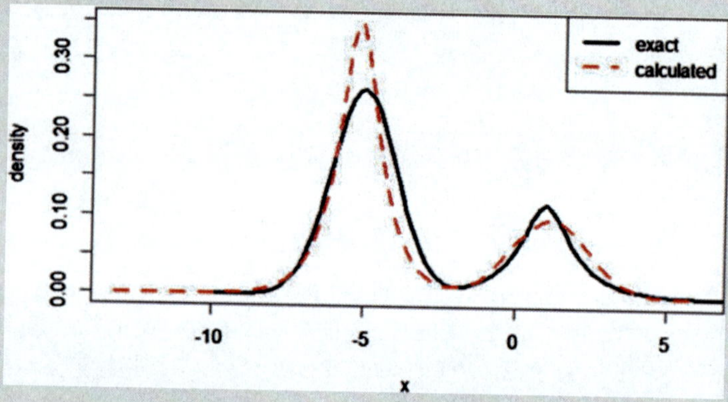

Fig. 5.34 PDF furnished by the implementation of (Zhu 2016)

Example 5.22 The algorithm EM can be combined with UQ representations, by considering the coefficients of the representation as the parameters to be determined. We shall illustrate such an approach using a simple interpolation of the PDFs of the components of the mixture by a polynomial in the situation of Example 5.20. In this case, $\boldsymbol{\theta}$ is the vector of the coefficients of the approximation of the PDF and we make no prior assumption on the distributions. We shall use as starting values, those furnished by `mixtools` with an erroneous prior – we assume that the distributions are Gaussian:

(continued)

Example 5.22 (continued)

```
library(pracma)
library(mixtools)
gm<-normalmixEM(xs,k=2,lambda=p_ini,mu=c(m_ini_1,m_ini_2),sigma
= c(sig_ini_1, sig_ini_2))
teta_ini = vector(length = 2, 'list')
degre = 9
epscol = 0
matA = exp1$matA(degre,xs)
matB = dnorm(xs,gm$mu[[1]],gm$sigma[[1]])
caux1 = lsqlincon(matA, matB,  A = -matA, b = rep(0,ndata))
pp1 = exp1$proj(caux1,xs)
auxx1 = sum(pp1)
teta_ini[[1]]= caux1/auxx1
matA = exp1$matA(degre,xs)
matB = dnorm(xs,gm$mu[[2]],gm$sigma[[2]])
caux2 = lsqlincon(matA, matB,  A = -matA, b = rep(0,ndata))
pp2 = exp1$proj(caux2,xs)
auxx2 = sum(pp2)
teta_ini[[2]]= caux2/auxx2
```

Notice that we used a constrained least squares procedure to get an interpolation that respects the nonnegativity of the PDF. The distribution is evaluated as follows:

```
xmin = min(xs)
xmax = max(xs)
source("polynom.R")
phi = polynom$new(xmin,xmax,1e-3)
source("expansion1D.R")
exp1 = expansion1D$new(phi)
distrib = vector(length = 2, 'list')
for (i in 1: length(distrib)){
  distrib[[i]] = function(x,teta){
    resu = exp1$proj(teta,x)
    return(resu)}}
```

(continued)

Example 5.22 (continued)

and the update of the parameters is made as follows:

```
update_teta = function(x,p){
  matA = exp1$matA(degre,x)
  matB = p
  csol = lsqlincon(matA, matB,  A = -matA, b = rep(0,length(p))
)
  ppp = exp1$proj(csol,x)
  aux = sum(ppp)
  teta_new = as.vector(csol/aux)
  return(teta_new)}
```

We obtain

```
p initial =  0.39  final =  0.410847323688906
q initial =  0.61  final =  0.589152676311094
m1 initial =  1.16744398288578  final =  1.45853449665608
m2 initial =  5.14009750077873  final =  5.0776733516031
s1 initial =  1.07224794644653  final =  1.3296707727723
s2 initial =  1.04327156092553  final =  1.2682672079083
```

The PDF generated is shown in Fig. 5.35.

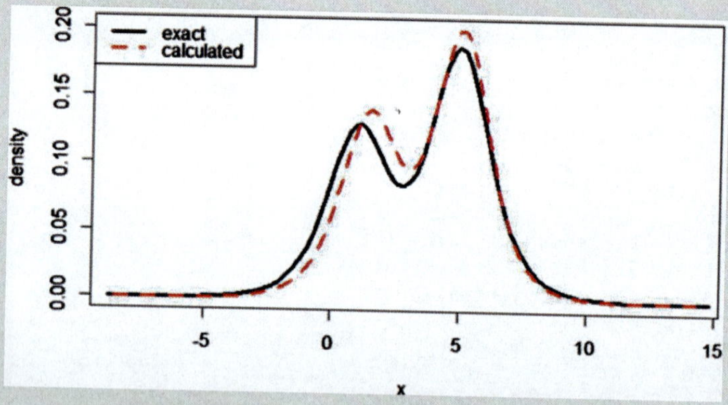

Fig. 5.35 PDF furnished by the implementation of (Zhu 2016), taking the coefficients of interpolation as parameters (situation of Example 5.20)

(continued)

Example 5.22 (continued)

For the situation in Example 5.21, we obtain

```
p initial =  0.68  final =  0.687527111301594
q initial =  0.32  final =  0.312472888698406
m1 initial =  -5.1634829044797  final =  -4.98526739677174
m2 initial =  1.11175698448645  final =  0.870796674531235
s1 initial =  0.989117445247671  final =  1.52217261881514
s2 initial =  1.27370316406965  final =  1.56652391848206
```

The PDF generated is shown in Fig. 5.36.

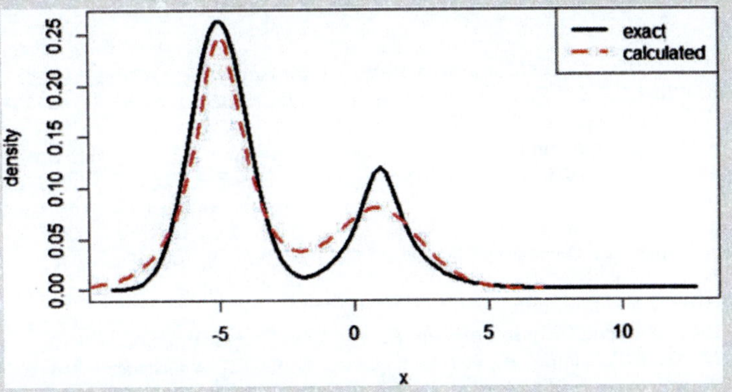

Fig. 5.36 PDF furnished by the implementation of (Zhu 2016), taking the coefficients of interpolation as parameters (situation of Example 5.21)

Exercises

1. The file `mixdata4.Rdata` contains a sample from a mixed distribution, formed by two gamma distributions $Gamma(\alpha_i, \beta_i)$. We have $\alpha_i = (E(X_i)/\sigma(X_i))^2$, $\beta_i = \alpha_i/E(X_i)$.
   (a) Run the EM algorithm to determine the parameters of the distributions and the probabilities p and q.
   (b) Run the EM algorithm with priors determined by interpolation to determine the parameters of the distributions and the probabilities p and q.
2. The file `mixdata5.Rdata` contains a sample from a mixed distribution, formed by two Chi-Squared distributions $\chi^2(\alpha_i)$. We have $\alpha_i = E(X_i) = \sigma^2(X_i)/2$.
   (c) Run the EM algorithm to determine the parameters of the distributions and the probabilities p and q.
   (d) Run the EM algorithm with priors determined by interpolation to determine the parameters of the distributions and the probabilities p and q.

(continued)

3. The file `mixdata6.Rdata` contains a sample from a mixed distribution, formed by two Logistic distributions *Logistic*(μ_i, s_i). We have $\mu_i = E(X_i)$, $s_i = \sigma(X_i)\sqrt{3}/\pi$.

   (e) Run the EM algorithm to determine the parameters of the distributions and the probabilities p and q.

   (f) Run the EM algorithm with priors determined by interpolation to determine the parameters of the distributions and the probabilities p and q.

References

Agliari, A., & Parisetti, C. C. (1986). A-g Reference Informative Prior: A Note on Zellner's g-Prior. *Journal of the Royal Statistical Society. Series D (The Statistician), 37*(3), pp. 271–275. doi: https://doi.org/10.2307/2348164

Akaike, H. (1998). Information theory and an extension of the maximum likelihood principle. In E. Parzen, K. Tanabe, & G. Kitagawa (Eds.), *Selected Papers of Hirotugu Akaike* (pp. 199–213). New York, NY, USA: Springer. doi:https://doi.org/10.1007/978-1-4612-1694-0_15

Albert, I., Donnet, S., Guihenneuc-Jouyaux, C., Low-Choy, S., Mengersen, K., & Rousseau, J. (2002). Combining Expert Opinions in Prior Elicitation. *Bayesian Analysis, 7*(3), pp. 503–532. doi:https://doi.org/10.1214/12-BA717

Ash, R. (1965, reedited 2003). *Information Theory*. New York, NY, USA: Dover.

Attiaoui, D., Doré, P., Martin, A., & Ben Yaghlane, B. (2012). A Distance between Continuous Belief Functions. In E. Hüllermeier, S. Link, T. Fober, & B. Seeger (Ed.), *Scalable Uncertainty Management: SUM 2012* (pp. 194–205). Marburg, Germany: Springer. doi:https://doi.org/10.1007/978-3-642-33362-0_15

Aurnhammer, C., & Frank, S. L. (2019). Evaluating information-theoretic measures of word prediction in naturalistic sentence reading. *Neuropsychologia, 134*. doi:https://doi.org/10.1016/j.neuropsychologia.2019.107198

Baldi, P., & Itti, L. (2010). Of bits and wows: A Bayesian theory of surprise with applications to attention. *Neural Networks, 23*(5), 649–666. doi:https://doi.org/10.1016/j.neunet.2009.12.007

Banner, K. M., Irvine, K. M., & Rodhouse, T. J. (2020, May). The use of Bayesian priors in Ecology: The good, the bad and the not great. *Methods in Ecology and Evolution, 11*(8), 882–889. doi:https://doi.org/10.1111/2041-210X.13407

Batou, A., & Soize, C. (2013). Calculation of Lagrange Multipliers in the Construction of Maximum Entropy Distributions in High Stochastic Dimension. *SIAM/ASA Journal on Uncertainty Quantification, 1*(1). doi:https://doi.org/10.1137/120901386

Bayarria, M. J., & Morales, J. (2003). Bayesian measures of surprise for outlier detection. *Journal of Statistical Planning and Inference, 111*(1–2), 3–22. doi:https://doi.org/10.1016/S0378-3758(02)00282-3

Benaglia, T., Chauveau, D., Hunter, D. R., & Young, D. S. (2009, October 21). mixtools: An R Package for Analyzing Finite Mixture Models. *Journal of Statistical Software, 32*(6), pp. 1–29. doi:https://doi.org/10.18637/jss.v032.i06

Benavoli, A. (2014). Belief function and multivalued mapping robustness in statistical estimation. *International Journal of Approximate Reasoning, 55*, 311–329. doi:https://doi.org/10.1016/j.ijar.2013.04.014

Beretta, G. P. (2008). Axiomatic Definition of Entropy for Nonequilibrium States. *International Journal of Thermodynamics, 11*(2), 39–48. doi:https://doi.org/10.5541/ijot.211

Berger, J. O. (1985). *Statistical Decision Theory and Bayesian Analysis*. New York, USA: Springer. doi:https://doi.org/10.1007/978-1-4757-4286-2

Berger, J. O., & Bernardo, J. M. (1989, March). Estimating a Product of Means: Bayesian Analysis with Reference Priors. *Journal of the American Statistical Association, 84*(405), pp. 200–207. doi:https://doi.org/10.2307/2289864

Berger, J. O., & Bernardo, J. M. (1992). On the development of reference priors. In J. M. Bernardo, J. O. Berger, A. P. Dawid, & A. F. Smith (Ed.), *Bayesian Statistics 4*. Clarendon Press.

Berger, J. O., Bernardo, J. M., & Sun, D. (2009, April). The formal definition of reference priors. *The Annals of Statistics, 37*(2), pp. 905–938. doi:https://doi.org/10.1214/07-AOS587

Boivin, C. (2022a, December 25). https://cran.r-project.org/web/packages/dst/vignettes/Captain_Example.html. Retrieved from Captain's Example: https://cran.r-project.org/web/packages/dst/vignettes/Captain_Example.html

Boivin, C. (2022b, 12 24). *Introduction to Belief Functions*. Retrieved from https://cran.r-project.org/web/packages/dst/vignettes/: https://cran.r-project.org/web/packages/dst/vignettes/Introduction_to_Belief_Functions.html#fn2

Boivin, C. (2022c, December 25). *Introduction to Belief Functions: The Monty Hall Game*. Retrieved 2022, from https://cran.microsoft.com/snapshot/2018-08-11/web/packages/dst/vignettes/Monty-hall-Example.html: https://cran.microsoft.com/snapshot/2018-08-11/web/packages/dst/vignettes/Monty-hall-Example.html

Boltzmann, L. E. (1866). Über die mechanische Bedeutung des zweiten Hauptsatzes der Wärmetheorie. *Wiener Berichte, 53*, 195–220. Retrieved from http://opacplus.bsb-muenchen.de/title/BV020135572/ft/bsb10133426?page=3

Boltzmann, L. E. (1877 traduction 2002). On the relationship between the second main theorem of mechanical heat theory and the probability calculation with respect to the results about the heat equilibrium. *Akademie der Wissenschaften in Wien Mathematisch-naturwissenschaftliche Klasse Sitzungsberichte, 2*(76), 373–435. Retrieved February 14, 2023, from http://users.polytech.unice.fr/~leroux/boltztrad.pdf

Boltzmann, L. E. (1896). *Vorlesungen ̈uber Gastheorie* (Vol. 1). Leipzig: Barth.

Boltzmann, L. E. (1995). *Lectures on Gas Theory*. (S. G. Brush, Trans.) New York, NY, USA: Dover.

Boyle, G. H. (n.d.). *modelingcommons.org/file/download/6101?file_id=3384*. Retrieved from http://modelingcommons.org/file/download/6101?file_id=3384

Brillouin, L. (1956). *Science and Information Theory*. New York, NY, USA: Dover.

Campagner, A., Ciucci, D., & Denœux, T. (2022). Belief Functions and Rough Sets: Suvey and New Insights. *International Journal of Approximate Reasoning, 143*, 92–215. doi:https://doi.org/10.1016/j.ijar.2022.01.011

Chadli, A., Khawla, B., Asma, M., & Fellag, H. (2017, April). Bayesian estimation of the Rayleigh distribution under different loss function. *Electronic Journal of Applied Statistical Analysis, 10*(1), pp. 50–64. doi:https://doi.org/10.1285/i20705948v10n1p50

Chakrabarti, C. G., & Indranil, C. (2005). Shannon entropy: axiomatic characterization and application. *International Journal of Mathematics and Mathematical Sciences*. doi:https://doi.org/10.1155/IJMMS.2005.2847

Chang, Y.-C., & Hung, W.-L. (2007). LINEX Loss Functions with Applications to Determining the Optimum Process Parameters. Quality & Quantity. *Quality & Quantity , 41*(2), pp. 291–301. doi:https://doi.org/10.1007/s11135-005-5425-3

Cheung, V. K., M.C., P., Meyer, L., Pearce, M. T., Haynes, J.-D., & Koelsch, S. (2019). Uncertainty and Surprise Jointly Predict Musical Pleasure and Amygdala, Hippocampus, and Auditory Cortex Activity. *Current Biology, 29*(23), 4084–4092.e4. doi:https://doi.org/10.1016/j.cub.2019.09.067

Clausius, R. J. (1850). Ueber die bewegende Kraft der Wärme und die Gesetze, welche sich daraus für die Wärmelehre selbst ableiten lassen. *Annalen der Physik*, 368–397 , 500–524.

Clausius, R. J. (1851 republished in 2009). On the Moving Force of Heat, and the Laws regarding the Nature of Heat itself which are deducible therefrom. *The London, Edinburgh, and Dublin Philosophical Magazine and Journal of Science, 2*(8), 1–21 , 102–119. doi:https://doi.org/10.1080/14786445108646819 , https://doi.org/10.1080/14786445108646840

Cobb, B. R., & Shenoy, P. (2006, April). On the plausibility transformation method for translating belief function models to probability models. *Journal of Approximate Reasoning, 41*(3), pp. 314–330. doi:https://doi.org/10.1016/j.ijar.2005.06.008

Consonni, G., Fouskakis, D., Liseo, B., & Ntzoufras, I. (2018). Prior Distributions for Objective Bayesian Analysis. *Bayesian Analysis, 13*(2), 627–679. doi:https://doi.org/10.1214/18-BA1103

Couso, I., Dubois, D., & Sanchez, L. (2014). *Random Sets and Random Fuzzy Sets as Ill-Perceived Random Variables.* Springer. doi:https://doi.org/10.1007/978-3-319-08611-8

Covarrubias-Pazaran, G. (2023, June 14). *sommer: Solving Mixed Model Equations in R.* Retrieved July 29, 2023, from https://cran.r-project.org/web/packages/sommer/: https://cran.r-project.org/web/packages/sommer/sommer.pdf

Csiszár, I. (2008). Axiomatic Characterizations of Information Measures. *Entropy, 10*(3), 261–273. doi:https://doi.org/10.3390/e10030261

Dale, A. I. (1982). Bayes or Laplace? An Examination of the Origin and Early Applications of Bayes' Theorem. *Archive for History of Exact Sciences, 27*(1), pp. 23–47.

Dale, A. I. (1999). *A History of Inverse Probability—From Thomas Bayes to Karl Pearson* (2nd. ed.). New York: Springer.

De Finetti, B. (2017). *Theory of Probability—A Critical Introductory Treatment.* (A. M. Smith, Trans.) UK: John Wiley and Sons.

DeGroot, M. H. (2004). *Optimal Statistical Decisions.* New Jersey: John Wiley & Sons.

Dempster, A. (1967, April). Upper and Lower Probabilities Induced by a Multivalued Mapping. *Ann. Math. Statist., 38*(2), pp. 325–339,. doi:https://doi.org/10.1214/aoms/1177698950

Denoeux, T. (2019, June). Decision-making with belief functions: A review. *International Journal of Approximate Reasoning, 109*, 87–110. doi:https://doi.org/10.1016/j.ijar.2019.03.009

Denoeux, T. (2021). *Uncertainty Analysis using Belief Functions: Applications to Statistical Inference and Pattern Recognition.* Université de technologie de Compiègne, Department of Computer Science, Compiègne, France. Retrieved December 20, 2022, from https://www.hds.utc.fr/~tdenoeux/dokuwiki/_media/en/book_bf.pdf

Denœux, T., Younes, Z., & Abdallah, F. (2010). Representing uncertainty on set-valued variables using belief functions. *Artificial Intelligence, 174*(7–8), 479–499. doi:https://doi.org/10.1016/j.artint.2010.02.002

Depaoli, S., Winter, S. D., & Visser, M. (2020). The Importance of Prior Sensitivity Analysis in Bayesian Statistics: Demonstrations Using an Interactive Shiny App. *Frontiers in Psychology, 11.* doi:https://doi.org/10.3389/fpsyg.2020.608045

Dezert, J., & Smarandache, F. (2008). A new probabilistic transformation of belief mass assignment. *Fusion 2008: International Conference on Information Fusion*, (pp. 1410–1417). Retrieved December 19, 2022, from https://hal.archives-ouvertes.fr/hal-00304319/document

Dezert, J., & Smarandache, F. (2009). Transformations of belief masses into subjective probabilities. In J. Dezert, & F. Smarandache (Eds.), *Advances and Applications of DSmT for Information Fusion* (pp. 85–136). Rehoboth: American Research Press . Retrieved December 19, 2022, from https://www.researchgate.net/publication/306413841_Transformations_of_belief_masses_into_subjective_probabilities

Dezert, J., Smarandache, F., & Daniel, M. (2004). A Generalized Pignistic Transformation. In J. Dezert, & F. Smarandache (Eds.), *Advances and Applications of DSmT for Information Fusion* (pp. 143–153). Rehoboth: American Research Press. Retrieved December 19, 2022, from https://www.onera.fr/sites/default/files/297/C023-Dezert-Fusion2004Stockholm.pdf

Diaconis, P., & Ylvisaker, D. (1979). Conjugate Priors for Exponential Families. *The Annals of Statistics, 7*(2), pp. 269–281. doi:https://doi.org/10.1214/aos/1176344611

Dieudonné, J. (1969). *Foundations of Modern Analysis.* London: Academic Press.

Doré, P. E., Fiche, A., & Martin, A. (2010). Models of belief functions — Impacts for patterns recognitions. *13th International Conference on Information Fusion*. Edinburgh, UK. doi:https://doi.org/10.1109/ICIF.2010.5711936

Doré, P., Martin, A., Abi-Zeid, I., Jousselme, A., & Maupin, P. (2011a, January). Belief functions induced by multimodalprobability density functions, an application to the search and rescue problem. *RAIRO—Operations Research, 44*(4), 323–343. doi:https://doi.org/10.1051/ro/2011001

Doré, P., Osswald, C., Martin, A., Jousselme, A., & Maupin, P. (2011b). Continuous belief functions to qualify sensors performances. In W. Liu (Ed.), *Symbolic and Quantitative Approaches to Reasoning with Uncertainty. ECSQARU 2011* (pp. 350–361). Belfast, Ireland: Springer. doi:https://doi.org/10.1007/978-3-642-22152-1_30

Dragulescu, A., & Yakovenko, V. (2000). Statistical mechanics of money. *Eur. Phys. J. B, 17*, 723–729. doi:https://doi.org/10.1007/s100510070114

Dubois, D., & Prade, H. (1986). A Set-Theoretic View of Belief Functions. *International Journal of General Systems, 12*(3), 193–226. doi:https://doi.org/10.1080/03081078608934937

Falconer, J. R., Frank, E., Polaschek, D. L., & Joshi, C. (2021). Methods for Eliciting Informative Prior Distributions: A Critical Review. *Decision Analysis, 19*(3), pp. 189–204. doi:https://doi.org/10.1287/deca.2022.0451

Falconer, J., Frank, E., Polaschek, D., & Joshi, C. (2022). Methods for Eliciting Informative Prior Distributions: A Critical Review. *Decision Analysis, 19*(3), 189–204. doi:https://doi.org/10.1287/deca.2022.0451

Fetz, T., & Oberguggenberger, M. (2016, November). Imprecise random variables, random sets, and Monte Carlo simulation. *78*, 252–264. doi:https://doi.org/10.1016/j.ijar.2016.06.012

Fiche, A., Martin, A., Cexus, J.-C., & Khenchaf, A. (2010). Continuous belief functions and α-stable distributions. *13th Conference on Information Fusion*. Edinburgh. doi:https://doi.org/10.1109/ICIF.2010.5711934

Fink, D. (1997). *A Compendium of Conjugate Priors*. Retrieved April 23, 2023, from https://www.johndcook.com/CompendiumOfConjugatePriors.pdf

Fisher, R. A. (1955). Statistical Methods and Scientific Induction. *Journal of the Royal Statistical Society, Series B, 17*(1), pp. 69–78. doi:https://doi.org/10.1111/j.2517-6161.1955.tb00180.x

Fort, H. (2022). *Forecasting with Maximum Entropy*. IOP Publishing Ltd.

Fox, C. R., & See, K. E. (2006). Belief and Preference in Decision Under Uncertainty. In D. Hardman, & L. Macchi (Eds.), *Thinking: Psychological Perspectives on Reasoning, Judgment and Decision Making*. John Wiley & Sons. doi:https://doi.org/10.1002/047001332X.ch14

Garthwaite, P. H., Kadane, J. B., & O'Hagan, A. (2005). Statistical Methods for Eliciting Probability Distributions. *Journal of the American Statistical Association, 100*(470), pp. 680–701. doi:https://doi.org/10.1198/016214505000000105

Gelman, A. (2002). Prior distribution. In A. H.-S. Piegorsch (Ed.), *Encyclopedia of Environmetrics*, (pp. 1634–1637). doi:https://doi.org/10.1002/9780470057339.vap039

Gelman, A., Simpson, D., & Betancourt, M. (2017). The Prior Can Often Only Be Understood in the Context of the Likelihood. *Entropy, 19*(10). doi:https://doi.org/10.3390/e19100555

Ghaderinezhad, F., & Ley, C. (2019). Quantification of the impact of priors in Bayesian statistics via Stein's Method. *146*, 206–212. doi:https://doi.org/10.1016/j.spl.2018.11.012

Ghaderinezhad, F., & Ley, C. (2020). On the Impact of the Choice of the Prior in Bayesian Statistics. In N. Tang (Ed.), *Bayesian Inference on Complicated Data* (pp. 1–14). IntechOpen. doi:https://doi.org/10.5772/intechopen.88994

Ghaderinezhad, F., Ley, C., & Serrien, B. (2022, October). The Wasserstein Impact Measure (WIM): A practical tool for quantifying prior impact in Bayesian statistics. *Computational Statistics and Data Analysis, 174*. doi:https://doi.org/10.1016/j.csda.2021.107352

Giambartolomei, G. (2015). *The Karhunen-Loeve Theorem*. Tesi di Laurea, Universita di Bologna, Bologna, Italy. Retrieved March 25, 2023, from https://amslaurea.unibo.it/10169/1/Giambartolomei_Giordano_Tesi.pdf

Gibbs, J. W. (1901 reprinted 1960). *Elementary Principles of Statistical Mechanics*. Dover.

Grassi, P. R., & Bartels, A. (2021). Magic, Bayes and wows: A Bayesian account of magic tricks. *Neuroscience & Biobehavioral Reviews, 126*, 515–527. doi:https://doi.org/10.1016/j.neubiorev.2021.04.001

Guan, J., & Bell, D. (1993). Discounting and Combination Operations in Evidential Reasoning. In D. Heckerman, & A. Mamdani (Ed.), *Ninth Conference on Uncertainty in Artificial Intelligence* (pp. 477–484). Washington, USA: Elsevier. doi:https://doi.org/10.1016/B978-1-4832-1451-1.50062-7

Halmos, P. (1978). *Measure theory.* Berlin, Heidelberg, New York: : Springer Verlag.

Halpern, J., & Fagin, R. (1992). Two views of belief: belief as generalized probability and belief as evidence. *Artificial Intelligence,, 54*, pp. 275–317.

Hansen, N. R. (2022, May 8). *Expectation maximization algorithms.* Retrieved July 27, 2023, from Computational Statistics with R: https://cswr.nrhstat.org/em.html

Harney, H. L. (2003). *Bayesian Inference.* Berlin, Germany: Springer. doi:https://doi.org/10.1007/978-3-662-06006-3

Hartley, R. (1928, July). Transmission of Information. *Bell System Technical Journal, 7*(3), 535–563.

He, Y. (2013). *Uncertainty Quantification and Data Fusion based on Dempster-Shafer Theory.* Florida State University, Mathematics. Florida State University Libraries.

Howson, C., & Urbach, P. (2006). *Scientific reasoning : the Bayesian approach* . Chicago: Open Court.

Hüllermeier, E., Kruse, R., & Hoffmann, F. (Eds.). (2010). Consonant continuous belief functions conflicts calculation. *IPMU'10: Proceedings of the Computational intelligence for knowledge-based systems design, and 13th international conference on Information processing and management of uncertainty* (pp. 706–715). Dortmund, Germany: Springer.

Hulse, A., Schumacher, B., & Westmoreland, M. D. (2018). Axiomatic Information Thermodynamics. *Entropy, 20*(4). doi:https://doi.org/10.3390/e20040237

Hurwich, C. M., & Tsai, C. L. (1989, June). Regression and Time Series Model Selection in Small Samples. *Biometrika, 76*(2), pp. 297–307. doi:https://doi.org/10.1093/biomet/76.2.297

Islam, A. F. (2011). *Loss functions, utility functions and Bayesian sample size determination.* Ph. D. Thesis, Queen Mary, University of London, London. Retrieved April 17, 2023, from https://core.ac.uk/download/pdf/30695406.pdf

Itti, L., & Baldi, P. (2009). Bayesian surprise attracts human attention. *Vision Research, 49*(10), 1295–1306. doi:https://doi.org/10.1016/j.visres.2008.09.007

Jaffray, J.-Y., & Wakker, P. (1993). Decision making with belief functions: Compatibility and incompatibility with the sure-thing principle. *Journal of Risk and Uncertainty, 7*(3), 255–271. doi:https://doi.org/10.1007/BF01079626

Jaynes, E. (1989). *Clearing up Mysteries—The Original Goal.* (J. Skilling, Ed.) Dordrecht: Springer. doi:https://doi.org/10.1007/978-94-015-7860-8_1

Jaynes, E. (2003). *Probability Theory: The Logic of Science.* Cambridge: Cambridge University Press.

Jaynes, E. T. (1957a, May). Information Theory and Statistical Mechanics. *Phys. Rev., 106*(4), 620–630. doi:https://doi.org/10.1103/PhysRev.106.620

Jaynes, E. T. (1957b, October). Information Theory and Statistical Mechanics. II. *Phys. Rev., 108*(2), 171–190. doi:https://doi.org/10.1103/PhysRev.108.171

Jaynes, E. T. (1965). Gibbs vs. Boltzmann Entropies. *American Journal of Physics, 33*(5), 391–398.

Jaynes, E. T. (1968a, September). Prior Probabilities. *IEEE Transactions on Systems Science and Cybernetics, 4*(3), 227–241. doi:https://doi.org/10.1109/TSSC.1968.300117

Jaynes, E. T. (1968b, September). Prior Probabilities. *IEEE Transactions on Systems Science and Cybernetics, 4*(3), pp. 227–241. doi:https://doi.org/10.1109/TSSC.1968.300117

Jeffreys, H. (1935, April). Some Tests of Significance, Treated by the Theory of Probability. *Mathematical Proceedings of the Cambridge Philosophical Society, 31*(2), pp. 203–222. doi:https://doi.org/10.1017/S030500410001330X

Jeffreys, H. (1939). *Theory of probability.* Oxford: University Press.

Jeffreys, H. (1946, September 24). An invariant form for the prior probability in estimation problems. *Proceedings of the Royal Society of London. Series A,, 10*(1007), pp. 453–461. doi: https://doi.org/10.1098/rspa.1946.0056

Johnson, S. R., Tomlinson, G. A., Hawker, G. A., Granton, J. T., & Feldman, B. M. (2010). Methods to elicit beliefs for Bayesian priors: a systematic review. *J Clin Epidemiol., 63*(4), pp. 355–369. doi:https://doi.org/10.1016/j.jclinepi.2009.06.003

Jones, D. E., Trangucci, R. N., & Chen, Y. (2022). Quantifying Observed Prior Impact. *Bayesian Analysis, 17*(3), 737–764. doi:https://doi.org/10.1214/21-BA1271

Kapur, J. N. (1989 revised 1993). *Maximum Entropy Models in Science and Technology.* New Delhi, India: Wiley Eastern.

Kass, R. E., & Raftery, A. E. (1995, June). Bayes Factors. *Journal of the American Statistical Association, 90*(430), pp. 773–795. doi:https://doi.org/10.2307/2291091

Kass, R. E., & Wasserman, L. (1996). The selection prior distribution by formal rules. *Journal of the American Statistical Association, 91*(435), 1343–1370. doi:https://doi.org/10.1080/01621459.1996.10477003

Khinchin, A. Y. (1957). *Mathematical Foundations of Information Theory.* New York, NY, USA: Dover.

Klopotek, M. A., & Wierzchon, S. T. (1998). A New Qualitative Rough-Set Approach to Modeling Belief Functions. In L. Polkowski, & A. Skowron (Ed.), *Rough Sets and Current Trends in Computing, First International Conference, RSCTC'98* (pp. 346–354). Warzsaw, Poland: Springer. doi:https://doi.org/10.1007/3-540-69115-4_47

Kojadinovic, I., Marichal, J.-L., & Roubens, M. (2005). An axiomatic approach to the definition of the entropy of a discrete Choquet capacity. *Information Sciences, 172*(1–2), 131–153. doi: https://doi.org/10.1016/j.ins.2004.05.011

Kolossa, A., Kopp, B., & Fingscheidt, T. (2015). A computational analysis of the neural bases of Bayesian inference. *NeuroImage, 106*, 222–237. doi:https://doi.org/10.1016/j.neuroimage.2014.11.00

Kullback, S. (1951 reed. 1969 reprint 1979). *Information Theory and Statistics.* New York: Wiley, reed. Dover.

Kullback, S., & Leibler, R. A. (1951, March). On Information and Sufficiency. *The Annals of Mathematical Statistics, 22*(1), 79–86. doi:https://doi.org/10.1214/aoms/1177729694

Laghmara, H., Laurain, T., Cudel, C., & Lauffenburger, J. P. (2020). Heterogeneous sensor data fusion for multiple object association using belief functions. *Information Fusion, 57*, 44–58. doi:https://doi.org/10.1016/j.inffus.2019.11.002

Laplace, P.-S. (1774). Memoire sur la probabilité des causes par les événements. *Memoires de Mathématique et de Physique, Presentés à l'Académie Royale des Sciences par divers Savans & lus dans ses Assemblées*, pp. 621–656.

Laplace, P.-S. (1891). *Oeuvres complètes* (Vol. 8). (A. d. Paris, Ed.) Paris: Gauthiers-Villars.

Laplace, P.-S. (1986). Memoir on the Probability of the Causes of Events. *Statistical Science, 1*(3), pp. 364–78. Retrieved from http://www.jstor.org/stable/2245476

Lavenda, B. H. (2001). Derivation of the Prior Distribution in Bayesian Analysis from the Principle of Statistical Equivalence. *Open Systems & Information Dynamics, 8*(2), 103–114. doi:https://doi.org/10.1023/A:1011910512406

Lazo, A. V., & Rathie, P. (1978). On the entropy of continuous probability distributions (Corresp.). *IEEE Transactions on Information Theory, 24*(1), 120–122. doi:https://doi.org/10.1109/tit.1978.1055832

Le Roux, Q. (2021, November 04). https://github.com/qlero/EMr. Retrieved July 27, 2023, from GitHub: https://github.com/qlero/EMr

Lee, J., Fan, Y., & Sisson, S. (2015). Bayesian threshold selection for extremal models using measures of surprise. *Computational Statistics and Data Analysi, 85*, 84–99. doi:https://doi.org/10.1016/j.csda.2014.12.004

Ley, C., Reinert, G., & Swan, Y. (2017). Distances between nested densities and a measure of the impact of the prior in Bayesian statistics. *The Annals of Applied Probability, 27*(1), 216–241. doi:https://doi.org/10.1214/16-AAP1202

Lian, C. (2017). *Information Fusion and Decision Making using Belief Functions.* Compiègne, France: Université de Technologie de Compiègne.

Liboff, R. L. (1974). Gibbs vs. Shannon entropies. *J Stat Phys, 11*, 343–357. doi:https://doi.org/10.1007/BF01009793

Liu, W. (2006, August). Analyzing the degree of conflict among belief functions. *Artificial Intelligence, 170* (11), 909–924. doi:https://doi.org/10.1016/j.artint.2006.05.002

Lu, Z., & Lou, W. (2022). Bayesian approaches to variable selection: a comparative study from practical perspectives. *The International Journal of Biostatistics, 18*(1), 83–108. doi:https://doi.org/10.1515/ijb-2020-0130

Martin, A. (2019). Conflict management in information fusion with belief functions. In E. Bossé, & G. Rogova (Eds.), *Information quality in information fusion and decision making* (pp. 79–97). Springer. doi:https://doi.org/10.1007/978-3-030-03643-0_4

Martin, R., Zhang, J., & Liu, C. (2010, June). Dempster–Shafer Theory and Statistical Inference with Weak Beliefs. *Statistical Science, 25*(1), 72–87. doi:https://doi.org/10.1214/10-STS322

Masnadi-Shirazi, H. (2011). *The design of Bayes consistent loss functions for classification.* Ph. D. Thesis, University of California, San Diego, San Diego. Retrieved April 17, 2023, from https://escholarship.org/uc/item/1cv1947c

Mercer, J. (1909). Functions of positive and negative type, and their connection the theory of integral equations. *Philosophical Transactions of the Royal Society of London. Series A., 209*(441–458), 415–446. doi:https://doi.org/10.1098/rsta.1909.0016

Mercier, D., Quost, B., & Denœux, T. (2005). Contextual Discounting of Belief Functions. In L. Godo (Ed.), *ECSQARU 2005: Symbolic and Quantitative Approaches to Reasoning with Uncertainty* (pp. 552–562). Barcelona, Spain: Springer. doi:https://doi.org/10.1007/11518655_47

Miranda, E., Couso, I., & Gil, P. (2005, July). Random sets as imprecise random variables. *Journal of Mathematical Analysis and Applications, 307*(1), 32–47. doi:https://doi.org/10.1016/j.jmaa.2004.10.022

Modirshanechi, A., Brea, J., & Gerstner, W. (2022). A taxonomy of surprise definitions. *Journal of Mathematical Psychology, 110*. doi:https://doi.org/10.1016/j.jmp.2022.102712

Morita, S., Thall, P. F., & Müller, P. (2010, July 1). Evaluating the Impact of Prior Assumptions in Bayesian Biostatistics. *Stat Biosci., 2*(1), 1–17. doi:https://doi.org/10.1007/s12561-010-9018-x

Muñoz-Cobo, J.-L., Mendizábal, R., Miquel, A., Berna, C., & Escrivá, A. (2017). Use of the Principles of Maximum Entropy and Maximum Relative Entropy for the Determination of Uncertain Parameter Distributions in Engineering Applications. *Entropy, 19*(9). doi:https://doi.org/10.3390/e19090486

Murphy, K. P. (2007). *Conjugate Bayesian analysis of the Gaussian distribution.* Retrieved April 23, 2023, from https://www.cs.ubc.ca/~murphyk/Papers/bayesGauss.pdf

Nambiar, K. K., Varma, P. K., & Saroch, V. (1992). An axiomatic definition of Shannon's entropy. *Appl. Math. Lett., 5*(4), 45–46. doi:https://doi.org/10.1016/0893-9659(92)90084-M

Nassar, M., Alotaibi, R., Okasha, H., & Wang, L. (2022). Bayesian Estimation Using Expected LINEX Loss Function: A Novel Approach with Applications. *Mathematics, 10*. doi:https://doi.org/10.3390/math10030436

Neyman, J., & Pearson, E. S. (1933). On the problem of the most efficient tests of statistical hypotheses. *Philosophical Transactions of the Royal Society A, 231*(694–706), pp. 289–337. doi:https://doi.org/10.1098/rsta.1933.0009

Nguyen, H. T. (1977). *On Random sets and Belief Functions.* University of California, Berkeley, EECS Department. Retrieved from http://www2.eecs.berkeley.edu/Pubs/TechRpts/1977/28879.html

Nguyen, H. T. (1978). On Random Sets and Belief Functions. *Journal of Mathematical Analysis and Applications, 65*, 531–542. doi:https://doi.org/10.1016/0022-247X(78)90161-0

Nguyen, H., & Wang, T. (1997). Belief Functions and Random Sets. In J. Goutsias, R. Mahler, & H. Nguyen (Eds.), *Random Sets* (pp. 243–255). Springer. doi:https://doi.org/10.1007/978-1-4612-1942-2_11

Ni, S., Lei, Y., & Tang, Y. (2020). Improved Base Belief Function-Based Conflict Data Fusion Approach Considering Belief Entropy in the Evidence Theory. *Entropy, 22*(8). doi:https://doi.org/10.3390/e22080801

O'Hara, R., & Sillanpää, M. J. (2009, March). A Review of Bayesian Variable Selection Methods: What, How and Which. *Bayesian Analysis, 4*(1), 85–118. doi:https://doi.org/10.1214/09-BA403

Ostwald, D., Spitzer, B., Guggenmos, M., Schmidt, T. T., Kiebel, S. J., & Blankenburg, F. (2012). Evidence for neural encoding of Bayesian surprise in human somatosensation. *NeuroImage, 62*(1), 177–188. doi:https://doi.org/10.1016/j.neuroimage.2012.04.050

Palm, G. (2023). *Novelty, Information and Surprise.* Germany: Springer-Verlag. doi:https://doi.org/10.1007/978-3-662-65875-8

Pearson, K. (1900). On the criterion that a given system of deviations from the probable in the case of a correlated system of variables is such that it can be reasonably supposed to have arisen from random sampling. *Philosophical Magazine Series 5, 50*(302), pp. 157–175. doi:https://doi.org/10.1080/14786440009463897

Pearson, K. (1904). On the theory of contingency and its relation to association and normal correlation. In B. Series, *Mathematical Contributions to the Theory of Evolution* (Vol. 1, pp. 1–35). Drapers' Company Research Memoirs . Retrieved July 1, 2023, from https://ia800207.us.archive.org/16/items/cu31924003064833/cu31924003064833.pdf

Planck, M. K. (1901). Über das Gesetz der Energieverteilung im Normalspektrum. *Annalen der Physik, 309*(3), 553–563. doi:https://doi.org/10.1002/andp.19013090310

Planck, M. K. (1914). *The Theory of Heat Radiation.* (M. Masius, Trans.) Philadelphia: P. Blakiston's Son & Co.

Press, S. J., & Tanur, J. M. (2001). *The Subjectivity of Scientists and the Bayesian Approach.* New York: John Wiley & Sons.

Quiroga-Martinez, D., Hansen, N., Højlund, A., Pearce, M., Brattico, E., & Vuust, P. (2020). Decomposing neural responses to melodic surprise in musicians and non-musicians: Evidence for a hierarchy of predictions in the auditory system. *NeuroImage*. doi:https://doi.org/10.1016/j.neuroimage.2020.116816

Rahman, A., Gao, J., D'Este, C., & Ahmed, S. E. (2016, September). An Assessment of the Effects of Prior Distributions on the Bayesian Predictive Inference. *International Journal of Statistics and Probability, 5*(5), 31–42. doi:https://doi.org/10.5539/ijsp.v5n5p31

Rao, C. R. (1948, January). Large sample tests of statistical hypotheses concerning several parameters with applications to problems of estimation. *Mathematical Proceedings of the Cambridge Philosophical Society, 44*(1), pp. 50–57 . doi:https://doi.org/10.1017/S0305004100023987

Rényi, A. (1961). On Measures of Entropy and Information. In J. Neyman (Ed.), *4th Berkeley Symposium on Mathematics, Statistics and Probability,. 1*, pp. 547–561. University of California Press.

Risti, B., & Smets, P. (2006). Belief function theory on the continuous space with an application to model based classification. In B. Bouchon-Meunier, G. Coletti, & R. R. Yager (Ed.), *Modern Information Processing: From Theory to Applications. IPMU'04* (pp. 11–24). Perugia, Italy.: Elsevier. doi:https://doi.org/10.1016/B978-044452075-3/50002-9

Robert, C. P. (1994). *The Bayesian Choice.* New York, NY, USA: Springer Verlag. doi:https://doi.org/10.1007/978-1-4757-4314-2

Ruggieri, A. (2020, October 2022). *Expectation-Maximisation.* Retrieved July 27, 2023, from GitHub: https://github.com/madlabunimib/Expectation-Maximisation/tree/master

Saravanan, R., & Levine, R. (2022). Surprisal analysis of diffusion processes. *Chemical Physics, 556*. doi:https://doi.org/10.1016/j.chemphys.2022.111450

Savchuk, V. P., & Tsokos, C. P. (2011). *Bayesian Theory and Methods with Applications.* Paris: Atlantis Press.

Schwarz, G. E. (1978). Estimating the dimension of a model. *Annals of Statistics,, 6*(2), 461–464. doi:https://doi.org/10.1214/aos/1176344136, MR 0468014

Shafer, G. (1976). *A Mathematical Theory of Evidence* . New Jersey: Princeton University Press.
Shafer, G. (1990). Perspectives on the Theory and Practice of Belief Functions. *International Journal of Approximate Reasoning, 4*(5-6), 323–362. doi:https://doi.org/10.1016/0888-613X(90)90012-Q
Shannon, C. E. (1948a). A Mathematical Theory of Communication. *The Bell System Technical Journal, 27*(3), 379–423. doi:https://doi.org/10.1002/j.1538-7305.1948.tb01338.x
Shannon, C. E. (1948b). A Mathematical Theory of Communication. *Bell System Technical Journal, 27*(4), 623–666. doi:https://doi.org/10.1002/j.1538-7305.1948.tb00917.x
Shannon, C. E. (n.d.). *A mathematical theory of comunication—Nokia Bell Labs*. Retrieved 2 13, 2023, from https://www.bell-labs.com/claude-shannon/assets/images/discoveries: https://www.bell-labs.com/claude-shannon/assets/images/discoveries/1948-04-21-a-mathematical-theory-of-communication-parts-I-and-carousel-01.pdf
Shore, J., & Johnson, R. (1980, January). Axiomatic derivation of the principle of maximum entropy and the principle of minimum cross-entropy. *IEEE Transactions on Information Theory, 26*(1), 26–37. doi:https://doi.org/10.1109/TIT.1980.1056144.
Smets, P. (1990). Constructing the Pignistic Probability Function in a Context of Uncertainty. *Machine Intelligence and Pattern Recognition, 10*, 29–39. doi:https://doi.org/10.1016/B978-0-444-88738-2.50010-5
Smets, P. (2000). Data fusion in the transferable belief model. *Proceedings of the Third International Conference on Information Fusion. 1*, pp. 21–33. Paris, France: IEEE. doi:https://doi.org/10.1109/IFIC.2000.862713.
Smets, P. (2005). Belief functions on real numbers. *International Journal of Approximate Reasoning, 40*, 181–223. doi:https://doi.org/10.1016/j.ijar.2005.04.001
Sohrab, S. H. (2014). Boltzmann entropy of thermodynamics versus Shannon entropy of information theory. *International Journal of Mechanics, 8*, 73–84. Retrieved February 15, 2023, from https://www.naun.org/main/NAUN/mechanics/2014/a182003-086.pdf
Soize, C. (2005, April). Random matrix theory for modeling uncertainties in computational mechanics. *Computer Methods in Applied Mechanics and Engineering, 194*(12–16), 1333–1366. doi:https://doi.org/10.1016/j.cma.2004.06.038
Soize, C. (2008). Construction of probability distributions in high dimension using the maximum entropy principle: Applications to stochastic processes, random fields and random matrices. *Int. J. Numer. Meth. Engng, 76*, 1583–1611. doi:https://doi.org/10.1002/nme.2385
Souza de Cursi, E. (2023). *Uncertainty Quantification using R*. Springer Cham.
Souza de Cursi, E., & Sampaio, R. (2015). *Uncertainty Quantification and Stochastic Modeling with Matlab*. ISTE/Elsevier. doi:https://doi.org/10.1016/C2014-0-04713-2
Stigler, S. M. (1982). Thomas Bayes's Bayesian Inference. *Journal of the Royal Statistical Society. Series A (General), 145*(2), pp. .250–258.
Strat, T. (1984). Continuous belief functions for evidential reasoning. *Proceedings of the 4th National Conference on Artificial Intelligence*. Austin, Texas. Retrieved December 18, 2022, from https://www.aaai.org/Papers/AAAI/1984/AAAI84-035.pdf
Strat, T. (1987). The Generation of Explanations within Evidential Reasoning Systems. In J. P. McDermott (Ed.), *10th. International Joint Conference on Artificial Intelligence (IJCAI)*, (pp. 1097–1104). Milan, Italy. Retrieved December 20, 2022, from https://www.ijcai.org/Proceedings/87-2/Papers/104.pdf
Strat, T. (1990). Decision analysis using belief functions. *International Journal of Approximate Reasoning, 4*(5-6), 391–417. doi:https://doi.org/10.1016/0888-613X(90)90014-S
Taillandier, P., & Therond, O. (2011). Use of the Belief Theory to formalize Agent DecisionMaking Processes : Application to cropping Plan Decision Making. *European Simulation and Modelling Conference*, (pp. 138–142). Guimaraes, Portugal. Retrieved 12 17, 2022, from https://hal.archives-ouvertes.fr/hal-00688405
Thrun, M. C. (2020, February 3). *Short Intro into Gaussian Mixture Models*. Retrieved from https://cran.r-project.org/web/packages/AdaptGauss: https://cran.r-project.org/web/packages/AdaptGauss/vignettes/AdaptGauss.html

Turing, A. M. (1941). *The Applications of Probability to Cryptography.* Bletchley Park. Available at Archive.org and https://www.nationalarchives.gov.uk/. Retrieved from https://archive.org/details/hw-25-37

Turing, A. M. (2015, May 26). *The Applications of Probability to Cryptography.* doi:https://doi.org/10.48550/arXiv.1505.04714

Wald, A. (1939). Contributions to the Theory of Statistical Estimation and Testing Hypotheses. *Annals of Mathematical Statistics, 10* (4), pp. 299–326. doi:https://doi.org/10.1214/aoms/1177732144

Wald, A. (1943, November). Tests of Statistical Hypotheses concerning several parameters when the number of observations is large. *Transactions of the American Mathematical Society, 54*(3), pp. 426–482. doi:https://doi.org/10.2307/1990256

Wang, L. (2008). *K arhunen-Loeve E xpansions and their Applications.* The London School of Economics and Political Science. Ann Arbor, MI, USA: ProQuest LLC. Retrieved March 25, 2023, from http://etheses.lse.ac.uk/2950/1/U615901.pdf

Wasserman, L. A. (1990, September). Belief functions and statistical inference. *Canadian Journal of Statistics, 18*(3), 183–196. doi:https://doi.org/10.2307/3315449

Wesner, J. S., & Pomeranz, J. P. (2021). Choosing priors in Bayesian ecological models by simulating from the prior predictive distribution. *Ecosphere, 12*(9). doi:https://doi.org/10.1002/ecs2.3739

Wilks, S. S. (1938, March). The Large-Sample Distribution of the Likelihood Ratio for Testing Composite Hypotheses. *Ann. Math. Statist., 9*(1), pp. 60–62. doi:https://doi.org/10.1214/aoms/1177732360

Winkler, R. L. (1966). *The assessment of prior distributions in Bayesian analysis.* Chicago, IL, USA: University of Chicago.

Winkler, R. L. (1967, September). The Assessment of Prior Distributions in Bayesian Analysis. *Journal of the American Statistical Association, 62*(319), 776–800. doi:https://doi.org/10.2307/2283671

Wu, D. (2023, January 11). *em: A Generic Function of the EM Algorithm for Finite Mixture Models in R.* Retrieved July 27, 2023, from https://rdrr.io/cran/em/: https://rdrr.io/cran/em/f/inst/doc/em_intro.pdf; https://cran.r-project.org/web/packages/em/vignettes/em_intro.pdf

Wu, W.-Z., & Mi, J.-S. (2008). An Interpretation of Belief Functions on Infinite Universes in the Theory of Rough Sets. In C. Chan, J. Grzymala-Busse, & W. P. Ziarko (Ed.), *Rough Sets and Current Trends in Computing. RSCTC 2008* (pp. 71–80). Akron, OH, USA: Springer. doi: https://doi.org/10.1007/978-3-540-88425-5_8

Yakovenko, V. M. (2010). Statistical Mechanics of Money, Debt, and Energy Comsumption. *Science and Culture, 76*(9–10), 430–436. doi:https://doi.org/10.48550/arXiv.1008.2179

Yakovenko, V. M., & Rosser, J. B. (2009, December 2). Statistical mechanics of money, wealth, and income. *Rev. Mod. Phys., 81*(4), 1703–1725. doi:https://doi.org/10.1103/RevModPhys.81.1703

Yao, Y. Y., & Lingras, P. J. (1998). Interpretations of Belief Functions in the Theory of Rough Sets. *Information Sciences, 104*(1–2), 81–106. doi:https://doi.org/10.1016/S0020-0255(97)00076-5

Yosida, K. (1995). *Functional Analysis.* Berlin Heidelberg: Springer-Verlag . doi:https://doi.org/10.1007/978-3-642-61859-8

Zanchini, E., & Beretta, G. P. (2008). Rigorous Axiomatic Definition of Entropy Valid Also for Non-Equilibrium States. In G. P. Beretta, A. Ghoniem, & G. Hatsopoulos (Ed.), *MEETING THE ENTROPY CHALLENGE: An International Thermodynamics Symposium in Honor and Memory of Professor Joseph H. Keenan. 1033.* Cambridge, MA, USA: AIP Conference Proceedings. doi:https://doi.org/10.1063/1.2979048

Zellner, A. (1986a, June). Bayesian Estimation and Prediction Using Asymmetric Loss Functions. *Journal of the American Statistical Association, 81*(394), pp. 446–451. doi:https://doi.org/10.2307/2289234

Zellner, A. (1986b). On Assessing Prior Distributions and Bayesian Regression Analysis with g Prior Distributions. In P. Goel, & A. Zellner (Eds.), *Bayesian Inference and Decision Techniques: Essays in Honor of Bruno de Finetti.* (pp. 233–243). Elsevier.

Zhang, H., & Deng, Y. (2020). Weighted belief function of sensor data fusion in engine fault diagnosis. *Soft Comput, 24*, 2329–2339. doi:https://doi.org/10.1007/s00500-019-04063-7

Zhang, Y.-Y. (2020). The Bayesian Posterior Estimators under Six Loss Functions for Unrestricted and Restricted Parameter Spaces. In N. Tang (Ed.), *Bayesian Inference on Complicated Data.* doi:https://doi.org/10.5772/intechopen.88587

Zhou, K., Martin, A., & Pan, Q. (2018). A belief combination rule for a large number of sources. . *Journal of Advances in Information Fusion, 13*(2). Retrieved December 2022, 20, from https://www.researchgate.net/deref/https%3A%2F%2Fhal.archives-ouvertes.fr%2Fhal-01883239

Zhu, H. (2016, November 6). *EM Algorithm Implementation.* Retrieved July 27, 2023, from RPubs: https://rpubs.com/H_Zhu/246450

Zondervan-Zwijnenburg, M., Peeters, M., Depaoli, S., & Van de Schoot, R. (2017). Where Do Priors Come From? Applying Guidelines to Construct Informative Priors in Small Sample Research. *Research in Human Development, 14*(4), 305–320. doi:https://doi.org/10.1080/15427609.2017.1370966

Chapter 6
Sequential Bayesian Estimation

In many situations, the evolution of a system is described by a state vector $\boldsymbol{x}$, which varies in time. Such a variation can be described by a function $t \to \boldsymbol{x}(t)$ associating with time $t \in (0, T)$ the state vector $\boldsymbol{x}(t)$ (continuous time formulation), but it is generally discretized and reduced to the determination of a sequence of states $\boldsymbol{x}^{(0)}$, $\boldsymbol{x}^{(1)}, \boldsymbol{x}^{(2)}, \ldots, \boldsymbol{x}^{(k)}, \boldsymbol{x}^{(k+1)}, \ldots$, where $\boldsymbol{x}^{(k)} = \boldsymbol{x}(t^{(k)})$, $t^{(k)} = (k-1)\Delta t$, with a time step system $\Delta t > 0$ (discrete time formulation). The evolution may be caused by internal changes or external actions, which are modeled by a function $t \to \boldsymbol{u}(t)$, which corresponds to the discretized values $\boldsymbol{u}^{(0)}, \boldsymbol{u}^{(1)}, \boldsymbol{u}^{(2)}, \ldots, \boldsymbol{u}^{(k)}, \boldsymbol{u}^{(k+1)}, \ldots$, where $\boldsymbol{u}^{(k)} = \boldsymbol{u}(t^{(k)})$. We are often interested in systems having their evolution governed by a state equation:

$$\boldsymbol{x}^{(k+1)} = \varphi\left(t^{(k)}, \boldsymbol{x}^{(k)}, \boldsymbol{u}^{(k)}\right). \tag{6.1}$$

Such an equation can be generated, for instance, by the discretization of a differential equation. In some situations, $\boldsymbol{x}^{(k)}$ collects information from the past, with a finite horizon: $\boldsymbol{x}^{(k)} = (\boldsymbol{s}^{(k)}, \boldsymbol{s}^{(k-1)}, \ldots, \boldsymbol{s}^{(k-p)})$.

In such a situation, we look for the estimation of $\boldsymbol{x}^{(k+1)}$ using these equations and some measured data – such a problem is known as the state estimation problem. The measured data generally contains measurement errors so that it is generally uncertain. To include the uncertainty, it is usual to introduce a random vector $\boldsymbol{w}^{(k)}$ and rewrite Eq. (6.1) as:

$$\boldsymbol{x}^{(k+1)} = f\left(t^{(k)}, \boldsymbol{x}^{(k)}, \boldsymbol{u}^{(k)}, \boldsymbol{w}^{(k)}\right). \tag{6.2}$$

The reader will find in the literature many approaches to solving this problem, such as, for instance – a few among many: Kalman filtering for linear systems (Kalman 1960a, b; Kalman and Bucy 1961; Tusell 2011), or nonlinear (Julier and

E. Souza de Cursi, *Uncertainty Quantification with R*, International Series in Operations Research & Management Science 352,
https://doi.org/10.1007/978-3-031-48208-3_6

Uhlmann 2004) with Gaussian Noise; Particle filtering (Del Moral 1996, 1997; Del Moral and Doucet 2014; Dahlin and Schön 2019), UQ approaches (Lopez et al. 2018).

Since $\boldsymbol{w}^{(k)}$ is a random variable, $\boldsymbol{x}^{(k+1)}$ is also a random variable so that we can look for its distribution and use it to determine an estimation of $\boldsymbol{x}^{(k+1)}$. Many of the existing methods are based on Bayesian principles, such as the Bayesian Update of a prior distribution. In this chapter, we examine some methods for the numerical determination of the distribution of $\boldsymbol{x}^{(1)},\boldsymbol{x}^{(2)},\ldots,\boldsymbol{x}^{(k)},\boldsymbol{x}^{(k+1)},\ldots$. In the general situation, $\boldsymbol{x}$ is not observed directly, but a variable $\boldsymbol{y}$ is measured, and we have an analogous equation for its evolution:

$$\boldsymbol{y}^{(k)} = g\left(t^{(k)}, \boldsymbol{x}^{(k)}, \boldsymbol{v}^{(k)}\right). \tag{6.3}$$

Here, analogously to Eq. (6.2), $\boldsymbol{v}$ is a random vector modeling measurement errors. In many practical applications, $\boldsymbol{y}^{(k)} = \boldsymbol{x}^{(k)} + \boldsymbol{v}^{(k)}$ (additive noise) or $\boldsymbol{y}^{(k)} = (1 + \boldsymbol{v}^{(k)})\boldsymbol{x}^{(k)}$ (multiplicative noise). In such a situation, the system can be modeled by the single Eq. (6.2) – then, we consider $\boldsymbol{v}^{(k)}$ as part of $\boldsymbol{w}^{(k)}$, i.e., $\tilde{\boldsymbol{w}}^{(k)} = \left(\boldsymbol{w}^{(k)}, \boldsymbol{v}^{(k)}\right)$ replaces $\boldsymbol{w}^{(k)}$ in Eq. (6.2).

The main idea in filtering is to determine the distribution $f(\boldsymbol{x}^{(k+1)}|\boldsymbol{y}^{(k+1)},\ldots,\boldsymbol{y}^{(k-h)})$, where $h \geq 0$ is a convenient horizon. This can be achieved by the Bayesian Update:

$$f\left(\boldsymbol{x}^{(k)}|\boldsymbol{y}^{(k)},\ldots,\boldsymbol{y}^{(k-h)}\right) = \frac{f\left(\boldsymbol{y}^{(k)}|\boldsymbol{x}^{(k)}\right)f\left(\boldsymbol{x}^{(k)}|\boldsymbol{y}^{(k-1)},\ldots,\boldsymbol{y}^{(k-h)}\right)}{f\left(\boldsymbol{y}^{(k)},\ldots,\boldsymbol{y}^{(k-h)}\right)}.$$

Prediction can be made using $f(\boldsymbol{x}^{(k+1)}|\boldsymbol{y}^{(k)},\ldots,\boldsymbol{y}^{(k-h)})$.

6.1 Monte-Carlo Approach

As shown in the preceding chapter, Bayesian estimation of a vector of parameters $\boldsymbol{\theta}$ using the data $\mathcal{Y}$ is carried by determining a posterior distribution $f_{po}(\boldsymbol{\theta}|\mathcal{Y})$ and using it to determine the value $\widehat{\boldsymbol{\theta}}$ that minimizes the mean value of a loss function $\ell\left(\boldsymbol{\theta},\widehat{\boldsymbol{\theta}}\right)$:

$$\widehat{\theta} = \arg\min\left\{J(\boldsymbol{\eta}) = \int f_{po}(\boldsymbol{\theta}|\mathcal{Y})\ell(\boldsymbol{\theta},\boldsymbol{\eta})d\boldsymbol{\theta}\right\}.$$

The evaluation of $J(\boldsymbol{\eta})$ requires the evaluation of an integral. In Chap. 5, the integration was carried by classical numerical methods such as Trapezoidal or Romberg integration. An alternative approach consists in using Monte-Carlo integration. Indeed, let $\Omega \subset \mathbb{R}^n, f: \Omega \to \mathbb{R}$, and

$$I = \int_{\Omega} f(\boldsymbol{x}) d\boldsymbol{x}.$$

Let U be a random vector, uniformly distributed on Ω. Then, its density is $f_U(\boldsymbol{x}) = 1/|\Omega|$, where $|\Omega|$ is the measure of Ω. Thus,

$$E(f(\boldsymbol{U})) = \int_{\Omega} f(\boldsymbol{x}) f_U(\boldsymbol{x}) d\boldsymbol{x} = \frac{I}{|\Omega|}, \tag{6.4}$$

and

$$I = |\Omega| E(f(\boldsymbol{U})) \approx \frac{|\Omega|}{N} \sum_{i=1}^{N} f\left(\boldsymbol{x}^{(i)}\right), \tag{6.5}$$

where $\mathcal{X} = \left(\boldsymbol{x}^{(1)}, \boldsymbol{x}^{(2)}, \ldots, \boldsymbol{x}^{(N)}\right)$ is a sample from the uniform distribution on Ω. This method can be improved by using an approach by importance sampling: let Z be a second random variable defined on Ω having as density f_Z (strictly positive on Ω), and

$$w(\boldsymbol{x}) = \frac{f_U(\boldsymbol{x})}{f_Z(\boldsymbol{x})}. \tag{6.6}$$

Then:

$$\int_{\Omega} f(\boldsymbol{x}) f_U(\boldsymbol{x}) d\boldsymbol{x} = \int_{\Omega} \frac{f_U(\boldsymbol{x})}{f_Z(\boldsymbol{x})} f(\boldsymbol{x}) f_Z(\boldsymbol{x}) d\boldsymbol{x} = \int_{\Omega} w(\boldsymbol{x}) f(\boldsymbol{x}) f_Z(\boldsymbol{x}) d\boldsymbol{x}. \tag{6.7}$$

Thus,

$$I = E(w(\mathbf{Z}) f(\mathbf{Z})) \approx \frac{|\Omega|}{N} \sum_{i=1}^{N} w\left(z^{(i)}\right) f\left(z^{(i)}\right), \tag{6.8}$$

where $\mathcal{Z} = \left(z^{(1)}, z^{(2)}, \ldots, z^{(N)}\right)$ is a sample from $\mathbf{Z}$. Ideally, $w(\boldsymbol{x}) \approx c$, independent from $\boldsymbol{x}$. In this case, $\int_{\Omega} f(\boldsymbol{x}) d\boldsymbol{x} = \int_{\Omega} w(\boldsymbol{x}) f_Z(\boldsymbol{x}) d\boldsymbol{x} = c \int_{\Omega} f_Z(\boldsymbol{x}) d\boldsymbol{x} = c$ since $\int_{\Omega} f_Z(\boldsymbol{x}) d\boldsymbol{x} = 1$.

As an alternative, we can consider a self-normalized version

$$I \approx \frac{|\Omega|}{N S_Z} \sum_{i=1}^{N} w\left(z^{(i)}\right) f\left(z^{(i)}\right), S_Z = \sum_{i=1}^{N} w\left(z^{(i)}\right). \tag{6.9}$$

Under this form, f_U, f_Z cannot be normalized – but we must be able to generate a sample from **Z**.

Example 6.1 Let us consider the function $f(x) = \exp(-|x - 1|/2)$ and $I = \int_{-5}^{5} f(x)dx$. The exact value is $I = \frac{2(-1-e+2e^3)}{e^3} \approx 3.6297552967910467$.

Let us use as importance distribution $N(1, 1)$:

```
a = -5
b = 5
f = function(x){
  if ( (x >= a) & (x <= b) ){
    exp(-abs(x-1)/2)
  }
  else{0}}
vecf = Vectorize(f)
vex = 2*(2*exp(3) - exp(1) - 1)/exp(3)
vtrap = int1$trapz1(f,a,b,100,1e-10)
vromb = int1$romb1(f,a,b,100,1e-10)
fu = function(x){1/(b-a)}
sig = 1
fz = function(x){dnorm(x,1,sig)}
w = function(x){fu(x)/fz(x)}
vecw = Vectorize(w)
ns = 1e5
nforget = 1e3
set.seed(NULL)
zs = rnorm(ns + nforget,1, sig)
zs = zs[(nforget+1):(ns + nforget)]
ws = vecw(zs)
fs = vecf(zs)
sn = mean(ws)
vs = (b-a)*mean(ws*fs)
vsu = vs/sn
```

An example of the result is the following one:

```
## [1] "Trapezoidal :  3.62975529672647"
## [1] "Romberg :  3.62974778470075"
## [1] "Monte-Carlo with N(1,1) :  3.59269117778257"
## [1] "... self-normalized :  3.63119367813341"
## [1] "... mean :  3.61194242795799"
```

The parameter `nforget` indicates the number of initial steps which are not considered in the evaluation of the integral. Its influence is illustrated in

(continued)

Example 6.1 (continued)

Tables 6.1 and 6.2: we evaluated the integral using `nforget` going from 0 to $1e4$ and determined the distribution of the estimated values.

Table 6.1 Example of analysis of the influence of `nforget`

Min	Max	Mean	Standard deviation
3.35288	3.51437	3.42529	0.04158

Statistics of the estimation of I for `nforget` going from 0 to 1e4 and $ns = 1e4$

Table 6.2 Example of analysis of the influence of `nforget`

Min	Max	Mean	Standard deviation
3.69040	3.72213	3.70035	0.00839

Statistics of the estimation of I for `nforget` going from 0 to 1e4 and $ns = 1e5$

The result of the Monte-Carlo evaluation is a variate of a random variable. For instance, we can repeat the evaluation to get a sample of the result and evaluate the distribution of the estimation of the integral, namely its mean and variance. The results of the runs are collected in a numeric vector and we use the sample generated by this way to determine its CDF and its PDF.

An example of result is shown in Fig. 6.1. We see that the PDF falls to zero very quickly (and the CDF goes to 1 very quickly) – such a behavior indicates the presence of outliers, i.e., a few values that are not significant since they are much larger to correspond to the real value of the integral. Since these values are just a few, the results remain convenient: the estimation is $I \approx 3.59$.

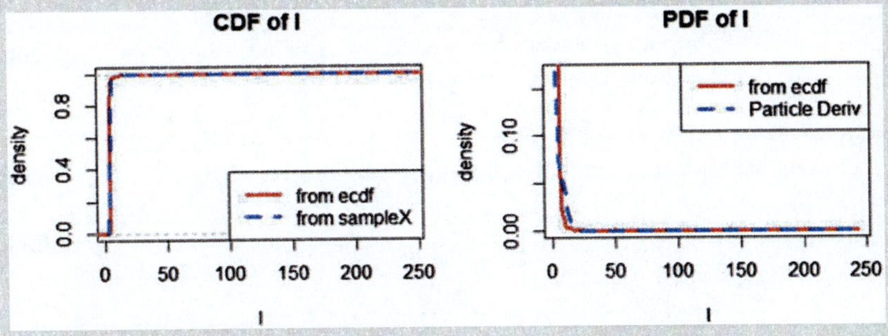

Fig. 6.1 Example of variability of the results and the existence of outliers

Of course, we can also eliminate outliers: for instance, we can consider that 1% of the results are outliers and eliminate them. Eliminating the 1% larger results leads to the results in Fig. 6.2. In this case, the estimation is $I \approx 3.49$.

(continued)

Example 6.1 (continued)

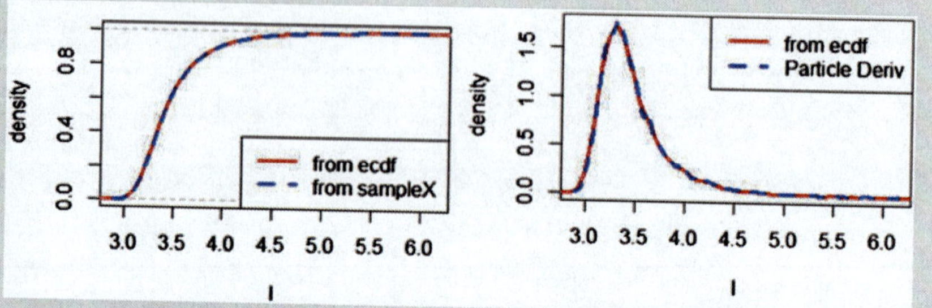

Fig. 6.2 Variability of the results after elimination of the 1% larger results, considered as outliers

Eliminating the 5% larger results, the estimation is $I \approx 3.43$.
The results also depend on *ns*, as shown in Table 6.3.

Table 6.3 Illustration of the variability in the Monte-Carlo evaluation of the integral

Number of realizations	Mean	Standard deviation	Confidence interval $\alpha = 0.05$
100	2.99	1.76	(2.64, 3.33)
1000	3.34	5.18	(2.95, 3.74)
10000	3.46	10.72	(3.25, 3.67)

Example 6.2 Let us consider the function $f(\boldsymbol{x}) = \exp(-\|\boldsymbol{x} - 1\|/2)$ and $I = \int_\Omega f(\boldsymbol{x})d\boldsymbol{x}, \Omega = \{\boldsymbol{x} \in \mathbb{R}^3 : \|\boldsymbol{x} - 1\| \leq 5\}$. We have $I = 4\pi(16 - 106/\exp(5/2)) \approx 91.7218152835168$.

Let us use as importance distribution $N(1, \sigma^2 \boldsymbol{Id})$, $\sigma = 3$.

(continued)

Example 6.2 (continued)

```
library("mvtnorm")

vect = function(f){
  ff = function(x){
    dd = dim(x)
    rrr = matrix(nrow = dd[[1]], ncol = 1)
    for ( i in 1: dd[[1]]){
      rrr[[i]] = f(x[i,])
    }
    return(rrr)
  }
  return(ff)}

source("ints.R")
int1 = ints$new()

ndim = 3
a = 0
b = 5

surf = 4*pi*(b^3 - a^3)/3
fu = function(x){1/surf}
f = function(x){
  r = sqrt(sum((x-1)^2))
  if ( r <= 5){
        rrr = exp(-0.5*r)}
  else{ rrr = 0}
  return(rrr)}
vecf = vect(f)
vex = 91.7218152835168
fr = function(r){4*pi*r^2*exp(-0.5*r)}
vtrap = int1$trapz1(fr,a,b,100,1e-10)
vromb = int1$romb1(fr,a,b,100,1e-10)

sig = ndim
fz = function(x){dmvnorm(x,rep(1,ndim),sig^2*diag(ndim))}

w = function(x){fu(x)/fz(x)}
vecw = vect(w)

set.seed(NULL)
ns = 1e5
nforget = 0
zs = rmvnorm(ns + nforget,rep(1,ndim),sig^2*diag(ndim))
ws = vecw(zs[(nforget+1):(ns +nforget), ])
fs = vecf(zs[(nforget+1):(ns +nforget), ])
vs = surf*mean(ws*fs)
```

(continued)

Example 6.2 (continued)

An example of result with:

```
[1] " Trapezoidal :  91.7218152834976"
[1] " Romberg :  91.7218152834956"
[1] " Monte-Carlo avec N(1, Id) =  91.6718266893683"
[1] " exact :  91.7218152835168"
```

As previously observed, the Monte-Carlo evaluation generates variates from a random variable. An example of distribution is shown in Fig. 6.3, obtained from a sample of $1E4$ variates, generated with $ns = 1E3$.

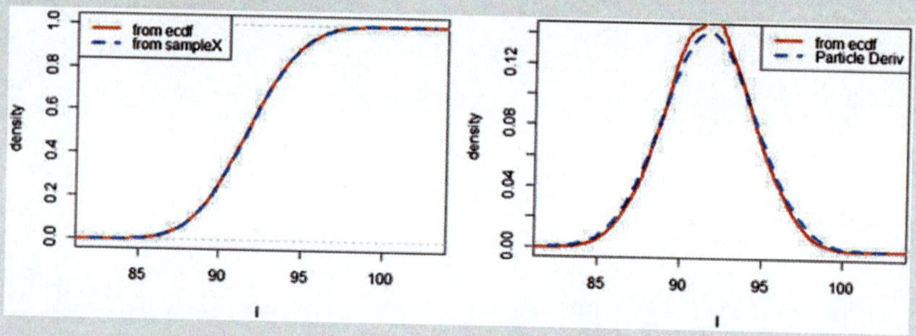

Fig. 6.3 Distribution of the estimation of the integral

The results also depend on *ns*, as shown in Table 6.4.

Table 6.4 Illustration of the variability with *ns* in the Monte-Carlo evaluation of the integral

Number of realizations	Mean	Standard deviation	Confidence interval $\alpha = 0.05$
100	99.26	77.52	(84.07, 114.46)
1000	90.35	82.66	(85.23, 95.47)
10000	90.41	83.13	(88.78, 92.04)

Exercises

1. Let $f(x) = 1_{(0,1)}(x) \sin(\pi x)$, where $1_{(0,1)}(x) = 1$, *if* $x \in (0, 1)$; $1_{(0,1)}(x) = 0$, *if* $\notin x$ (0, 1). Evaluate $I = E(f(X))$ using a Monte-Carlo method, when
   (a) X is uniformly distributed on (0,1).
   (b) X is Poisson distributed $\mathcal{P}(1)$.
   (c) X is Laplace distributed $\mathcal{L}\left(\frac{1}{2}, 1\right)$.
   (d) X is Gaussian $N(0, 1)$.
   - To generate an importance function, you can consider a polynomial expansion of f.

(continued)

2. Let $f(x) = 1_{(0,1)}(x)\left(\sin(\pi x) - \pi x + \frac{\pi}{2}\right)$, where $1_{(0,1)}(x) = 1$, *if* $x \in (0,1)$; $1_{(0,1)}(x) = 0$, *if* $\notin x\ (0,1)$. Evaluate $I = E(f(X))$ using a Monte-Carlo method, when X is uniformly distributed on (0,1).
3. Let $f(x) = 1_{(0,1)}(x) \exp(x)$, where $1_{(0,1)}(x) = 1$, *if* $x \in (0,1)$; $1_{(0,1)}(x) = 0$, *if* $\notin x\ (0,1)$. Evaluate $I = E(f(X))$ using a Monte-Carlo method, when

   (a) X is uniformly distributed on (0,1).
   (b) X is Poisson distributed $\mathcal{P}(1)$.
   (c) X is Laplace distributed $\mathcal{L}\left(\frac{1}{2}, 1\right)$.
   (d) X is Gaussian $N(0, 1)$.

   - To generate an importance function, you can consider a polynomial expansion of f.
4. Let $f(x) = 1_{(0,+\infty)}(x)\, x^2 \exp(-x)$, where $1_{(0,+\infty)}(x) = 1$, *if* $x \in (0,+\infty)$; $1_{(0,1)}(x) = 0$, *if* $\notin x\ (0,+\infty)$. Evaluate $I = E(f(X))$ using a Monte-Carlo method, when

   (a) X is uniformly distributed on (0,10).
   (b) X is Poisson distributed $\mathcal{P}(1)$.
   (c) X is Laplace distributed $\mathcal{L}(1, 1)$.
   (d) X is Gaussian $N(0, 1)$.

   - To generate an importance function, you can consider a polynomial expansion of f.
5. Let $f(x) = 1_{(0,+\infty)}(x)\sqrt{x}\exp(-x)$, where $1_{(0,+\infty)}(x) = 1$, *if* $x \in (0,+\infty)$; $1_{(0,1)}(x) = 0$, *if* $\notin x\ (0,+\infty)$. Evaluate $I = E(f(X))$ using a Monte-Carlo method, when

   (a) X is uniformly distributed on (0,10).
   (b) X is Poisson distributed $\mathcal{P}(1)$.
   (c) X is Laplace distributed $\mathcal{L}(1, 1)$.
   (d) X is Gaussian $N(0, 1)$.

   - To generate an importance function, you can consider a polynomial expansion of f.
6. Let $f(x) = 1_{(0,2\pi)}(x)\frac{1}{x^2 + \sin^2(2x)}$, where $1_{(0,2\pi)}(x) = 1$, *if* $x \in (0, 2\pi)$; $1_{(0,2\pi)}(x) = 0$, *if* $\notin x\ (0, 2\pi)$. Evaluate $I = E(f(X))$ using a Monte-Carlo method, when

   (a) X is uniformly distributed on $(0, 2\pi)$.
   (b) X is Poisson distributed $\mathcal{P}(1)$.
   (c) X is Laplace distributed $\mathcal{L}(1, 1)$.
   (d) X is Gaussian $N(0, 1)$.

   - To generate an importance function, you can consider a polynomial expansion of f.

6.2 Metropolis-Hastings Algorithm

The generation of random points may be difficult if the target density f_Z does not correspond to a standard distribution, implemented in R. To solve this difficulty, we must use a convenient algorithm, such as, for instance, the rejection method (see, for instance, (Robert and Casella 2004, 2010) or the popular Metropolis-Hastings algorithm, introduced in (Metropolis et al. 1953; Hastings 1970).

The Metropolis-Hastings algorithm furnishes a fast method to generate random vectors corresponding to a given density f_Z by simulating a random walk, whose moves are made according to a simplified metropolis dynamics.

The method is based on the classical approach of perturbation followed by decision of acceptation or rejection: given the actual point $\boldsymbol{x}^{(n)}$, we generate a trial point $\boldsymbol{x}_t$. Then, a decision is taken about the acceptation or rejection of the point $\boldsymbol{x}_t$: if $\boldsymbol{x}_t$ is accepted, then $\boldsymbol{x}^{(n+1)} = \boldsymbol{x}_t$, else $\boldsymbol{x}^{(n+1)} = \boldsymbol{x}^{(n)}$. The trial point $\boldsymbol{x}_t$ is generated using a distribution $p(\boldsymbol{x}_t|\boldsymbol{x})$ and the decision is taken according to a probability of acceptation:

$$p_{\text{accept}} = \min\left\{1, \frac{f_Z(x)}{f_Z(x_t)} \frac{p(\boldsymbol{x}_t|\boldsymbol{x})}{p(\boldsymbol{x}|\boldsymbol{x}_t)}\right\}.$$

To take the decision, a random number $u \sim U(0, 1)$ is generated and $\boldsymbol{x}_t$ is accepted if $u < p_{accept}$. A simple way to generate a trial point consists in generating a perturbation $\boldsymbol{dx}^{(n)}$, according to a density $q(\boldsymbol{dx})$ so that $\boldsymbol{x}_t = \boldsymbol{x}^{(n)} + \boldsymbol{dx}^{(n)}$. In this case, $p(\boldsymbol{x}_t|\boldsymbol{x}) = q(\boldsymbol{dx})$, $p(\boldsymbol{x}|\boldsymbol{x}_t) = q(-\boldsymbol{dx})$. For a symmetric distribution, $p(\boldsymbol{x}_t|\boldsymbol{x}) = p(\boldsymbol{x}|\boldsymbol{x}_t)$ so that

$$p_{\text{accept}} = \min\left\{1, \frac{f_Z(x)}{f_Z(x_t)}\right\}.$$

The reader can find on the net many implementations of the basic Metropolis-Hastings algorithm (for instance, (Navarro 2023; The Metropolis Algorithm; Neel 2014; Law 2019), packages `ErlisR`, `EMC`, `mcmc`, ...). Namely, you can find code at (Robert 2016; Dahlin and Schön 2019). An example of implementation, using a Gaussian q is the following:

```
mhdyn = function(x,pert,fz) {
  xt = pert(x)
  p_accept = min(1, fz(xt) / fz(x))
  u = runif(1)
  if(u < p_accept) {
    resu = xt
  } else {
    resu = x}
  return(resu)}
pert_gauss = function(x,sig){
  dx = rnorm(length(x),0,sig)
  return(x+dx)}
hmsampler = function(x0,ns,pert,nforget,fz) {
  xs = matrix(nrow = ns+nforget,ncol = length(x0))
  xs[1, ] = x0
  for(n in 2:(ns+nforget)) {
    xs[n,] = mhdyn(xs[n-1, ],pert,fz)
  }
  return(xs)}
```

Example 6.3 Let us consider the function $f(x) = \exp(-|x - 1|/2)$, and $f_Z(x) = f(x)/\int f(s)ds$ We can use the Metropolis-Hastings (MH) algorithm to generate a sample from f_Z as follows:

```
f = function(x) {
  exp(-abs(x-1)/2)}

ns = 10000
nforget = 0
pert = function(x){pert_gauss(x,1)}
set.seed(NULL)
x0 = rnorm(1)
xs = hmsampler(x0,ns,pert,nforget,f)
```

We use the nonnormalized f since

$$\frac{f_Z(x)}{f_Z(x_t)} = \frac{f(x)}{f(x_t)}.$$

(continued)

Example 6.3 (continued)

An example of result is shown in Fig. 6.4.

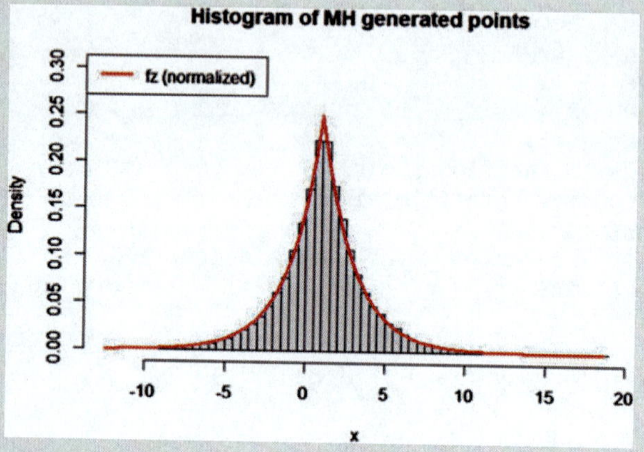

Fig. 6.4 Example of result furnished by the Metropolis-Hastings algorithm

We can limit the analysis to an interval (a, b) by modifying the definition of f_Z as follows:

```
f = function(x) {
  if ( (x >= a) & (x <= b) ){
  exp(-abs(x-1)/2)}
  else{0}}
```

An example of the result is shown in Fig. 6.5, with $a = -5$, $b = 5$.

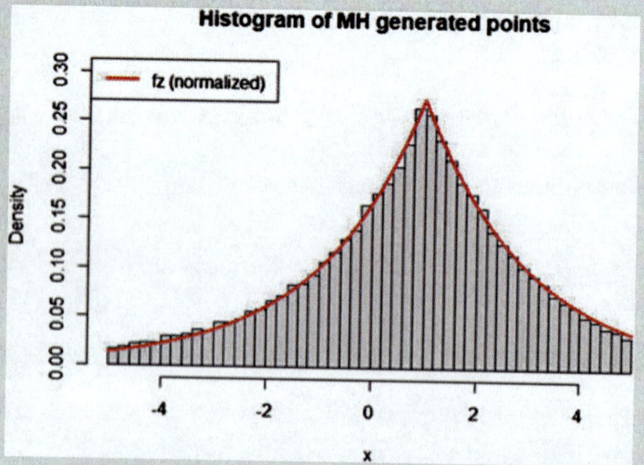

Fig. 6.5 Example of result furnished by the Metropolis-Hastings algorithm, with points generated only on the interval (-5,5)

(continued)

Example 6.3 (continued)

We can estimate

$$I = \int_a^b f(x)dx = \int_a^b \frac{f(x)}{f_Z(x)} f_Z(x)dx$$

as follows:

```
F = ecdf(xs)  # empirical CDF of the points generated
xk = knots(F)  # nodes
df = density(xk)  # fz is the density associated to F
xdf = df$x  # abscissa
ydf = df$y  # values of fz
vf = Vectorize(f)
vs = mean( vf(xdf)/ydf) # mean of f(x)/fz(x)
```

We obtain

```
## [1] "estimation of the integral =  3.44063038607349"
```

Of course, it is more efficient to use an importance distribution. Let us use $f_Z(x) = (1 - |x - 1|/6)/5$ *on* $(-5, 5)$:

```
a = -5
b = 5
f = function(x) {
  if ( (x >= a) & (x <= b) ){
    exp(-abs(x-1)/2)
  }
  else{0}}
fz = function(x) {
  if ( (x >= a) & (x <= b) ){
  (1 - abs(x-1)/6)/5
  }
  else{0}}
```

(continued)

Example 6.3 (continued)

In this case, the estimation of the integral is made as follows:

```
w = function(x){f(x)/fz(x)}
vw = Vectorize(w)
vs = mean( vw(xs))
```

We obtain:

```
## [1] "estimation of the integral =  3.20602349008448"
```

The distribution of the points generated is shown in Fig. 6.6.

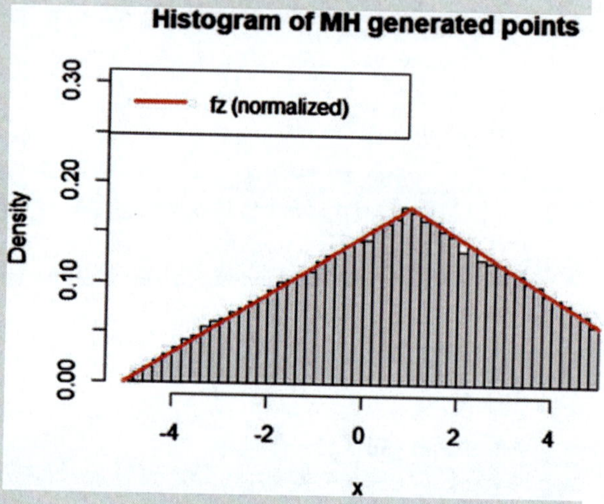

Fig. 6.6 Example of result furnished by the Metropolis-Hastings algorithm, with points generated only on the interval (-5,5) and $fz(x) = (1 - |x - 1|/6)/5$

Analogously to the preceding, MH estimates are random variables and present a variability according to the parameters chosen. For instance, an illustration of the variability with *ns* is shown in Table 6.5.

Table 6.5 Illustration of the variability with *ns* in the Monte-Carlo evaluation of the integral

Number of realizations	Mean	Standard deviation	Confidence interval $\alpha = 0.05$
100	3.19	0.89	(3.25, 3.43)
1000	3.21	0.91	(3.15, 3.27)
10000	3.20	1.1	(3.18, 3.22)

An example of distribution of the estimation is shown in Fig. 6.7, obtained from a sample of $1E4$ variates, generated with $ns = 1E4$. The mean is about $I \approx 3.2$.

(continued)

Example 6.3 (continued)

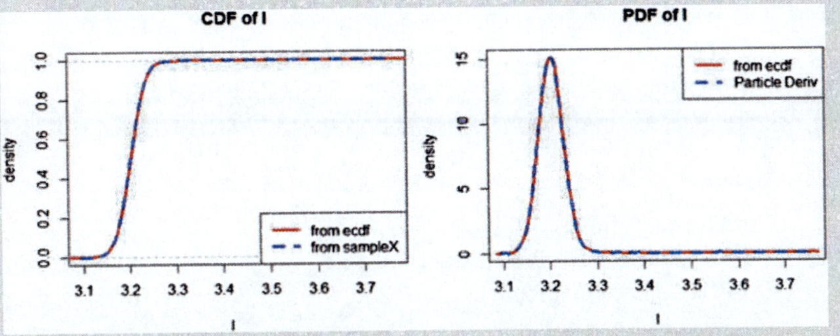

Fig. 6.7 Distribution of the estimation of I by MH, from a sample of size 1E4

Again, the CDF and the PDF show the existence of outliers. As in the preceding example, we can eliminate 1% of the extreme values, a half at left, a half at right. The resulting distribution appears in Fig. 6.8. The mean is yet about $I \approx 3.2$.

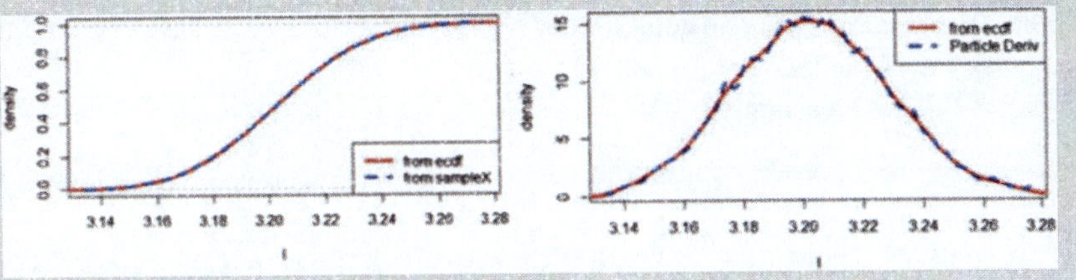

Fig. 6.8 Distribution of the estimation of I by MH, from a sample of size 1E4

Exercises

1. Let $f(x) = 1_{(0,\,1)}(x) \sin(\pi x)$, where $1_{(0,\,1)}(x) = 1$, *if* $x \in (0, 1)$; $1_{(0,\,1)}(x) = 0$, *if* $\notin x\ (0, 1)$. Evaluate $I = E(f(X))$ using a Metropolis-Hastings method, when

   (a) X is uniformly distributed on (0,1).
   (b) X is Poisson distributed $\mathcal{P}(1)$.
   (c) X is Laplace distributed $\mathcal{L}\left(\frac{1}{2}, 1\right)$.
   (d) X is Gaussian $N(0, 1)$.

(continued)

- To generate an importance function, you can consider a polynomial expansion of *f*.

2. Let $f(x)=1_{(0,1)}(x)\left(\sin(\pi x)-\pi x+\frac{\pi}{2}\right)$, where $1_{(0,1)}(x)=1$, *if* $x\in(0,1)$; $1_{(0,1)}(x)=0$, *if* $\notin x\ (0,1)$. Evaluate $I=E(f(X))$ using a Metropolis-Hastings method, when X is uniformly distributed on (0,1).
3. Let $f(x)=1_{(0,1)}(x)\exp(x)$, where $1_{(0,1)}(x)=1$, *if* $x\in(0,1)$; $1_{(0,1)}(x)=0$, *if* $\notin x\ (0,1)$. Evaluate $I=E(f(X))$ using a Metropolis-Hastings method, when

   (a) X is uniformly distributed on (0,1).
   (b) X is Poisson distributed $\mathcal{P}(1)$.
   (c) X is Laplace distributed $\mathcal{L}\left(\frac{1}{2},1\right)$.
   (d) X is Gaussian $N(0,1)$.

   - To generate an importance function, you can consider a polynomial expansion of *f*.

4. Let $f(x)=1_{(0,+\infty)}(x)\,x^2\exp(-x)$, where $1_{(0,+\infty)}(x)=1$, *if* $x\in(0,+\infty)$; $1_{(0,1)}(x)=0$, *if* $\notin x\ (0,+\infty)$. Evaluate $I=E(f(X))$ using a Metropolis-Hastings method, when

   (a) X is uniformly distributed on (0,10).
   (b) X is Poisson distributed $\mathcal{P}(1)$.
   (c) X is Laplace distributed $\mathcal{L}(1,1)$.
   (d) X is Gaussian $N(0,1)$.

   - To generate an importance function, you can consider a polynomial expansion of *f*.

5. Let $f(x)=1_{(0,+\infty)}(x)\sqrt{x}\exp(-x)$, where $1_{(0,+\infty)}(x)=1$, *if* $x\in(0,+\infty)$; $1_{(0,1)}(x)=0$, *if* $\notin x\ (0,+\infty)$. Evaluate $I=E(f(X))$ using a Metropolis-Hastings method, when

   (a) X is uniformly distributed on (0,10).
   (b) X is Poisson distributed $\mathcal{P}(1)$.
   (c) X is Laplace distributed $\mathcal{L}(1,1)$.
   (d) X is Gaussian $N(0,1)$.

   - To generate an importance function, you can consider a polynomial expansion of *f*.

6. Let $f(x)=1_{(0,2\pi)}(x)\frac{1}{x^2+\sin^2(2x)}$, where $1_{(0,2\pi)}(x)=1$, *if* $x\in(0,2\pi)$; $1_{(0,2\pi)}(x)=0$, *if* $\notin x\ (0,2\pi)$. Evaluate $I=E(f(X))$ using a Metropolis-Hastings method, when

   (a) X is uniformly distributed on $(0,2\pi)$.
   (b) X is Poisson distributed $\mathcal{P}(1)$.

(continued)

(c) X is Laplace distributed $\mathcal{L}(1, 1)$.
(d) X is Gaussian $N(0, 1)$.

- To generate an importance function, you can consider a polynomial expansion of f.

6.3 Kalman Filtering

As explained in the introduction, filtering consists in determining an estimation for the states of a system having its evolution given by recurrence equations, such as (6.2). Noise $\boldsymbol{w}$ is modeled as a random variable so that $\boldsymbol{x}$ is also a random variable. The main idea in filtering is to estimate the distribution of $\boldsymbol{x}^{(k)}$ by using equations (6.2) and (6.3) and the observations $\boldsymbol{y}^{(k)}, \ldots, \boldsymbol{y}^{(1)}$. In the particular case where the system is linear and the noise is addictive, we have

$$\boldsymbol{x}^{(k+1)} = \boldsymbol{A}^{(k)}\boldsymbol{x}^{(k)} + \boldsymbol{B}^{(k)}\boldsymbol{u}^{(k)} + \boldsymbol{w}^{(k)}, \boldsymbol{y}^{(k)} = \boldsymbol{C}^{(k)}\boldsymbol{x}^{(k)} + \boldsymbol{v}^{(k)}. \tag{6.10}$$

Let us assume that both the noises are Gaussian:

$$\boldsymbol{w}^{(k)} \sim \boldsymbol{N}\left(\boldsymbol{0}, \boldsymbol{Q}^{(k)}\right), \quad \boldsymbol{v}^{(k)} \sim \boldsymbol{N}\left(\boldsymbol{0}, \boldsymbol{R}^{(k)}\right). \tag{6.11}$$

Assume in addition that the noises $\boldsymbol{w}$, $\boldsymbol{v}$ are independent. Let $\widehat{\boldsymbol{x}}^{(k)}$ be an estimator for $\boldsymbol{x}^{(k)}$. In such a situation, filtering can be achieved by a Kalman filter – the reader can find in the literature many derivations of the basic equations of a Kalman Filter: for instance, (Grewal and Andrews 2001; Simons 2006; Gibbs 2011; Faragher 2012; Chui and Chen 2017). The basic idea is to look for $\widehat{\boldsymbol{x}}^{(k)}$ that minimizes the error $E\left(\left\|\widehat{\boldsymbol{x}}^{(k+1)} - \boldsymbol{x}^{(k+1)}\right\|^2\right)$. Adopting the Bayesian point of view, we can consider a prior

$$\widehat{\boldsymbol{x}_p}^{(k)} = \boldsymbol{A}^{(k-1)}\boldsymbol{x}^{(k-1)} + \boldsymbol{B}^{(k-1)}\boldsymbol{u}^{(k-1)} \tag{6.12}$$

to be corrected. We can look for a correction having the form

$$\widehat{\boldsymbol{x}}^{(k)} = \widehat{\boldsymbol{x}_p}^{(k)} + \boldsymbol{K}^{(k)}\widehat{\boldsymbol{v}}^{(k)} = \widehat{\boldsymbol{x}_p}^{(k)} + \boldsymbol{K}^{(k)}\left(\boldsymbol{y}^{(k)} - \boldsymbol{C}^{(k)}\widehat{\boldsymbol{x}_p}^{(k)}\right), \tag{6.13}$$

where $\boldsymbol{K}^{(k)}$ is a matrix to be determined. By determining the least-squares solution abovementioned, we obtain the Kalman gain matrix:

$$\boldsymbol{K}^{(k)} = \boldsymbol{P}^{(k)}\left(\boldsymbol{C}^{(k)}\right)^t\left(\boldsymbol{C}^{(k)}\boldsymbol{P}^{(k)}\left(\boldsymbol{C}^{(k)}\right)^t + \boldsymbol{R}^{(k)}\right)^{-1}, \tag{6.14}$$

where $\boldsymbol{P}^{(k)}$ is the covariance matrix of the error $\boldsymbol{e}^{(k)} = \widehat{\boldsymbol{x}}^{(k)} - \boldsymbol{x}^{(k)}$. We have, on the one hand,

$$\boldsymbol{P}^{(k)} = \boldsymbol{A}^{(k-1)}\boldsymbol{P}^{(k-1)}\left(\boldsymbol{A}^{(k-1)}\right)^t + \boldsymbol{Q}^{(k-1)}, \tag{6.15}$$

and, on the other hand,

$$\boldsymbol{P}^{(k)} = \left(Id - \boldsymbol{K}^{(k)}\boldsymbol{A}^{(k)}\right)\boldsymbol{P}^{(k-1)}. \tag{6.16}$$

The first equation is used as prior for $\boldsymbol{P}^{(k)}$, which is updated using the second one, after the determination of $\boldsymbol{K}^{(k)}$. Some algorithms use these equations iteratively, making loops involving Eqs. 6.14, 6.15 and 6.16. In a basic algorithm, the calculations can be made sequentially as follows:

- Initialization: $\boldsymbol{x}^{(0)}$ and $\boldsymbol{P}^{(0)}$ are known.
- For $k \geq 0$,
  - Determine $\widehat{\boldsymbol{x}_p}^{(k)}$
  - Determine the a priori estimation of $\boldsymbol{P}^{(k)}$: $\widehat{\boldsymbol{P}}^{(k)} = \boldsymbol{A}^{(k-1)}\boldsymbol{P}^{(k-1)}\left(\boldsymbol{A}^{(k-1)}\right)^t + \boldsymbol{Q}^{(k-1)}$
  - Determine $\boldsymbol{K}^{(k)} = \widehat{\boldsymbol{P}}^{(k)}\left(\boldsymbol{C}^{(k)}\right)^t\left(\boldsymbol{C}^{(k)}\widehat{\boldsymbol{P}}^{(k)}\left(\boldsymbol{C}^{(k)}\right)^t + \boldsymbol{R}^{(k)}\right)^{-1}$
  - Update $\widehat{\boldsymbol{x}}^{(k)} = \widehat{\boldsymbol{x}_p}^{(k)} + \boldsymbol{K}^{(k)}\left(\boldsymbol{y}^{(k)} - \boldsymbol{C}^{(k)}\widehat{\boldsymbol{x}_p}^{(k)}\right)$
  - Update $\boldsymbol{P}^{(k)} = \left(\boldsymbol{Id} - \boldsymbol{K}^{(k)}\boldsymbol{A}^{(k)}\right)\widehat{\boldsymbol{P}}^{(k-1)}$
  - Increase k

You can find many implementations of Kalman Filters in the literature and in the net. See, for instance, (LearningMachines 2020; Tusell 2011). There are also instructions and packages in R, such as `pomp`, `FKF`, `dlm`, `KFAS`, `MARSS`, `kalmanfilter` (see Hubbard 2023), and the instructions `KalmanRun`, `KalmanSmooth`, `KalmanForecast` of the standard package `stats`. A basic implementation is the following:

```
kf = function(A,B,C,y,u,v,w,t,x0,P0,Q,R){
  n0 = length(x0)
  nt = length(t)
  nw = length(w)
  xf = matrix(nrow = n0, ncol = nt)
  xf[, 1] = matrix(x0, ncol = 1)
  Rinv = solve(R)
  At = t(A)
  Ct = t(C)
  Pkm1 = P0
  for (k in 2:nt){
    xkm1 = xf[ ,k-1]
    xp = A %*% xf[ ,k-1] + B %*% u[ , k-1]
    Pp = A %*% Pkm1 %*% At + Q
    Minv = solve(C*Pp*Ct + R)
    Kg = Pp %*% Ct %*% Minv
    xk = xkm1 + Kg%*% (y[ ,k] - C %*% xp)
    Pk = (diag(length(xkm1)) - Kg*A)*Pp
    xf[ , k] = xk
    Pkm1 = Pk
  }
  return(xf)
}
```

Example 6.4 Let us consider the following situation (inspired from (Kalman Filtering 2023)

$$\boldsymbol{A} = \begin{pmatrix} 1.2269 & -0.4940 & 0.1129 \\ 1 & 0 & 0 \\ 0 & 1 & 0 \end{pmatrix}, \boldsymbol{B} = \begin{pmatrix} -0.382 \\ 0.5919 \\ 0.5191 \end{pmatrix},$$

$$\boldsymbol{C} = \begin{pmatrix} 1 & 0 & 0 \\ 0 & 0 & 0 \\ 0 & 0 & 0 \end{pmatrix}, \boldsymbol{Q} = \sigma_W^2 \begin{pmatrix} B_1^2 & 0 & 0 \\ 0 & B_2^2 & 0 \\ 0 & 0 & B_3^2 \end{pmatrix}, R = \sigma_v^2 \boldsymbol{Id}.$$

In this case, only the component 1 is observed. We consider $t = (0, 1, \ldots, 100)$, $u(t) = \sin\,(2\pi t/100)$, $x^{(0)} = (0.6, 0.5, 0.5)$, and $P^{(0)} = Q$. We generate a trajectory using the code below:

(continued)

Example 6.4 (continued)

```
kfsim = function(A,B,C,u,v,w,t,x0){
  n0 = length(x0)
  nt = length(t)
  nw = length(w)
  xs = matrix(nrow = n0, ncol = nt)
  ys = matrix(nrow = n0, ncol = nt)
  xs[, 1] = matrix(x0, ncol = 1)
  y0 = C %*% xs[ ,1] + v[ , 1]
  ys[, 1] = matrix(y0, ncol = 1)
  for (k in 2:nt){
    xs[ ,k] = A %*% xs[ ,k-1] + B %*% u[ , k-1] + w[ , k-1]
    ys[ ,k] = C %*% xs[ , k] + v[ , k]
  }
  sol = list(xs, ys)
  names(sol) = c("x","y")
  return(sol)}

sol1 = kfsim(A,B,C,u,v,w,t,x0)
xo = sol1$x
yo = sol1$y
```

Then, we call the basic Kalman Filter:

```
xf = kf(A,B,C,yo,u,v,w,t,x0,P0,Q,R)
```

An example of result for $\sigma_w = 0.25$, $\sigma_v = 0.5$ is shown in Fig. 6.9. The points "`data`" correspond to the observations y. An example with $\sigma_w = 0.5$, $\sigma_v = 0.5$ is shown in Fig. 6.10.

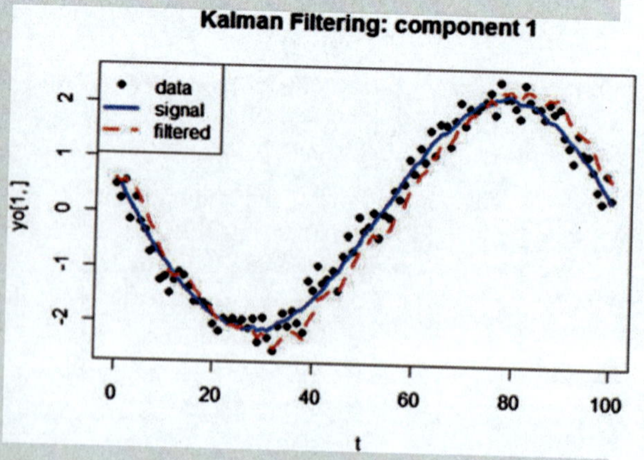

Fig. 6.9 Example of result furnished by the Kalman filter, with $\sigma_w = 0.25$, $\sigma_v = 0.5$

(continued)

Example 6.4 (continued)

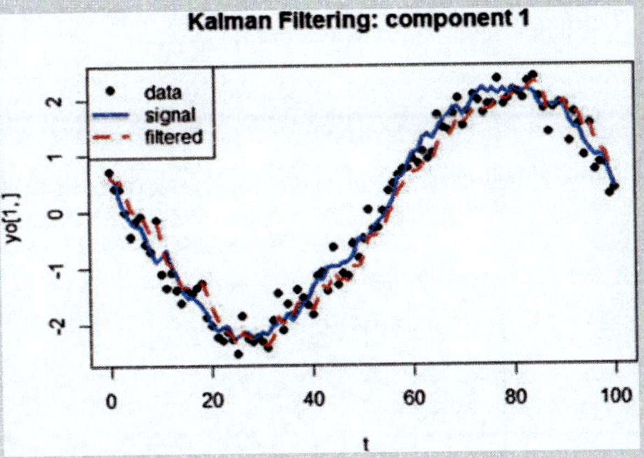

Fig. 6.10 Example of result furnished by the Kalman filter, with $\sigma_w = 0.5, \sigma_v = 0.5$

Example 6.5 Let us consider the differential equation

$$\frac{d}{dt}\begin{pmatrix} x_1 \\ x_2 \end{pmatrix} = \begin{pmatrix} 0 & 1 \\ -1 & 0 \end{pmatrix}\begin{pmatrix} x_1 \\ x_2 \end{pmatrix} + \begin{pmatrix} w_1 \\ w_2 \end{pmatrix}.$$

The equation is discretized in time on the interval $(0, T)$, by Euler's method, with a step Δt so that

$$\begin{pmatrix} x_1 \\ x_2 \end{pmatrix}^{(k+1)} = \begin{pmatrix} 1 & \Delta t \\ -\Delta t & 1 \end{pmatrix}\begin{pmatrix} x_1 \\ x_2 \end{pmatrix}^{(k)} + \Delta t\begin{pmatrix} w_1 \\ w_2 \end{pmatrix}^{(k)}.$$

Assume that the observation is $y = x + v$ so that

$$\begin{pmatrix} y_1 \\ y_2 \end{pmatrix}^{(k)} = \begin{pmatrix} 1 & 0 \\ 0 & 1 \end{pmatrix}\begin{pmatrix} x_1 \\ x_2 \end{pmatrix}^{(k)} + \begin{pmatrix} v_1 \\ v_2 \end{pmatrix}^{(k)}$$

Let us consider $\boldsymbol{w} \sim N(0, \sigma_w^2 \boldsymbol{Id})$, $\boldsymbol{v} \sim N(0, \sigma_v^2 \boldsymbol{Id})$. Then

(continued)

Example 6.5 (continued)

$$A = \begin{pmatrix} 1 & \Delta t \\ -\Delta t & 1 \end{pmatrix}, B = 0, C = Id, Q = \Delta t^2 \sigma_W^2 \, Id, R = \sigma_v^2 Id.$$

We consider $t = \left(i \frac{2\pi}{npt}, 0 \leq i \leq npt \right)$, $x^{(0)} = (0, 1)$, and $P^{(0)} = Q$.
An example of result for $\sigma_w = 0.4$, $\sigma_v = 0.2$ is shown in Figs. 6.11 and 6.12.

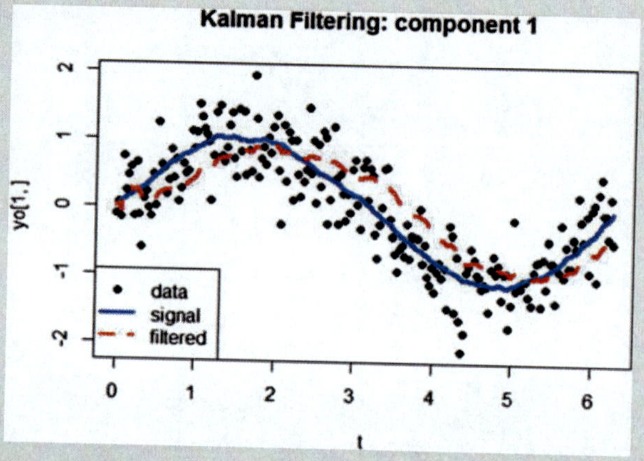

Fig. 6.11 Example of result furnished by the Kalman filter, with $\sigma_w = 0.4$, $\sigma_v = 0.2$

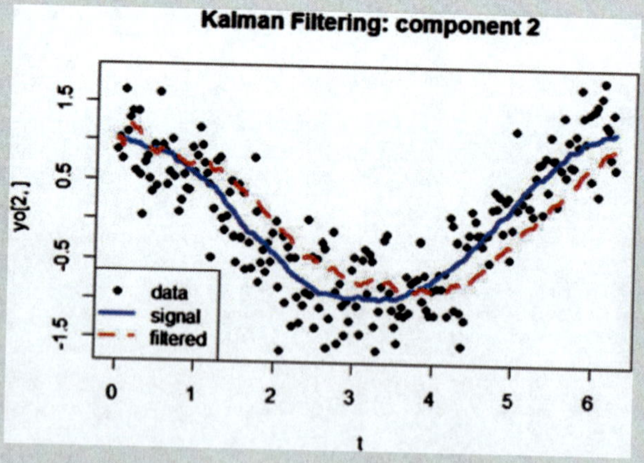

Fig. 6.12 Example of result furnished by the Kalman filter, with $\sigma_w = 0.4$, $\sigma_v = 0.2$

Example 6.6 Let us consider the situation where

$$\begin{pmatrix} x_1 \\ x_2 \end{pmatrix}^{(k+1)} = A \begin{pmatrix} x_1 \\ x_2 \end{pmatrix}^{(k)} + \Delta t \begin{pmatrix} w_1 \\ w_2 \end{pmatrix}^{(k)},$$

with

$$A = \begin{pmatrix} 1 & -\frac{\Delta t}{2} \\ \frac{\Delta t}{2} & 1 \end{pmatrix}^{-1} \begin{pmatrix} 1 & \frac{\Delta t}{2} \\ -\frac{\Delta t}{2} & 1 \end{pmatrix}.$$

The observation is $y = x + v$. Let $\boldsymbol{w} \sim N(0, \sigma_w^2 \boldsymbol{Id})$, $\boldsymbol{v} \sim N(0, \sigma_v^2 \boldsymbol{Id})$, $\boldsymbol{B} = \boldsymbol{0}, \boldsymbol{C} = \boldsymbol{Id}, \boldsymbol{R} = \sigma_v^2 \boldsymbol{Id}$,

$$Q = \frac{\Delta t^2 \sigma_W^2}{1 + \frac{\Delta t^2}{4}} \begin{pmatrix} 1 & \frac{\Delta t}{2} \\ -\frac{\Delta t}{2} & 1 \end{pmatrix}.$$

We consider $t = (i\Delta t, 0 \leq i \leq npt)$, $\Delta t = 2\pi/npt$, $\boldsymbol{x}^{(0)} = (0, 1)$, and $\boldsymbol{P}^{(0)} = \boldsymbol{Q}$. An example of result for $\sigma_w = 0.4$, $\sigma_v = 0.2$, $npt = 200$ is shown in Figs. 6.13 and 6.14.

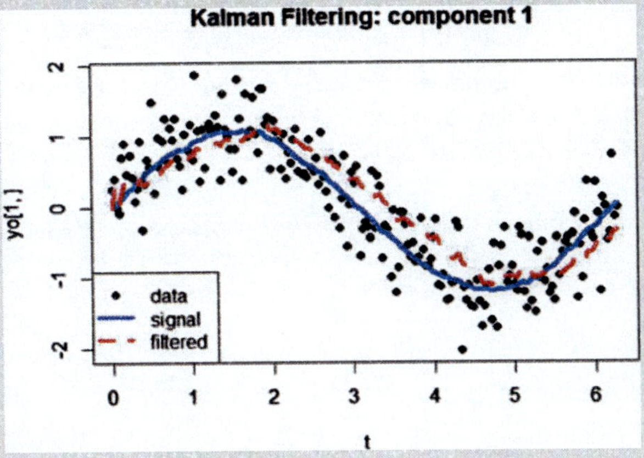

Fig. 6.13 Example of result furnished by the Kalman filter, with $\sigma_w = 0.4$, $\sigma_v = 0.2$

(continued)

Example 6.6 (continued)

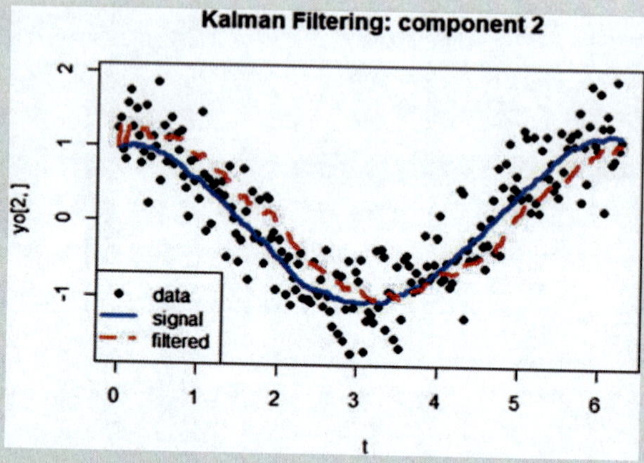

Fig. 6.14 Example of result furnished by the Kalman filter, with $\sigma_w = 0.4$, $\sigma_v = 0.2$

Example 6.7 Let us consider the differential equation

$$\frac{d}{dt}\begin{pmatrix} x_1 \\ x_2 \end{pmatrix} = \begin{pmatrix} 0 & 1 \\ -g/\ell & 0 \end{pmatrix}\begin{pmatrix} x_1 \\ x_2 \end{pmatrix} + \begin{pmatrix} w_1 \\ w_2 \end{pmatrix}.$$

The equation is discretized in time on the interval $(0, T)$, by Euler's method, with a step Δt, and the observation is $y = x + v$, as in the preceding example.

We consider $t = (i\Delta t, 0 \leq i \leq npt)$, $\Delta t = 2\pi/npt$, $\mathbf{x}^{(0)} = \left(\frac{\pi}{8}, 0\right)$, $g = 9.81$, $\ell = 2$, and $\boldsymbol{P}^{(0)} = \boldsymbol{Q}$.

An example of result for $\sigma_w = 0.5$, $\sigma_v = 0.2$ is shown is shown in Figs. 6.15 and 6.16.

(continued)

Example 6.7 (continued)

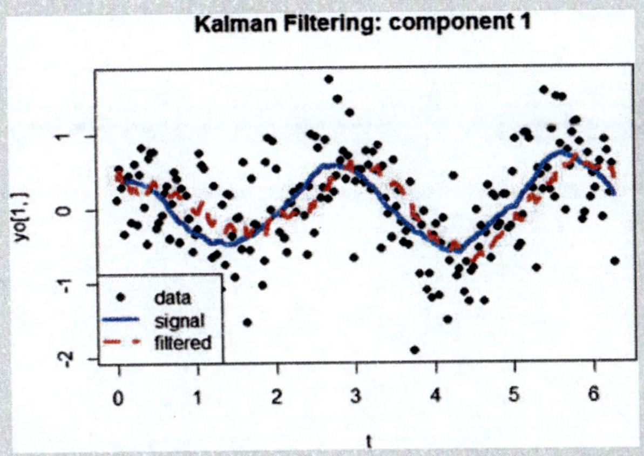

Fig. 6.15 Example of result furnished by the Kalman filter, with $\sigma_w = 0.5$, $\sigma_v = 0.2$

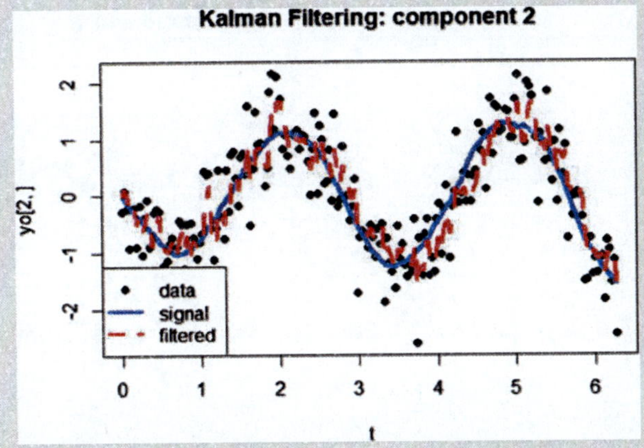

Fig. 6.16 Example of result furnished by the Kalman filter, with $\sigma_w = 0.5$, $\sigma_v = 0.2$

Example 6.8 Let us consider the situation where

$$\begin{pmatrix} x_1 \\ x_2 \end{pmatrix}^{(k+1)} = A \begin{pmatrix} x_1 \\ x_2 \end{pmatrix}^{(k)} + \Delta t \begin{pmatrix} w_1 \\ w_2 \end{pmatrix}^{(k)},$$

with

(continued)

Example 6.8 (continued)

$$A = \begin{pmatrix} 1 & -\frac{\Delta t}{2} \\ \frac{g\Delta t}{2\ell} & 1 \end{pmatrix}^{-1} \begin{pmatrix} 1 & \frac{\Delta t}{2} \\ -\frac{g\Delta t}{2\ell} & 1 \end{pmatrix}.$$

The observation is $y = x + v$. Let $\boldsymbol{w} \sim N(0, \sigma_w^2 \boldsymbol{Id})$, $\boldsymbol{v} \sim N(0, \sigma_v^2 \boldsymbol{Id})$, $\boldsymbol{B} = \boldsymbol{0}, \boldsymbol{C} = \boldsymbol{Id}, \boldsymbol{R} = \sigma_v^2 \boldsymbol{Id}$,

$$\boldsymbol{Q} = \frac{4\ell \Delta t^2 \sigma_W^2}{4\ell + g\Delta t^2} \begin{pmatrix} 1 & \frac{\Delta t}{2} \\ -\frac{g\Delta t}{2\ell} & 1 \end{pmatrix}.$$

We consider $t = (i\Delta t, 0 \leq i \leq npt), \Delta t = 2\pi / npt, \boldsymbol{x}^{(0)} = \left(\frac{\pi}{8}, 0\right)$, $g = 9.81$, $\ell = 2$, and $\boldsymbol{P}^{(0)} = \boldsymbol{Q}$. An example of result for $\sigma_w = 0.4$, $\sigma_v = 0.2$ is shown in Figs. 6.17 and 6.18.

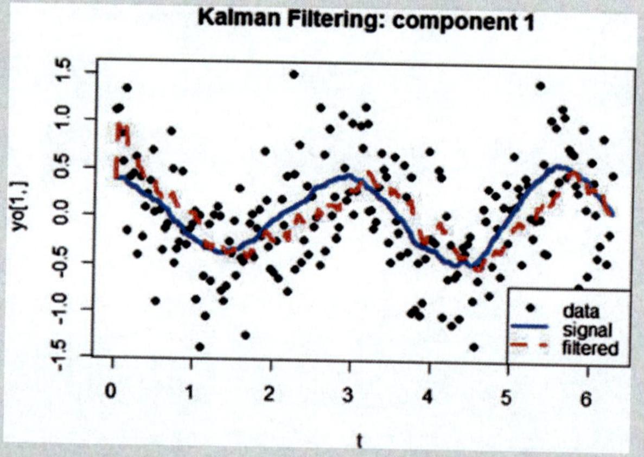

Fig. 6.17 Example of result furnished by the Kalman filter, with $\sigma_w = 0.4$, $\sigma_v = 0.2$

(continued)

Example 6.8 (continued)

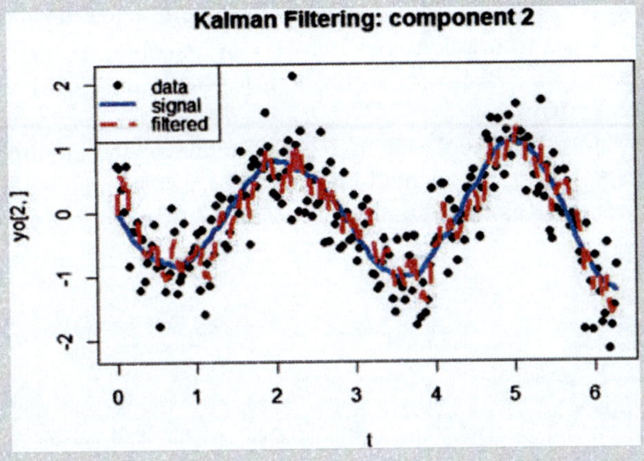

Fig. 6.18 Example of result furnished by the Kalman filter, with $\sigma_w = 0.4$, $\sigma_v = 0.2$

The reader can find in the literature many variations and generalizations about the basic Kalman Filter, namely extensions to nonlinear state equations (Extended Kalman Filter) and improvements (as, for instance, Unscented Kalman Filter, Ensemble Kalman Filter). The Extended Kalman Filter (EKF) is based on successive linearizations of the state equation:

$$\boldsymbol{f}\left(t^{(k)}, \boldsymbol{x}^{(k)}, \boldsymbol{u}^{(k)}, \boldsymbol{w}^{(k)}\right) \approx \boldsymbol{A}^{(k)}\boldsymbol{x}^{(k)} + \boldsymbol{B}^{(k)}\boldsymbol{u}^{(k)} + \boldsymbol{D}^{(k)}\boldsymbol{w}^{(k)} + \boldsymbol{E}^{(k)}. \tag{6.17}$$

$$\boldsymbol{g}\left(t^{(k)}, \boldsymbol{x}^{(k)}, \boldsymbol{v}^{(k)}\right) \approx \boldsymbol{C}^{(k)}\boldsymbol{x}^{(k)} + \boldsymbol{F}^{(k)}\boldsymbol{v}^{(k)} + \boldsymbol{G}^{(k)}. \tag{6.18}$$

The design of a particular EKF depends upon the linearization chosen. Indeed, many linearizations can be found in the literature. For instance, let $\boldsymbol{a}^{(k)} = \left(\bar{\boldsymbol{x}}^{(k)}, \boldsymbol{u}^{(k)}, \boldsymbol{w}^{(k)}\right)$, $\boldsymbol{b}^{(k)} = \left(\bar{\boldsymbol{x}}^{(k)}, \boldsymbol{v}^{(k)}\right)$. If a Taylor's expansion around $\boldsymbol{a}^{(k-1)}$ and $\boldsymbol{b}^{(k-1)}$ is used for the linearization,

$$\boldsymbol{A}^{(k)} = \frac{\partial \boldsymbol{f}}{\partial \boldsymbol{x}}\left(t^{(k)}, \boldsymbol{a}^{(k-1)}\right), \quad \boldsymbol{B}^{(k)} = \frac{\partial \boldsymbol{f}}{\partial \boldsymbol{u}}\left(t^{(k)}, \boldsymbol{a}^{(k-1)}\right),$$

$$\boldsymbol{C}^{(k)} = \frac{\partial \boldsymbol{g}}{\partial \boldsymbol{x}}\left(t^{(k)}, \boldsymbol{b}^{(k)}\right), \quad \boldsymbol{F}^{(k)} = \frac{\partial \boldsymbol{g}}{\partial \boldsymbol{v}}\left(t^{(k)}, \boldsymbol{b}^{(k)}\right),$$

$$\boldsymbol{E}^{(k)} = \boldsymbol{f}\left(t^{(k)}, \boldsymbol{a}^{(k-1)}\right) - \boldsymbol{A}^{(k)}\bar{\boldsymbol{x}}^{(k-1)} - \boldsymbol{B}^{(k)}\boldsymbol{u}^{(k-1)} - \boldsymbol{D}^{(k)}\boldsymbol{w}^{(k-1)},$$

$$\boldsymbol{G}^{(k)} = \boldsymbol{g}\left(t^{(k)}, \boldsymbol{b}^{(k-1)}\right) - \boldsymbol{C}^{(k)}\,\bar{\boldsymbol{x}}^{(k-1)} - \boldsymbol{F}^{(k)}\boldsymbol{v}^{(k-1)}.$$

Choosing $\bar{\boldsymbol{x}}^{(k-1)} = \boldsymbol{x}^{(k-1)}$ leads to explicit equations, while choosing $\bar{\boldsymbol{x}}^{(k-1)} = \widehat{\boldsymbol{x}}^{(k)}$ leads to implicit ones, which can be solved by prediction-correction or internal iterations. Whatever the choice, Eqs. (6.17) and (6.18) can be rewritten to take the form (6.10).

Linearization can be replaced by the estimation of the conditional distribution $f(\boldsymbol{x}^{(k+1)}|\boldsymbol{x}^{(k)}, \boldsymbol{w}^{(k)}, \boldsymbol{v}^{(k)})$, by adapted methods. For instance, sigma-point estimations lead to Unscented Kalman Filters, while Monte-Carlo estimations lead to Particle Filters.

Exercises

1. Let

- $\begin{pmatrix} X_1^{(k+1)} \\ X_2^{(k+1)} \end{pmatrix} = \begin{pmatrix} 1 & \Delta t \\ -\omega^2\Delta t & 1 \end{pmatrix}\begin{pmatrix} X_1^{(k)} \\ X_2^{(k)} \end{pmatrix} + \begin{pmatrix} 0 \\ u^{(k)}\Delta t \end{pmatrix} + \Delta t\begin{pmatrix} w_1 \\ w_2 \end{pmatrix}.$
- Only the first variable is observed: $y_1^{(k)} = x_1^{(k)} + v^{(k)}$. Assume that the observations are made for $1 \le k \le 100$, with $\Delta t = 0.1$, and, on the one hand, all the noises w_1, w_2, v are independent Gaussian $N(0, \sigma)$, with $\sigma_w = 3\sigma_v$; on the other hand, $u^{(k)} = k\Delta t \sin(k\Delta t)$.

  (a) Generate a trajectory $(\boldsymbol{X}^{(k)}, \boldsymbol{Y}^{(k)})$ of the system
  (b) Apply the Kalman Filter to the data generated
  (c) Draw a graph comparing the filtered values of $X_1^{(k)}$ and the real ones

- Test different values of σ_v.

2. Let

- $\begin{pmatrix} X_1^{(k+1)} \\ X_2^{(k+1)} \end{pmatrix} = \begin{pmatrix} 1 & \Delta t \\ 0 & 1 \end{pmatrix}\begin{pmatrix} X_1^{(k)} \\ X_2^{(k)} \end{pmatrix} + \begin{pmatrix} 0 \\ u^{(k)}\Delta t \end{pmatrix} + \Delta t\begin{pmatrix} w_1 \\ w_2 \end{pmatrix}.$
- Only the first variable is observed: $y_1^{(k)} = x_1^{(k)} + v^{(k)}$. Assume that the observations are made for $1 \le k \le 100$, with $\Delta t = 0.1$, and, on the one hand, all the noises w_1, w_2, v are independent Gaussian $N(0, \sigma)$, with $\sigma_w = 3\sigma_v$; on the other hand, $u^{(k)} = k\Delta t \sin(k\Delta t)$.

  (a) Generate a trajectory $(\boldsymbol{X}^{(k)}, \boldsymbol{Y}^{(k)})$ of the system
  (b) Apply the Kalman Filter to the data generated
  (c) Draw a graph comparing the filtered values of $X_1^{(k)}$ and the real ones

- Test different values of σ_v.

(continued)

3. Let

- $$\begin{pmatrix} X_1^{(k+1)} \\ X_2^{(k+1)} \end{pmatrix} = \begin{pmatrix} \cos(\theta) & -\sin(\theta) \\ \sin(\theta) & \cos(\theta) \end{pmatrix} \begin{pmatrix} X_1^{(k)} \\ X_2^{(k)} \end{pmatrix} + \begin{pmatrix} w_1 \\ w_2 \end{pmatrix}.$$
- Both the variables are observed: $y_1^{(k)} = x_1^{(k)} + v_1^{(k)}$, $y_2^{(k)} = x_2^{(k)} + v_2^{(k)}$. Assume that the observations are made for $1 \le k \le 100$, and, on the one hand, all the noises w_1, w_2, v_1, v_2 are independent Gaussian $N(0, \sigma)$, with $\sigma_w = 3\sigma_v$; on the other hand, $\theta = \pi/16$.

   (a) Generate a trajectory $(\boldsymbol{X}^{(k)}, \boldsymbol{Y}^{(k)})$ of the system
   (b) Apply the Kalman Filter to the data generated
   (c) Draw a graph comparing the filtered values of $X_1^{(k)}$ and the real ones

- Test different values of σ_v. Test the case where $\theta = \pi/4$.

4. Let

- $$\begin{pmatrix} X_1^{(k+1)} \\ X_2^{(k+1)} \\ X_3^{(k+1)} \\ X_4^{(k+1)} \end{pmatrix} = \begin{pmatrix} 1 & 0 & \Delta t & 0 \\ 0 & 1 & 0 & \Delta t \\ 0 & 0 & 1 & 0 \\ 0 & 0 & 0 & 1 \end{pmatrix} \begin{pmatrix} X_1^{(k)} \\ X_2^{(k)} \\ X_3^{(k)} \\ X_4^{(k)} \end{pmatrix} + \begin{pmatrix} w_1 \\ w_2 \\ w_3 \\ w_4 \end{pmatrix}.$$
- Only the two first variables are observed: $y_1^{(k)} = x_1^{(k)} + v_1^{(k)}$, $y_2^{(k)} = x_2^{(k)} + v_2^{(k)}$. Assume that the observations are made for $1 \le k \le 100$, and, on the one hand, all the noises w_i, v_j are independent Gaussian $N(0, \sigma)$, with $\sigma_w = \sigma_v$.

   (a) Generate a trajectory $(\boldsymbol{X}^{(k)}, \boldsymbol{Y}^{(k)})$ of the system
   (b) Apply the Kalman Filter to the data generated
   (c) Draw a graph comparing the filtered values of $X_1^{(k)}$ and the real ones

- Test different values of σ_v.

6.4 Particle Filtering

As mentioned above, particle filters look for an estimation of the conditional distribution $f(\boldsymbol{x}^{(k+1)} | \boldsymbol{x}^{(k)}, \boldsymbol{w}^{(k)}, \boldsymbol{v}^{(k)})$. Once the estimation is generated, it is used to determine an estimation for $\boldsymbol{x}^{(k+1)}$, such as, for instance, the mean of the distribution. Implementations in R of particle filters can be found on the net and in the R packages, such as, for instance, `pomp`, `sfmsb`, `SMC`, `LaplacesDemon`, `nimbleSMC`, `NIMBLE`. You can find also tutorials in the literature – for instance, (Arulampalam et al. 2002; de Freitas 2002; Doucet and Johansen 2009; Künsch 2013; Speekenbrink 2016; Dahlin and Schön 2019; Elfring et al. 2021).

Particle filters are based on the observation that (6.2) and (6.3) defines a Hidden Markov Model, where the state $(k + 1)$ depends only upon the state (k) so that the condition PDFs verify (below, f denotes a PDF and not the state equation)

$$f\left(\boldsymbol{x}^{(k)}|\boldsymbol{x}^{(k-1)},\ldots,\boldsymbol{x}^{(0)}\right)=f\left(\boldsymbol{x}^{(k)}|\boldsymbol{x}^{(k-1)}\right). \tag{6.19}$$

Thus, the PDF of $(\boldsymbol{x}^{(0)},\ldots,\boldsymbol{x}^{(k)})$ satisfies

$$f\left(\boldsymbol{x}^{(0)},\ldots,\boldsymbol{x}^{(k)}\right)=f\left(\boldsymbol{x}^{(0)},\ldots,\boldsymbol{x}^{(k-1)}\right)f\left(\boldsymbol{x}^{(k)}|\boldsymbol{x}^{(k-1)}\right),$$

and

$$f\left(\boldsymbol{x}^{(0)},\ldots,\boldsymbol{x}^{(k)}\right)=f\left(\boldsymbol{x}^{(k)}|\boldsymbol{x}^{(k-1)}\right)f\left(\boldsymbol{x}^{(k-1)}|\boldsymbol{x}^{(k-2)}\right)\ldots f\left(\boldsymbol{x}^{(1)}|\boldsymbol{x}^{(0)}\right)f\left(\boldsymbol{x}^{(0)}\right),$$

i.e.,

$$f\left(\boldsymbol{x}^{(0)},\ldots,\boldsymbol{x}^{(k)}\right)=f\left(\boldsymbol{x}^{(0)}\right)\prod_{i=1}^{k}f\left(\boldsymbol{x}^{(i)}|\boldsymbol{x}^{(i-1)}\right). \tag{6.20}$$

Analogously,

$$f\left(\boldsymbol{y}^{(k)}|\boldsymbol{x}^{(k)},\ldots,\boldsymbol{x}^{(0)}\right)=f\left(\boldsymbol{y}^{(k)}|\boldsymbol{x}^{(k)}\right). \tag{6.21}$$

Therefore,

$$f\left(\boldsymbol{y}^{(0)},\ldots,\boldsymbol{y}^{(k)}|\boldsymbol{x}^{(0)},\ldots,\boldsymbol{x}^{(k)}\right)=f\left(\boldsymbol{y}^{(k)}|\boldsymbol{x}^{(k)}\right)\ldots f\left(\boldsymbol{y}^{(0)}|\boldsymbol{x}^{(0)}\right)=\prod_{i=1}^{k}f\left(\boldsymbol{y}^{(i)}|\boldsymbol{x}^{(i)}\right),$$

i.e.,

$$f\left(\boldsymbol{y}^{(0)},\ldots,\boldsymbol{y}^{(k)}|\boldsymbol{x}^{(0)},\ldots,\boldsymbol{x}^{(k)}\right)=\prod_{i=0}^{k}f\left(\boldsymbol{y}^{(i)}|\boldsymbol{x}^{(i)}\right). \tag{6.22}$$

Since

$$f\left(\boldsymbol{x}^{(0)},\ldots,\boldsymbol{x}^{(k)}|\boldsymbol{y}^{(0)},\ldots,\boldsymbol{y}^{(k)}\right)=\frac{f\left(\boldsymbol{y}^{(0)},\ldots,\boldsymbol{y}^{(k)}|\boldsymbol{x}^{(0)},\ldots,\boldsymbol{x}^{(k)}\right)}{f\left(\boldsymbol{y}^{(0)},\ldots,\boldsymbol{y}^{(k)}\right)}f\left(\boldsymbol{x}^{(0)},\ldots,\boldsymbol{x}^{(k)}\right),$$

We have, from Eqs. (6.20) and (6.22),

$$f\left(\boldsymbol{x}^{(0)},\ldots,\boldsymbol{x}^{(k)}|\boldsymbol{y}^{(0)},\ldots,\boldsymbol{y}^{(k)}\right)=f\left(\boldsymbol{x}^{(0)}\right)\frac{\prod_{i=1}^{k}f\left(\boldsymbol{x}^{(i)}|\boldsymbol{x}^{(i-1)}\right)\prod_{i=0}^{k}f\left(\boldsymbol{y}^{(i)}|\boldsymbol{x}^{(i)}\right)}{f\left(\boldsymbol{y}^{(0)},\ldots,\boldsymbol{y}^{(k)}\right)}.$$

Thus, we can evaluate $p^{(k)}=f(\boldsymbol{x}^{(0)},\ldots,\boldsymbol{x}^{(k)}|\boldsymbol{y}^{(0)},\ldots,\boldsymbol{y}^{(k)})$ recursively using the update equation:

$$p^{(k)}=c^{(k)}p^{(k-1)},c^{(k)}=\frac{f\left(\boldsymbol{y}^{(k)}|\boldsymbol{x}^{(k)}\right)f\left(\boldsymbol{x}^{(k)}|\boldsymbol{x}^{(k-1)}\right)}{f\left(\boldsymbol{y}^{(k)}|\boldsymbol{y}^{(k-1)},\ldots,\boldsymbol{y}^{(0)}\right)}. \quad (6.23)$$

Equation (6.23) can be combined to importance sampling: let $f_{imp}^{(k)}$ be the importance distribution and assume that

$$f_{imp}^{(k)}\left(\boldsymbol{x}^{(0)},\ldots,\boldsymbol{x}^{(k)}\right)=f_{imp}^{(k)}\left(\boldsymbol{x}^{(k)}|\boldsymbol{x}^{(0)},\ldots,\boldsymbol{x}^{(k-1)}\right)f_{imp}^{(k-1)}\left(\boldsymbol{x}^{(0)},\ldots,\boldsymbol{x}^{(k-1)}\right). \quad (6.24)$$

An example of distribution verifying (6.24) is a multivariate Gaussian distribution having a diagonal covariance matrix. Let $q^{(k)}=f_{imp}^{(k)}\left(\boldsymbol{x}^{(0)},\ldots,\boldsymbol{x}^{(k)}\right)$, $r^{(k)}=f_{imp}^{(k)}\left(\boldsymbol{x}^{(k)}|\boldsymbol{x}^{(0)},\ldots,\boldsymbol{x}^{(k-1)}\right)$, then

$$\frac{p^{(k)}}{q^{(k)}}=c^{(k)}\frac{p^{(k-1)}}{q^{(k)}}=\frac{c^{(k)}}{r^{(k)}}\frac{p^{(k-1)}}{q^{(k-1)}}. \quad (6.25)$$

Thus, the weights $wgt^{(k)}=p^{(k)}/r^{(k)}$ can be recursively determined by an equation analogous to (6.23):

$$wgt^{(k)}=cwgt^{(k)}wgt^{(k-1)},cwgt^{(k)}=\frac{c^{(k)}}{r^{(k)}}. \quad (6.26)$$

A basic Particle Filtering Algorithm is the following:

- Initialization: Choose an importance distribution satisfying (6.24). Generate a sample $S^{(0)}=\left\{\left(\boldsymbol{x}^{(0,i)},\boldsymbol{y}^{(0,i)}\right),1\leq i\leq ns\right\}$ of ns variates from the initial distribution and determine the corresponding weights $wgt^{(0,i)}$. Normalize the weights so that $\sum_{i=1}^{ns}wgt^{(0,i)}=1$.

- For $k \geq 0$,
  - Generate a sample $\mathcal{T}^{(k)} = \{ (\boldsymbol{x}^{(k,i)}, \boldsymbol{y}^{(k,i)}), 1 \leq i \leq ns \}$ using $f^{(k)}_{imp}(\boldsymbol{x}^{(k)} | \boldsymbol{x}^{(0)}, \ldots, \boldsymbol{x}^{(k-1)})$.
  - Use $\mathcal{T}^{(k)}$ to generate
    $\mathcal{S}^{(k)} = \{ ((\boldsymbol{x}^{(0,i)}, \boldsymbol{y}^{(0,i)}), \ldots, (\boldsymbol{x}^{(k,i)}, \boldsymbol{y}^{(k,i)})), 1 \leq i \leq ns \}$.
  - Determine the corresponding weights $wgt^{(k,i)}$, using the update equation (6.26). Normalize the weights so that $\sum_{i=1}^{ns} wgt^{(k,i)} = 1$.
  - Estimate $\widehat{\boldsymbol{x}}^{(k)}$ using the distribution determined.
  - Increase k.

Many algorithms in the literature include a supplementary of resampling: the sample is reduced to increase the significance of the points having larger weights and eliminate points with small ones. After resampling, the sample is often set to be equiprobable (equal weights for all the elements). A code in R for the implementation of a Particle Filter is given in (Dahlin and Schön 2019; Dahlin 2023). These codes include resampling.

Example 6.9 Let us consider the situation where

$$x^{(k+1)} = \min\left(1, x^{(k)} - \Delta t\sqrt{1 - (x^{(k)})^2} + \Delta t w^{(k)}\right).$$

The observation is $y = \min(1, x + v)$. Let $\boldsymbol{w} \sim N(0, \sigma_w^2 \boldsymbol{Id})$, $\boldsymbol{v} \sim N(0, \sigma_v^2 \boldsymbol{Id})$. We consider $t = (i\Delta t, 0 \leq i \leq npt)$, $\Delta t = 10/npt$ $x^{(0)} = 0$. An example of result for $\sigma_w = 0.4$, $\sigma_v = 0.4$, $npt = 200$ is shown in Fig. 6.19.

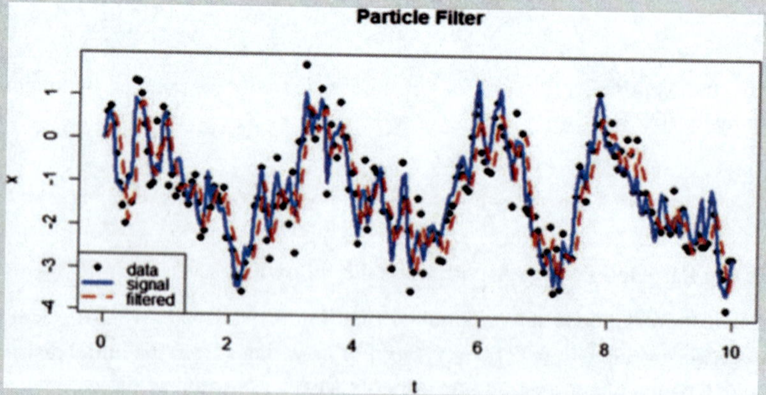

Fig. 6.19 Example of result furnished by the particle filtering, with $\sigma_w = 0.4$, $\sigma_v = 0.4$, using the code in (Dahlin and Schön 2019; Dahlin 2023)

Example 6.10 Let us consider the situation where

$$x^{(k+1)} = \min\left(1, x^{(k)} + \Delta t\sqrt{1-(x^{(k)})^2} + \Delta t w^{(k)}\right).$$

The observation is $y = \min(1, x + v)$. Let $\boldsymbol{w} \sim N(0, \sigma_w^2 \boldsymbol{Id})$, $\boldsymbol{v} \sim N(0, \sigma_v^2 \boldsymbol{Id})$. We consider $t = (i\Delta t, 0 \leq i \leq npt)$, $\Delta t = 10/npt$ $x^{(0)} = 0$. An example of result for $\sigma_w = 0.4$, $\sigma_v = 0.4$, $npt = 200$ is shown in Fig. 6.20.

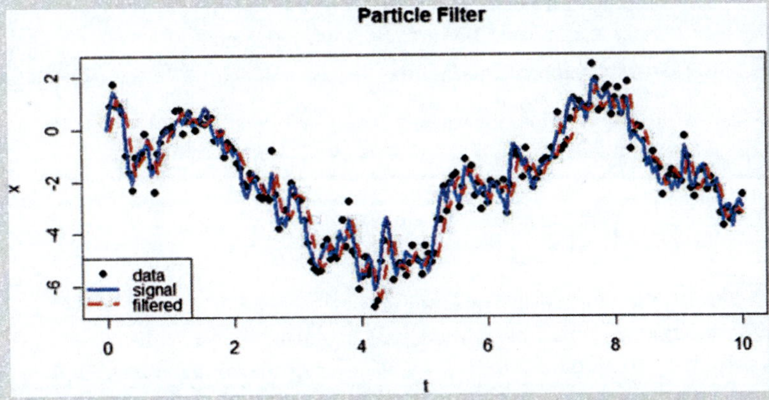

Fig. 6.20 Example of result furnished by the particle filtering, with $\sigma_w = 0.4$, $\sigma_v = 0.4$, using the code in (Dahlin and Schön 2019; Dahlin 2023)

Exercises

1. Let

- $$\begin{pmatrix} X_1^{(k+1)} \\ X_2^{(k+1)} \end{pmatrix} = \begin{pmatrix} 1 & \Delta t \\ -\omega^2\Delta t & 1 \end{pmatrix}\begin{pmatrix} X_1^{(k)} \\ X_2^{(k)} \end{pmatrix} + \begin{pmatrix} 0 \\ u^{(k)}\Delta t \end{pmatrix} + \Delta t\begin{pmatrix} w_1 \\ w_2 \end{pmatrix}.$$
- Only the first variable is observed: $y_1^{(k)} = x_1^{(k)} + v^{(k)}$. Assume that the observations are made for $1 \leq k \leq 100$, with $\Delta t = 0.1$, and, on the one hand, all the noises w_1, w_2, v are independent Gaussian $N(0, \sigma)$, with $\sigma_w = 3\sigma_v$; on the other hand, $u^{(k)} = k\Delta t \sin(k\Delta t)$.

  (a) Generate a trajectory $(\boldsymbol{X}^{(k)}, \boldsymbol{Y}^{(k)})$ of the system
  (b) Apply the Particle Filter to the data generated
  (c) Draw a graph comparing the filtered values of $X_1^{(k)}$ and the real ones

- Test different values of σ_v.

(continued)

2. Let

 - $$\begin{pmatrix} X_1^{(k+1)} \\ X_2^{(k+1)} \end{pmatrix} = \begin{pmatrix} 1 & \Delta t \\ 0 & 1 \end{pmatrix} \begin{pmatrix} X_1^{(k)} \\ X_2^{(k)} \end{pmatrix} + \begin{pmatrix} 0 \\ u^{(k)}\Delta t \end{pmatrix} + \Delta t \begin{pmatrix} w_1 \\ w_2 \end{pmatrix}.$$
 - Only the first variable is observed: $y_1^{(k)} = x_1^{(k)} + v^{(k)}$. Assume that the observations are made for $1 \le k \le 100$, with $\Delta t = 0.1$, and, on the one hand, all the noises w_1, w_2, v are independent Gaussian $N(0, \sigma)$, with $\sigma_w = 3\sigma_v$; on the other hand, $u^{(k)} = k\Delta t \sin(k\Delta t)$.

   (a) Generate a trajectory $(\boldsymbol{X}^{(k)}, \boldsymbol{Y}^{(k)})$ of the system
   (b) Apply the Particle Filter to the data generated
   (c) Draw a graph comparing the filtered values of $X_1^{(k)}$ and the real ones
 - Test different values of σ_v.

3. Let

 - $$\begin{pmatrix} X_1^{(k+1)} \\ X_2^{(k+1)} \end{pmatrix} = \begin{pmatrix} \cos(\theta) & -\sin(\theta) \\ \sin(\theta) & \cos(\theta) \end{pmatrix} \begin{pmatrix} X_1^{(k)} \\ X_2^{(k)} \end{pmatrix} + \begin{pmatrix} w_1 \\ w_2 \end{pmatrix}.$$
 - Both the variables are observed: $y_1^{(k)} = x_1^{(k)} + v_1^{(k)}$, $y_2^{(k)} = x_2^{(k)} + v_2^{(k)}$. Assume that the observations are made for $1 \le k \le 100$, and, on the one hand, all the noises w_1, w_2, v_1, v_2 are independent Gaussian $N(0, \sigma)$, with $\sigma_w = 3\sigma_v$; on the other hand, $\theta = \pi/16$.

   (a) Generate a trajectory $(\boldsymbol{X}^{(k)}, \boldsymbol{Y}^{(k)})$ of the system
   (b) Apply the Particle Filter to the data generated
   (c) Draw a graph comparing the filtered values of $X_1^{(k)}$ and the real ones
 - Test different values of σ_v. Test the case where $\theta = \pi/4$.

4. Let

 - $$\begin{pmatrix} X_1^{(k+1)} \\ X_2^{(k+1)} \\ X_3^{(k+1)} \\ X_4^{(k+1)} \end{pmatrix} = \begin{pmatrix} 1 & 0 & \Delta t & 0 \\ 0 & 1 & 0 & \Delta t \\ 0 & 0 & 1 & 0 \\ 0 & 0 & 0 & 1 \end{pmatrix} \begin{pmatrix} X_1^{(k)} \\ X_2^{(k)} \\ X_3^{(k)} \\ X_4^{(k)} \end{pmatrix} + \begin{pmatrix} w_1 \\ w_2 \\ w_3 \\ w_4 \end{pmatrix}.$$
 - Only the two first variables are observed: $y_1^{(k)} = x_1^{(k)} + v_1^{(k)}$, $y_2^{(k)} = x_2^{(k)} + v_2^{(k)}$. Assume that the observations are made for $1 \le k \le 100$, and, on the one hand, all the noises w_i, v_j are independent Gaussian $N(0, \sigma)$, with $\sigma_w = \sigma_v$.

(continued)

(a) Generate a trajectory $(\boldsymbol{X}^{(k)}, \boldsymbol{Y}^{(k)})$ of the system
(b) Apply the Particle Filter to the data generated
(c) Draw a graph comparing the filtered values of $X_1^{(k)}$ and the real ones

- Test different values of σ_v.

6.5 Determination of the Distribution of the Noise

Uncertainty Quantification can be used to determine the distribution of the noises $\boldsymbol{v}, \boldsymbol{w}$, whenever it is possible to generate a sample for the variable under analysis. For instance,

- Determination of the distribution of the noise $\boldsymbol{v}^{(k)}$ can be made when a sample of $(\boldsymbol{x}^{(k)}, \boldsymbol{y}^{(k)})$ is available – or a trajectory $(\boldsymbol{x}^{(0)}, \boldsymbol{y}^{(0)}), \ldots, (\boldsymbol{x}^{(n)}, \boldsymbol{y}^{(n)})$ is available and the distribution of $\boldsymbol{v}^{(k)}$ is the same for $0 \leq k \leq n$.
- Determination of the noise $\boldsymbol{w}^{(k)}$ can be made when a sample of $(\boldsymbol{x}^{(k)}, \boldsymbol{x}^{(k+1)})$ is available – or a trajectory $\boldsymbol{x}^{(0)}, \ldots, \boldsymbol{x}^{(n)}$ is available and the distribution of $\boldsymbol{w}^{(k)}$ is the same for $0 \leq k \leq n$.

We shall illustrate the approach in the sequel.

Example 6.11 Let us consider a system governed by the equation:

$$x^{(k+1)} = Ax^{(k)} + Bu^{(k)} + Dw^{(k)}.$$

Assume that the matrices A and B are those presented in Example 6.4, and

$$D = \frac{1}{4}\begin{pmatrix} 2 & 1 & 0 \\ 0 & \sqrt{5} & 0 \\ 0 & 0 & 4 \end{pmatrix}.$$

$w = (w_1, w_2, w_3)'$ is a vector of independent variables having unknown distributions. To determine the distribution of w, we have a trajectory $x^{(0)}, \ldots, x^{(kmax)}$ in file `UQ1.Rdata`. This data was generated as follows:

(continued)

Example 6.11 (continued)

```
x0 = matrix(c(0.6,0.5,0.5), ncol = 1)
sigw = 0.5
sigv = 0.5
A = matrix( c(1.1269, -0.4940,0.1129,1,0,0 ,0,1,0), nrow = 3, by
row=TRUE)
B = matrix(c(-0.3832, 0.5919, 0.5191), ncol=1)
C = diag(length(x0))
C[[2,2]] = 0
C[[3, 3]] = 0
t = seq(0,100)
u = matrix(sin(2*pi*t/100), nrow = 1)
matD = 0.25*matrix(c(2, 1, 0, 0, sqrt(5), 0, 0, 0, 4), ncol = 3,
byrow = TRUE)
sig = matrix(c(1,0.5,1.5),nrow = 3)
S = sigw*sig
dx12 = matrix(NA, nrow = 2, ncol = length(t)-1)
for (i in 1: (length(t)-1)){
  dx12[ ,i] = S[1:2] * matrix(rnorm(2), ncol = 1)
}
library(extraDistr)
dx3 = rlaplace(length(t)-1,0,sigw)
xs = matrix(NA,nrow= 3, ncol = length(t))
xs[ , 1] = x0
for (k in 2: length(t)){
  xs[ ,k] = A %*% matrix(xs[ , k-1],ncol=1) + B*u[[k-1]] + matD
%*% matrix(c(dx12[ ,k-1],dx3[[k-1]]), ncol = 1)}
save(list=c("A","B","C",”D”,"t","u","xs"), file = "UQ1.Rdata")
```

Thus (w_1, w_2) is a Gaussian vector having mean zero, while w_3 is Laplace distributed. We load the data and we determine a sample of w as follows:

```
load(file = "UQ1.Rdata")
wo = matrix(nrow = 3, ncol = length(t) - 1)
for (k in 1: (length(t)-1)){
  wo[, k] = xs[ ,k+1] - A %*% xs[ , k] - B*u[[k]]}
```

We start by determining an UQ representation of the variables. For instance, for w_1:

(continued)

Example 6.11 (continued)

```
eps = 1e-8
k = 3
u1 = sort(rnorm(length(t)-1))
a1 = min(u1)
b1 = max(u1)
source("polynom.R")
source("expansion1D.R")
phi1 = polynom$new(a1,b1,1e-3)
exp1 = expansion1D$new(phi1)
aux1 = sort(wo[1, ],index.return = TRUE)
Zs1 = aux1$x # ordering to create a correlation
idZ1 = aux1$i
cZ1 = exp1$colloc(k,eps,u1,Zs1)
pZ1 = exp1$proj(cZ1,u1)
p1a = numeric(length = length(pZ1))
p1a[idZ1] = pZ1  # come back to initial order
```

Once the representation is determined, we can use it to generate a large sample of ns variates from w_1 : the computational cost is the cost of the evaluation of the polynomial representing the variable. The results for $ns = 1E5$ are shown in Figs. 6.21 and 6.22. The RMS errors were $5E - 2$. The PDF was evaluated by particle derivative of the empirical CDF, using the class `sampleX.R`.

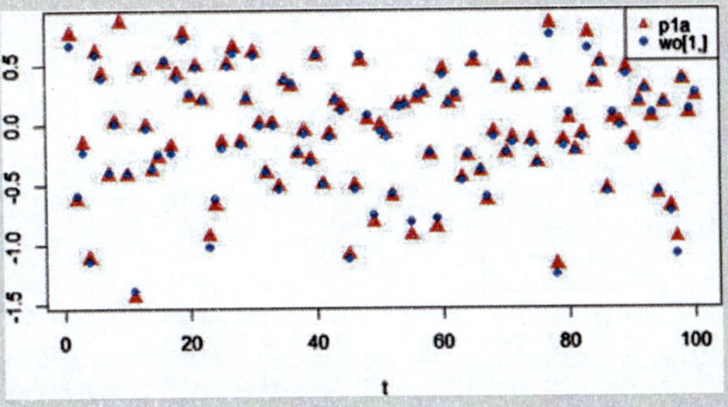

Fig. 6.21 UQ representation of w – component 1

(continued)

Example 6.11 (continued)

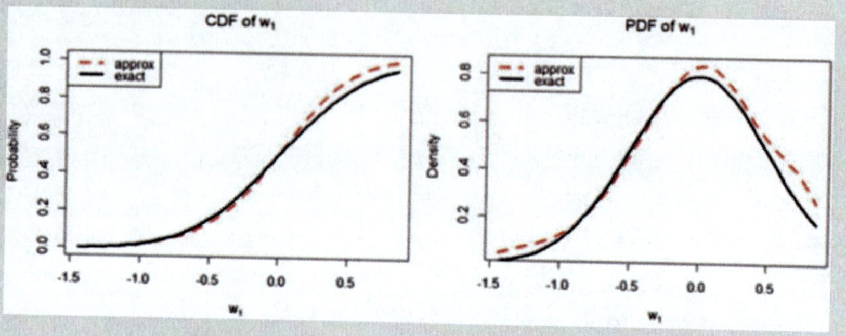

Fig. 6.22 CDF and PDF generated by the representation of w – component 1

The same approach can be applied to the other components, with similar results. Examples of results are shown in Figs. 6.23 and 6.24. We can verify the quality of the result by generating several trajectories and determining the mean value.

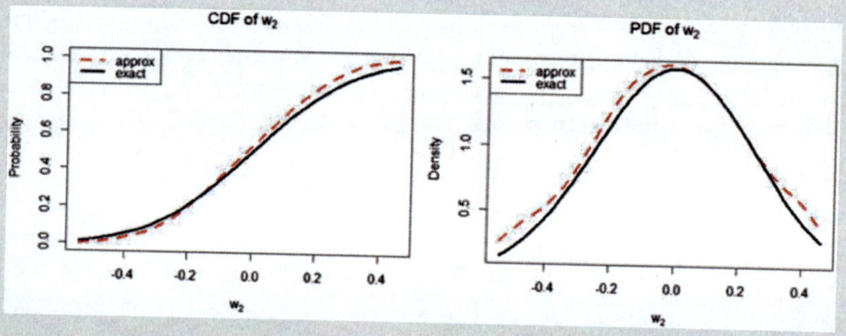

Fig. 6.23 CDF and PDF generated by the representation of w – component 2

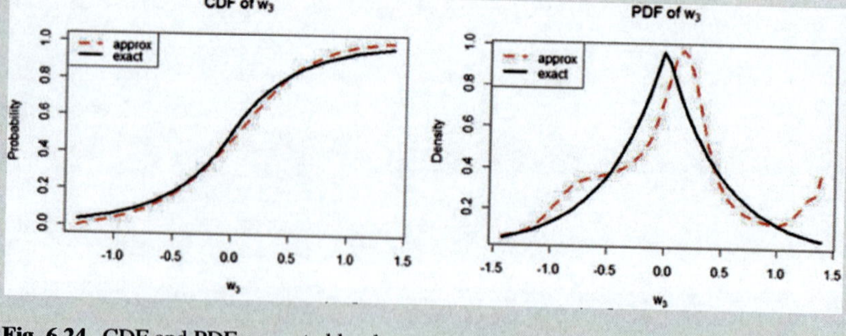

Fig. 6.24 CDF and PDF generated by the representation of w – component 3

(continued)

Example 6.11 (continued)

It is expected that the result will be the unnoisy evolution of the system. We can generate such a sample as follows:

```
xcumul = matrix(0, nrow = length(x0), ncol = length(t))
w = matrix(0, nrow = length(x0), ncol = length(t))
nsxmax = 1e3
for (nsx in 1: nsxmax){
  auxs = rnorm(2*length(t))
  ind1 = which( (auxs >= a1) & (auxs <= b1))
  us = auxs[ind1]
  p1s = exp1$proj(cZ1,us)
  w[1 ,] = p1s[1:length(t)]
  auxs = rnorm(2*length(t))
  ind2 = which( (auxs >= a2) & (auxs <= b2))
  us = auxs[ind2]
  p2s = exp2$proj(cZ2,us)
  w[2 ,] = p2s[1:length(t)]
  auxs = rnorm(2*length(t))
  ind3 = which( (auxs >= a3) & (auxs <= b3))
  us = auxs[ind3]
  p3s = exp3$proj(cZ3,us)
  w[3 ,] = p3s[1:length(t)]
xsb = matrix(NA,nrow= 3, ncol = length(t))
xsb[ , 1] = x0
for (k in 2: length(t)){
  xsb[ ,k] = A %*% matrix(xsb[ , k-1],ncol=1) + B*u[[k-1]] + mat
D %*% w[ ,k-1]
}
xcumul = xcumul + xsb}
xest = xcumul/nsxmax
```

The results for a sample of size $ns = 1E3$ are shown in Figs. 6.25, 6.26 and 6.27. As expected, the mean evolution is close to the unnoisy one.

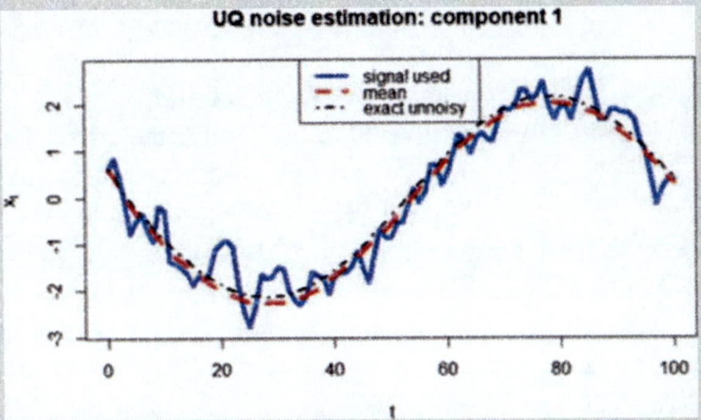

Fig. 6.25 Evaluation of the quality of the representation of the noise w – component 1

(continued)

Example 6.11 (continued)

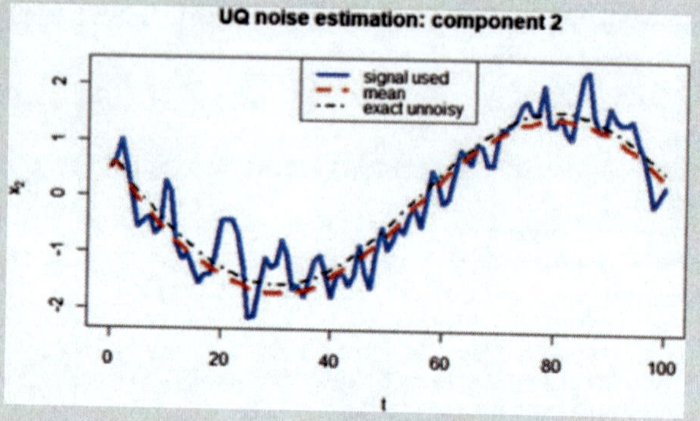

Fig. 6.26 Evaluation of the quality of the representation of the noise w – component 2

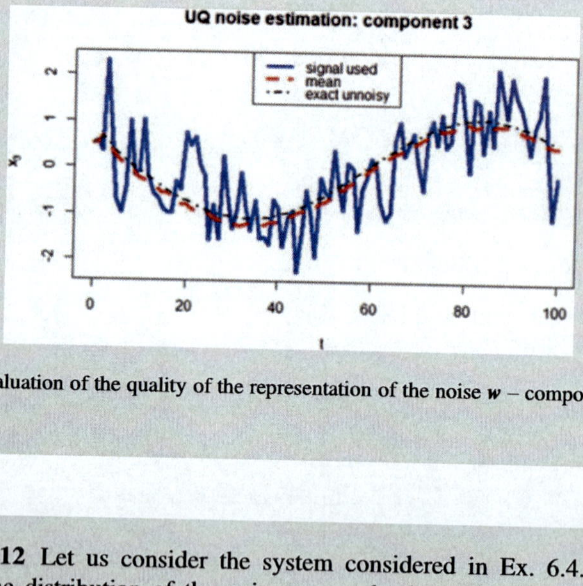

Fig. 6.27 Evaluation of the quality of the representation of the noise w – component 3

Example 6.12 Let us consider the system considered in Ex. 6.4. Let us determine the distribution of the noise v_1 – only the first component y_1 is observed. Assume that the data consist in the observation of one trajectory of the system, as in Ex. 6.4. We start by generating a sample from v_1 by taking $v_1^{(k)} = y_1^{(k)} - x_1^{(k)}$. Then, we determine an UQ representation of v_1. For instance, let us use a Laplace's variable to determine the representation:

(continued)

Example 6.12 (continued)

```
load(file = "UQ2.Rdata")
vo = yo[1, ] - xo[1, ]
eps = 0
k = 5
library(extraDistr)
u1 = sort(rlaplace(length(vo)))
a1 = min(u1)
b1 = max(u1)
source("polynom.R")
source("expansion1D.R")
phi1 = polynom$new(a1,b1,1e-3)
exp1 = expansion1D$new(phi1)
aux1 = sort(vo,index.return = TRUE)
Zs1 = aux1$x # ordering to create a correlation
idZ1 = aux1$i
cZ1 = exp1$colloc(k,eps,u1,Zs1)
pZ1 = exp1$proj(cZ1,u1)
p1a = numeric(length = length(pZ1))
p1a[idZ1] = pZ1  # come back to initial order
```

As in the preceding example, the representation is used to generate a large sample of ns variates from v_1. The results for $ns = 1E5$ are shown in Figs. 6.28 and 6.29. The PDF was evaluated by particle derivative of the empirical CDF, using the class `sampleX.R`.

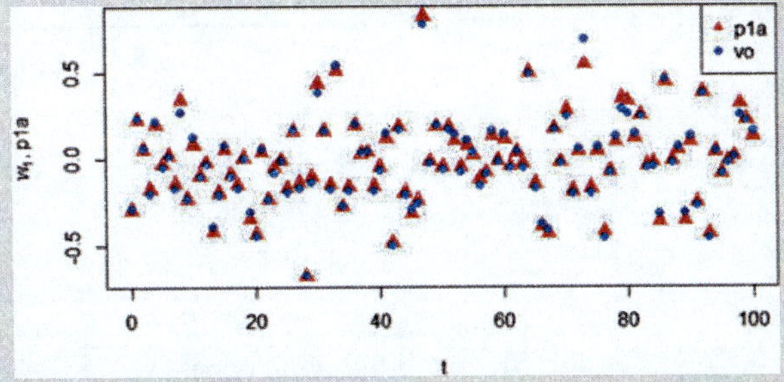

Fig. 6.28 UQ representation of v_1

(continued)

Example 6.12 (continued)

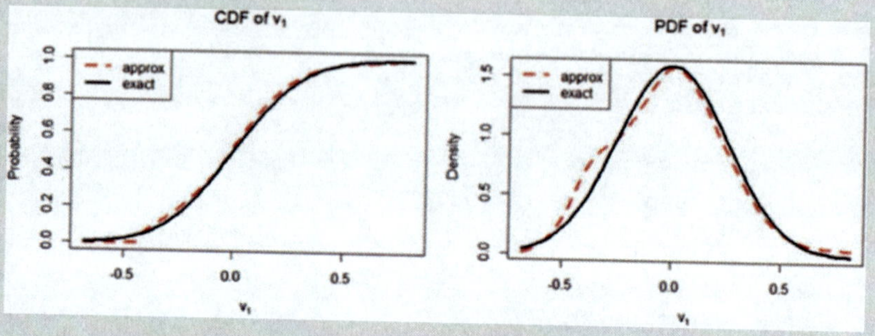

Fig. 6.29 CDF and PDF generated by the representation of v_1

Again, we can test the quality of the representation of the noise, as shown in the preceding example. A result is shown in Fig. 6.30.

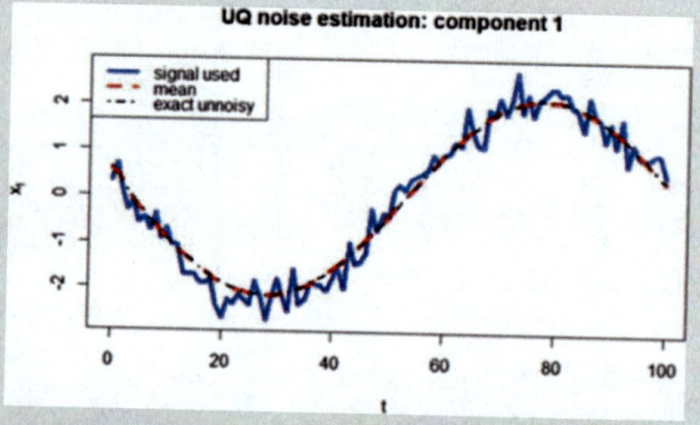

Fig. 6.30 Evaluation of the quality of the representation of the noise $\boldsymbol{v}$ – component 3

Example 6.13 Let us consider the differential equation

$$\frac{dx}{dt} = \exp(-x), x(0) = 0.$$

(continued)

Example 6.13 (continued)

This equation has as analytical solution $x = \log((1/)$. Let us consider a discretization by Euler's method and the addition of an error:

$$x^{(k+1)} = \left(1 + w^{(k)}\right)\left(x^{(k)} + \Delta t \exp\left(-x^{(k)}\right)\right)$$

Let the observation be

$$y^{(k)} = \left(1 + v^{(k)}\right)x^{(k)}.$$

We generate the data as follows:

```
x0 = 0
sigw = 0.05
sigv = 0.05
tmax = 1
nt = 100
dt = tmax/nt
t = seq(0,tmax,length.out = nt+1)
f0 = function(t,x){x + dt*exp(-x)}
f = function(t,x,w){(1+w)*f0(t,x)}
g = function(t,x,v){x + v*x}
w = sigw*rnorm(length(t))
v = sigv*rnorm(length(t))
xo = numeric(length = nt +1)
yo = numeric(length = nt +1)
xo[[1]] = x0
yo[[1]] = g(x0,0,v[[1]])
for (i in 1:nt){
  xo[[i+1]] = f(t[[i]],xo[[i]],w[[i]])
  yo[[i+1]] = g(t[[i+1]],xo[[i+1]],v[[i+1]])}
save(list=c("xo","yo","x0","t","f0","f","g","dt","nt","tmax"), f
ile = "UQ3.Rdata")
```

(continued)

Example 6.13 (continued)

Assume that we do not know that errors are of multiplicative type and we still use additive models:

```
# distribution of v
vo = yo - xo
u1 = sort(rnorm(length(vo)))
a1 = min(u1)
b1 = max(u1)
source("polynom.R")
source("expansion1D.R")
phi1 = polynom$new(a1,b1,1e-3)
exp1 = expansion1D$new(phi1)
kv = 3
auxv = sort(vo,index.return = TRUE)
Zv1 = auxv$x # ordering to create a correlation
idv1 = auxv$ix
cv = exp1$colloc(kv,0,u1,Zv1)

# distribution of w
wo = numeric(length = length(t)-1)
for (k in 2: length(t)){
  wo[[k-1]] = xo[[k]] - f0(t[[k-1]],xo[[k-1]])}
u2 = sort(rnorm(length(wo)))
a2 = min(u2)
b2 = max(u2)
source("polynom.R")
source("expansion1D.R")
phi2 = polynom$new(a2,b2,1e-3)
exp2 = expansion1D$new(phi2)
kw = 7
auxw = sort(wo,index.return = TRUE)
Zw2 = auxw$x # ordering to create a correlation
idw2 = auxw$ix
cw = exp2$colloc(kw,0,u2,Zw2)
```

The results are shown in Fig. 6.31.

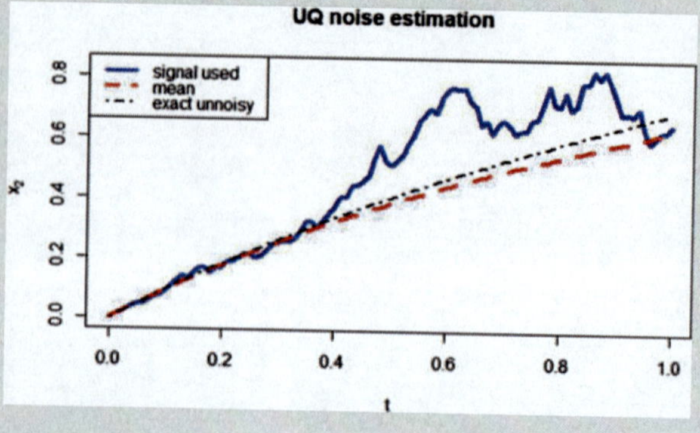

Fig. 6.31 UQ noise estimation using an additive model for multiplicative noise

(continued)

Example 6.13 (continued)

Examples of results using the correct multiplicative model are shown in Figs. 6.32, 6.33 and 6.34.

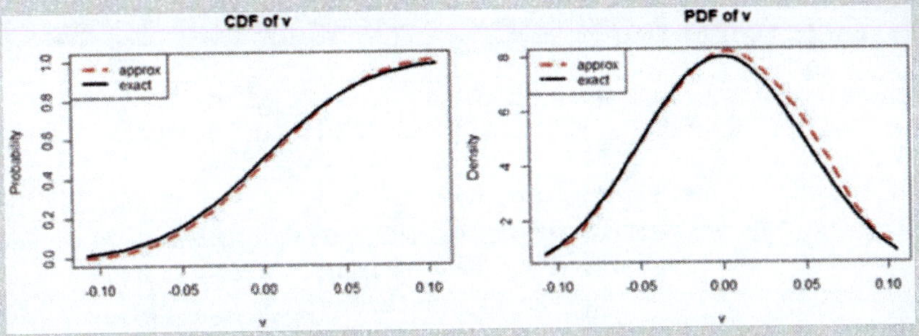

Fig. 6.32 CDF and PDF generated by the representation of $\boldsymbol{v}$, using the multiplicative model

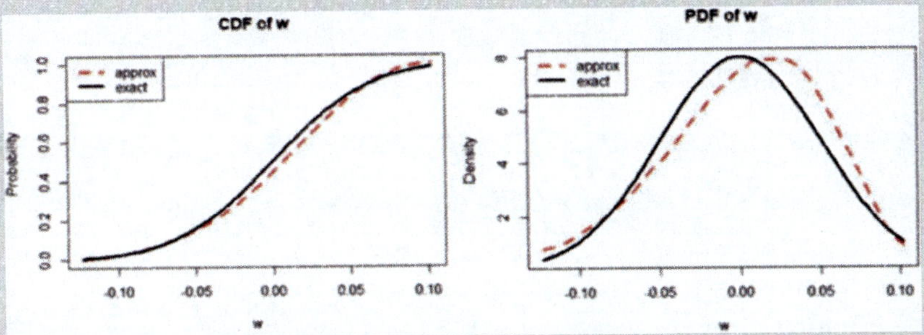

Fig. 6.33 CDF and PDF generated by the representation of $\boldsymbol{w}$, using a multiplicative model

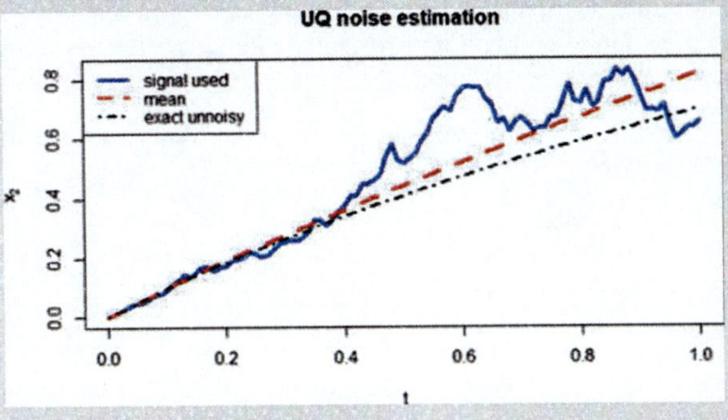

Fig. 6.34 UQ noise estimation using an additive model for multiplicative noise

We can also recursively determine a representation of $\boldsymbol{x}^{(k)}$, which can be used for noise estimation, filtering, or prediction. The reader can find this approach in (Lopez et al. 2016, 2018, 2019).

Exercises

1. Let

- $$\begin{pmatrix} X_1^{(k+1)} \\ X_2^{(k+1)} \end{pmatrix} = \begin{pmatrix} 1 & \Delta t \\ -\omega^2 \Delta t & 1 \end{pmatrix} \begin{pmatrix} X_1^{(k)} \\ X_2^{(k)} \end{pmatrix} + \begin{pmatrix} 0 \\ u^{(k)} \Delta t \end{pmatrix} + \Delta t \begin{pmatrix} w_1 \\ w_2 \end{pmatrix}.$$
- Only the first variable is observed: $y_1^{(k)} = x_1^{(k)} + v^{(k)}$. Assume that the observations are made for $1 \le k \le 100$, with $\Delta t = 0.1$, and, on the one hand, all the noises w_1, w_2, v are independent Gaussian $N(0, \sigma)$, with $\sigma_w = 3\sigma_v$; on the other hand, $u^{(k)} = k\Delta t \sin(k\Delta t)$.

  (a) Generate a trajectory $(\boldsymbol{X}^{(k)}, \boldsymbol{Y}^{(k)})$ of the system
  (b) Determine a representation of the noises using a Gaussian variable $N(0, 1)$
  (c) Determine a representation of the noise using a Uniform Variable $U(0, 1)$

- Test different values of σ_v.

2. Let

- $$\begin{pmatrix} X_1^{(k+1)} \\ X_2^{(k+1)} \end{pmatrix} = \begin{pmatrix} 1 & \Delta t \\ 0 & 1 \end{pmatrix} \begin{pmatrix} X_1^{(k)} \\ X_2^{(k)} \end{pmatrix} + \begin{pmatrix} 0 \\ u^{(k)} \Delta t \end{pmatrix} + \Delta t \begin{pmatrix} w_1 \\ w_2 \end{pmatrix}.$$
- Only the first variable is observed: $y_1^{(k)} = x_1^{(k)} + v^{(k)}$. Assume that the observations are made for $1 \le k \le 100$, with $\Delta t = 0.1$, and, on the one hand, all the noises w_1, w_2, v are independent Gaussian $N(0, \sigma)$, with $\sigma_w = 3\sigma_v$; on the other hand, $u^{(k)} = k\Delta t \sin(k\Delta t)$.

  (a) Generate a trajectory $(\boldsymbol{X}^{(k)}, \boldsymbol{Y}^{(k)})$ of the system
  (b) Determine a representation of the noises using a Gaussian variable $N(0, 1)$
  (c) Determine a representation of the noise using a Uniform Variable $U(0, 1)$

- Test different values of σ_v.

(continued)

3. Let

- $$\begin{pmatrix} X_1^{(k+1)} \\ X_2^{(k+1)} \end{pmatrix} = \begin{pmatrix} \cos(\theta) & -\sin(\theta) \\ \sin(\theta) & \cos(\theta) \end{pmatrix} \begin{pmatrix} X_1^{(k)} \\ X_2^{(k)} \end{pmatrix} + \begin{pmatrix} w_1 \\ w_2 \end{pmatrix}.$$
- Both the variables are observed: $y_1^{(k)} = x_1^{(k)} + v_1^{(k)}$, $y_2^{(k)} = x_2^{(k)} + v_2^{(k)}$. Assume that the observations are made for $1 \leq k \leq 100$, and, on the one hand, all the noises w_1, w_2, v_1, v_2 are independent Gaussian $N(0, \sigma)$, with $\sigma_w = 3\sigma_v$; on the other hand, $\theta = \pi/16$.
  (a) Generate a trajectory $(\boldsymbol{X}^{(k)}, \boldsymbol{Y}^{(k)})$ of the system
  (b) Determine a representation of the noises using a Gaussian variable $N(0, 1)$
  (c) Determine a representation of the noise using a Uniform Variable $U(0, 1)$
- Test different values of σ_v. Test the case where $\theta = \pi/4$.

4. Let

- $$\begin{pmatrix} X_1^{(k+1)} \\ X_2^{(k+1)} \\ X_3^{(k+1)} \\ X_4^{(k+1)} \end{pmatrix} = \begin{pmatrix} 1 & 0 & \Delta t & 0 \\ 0 & 1 & 0 & \Delta t \\ 0 & 0 & 1 & 0 \\ 0 & 0 & 0 & 1 \end{pmatrix} \begin{pmatrix} X_1^{(k)} \\ X_2^{(k)} \\ X_3^{(k)} \\ X_4^{(k)} \end{pmatrix} + \begin{pmatrix} w_1 \\ w_2 \\ w_3 \\ w_4 \end{pmatrix}.$$
- Only the two first variables are observed: $y_1^{(k)} = x_1^{(k)} + v_1^{(k)}$, $y_2^{(k)} = x_2^{(k)} + v_2^{(k)}$. Assume that the observations are made for $1 \leq k \leq 100$, and, on the one hand, all the noises w_i, v_j are independent Gaussian $N(0, \sigma)$, with $\sigma_w = \sigma_v$.
  (a) Generate a trajectory $(\boldsymbol{X}^{(k)}, \boldsymbol{Y}^{(k)})$ of the system
  (b) Determine a representation of the noises using a Gaussian variable $N(0, 1)$
  (c) Determine a representation of the noise using a Uniform Variable $U(0, 1)$
- Test different values of σ_v.

6.6 Bayesian Optimization

Sequential Bayesian Estimation generates an algorithm for Global Optimization, usually referred as "Bayesian Optimization." The model problem to be solved is:

$$\boldsymbol{x}^* = \arg\min\{f(\boldsymbol{x}) : \boldsymbol{x} \in \mathbb{R}^n\}.$$

Here, $f : \mathbb{R}^n \rightarrow \mathbb{R}$ is the objective function to be minimized.

Bayesian Optimization is based on the basic principle of Bayesian Update. A typical iteration of a Bayesian Optimization Algorithm (BOA) reads as

1. Generate a surrogate model of the objective function f, under the form of a Stochastic Process
2. Select a new point
3. Update the Stochastic Process using the Bayesian Update

To select a new point, a BOA uses an auxiliary optimization problem, involving an acquisition function a, destined to be maximized. A typical BOA reads as

1. Initialization: select n_0 initial points. Evaluate the objective function f in these points and set a stochastic process (for instance, Gaussian) describing these values of f. Select an acquisition function a, destined to generate new points.
2. At iteration $k > 0$: select the point $x^{(k+1)}$ that maximizes the acquisition function, evaluate $f(x^{(k+1)})$, update the stochastic process using this new value.
3. Increment k and test for the end of iterations. If the stopping conditions are satisfied, stop and return the best value determined along the iterations. Otherwise, go to step 2.

The reader can find in the literature and on internet number of tutorials and implementations of this approach. For instance, the packages `laGP`, `mlrMBO` and `rBayesianOptimization` in R; the tutorials (Brochu et al. 2010; Frazier 2018; Popov 2019; Adyatama 2019; Ahmed and Prince 2020; Community 2023).

Among the popular acquisition functions, we can cite the expected improvement: let $f^{(k)}$ be the best value observed in the past, at step $k > 0$, and $G^{(k)}$ be the distribution of the Stochastic Process at step $k > 0$. Then, the expected improvement is

$$a_{EI}(x) = E\left(u_{EI}(x)|G^{(k)}\right), u_{EI}(x) = \max\left\{0, f^{(k)} - f(x)\right\}.$$

A second popular acquisition function is the probability of improvement:

$$a_{POI}(x) = P\left(f(x) < f^{(k)}|G^{(k)}\right).$$

Notice that

$$a_{POI}(x) = E\left(u_{POI}(x)|G^{(k)}\right), \quad u_{POI}(x) = \begin{cases} 0, \textit{if } f(x) \geq f^{(k)} \\ 1, \textit{if } f(x) \leq f^{(k)} \end{cases}.$$

The reader will find in the literature other acquisition functions (Wilson et al. 2018; Gan et al. 2021) and associated numerical methods (Pawar and Warbhe 2021).

In terms of Bayesian Update, the choice of a Gaussian Process simplifies the procedure since the Bayesian Conjugate of a Gaussian Process is a Gaussian Process yet so that the update corresponds to a simple modification of the parameters defining the process. In addition, the maximum of a_{EI} and a_{POI} can be easily determined. For instance, if $G^{(k)} \approx N(m^{(k)}(x), \sigma^{(k)}(x))$, then

$$a_{EI}(x) = \sigma^{(k)}(x)(z(x)\Phi(z(x)) + \varphi(z(x))),$$

where Φ is the CDF of $N(0,1)$, φ is the associated PDF and

$$z(x) = \frac{f^{(k)} - m^{(k)}(x)}{\sigma^{(k)}(x)}.$$

Analogously,

$$a_{POI}(x) = \Phi(z(x)).$$

A less popular, but easy to calculate acquisition functions are the lower confidence bound and the upper confidence bound .

$$a_{LCB}(x) = m^{(k)}(x) - z_\alpha \sigma^{(k)}(x), \quad a_{UCB}(x) = m^{(k)}(x) + z_\alpha \sigma^{(k)}(x).$$

In practice, z_α is a parameter and not a p-value associated with a confidence interval.

The Gaussian stochastic process $G^{(k)}$ is generally defined by assuming that $Y = (f(x^{(1)}, \ldots, x^{(k)})$ is a Multivariate Gaussian Process $N(\boldsymbol{m}^{(k)}, \boldsymbol{\Sigma}^{(k)})$, having as covariance matrix $\Sigma_{ij}^{(k)} = K(x_i, x_j)$, where K is a correlation function (or kernel), while $m_i^{(k)}(x)$ is a mean function. Popular kernels take the form $K(x_i, x_j) = k(\|x_i - x_j\|)$. For instance, the Exponential Kernel, with

$$k(r) = \alpha_0 \exp(-r^\nu),$$

and the Matérn Kernel

$$k(r) = \alpha_0 \frac{2^{1-\nu}}{\Gamma(\nu)} \left(\frac{r}{\ell}\sqrt{2\nu}\right)^\nu K_\nu\left(\frac{r}{\ell}\sqrt{2\nu}\right).$$

Here, Γ is the Gamma function, and K_ν is the modified Bessel function of second kind. ν, α_0, ℓ are strictly positive parameters to be defined by the user. A typical mean function is generated by an expansion using a basis $\{\psi_i : i \in \mathbb{N}\}$:

$$m^{(k)}(x) = \sum_{i=0}^{n} \beta_i \psi_i(x).$$

Example 6.14 Let us consider the Rastrigin function

$$f(x) = n + \sum_{i=1}^{n} \left((x_i - i)^2 - \cos(2\pi(x_i - i)) \right).$$

Under R, we write this function as

```
frastr = function(x){length(x) + sum(x^2 - cos(2*pi*x))}
```

Let us use the package `mlrMBO` to find an approximated solution in dimension 3, using the acquisition by probability of improvement. The search is made on the region $(-100, 100)^3$. Using the acquisition by expected improvement (`makeMBOInfillCritEI()`):

```
library(mlrMBO)
library(smoof)

obj.fun = makeSingleObjectiveFunction(
  name = "DecenteredRastrigin",
  fn = function(x) {
    aux = x - seq(1: length(x))
    r1 =  aux^2 - cos(2*pi*aux)
    r = length(x) + sum(r1)
    return(r)},
  par.set = makeParamSet(
    makeNumericVectorParam("x", len = 3, lower = -100, upper = 1
00) ),
  minimize = TRUE)

des = generateDesign(n = 20, par.set = getParamSet(obj.fun), fun
= lhs::randomLHS)

control = makeMBOControl()
control = setMBOControlTermination(control, iters = 200)
control = setMBOControlInfill(control, crit = makeMBOInfillCritE
I())

surr.km = makeLearner("regr.km", predict.type = "se", covtype =
"matern3_2", control = list(trace = FALSE))

sol = mbo(obj.fun, design = des, learner = surr.km, control = co
ntrol, show.info = TRUE))
```

(continued)

Example 6.14 (continued)

Below, an example of result (the exact solution is (1,2,3)):

```
[1] "solution found :  c(0.869467597285984, 1.9807269478774,
2.94851143193604)"
```

Let us use the Lower Confidence Bound (`makeMBOInfillCritCB()`):

```
ctrl = setMBOControlInfill(ctrl, crit = makeMBOInfillCritCB())
```

An example of result is

```
[1] "solution found :  c(1.08976812759066, 2.00241992579541,
2.96630677024368)"
```

Example 6.15 Let us consider again the Rastrigin function

$$f(x) = n + \sum_{i=1}^{n} \left((x_i - i)^2 - \cos(2\pi(x_i - i)) \right).$$

Under R, you can easily generate your own simple BOA, making a minimization coordinate by coordinate, with acquisition by expected improvement:

```
library(pracma)
spam = function(x,f){
  dx = dim(x)
  v = matrix(0, nrow = dx[[1]], ncol = dx[[2]])
  for (i in 1: length(x)){
    v[[i]] = f(x[[i]])}
  return(v)}

spam2 = function(x,y,f){
  v = matrix(0, length(x),length(y))
  for (i in 1: length(x)){
    for (j in 1: length(y)){
      v[[i,j]] = f(x[[i]],y[[j]])}}
  return(v)}
```

(continued)

Example 6.15 (continued)

```
nitextmax = 100
x_acnd = rnorm(ndim)
f_acnd = fobjnd(x_acnd)
for (nitext in 1: nitextmax){
  for (i in 1: ndim){
    x_ini = matrix(c(a + (b-a)*rnorm(5), x_acnd[[i]]), nrow = 1)
    fobj = function(xxi){
      xxx = x_acnd
      xxx[[i]] = xxi
      r = fobjnd(xxx)
      return(r)}
    x_g = matrix(seq(a,b,length.out = 1001), nrow = 1)
    f_g = spam(x_g, fobj)
    mu = function(x){x}
    x_ac = matrix(x_ini, nrow = 1)
    f_ac = spam(x_ac,fobj)
    ind = which.min(f_ac)
    fstar_n = f_ac[[ind]]
    xstar_n = x_ac[[ind]]

    for (nit in 1: nitmax){
      Sig = function(x) {spam2(x_ac,x,kernel)}
      aux =  spam2(x_ac, x_ac,kernel)
      C = solve(aux + epsi*diag(length(x_ac)))
      m = spam(x_ac, mu)
      mu_n = function(x) {Sig(x) %*% C %*% (f_ac - m) + mu(x)}
      s_n = function(x){kernel(x,x) - t(Sig(x)) %*% C %*% t(spam
2(x,x_ac,kernel))}
```

Using

```
      delta_n = function(x) {mu(x) - fstar_n}
      EI_n = function(x) {max(delta_n(x), 0) + s_n(x)*dnorm(delt
a_n(x)/s_n(x)) - abs(delta_n(x))*pnorm(delta_n(x)/s_n(x))}
      negEI_n = function(x){ -EI_n(x)}
      x_new = fminbnd(negEI_n,a,b)$xmin
      x_ac = matrix(c(x_ac, x_new), nrow = 1)
      f_ac = matrix(c(f_ac, fobj(x_new)), nrow = 1)
      ind = which.min(f_ac)
      fstar_n = f_ac[[ind]]
      xstar_n = x_ac[[ind]]}
    x_acnd[[i]] = xstar_n
    f_acnd = fstar_n
}}
```

(continued)

Example 6.15 (continued)

An example of result is the following one – recall that the exact solution is (1,2,3):

```
fobjnd = function(x){frastr(x)}
```

This algorithm can be easily hybridized with another one. For instance, let us introduce Stochastic Perturbations of the Gradient Descent (see, for instance, (Pogu and Souza de Cursi 1994)):

```
[1] "solution found :  c(1.06413260420848, 2.12461726614363,
3.18524080943234)"
```

An example of result is the following:

```
for (nitext in 1: nitextmax){
  sigpert = sigp/sqrt(log(nitext + 1))
  xt = x_acnd
  gf = gradfobjnd(xt)
  xt = xt - pas*gf
  ft = fobjnd(xt)
  for (npert in 1: npertmax){
    dx = sigpert*rnorm(length(x_acnd))
    xtt = xt + dx
    ftt = fobjnd(xtt)
    if (ftt < ft){
      ft = ftt
      xt = xtt}}
   if (ft < f_acnd){
```

(continued)

Example 6.15 (continued)

```
    f_acnd = ft
    x_acnd = xt}
    for (i in 1: ndim){
      x_ini = matrix(c(a + (b-a)*rnorm(5), x_acnd[[i]]), nrow =
1)
      fobj = function(xxi){
        xxx = x_acnd
        xxx[[i]] = xxi
        r = fobjnd(xxx)
        return(r)}
      gradfobj = function(xxi){
        xxx = x_acnd
        xxx[[i]] = xxi
        auxg = gradfobjnd(xxx)
        rg = auxg[[i]]
        return(rg)}
      x_g = matrix(seq(a,b,length.out = 1001), nrow = 1)
      f_g = spam(x_g, fobj)
      mu = function(x){x}
      x_ac = matrix(x_ini, nrow = 1)
      f_ac = spam(x_ac,fobj)
      ind = which.min(f_ac)
      fstar_n = f_ac[[ind]]
      xstar_n = x_ac[[ind]]

      for (nit in 1: nitmax){
        Sig = function(x) {spam2(x_ac,x,kernel)}
        aux =  spam2(x_ac, x_ac,kernel)
        C = solve(aux + epsi*diag(length(x_ac)))
        m = spam(x_ac, mu)
        mu_n = function(x) {Sig(x) %*% C %*% (f_ac - m) + mu(x)}
        s_n = function(x){kernel(x,x) - t(Sig(x)) %*% C %*%
t(spam2(x,x_ac,kernel))}
        delta_n = function(x) {mu(x) - fstar_n}
        EI_n = function(x) {max(delta_n(x), 0) +
s_n(x)*dnorm(delta_n(x)/s_n(x)) -
abs(delta_n(x))*pnorm(delta_n(x)/s_n(x))}
        negEI_n = function(x){ -EI_n(x)}
        x_new = fminbnd(negEI_n,a,b)$xmin

        x_ac = matrix(c(x_ac, x_new), nrow = 1)
        f_ac = matrix(c(f_ac, fobj(x_new)), nrow = 1)
        ind = which.min(f_ac)
        fstar_n = f_ac[[ind]]
        xstar_n = x_ac[[ind]]
      }
      x_acnd[[i]] = xstar_n
      f_acnd = fstar_n
      xt = x_acnd
      gf = gradfobjnd(xt)
```

(continued)

Example 6.15 (continued)

```
xt = xt - pas*gf
ft = fobjnd(xt)
for (npert in 1: npertmax){
  dx = sigpert*rnorm(length(x_acnd))
  xtt = xt + dx
  ftt = fobjnd(xtt)
  if (ftt < ft){
    ft = ftt
    xt = xtt}}
if (ft < f_acnd){
  f_acnd = ft
  x_acnd = xt}
print(paste('*** i = ', i,', optimum = ',f_acnd))}}
```

```
[1] "solution found :   c(0.98942728950275, 2.00003809225827,
3.00184302446216)"
```

Exercises

1. Consider the Griewank's decentered function

$$f(x) = 1 + \frac{1}{4000}\sum_{i=1}^{n}(x_i - i)^2 - \prod_{i=1}^{n}\cos\left(\frac{x_i - i}{\sqrt{i}}\right).$$

   - *Use* a BOA to determine its point of minimum on $(-500,500)^n$, for $n = 3, 5, 10$.

2. Consider the Ackley's decentered function

$$f(x) = \exp(1) + 20 - \exp\left(-\frac{2}{10}\sqrt{\frac{1}{n}\sum_{i=1}^{n}(x_i - i)^2}\right) - \exp\left(\frac{1}{n}\sum_{i=1}^{n}\cos\left(\frac{x_i - i}{2\pi}\right)\right).$$

   - *Use* a BOA to determine its point of minimum on $(-30, 30)^n$, for $n = 3$, 5, 10.

(continued)

3. Consider the Schweffel's decentered function

$$f(x) = 420.9829n - \sum_{i=1}^{n} (x_i - i) \sin\left(\sqrt{|x_i - i|}\right).$$

- Use a BOA to determine its point of minimum on $(-500,500)^n$, for $n = 3, 5, 10$.

4. Consider the Rosenbrock's modified function

$$f(x) = \sum_{i=1}^{n} \left(\left(\frac{x_{i+1}}{i+1} - \left(\frac{x_i}{i} - 1\right)^2 \right)^2 + \frac{1}{100}\left(\frac{x_i}{i} - 1\right)^2 \right).$$

- *Use* a BOA to determine its point of minimum on $(-500,500)^n$, for $n = 3, 5, 10$.

References

Adyatama, A. (2019, December 29). *Bayesian Optimization*. Retrieved from RPubs: https://rpubs.com/Argaadya/bayesian-optimization

Agliari, A., & Parisetti, C. C. (1986). A-g Reference Informative Prior: A Note on Zellner's g-Prior. *Journal of the Royal Statistical Society. Series D (The Statistician), 37*(3), pp. 271–275. doi: https://doi.org/10.2307/2348164

Ahmed, M., & Prince, S. (2020, June 3). *Bayesian optimization*. Retrieved from https://www.borealisai.com/: https://www.borealisai.com/research-blogs/tutorial-8-bayesian-optimization/

Akaike, H. (1998). Information theory and an extension of the maximum likelihood principle. In E. Parzen, K. Tanabe, & G. Kitagawa (Eds.), *Selected Papers of Hirotugu Akaike* (pp. 199–213). New York, NY, USA: Springer. doi:https://doi.org/10.1007/978-1-4612-1694-0_15

Albert, I., Donnet, S., Guihenneuc-Jouyaux, C., Low-Choy, S., Mengersen, K., & Rousseau, J. (2002). Combining Expert Opinions in Prior Elicitation. *Bayesian Analysis, 7*(3), pp. 503–532. doi:https://doi.org/10.1214/12-BA717

Arulampalam, M., Maskell, S., Gordon, N., & and Clapp, T. (2002, February). A Tutorial on Particle Filters for Online nonlinear/non-Gaussian Bayesian tracking. *IEEE Transactions on Signal Processing, 50*(2), pp. 174–188. doi:https://doi.org/10.1109/78.978374

Ash, R. (1965, reedited 2003). *Information Theory*. New York, NY, USA: Dover.

Attiaoui, D., Doré, P., Martin, A., & Ben Yaghlane, B. (2012). A Distance between Continuous Belief Functions. In E. Hüllermeier, S. Link, T. Fober, & B. Seeger (Ed.), *Scalable Uncertainty Management: SUM 2012* (pp. 194–205). Marburg, Germany: Springer. doi:https://doi.org/10.1007/978-3-642-33362-0_15

Aurnhammer, C., & Frank, S. L. (2019). Evaluating information-theoretic measures of word prediction in naturalistic sentence reading. *Neuropsychologia, 134*. doi:https://doi.org/10.1016/j.neuropsychologia.2019.107198

Baldi, P., & Itti, L. (2010). Of bits and wows: A Bayesian theory of surprise with applications to attention. *Neural Networks, 23*(5), 649–666. doi:https://doi.org/10.1016/j.neunet.2009.12.007

Banner, K. M., Irvine, K. M., & Rodhouse, T. J. (2020, May). The use of Bayesian priors in Ecology: The good, the bad and the not great. *Methods in Ecology and Evolution, 11*(8), 882–889. doi:https://doi.org/10.1111/2041-210X.13407

Batou, A., & Soize, C. (2013). Calculation of Lagrange Multipliers in the Construction of Maximum Entropy Distributions in High Stochastic Dimension. *SIAM/ASA Journal on Uncertainty Quantification, 1*(1). doi:https://doi.org/10.1137/120901386

Bayarria, M. J., & Morales, J. (2003). Bayesian measures of surprise for outlier detection. *Journal of Statistical Planning and Inference, 111*(1–2), 3 22. doi:https://doi.org/10.1016/S0378-3758(02)00282-3

Benavoli, A. (2014). Belief function and multivalued mapping robustness in statistical estimation. *International Journal of Approximate Reasoning, 55*, 311–329. doi:https://doi.org/10.1016/j.ijar.2013.04.014

Beretta, G. P. (2008). Axiomatic Definition of Entropy for Nonequilibrium States. *International Journal of Thermodynamics, 11*(2), 39–48. doi:https://doi.org/10.5541/ijot.211

Berger, J. O. (1985). *Statistical Decision Theory and Bayesian Analysis*. New York, USA: Springer. doi:https://doi.org/10.1007/978-1-4757-4286-2

Berger, J. O., & Bernardo, J. M. (1989, March). Estimating a Product of Means: Bayesian Analysis with Reference Priors. *Journal of the American Statistical Association, 84*(405), pp. 200–207. doi:https://doi.org/10.2307/2289864

Berger, J. O., & Bernardo, J. M. (1992). On the development of reference priors. In J. M. Bernardo, J. O. Berger, A. P. Dawid, & A. F. Smith (Ed.), *Bayesian Statistics 4*. Clarendon Press.

Berger, J. O., Bernardo, J. M., & Sun, D. (2009, April). The formal definition of reference priors. *The Annals of Statistics, 37*(2), pp. 905–938. doi:https://doi.org/10.1214/07-AOS587

Boivin, C. (2022a, December 25). https://cran.r-project.org/web/packages/dst/vignettes/Captain_Example.html. Retrieved from Captain's Example: https://cran.r-project.org/web/packages/dst/vignettes/Captain_Example.html

Boivin, C. (2022b, 12 24). *Introduction to Belief Functions*. Retrieved from https://cran.r-project.org/web/packages/dst/vignettes/: https://cran.r-project.org/web/packages/dst/vignettes/Introduction_to_Belief_Functions.html#fn2

Boivin, C. (2022c, December 25). *Introduction to Belief Functions: The Monty Hall Game*. Retrieved 2022, from https://cran.microsoft.com/snapshot/2018-08-11/web/packages/dst/vignettes/Monty-hall-Example.html: https://cran.microsoft.com/snapshot/2018-08-11/web/packages/dst/vignettes/Monty-hall-Example.html

Boltzmann, L. E. (1866). Über die mechanische Bedeutung des zweiten Hauptsatzes der Wärmetheorie. *Wiener Berichte, 53*, 195–220. Retrieved from http://opacplus.bsb-muenchen.de/title/BV020135572/ft/bsb10133426?page=3

Boltzmann, L. E. (1877 traduction 2002). On the relationship between the second main theorem of mechanical heat theory and the probability calculation with respect to the results about the heat equilibrium. *Akademie der Wissenschaften in Wien Mathematisch-naturwissenschaftliche Klasse Sitzungsberichte*, 2(76), 373–435. Retrieved February 14, 2023, from http://users.polytech.unice.fr/~leroux/boltztrad.pdf

Boltzmann, L. E. (1896). *Vorlesungen �uber Gastheorie* (Vol. 1). Leipzig: Barth.

Boltzmann, L. E. (1995). *Lectures on Gas Theory*. (S. G. Brush, Trans.) New York, NY, USA: Dover.

Boyle, G. H. (n.d.). *modelingcommons.org/file/download/6101?file_id=3384*. Retrieved from http://modelingcommons.org/file/download/6101?file_id=3384

Brillouin, L. (1956). *Science and Information Theory*. New York, NY, USA: Dover.

Brochu, E., Cora, V., & de Freitas, N. (2010). *A Tutorial on Bayesian Optimization of Expensive Cost Functions, with Application to Active User Modeling and Hierarchical Reinforcement Learning*. Retrieved July 25, 2023, from https://arxiv.org/pdf/1012.2599.pdf

Campagner, A., Ciucci, D., & Denœux, T. (2022). Belief Functions and Rough Sets: Suvey and New Insights. *International Journal of Approximate Reasoning, 143*, 92–215. doi:https://doi.org/10.1016/j.ijar.2022.01.011

Chadli, A., Khawla, B., Asma, M., & Fellag, H. (2017, April). Bayesian estimation of the Rayleigh distribution under different loss function. *Electronic Journal of Applied Statistical Analysis, 10*(1), pp. 50–64. doi:https://doi.org/10.1285/i20705948v10n1p50

Chakrabarti, C. G., & Indranil, C. (2005). Shannon entropy: axiomatic characterization and application. *International Journal of Mathematics and Mathematical Sciences*. doi:https://doi.org/10.1155/IJMMS.2005.2847

Chang, Y.-C., & Hung, W.-L. (2007). LINEX Loss Functions with Applications to Determining the Optimum Process Parameters. Quality & Quantity. *Quality & Quantity , 41*(2), pp. 291–301. doi:https://doi.org/10.1007/s11135-005-5425-3

Cheung, V. K., M.C., P., Meyer, L., Pearce, M. T., Haynes, J.-D., & Koelsch, S. (2019). Uncertainty and Surprise Jointly Predict Musical Pleasure and Amygdala, Hippocampus, and Auditory Cortex Activity. *Current Biology, 29*(23), 4084–4092.e4. doi:https://doi.org/10.1016/j.cub.2019.09.067

Chui, C. K., & Chen, G. (2017). *Kalman Filtering*. Springer International Publishing AG. doi:https://doi.org/10.1007/978-3-319-47612-4

Clausius, R. J. (1850). Ueber die bewegende Kraft der Wärme und die Gesetze, welche sich daraus für die Wärmelehre selbst ableiten lassen. *Annalen der Physik*, 368–397 , 500–524.

Clausius, R. J. (1851 republished in 2009). On the Moving Force of Heat, and the Laws regarding the Nature of Heat itself which are deducible therefrom. *The London, Edinburgh, and Dublin Philosophical Magazine and Journal of Science, 2*(8), 1–21, 102–119. doi:https://doi.org/10.1080/14786445108646819, https://doi.org/10.1080/14786445108646840

Cobb, B. R., & Shenoy, P. (2006, April). On the plausibility transformation method for translating belief function models to probability models. *Journal of Approximate Reasoning, 41*(3), pp. 314–330. doi:https://doi.org/10.1016/j.ijar.2005.06.008

Community, T. R. (2023, 7 11). *Bayesian Optimization*. Retrieved from Tidy Modeling with R Book Club: https://r4ds.github.io/bookclub-tmwr/bayesian-optimization.html

Consonni, G., Fouskakis, D., Liseo, B., & Ntzoufras, I. (2018). Prior Distributions for Objective Bayesian Analysis. *Bayesian Analysis, 13*(2), 627–679. doi:https://doi.org/10.1214/18-BA1103

Couso, I., Dubois, D., & Sanchez, L. (2014). *Random Sets and Random Fuzzy Sets as Ill-Perceived Random Variables*. Springer. doi:https://doi.org/10.1007/978-3-319-08611-8

Csiszár, I. (2008). Axiomatic Characterizations of Information Measures. *Entropy, 10*(3), 261–273. doi:https://doi.org/10.3390/e10030261

Dahlin, J. (2023, July 20). *pmhtutorial: Minimal Working Examples for Particle Metropolis-Hastings*. Retrieved from https://rdrr.io/cran: https://rdrr.io/cran/pmhtutorial/

Dahlin, J., & Schön, T. B. (2019, March 26). Getting Started with Particle Metropolis-Hastings for Inference in Nonlinear Dynamical Models. *Journal of Statistical Software, Code Snippets, 88*(2), pp. 1–41. doi:https://doi.org/10.18637/jss.v088.c02

Dale, A. I. (1982). Bayes or Laplace? An Examination of the Origin and Early Applications of Bayes' Theorem. *Archive for History of Exact Sciences, 27*(1), pp. 23–47.

Dale, A. I. (1999). *A History of Inverse Probability—From Thomas Bayes to Karl Pearson* (2nd. ed.). New York: Springer.

De Finetti, B. (2017). *Theory of Probability—A Critical Introductory Treatment*. (A. M. Smith, Trans.) UK: John Wiley and Sons.

de Freitas, N. (2002). *Software for Research Purposes*. Retrieved from http://www.cs.ox.ac.uk/people/nando.defreitas/: https://www.cs.ubc.ca/~nando/software.html

DeGroot, M. H. (2004). *Optimal Statistical Decisions*. New Jersey: John Wiley & Sons.

Del Moral, P. (1996). Non Linear Filtering: Interacting Particle Solution. *Markov Processes and Related Fields, 2*(4), 555–580.

Del Moral, P. (1997, September). Filtrage non-linéaire par systèmes de particules en interaction. *Comptes Rendus de l'Académie des Sciences—Series I—Mathématiques, 325*(6), 653–658. doi:https://doi.org/10.1016/S0764-4442(97)84778-7

Del Moral, P., & Doucet, A. (2014, January 2014). Particle methods: An introduction with applications. *ESAIM Proceedings, 44*, 1–46. doi:https://doi.org/10.1051/proc/201444001

Dempster, A. (1967, April). Upper and Lower Probabilities Induced by a Multivalued Mapping. *Ann. Math. Statist., 38*(2), pp. 325–339,. doi:https://doi.org/10.1214/aoms/1177698950

Denoeux, T. (2019, June). Decision-making with belief functions: A review. *International Journal of Approximate Reasoning, 109*, 87–110. doi:https://doi.org/10.1016/j.ijar.2019.03.009

Denoeux, T. (2021). *Uncertainty Analysis using Belief Functions: Applications to Statistical Inference and Pattern Recognition.* Université de technologie de Compiègne, Department of Computer Science, Compiègne, France. Retrieved December 20, 2022, from https://www.hds.utc.fr/~tdenoeux/dokuwiki/_media/en/book_bf.pdf

Denœux, T., Younes, Z., & Abdallah, F. (2010). Representing uncertainty on set-valued variables using belief functions. *Artificial Intelligence, 174*(7–8), 479–499. doi:https://doi.org/10.1016/j.artint.2010.02.002

Depaoli, S., Winter, S. D., & Visser, M. (2020). The Importance of Prior Sensitivity Analysis in Bayesian Statistics: Demonstrations Using an Interactive Shiny App. *Frontiers in Psychology, 11*. doi:https://doi.org/10.3389/fpsyg.2020.608045

Dezert, J., & Smarandache, F. (2008). A new probabilistic transformation of belief mass assignment. *Fusion 2008: International Conference on Information Fusion*, (pp. 1410–1417). Retrieved December 19, 2022, from https://hal.archives-ouvertes.fr/hal-00304319/document

Dezert, J., & Smarandache, F. (2009). Transformations of belief masses into subjective probabilities. In J. Dezert, & F. Smarandache (Eds.), *Advances and Applications of DSmT for Information Fusion* (pp. 85–136). Rehoboth: American Research Press . Retrieved December 19, 2022, from https://www.researchgate.net/publication/306413841_Transformations_of_belief_masses_into_subjective_probabilities

Dezert, J., Smarandache, F., & Daniel, M. (2004). A Generalized Pignistic Transformation. In J. Dezert, & F. Smarandache (Eds.), *Advances and Applications of DSmT for Information Fusion* (pp. 143–153). Rehoboth: American Research Press. Retrieved December 19, 2022, from https://www.onera.fr/sites/default/files/297/C023-Dezert-Fusion2004Stockholm.pdf

Diaconis, P., & Ylvisaker, D. (1979). Conjugate Priors for Exponential Families. *The Annals of Statistics, 7*(2), pp. 269–281. doi:https://doi.org/10.1214/aos/1176344611

Dieudonné, J. (1969). *Foundations of Modern Analysis*. London: Academic Press.

Doré, P. E., Fiche, A., & Martin, A. (2010). Models of belief functions—Impacts for patterns recognitions. *13th International Conference on Information Fusion.* Edinburgh, UK. doi:https://doi.org/10.1109/ICIF.2010.5711936

Doré, P., Martin, A., Abi-Zeid, I., Jousselme, A., & Maupin, P. (2011a, January). Belief functions induced by multimodalprobability density functions, an application to the search and rescue problem. *RAIRO—Operations Research, 44*(4), 323–343. doi:https://doi.org/10.1051/ro/2011001

Doré, P., Osswald, C., Martin, A., Jousselme, A., & Maupin, P. (2011b). Continuous belief functions to qualify sensors performances. In W. Liu (Ed.), *Symbolic and Quantitative Approaches to Reasoning with Uncertainty. ECSQARU 2011* (pp. 350–361). Belfast, Ireland: Springer. doi:https://doi.org/10.1007/978-3-642-22152-1_30

Doucet, A., & Johansen, A. (2009). A Tutorial on Particle Filtering and Smoothing: Fifteen Years Later. In D. Crisan, & B. Rozovskii (Eds.), *The Oxford Handbook of Nonlinear Filtering* (pp. 656–704). New York: Oxford University Press.

Dragulescu, A., & Yakovenko, V. (2000). Statistical mechanics of money. *Eur. Phys. J. B, 17*, 723–729. doi:https://doi.org/10.1007/s100510070114

Dubois, D., & Prade, H. (1986). A Set-Theoretic View of Belief Functions. *International Journal of General Systems, 12*(3), 193–226. doi:https://doi.org/10.1080/03081078608934937

Elfring, J., Torta, E., & van de Molengraft, R. (2021, January 9). Particle Filters: A Hands-On Tutorial. *Sensors, 21*(2). doi:https://doi.org/10.3390/s21020438

Falconer, J. R., Frank, E., Polaschek, D. L., & Joshi, C. (2021). Methods for Eliciting Informative Prior Distributions: A Critical Review. *Decision Analysis, 19*(3), pp. 189–204. doi:https://doi.org/10.1287/deca.2022.0451

Falconer, J., Frank, E., Polaschek, D., & Joshi, C. (2022). Methods for Eliciting Informative Prior Distributions: A Critical Review. *Decision Analysis, 19*(3), 189–204. doi:https://doi.org/10.1287/deca.2022.0451

Faragher, R. (2012, September). Understanding the Basis of the Kalman Filter Via a Simple and Intuitive Derivation. *IEEE Signal Processing Magazine, 29*(5), pp. 128–132. doi:https://doi.org/10.1109/MSP.2012.2203621

Fetz, T., & Oberguggenberger, M. (2016, November). Imprecise random variables, random sets, and Monte Carlo simulation. *78*, 252–264. doi:https://doi.org/10.1016/j.ijar.2016.06.012

Fiche, A., Martin, A., Cexus, J.-C., & Khenchaf, A. (2010). Continuous belief functions and α-stable distributions. *13th Conference on Information Fusion.* Edinburgh. doi:https://doi.org/10.1109/ICIF.2010.5711934

Fink, D. (1997). *A Compendium of Conjugate Priors.* Retrieved April 23, 2023, from https://www.johndcook.com/CompendiumOfConjugatePriors.pdf

Fisher, R. A. (1955). Statistical Methods and Scientific Induction. *Journal of the Royal Statistical Society, Series B, 17*(1), pp. 69–78. doi:https://doi.org/10.1111/j.2517-6161.1955.tb00180.x

Fort, H. (2022). *Forecasting with Maximum Entropy.* IOP Publishing Ltd.

Fox, C. R., & See, K. E. (2006). Belief and Preference in Decision Under Uncertainty. In D. Hardman, & L. Macchi (Eds.), *Thinking: Psychological Perspectives on Reasoning, Judgment and Decision Making.* John Wiley & Sons. doi:https://doi.org/10.1002/047001332X.ch14

Frazier, P. I. (2018). *A Tutorial on Bayesian Optimization.* Retrieved July 25, 2023, from https://arxiv.org/pdf/1807.02811.pdf

Gan, W., Ji, Z., & Liang, Y. (2021). Acquisition Functions in Bayesian Optimization. *2nd International Conference on Big Data & Artificial Intelligence & Software Engineering (ICBASE),* (pp. 129–135). Zhuhai, China. doi:https://doi.org/10.1109/ICBASE53849.2021.00032.

Garthwaite, P. H., Kadane, J. B., & O'Hagan, A. (2005). Statistical Methods for Eliciting Probability Distributions. *Journal of the American Statistical Association, 100*(470), pp. 680–701. doi:https://doi.org/10.1198/016214505000000105

Gelman, A. (2002). Prior distribution. In A. H.-S. Piegorsch (Ed.), *Encyclopedia of Environmetrics,* (pp. 1634–1637). doi:https://doi.org/10.1002/9780470057339.vap039

Gelman, A., Simpson, D., & Betancourt, M. (2017). The Prior Can Often Only Be Understood in the Context of the Likelihood. *Entropy, 19*(10). doi:https://doi.org/10.3390/e19100555

Ghaderinezhad, F., & Ley, C. (2019). Quantification of the impact of priors in Bayesian statistics via Stein's Method. *146*, 206–212. doi:https://doi.org/10.1016/j.spl.2018.11.012

Ghaderinezhad, F., & Ley, C. (2020). On the Impact of the Choice of the Prior in Bayesian Statistics. In N. Tang (Ed.), *Bayesian Inference on Complicated Data* (pp. 1–14). IntechOpen. doi:https://doi.org/10.5772/intechopen.88994

Ghaderinezhad, F., Ley, C., & Serrien, B. (2022, October). The Wasserstein Impact Measure (WIM): A practical tool for quantifying prior impact in Bayesian statistics. *Computational Statistics and Data Analysis, 174.* doi:https://doi.org/10.1016/j.csda.2021.107352

Giambartolomei, G. (2015). *The Karhunen-Loeve Theorem.* Tesi di Laurea, Universita di Bologna, Bologna, Italy. Retrieved March 25, 2023, from https://amslaurea.unibo.it/10169/1/Giambartolomei_Giordano_Tesi.pdf

Gibbs, B. P. (2011). *Advanced Kalman Filtering, Least-Squares and Modeling.* New Jersey, USA: John Wiley & Sons. doi:https://doi.org/10.1002/9780470890042

Gibbs, J. W. (1901 reprinted 1960). *Elementary Principles of Statistical Mechanics.* Dover.

Grassi, P. R., & Bartels, A. (2021). Magic, Bayes and wows: A Bayesian account of magic tricks. *Neuroscience & Biobehavioral Reviews, 126*, 515–527. doi:https://doi.org/10.1016/j.neubiorev.2021.04.001

Grewal, M. S., & Andrews, A. P. (2001). *Kalman Filtering: Theory and Practice using Matlab.* New York, USA: John Wiley & Sons. doi:https://doi.org/10.1002/9780470377819

Guan, J., & Bell, D. (1993). Discounting and Combination Operations in Evidential Reasoning. In D. Heckerman, & A. Mamdani (Ed.), *Ninth Conference on Uncertainty in Artificial Intelligence* (pp. 477–484). Washington, USA: Elsevier. doi:https://doi.org/10.1016/B978-1-4832-1451-1.50062-7

Halmos, P. (1978). *Measure theory.* Berlin, Heidelberg, New York: : Springer Verlag.

Halpern, J., & Fagin, R. (1992). Two views of belief: belief as generalized probability and belief as evidence. *Artificial Intelligence,, 54*, pp. 275–317.

Harney, H. L. (2003). *Bayesian Inference.* Berlin, Germany: Springer. doi:https://doi.org/10.1007/978-3-662-06006-3

Hartley, R. (1928, July). Transmission of Information. *Bell System Technical Journal, 7*(3), 535–563.

Hastings, W. (1970, April). Monte Carlo sampling methods using Markov chains and their applications. *Biometrika, 57*(1), 97–109. doi:https://doi.org/10.1093/biomet/57.1.97

He, Y. (2013). *Uncertainty Quantification and Data Fusion based on Dempster-Shafer Theory.* Florida State University, Mathematics. Florida State University Libraries.

Howson, C., & Urbach, P. (2006). *Scientific reasoning : the Bayesian approach* . Chicago: Open Court.

Hubbard, A. (2023, February 22). *Kalman Filter for State Space Models.* Retrieved July 16, 2023, from Kalman Filter for State Space Models: https://cran.r-project.org/web/packages/kalmanfilter/vignettes/kalmanfilter_vignette.html

Hüllermeier, E., Kruse, R., & Hoffmann, F. (Eds.). (2010). Consonant continuous belief functions conflicts calculation. *IPMU'10: Proceedings of the Computational intelligence for knowledge-based systems design, and 13th international conference on Information processing and management of uncertainty* (pp. 706–715). Dortmund, Germany: Springer.

Hulse, A., Schumacher, B., & Westmoreland, M. D. (2018). Axiomatic Information Thermodynamics. *Entropy, 20*(4). doi:https://doi.org/10.3390/e20040237

Hurwich, C. M., & Tsai, C. L. (1989, June). Regression and Time Series Model Selection in Small Samples. *Biometrika, 76*(2), pp. 297–307. doi:https://doi.org/10.1093/biomet/76.2.297

Islam, A. F. (2011). *Loss functions, utility functions and Bayesian sample size determination.* Ph. D. Thesis, Queen Mary, University of London, London. Retrieved April 17, 2023, from https://core.ac.uk/download/pdf/30695406.pdf

Itti, L., & Baldi, P. (2009). Bayesian surprise attracts human attention. *Vision Research, 49*(10), 1295–1306. doi:https://doi.org/10.1016/j.visres.2008.09.007

Jaffray, J.-Y., & Wakker, P. (1993). Decision making with belief functions: Compatibility and incompatibility with the sure-thing principle. *Journal of Risk and Uncertainty, 7*(3), 255–271. doi:https://doi.org/10.1007/BF01079626

Jaynes, E. (1989). *Clearing up Mysteries—The Original Goal.* (J. Skilling, Ed.) Dordrecht: Springer. doi:https://doi.org/10.1007/978-94-015-7860-8_1

Jaynes, E. (2003). *Probability Theory: The Logic of Science.* Cambridge: Cambridge University Press.

Jaynes, E. T. (1957a, May). Information Theory and Statistical Mechanics. *Phys. Rev., 106*(4), 620–630. doi:https://doi.org/10.1103/PhysRev.106.620

Jaynes, E. T. (1957b, October). Information Theory and Statistical Mechanics. II. *Phys. Rev., 108*(2), 171–190. doi:https://doi.org/10.1103/PhysRev.108.171

Jaynes, E. T. (1965). Gibbs vs. Boltzmann Entropies. *American Journal of Physics, 33*(5), 391–398.

Jaynes, E. T. (1968a, September). Prior Probabilities. *IEEE Transactions on Systems Science and Cybernetics, 4*(3), 227–241. doi:https://doi.org/10.1109/TSSC.1968.300117

Jaynes, E. T. (1968b, September). Prior Probabilities. *IEEE Transactions on Systems Science and Cybernetics, 4*(3), pp. 227–241. doi:https://doi.org/10.1109/TSSC.1968.300117

Jeffreys, H. (1935, April). Some Tests of Significance, Treated by the Theory of Probability. *Mathematical Proceedings of the Cambridge Philosophical Society, 31*(2), pp. 203–222. doi:https://doi.org/10.1017/S030500410001330X

Jeffreys, H. (1939). *Theory of probability.* Oxford: University Press.

Jeffreys, H. (1946, September 24). An invariant form for the prior probability in estimation problems. *Proceedings of the Royal Society of London. Series A,, 10*(1007), pp. 453–461. doi: https://doi.org/10.1098/rspa.1946.0056

Johnson, S. R., Tomlinson, G. A., Hawker, G. A., Granton, J. T., & Feldman, B. M. (2010). Methods to elicit beliefs for Bayesian priors: a systematic review. *J Clin Epidemiol., 63*(4), pp. 355–369. doi:https://doi.org/10.1016/j.jclinepi.2009.06.003

Jones, D. E., Trangucci, R. N., & Chen, Y. (2022). Quantifying Observed Prior Impact. *Bayesian Analysis, 17*(3), 737–764. doi:https://doi.org/10.1214/21-BA1271

Julier, S., & Uhlmann, J. (2004, March). Unscented Filtering and Nonlinear Estimation. *Proceedings of the IEEE. 92 (3): 401–422, 92*(3), 401–422. doi:https://doi.org/10.1109/JPROC.2003.823141

Kalman Filtering. (2023). Retrieved July 17, 2023, from Mathworks Help Center: https://fr.mathworks.com/help/control/ug/kalman-filtering.html

Kalman, R. (1960a). Contributions to the Theory of Optimal Control. *Boletín de la Sociedad de Matematica Mexicana, 5*(2), 102–119. Retrieved July 10, 2023, from https://www.ee.iitb.ac.in/~belur/ee640/optimal-classic-paper.pdf

Kalman, R. (1960b, March 1). A New Approach to Linear Filtering and Prediction Problems. *Journal of Basic Engineering, 82*(1), 35–45. doi:https://doi.org/10.1115/1.3662552

Kalman, R., & Bucy, R. (1961, March 1). New Results in Linear Filtering and Prediction Theory. *Journal of Basic Engineering, 83*(1), 95–108. doi:https://doi.org/10.1115/1.3658902

Kapur, J. N. (1989 revised 1993). *Maximum Entropy Models in Science and Technology.* New Delhi, India: Wiley Eastern.

Kass, R. E., & Raftery, A. E. (1995, June). Bayes Factors. *Journal of the American Statistical Association, 90*(430), pp. 773–795. doi:https://doi.org/10.2307/2291091

Kass, R. E., & Wasserman, L. (1996). The selection prior distribution by formal rules. *Journal of the American Statistical Association, 91*(435), 1343–1370. doi:https://doi.org/10.1080/01621459.1996.10477003

Khinchin, A. Y. (1957). *Mathematical Foundations of Information Theory.* New York, NY, USA: Dover.

Klopotek, M. A., & Wierzchon, S. T. (1998). A New Qualitative Rough-Set Approach to Modeling Belief Functions. In L. Polkowski, & A. Skowron (Ed.), *Rough Sets and Current Trends in Computing, First International Conference, RSCTC'98* (pp. 346–354). Warzsaw, Poland: Springer. doi:https://doi.org/10.1007/3-540-69115-4_47

Kojadinovic, I., Marichal, J.-L., & Roubens, M. (2005). An axiomatic approach to the definition of the entropy of a discrete Choquet capacity. *Information Sciences, 172*(1–2), 131–153. doi: https://doi.org/10.1016/j.ins.2004.05.011

Kolossa, A., Kopp, B., & Fingscheidt, T. (2015). A computational analysis of the neural bases of Bayesian inference. *NeuroImage, 106*, 222–237. doi:https://doi.org/10.1016/j.neuroimage.2014.11.00

Kullback, S. (1951 reed. 1969 reprint 1979). *Information Theory and Statistics.* New York: Wiley, reed. Dover.

Kullback, S., & Leibler, R. A. (1951, March). On Information and Sufficiency. *The Annals of Mathematical Statistics, 22*(1), 79–86. doi:https://doi.org/10.1214/aoms/1177729694

Künsch, H. (2013, September). Particle filters. *Bernoulli 19 (4), 19*(4), pp. 1391–1403. doi:https://doi.org/10.3150/12-BEJSP07

Laghmara, H., Laurain, T., Cudel, C., & Lauffenburger, J. P. (2020). Heterogeneous sensor data fusion for multiple object association using belief functions. *Information Fusion, 57*, 44–58. doi:https://doi.org/10.1016/j.inffus.2019.11.002

Laplace, P.-S. (1774). Memoire sur la probabilité des causes par les événements. *Memoires de Mathématique et de Physique, Presentés à l'Académie Royale des Sciences par divers Savans & lus dans ses Assemblées*, pp. 621–656.

Laplace, P.-S. (1891). *Oeuvres complètes* (Vol. 8). (A. d. Paris, Ed.) Paris: Gauthiers-Villars.

Laplace, P.-S. (1986). Memoir on the Probability of the Causes of Events. *Statistical Science, 1*(3), pp. 364–78. Retrieved from http://www.jstor.org/stable/2245476

Lavenda, B. H. (2001). Derivation of the Prior Distribution in Bayesian Analysis from the Principle of Statistical Equivalence. *Open Systems & Information Dynamics, 8*(2), 103–114. doi:https://doi.org/10.1023/A:1011910512406

Law, J. (2019, February 11). *Efficient Markov chain Monte Carlo in R with Rcpp.* Retrieved July 27, 2023, from Bayesian Statistics and Functional Programming: https://jonnylaw.rocks/posts/2019-02-11-metropolis_r/

Lazo, A. V., & Rathie, P. (1978). On the entropy of continuous probability distributions (Corresp.). *IEEE Transactions on Information Theory, 24*(1), 120–122. doi:https://doi.org/10.1109/tit.1978.1055832

LearningMachines. (2020, July 7). *Kalman Filter as a Form of Bayesian Updating.* Retrieved July 16, 2023, from R-bloggers: https://www.r-bloggers.com/2020/07/kalman-filter-as-a-form-of-bayesian-updating/

Lee, J., Fan, Y., & Sisson, S. (2015). Bayesian threshold selection for extremal models using measures of surprise. *Computational Statistics and Data Analysi, 85*, 84–99. doi:https://doi.org/10.1016/j.csda.2014.12.004

Ley, C., Reinert, G., & Swan, Y. (2017). Distances between nested densities and a measure of the impact of the prior in Bayesian statistics. *The Annals of Applied Probability, 27*(1), 216–241. doi:https://doi.org/10.1214/16-AAP1202

Lian, C. (2017). *Information Fusion and Decision Making using Belief Functions.* Compiègne, France: Université de Technologie de Compiègne.

Liboff, R. L. (1974). Gibbs vs. Shannon entropies. *J Stat Phys, 11*, 343–357. doi:https://doi.org/10.1007/BF01009793

Liu, W. (2006, August). Analyzing the degree of conflict among belief functions. *Artificial Intelligence, 170* (11), 909–924. doi:https://doi.org/10.1016/j.artint.2006.05.002

Lopez, R., Cursi, J., & Carlon, A. (2018). A state estimation approach based on stochastic expansions. *Comp. Appl. Math., 37*, 3399–3430. doi:https://doi.org/10.1007/s40314-017-0515-0

Lopez, R., Cursi, J., & Carlon, A. (2019, July). A New Approach for State Estimation. In F. Canavero (Ed.), *Uncertainty Modeling for Engineering Applications* (Vol. 37, pp. 41–54). Springer, Cham. doi:https://doi.org/10.1007/978-3-030-04870-9_3

Lopez, R., Souza de Cursi, J., & Carlon, A. (2016). State estimation based on stochastic polynomials and variational approximation. In D. Rade, & A. M. Lima (Ed.), *3rd International Symposium on Uncertainty Quantification and Stochastic Modeling.* Maresias, SP, Brazil. doi:https://doi.org/10.20906/CPS/USM-2016-0005

Lu, Z., & Lou, W. (2022). Bayesian approaches to variable selection: a comparative study from practical perspectives. *The International Journal of Biostatistics, 18*(1), 83–108. doi:https://doi.org/10.1515/ijb-2020-0130

Martin, A. (2019). Conflict management in information fusion with belief functions. In E. Bossé, & G. Rogova (Eds.), *Information quality in information fusion and decision making* (pp. 79–97). Springer. doi:https://doi.org/10.1007/978-3-030-03643-0_4

Martin, R., Zhang, J., & Liu, C. (2010, June). Dempster–Shafer Theory and Statistical Inference with Weak Beliefs. *Statistical Science, 25*(1), 72–87. doi:https://doi.org/10.1214/10-STS322

Masnadi-Shirazi, H. (2011). *The design of Bayes consistent loss functions for classification.* Ph. D. Thesis, University of California, San Diego, San Diego. Retrieved April 17, 2023, from https://escholarship.org/uc/item/1cv1947c

Mercer, J. (1909). Functions of positive and negative type, and their connection the theory of integral equations. *Philosophical Transactions of the Royal Society of London. Series A., 209*(441–458), 415–446. doi:https://doi.org/10.1098/rsta.1909.0016

Mercier, D., Quost, B., & Denœux, T. (2005). Contextual Discounting of Belief Functions. In L. Godo (Ed.), *ECSQARU 2005: Symbolic and Quantitative Approaches to Reasoning with*

Uncertainty (pp. 552–562). Barcelona, Spain: Springer. doi:https://doi.org/10.1007/11518655_47

Metropolis, N., Rosenbluth, A. W., Rosenbluth, M. N., Teller, A. H., & Teller, E. (1953, June). Equation of State Calculations by Fast Computing Machines. *Journal of Chemical Physics, 21*(6), 1087–1092. doi:https://doi.org/10.1063/1.1699114

Miranda, E., Couso, I., & Gil, P. (2005, July). Random sets as imprecise random variables. *Journal of Mathematical Analysis and Applications, 307*(1), 32–47. doi:https://doi.org/10.1016/j.jmaa.2004.10.022

Modirshanechi, A., Brea, J., & Gerstner, W. (2022). A taxonomy of surprise definitions. *Journal of Mathematical Psychology, 110*. doi:https://doi.org/10.1016/j.jmp.2022.102712

Morita, S., Thall, P. F., & Müller, P. (2010, July 1). Evaluating the Impact of Prior Assumptions in Bayesian Biostatistics. *Stat Biosci., 2*(1), 1–17. doi:https://doi.org/10.1007/s12561-010-9018-x

Muñoz-Cobo, J.-L., Mendizábal, R., Miquel, A., Berna, C., & Escrivá, A. (2017). Use of the Principles of Maximum Entropy and Maximum Relative Entropy for the Determination of Uncertain Parameter Distributions in Engineering Applications. *Entropy, 19*(9). doi:https://doi.org/10.3390/e19090486

Murphy, K. P. (2007). *Conjugate Bayesian analysis of the Gaussian distribution.* Retrieved April 23, 2023, from https://www.cs.ubc.ca/~murphyk/Papers/bayesGauss.pdf

Nambiar, K. K., Varma, P. K., & Saroch, V. (1992). An axiomatic definition of Shannon's entropy. *Appl. Math. Lett., 5*(4), 45–46. doi:https://doi.org/10.1016/0893-9659(92)90084-M

Nassar, M., Alotaibi, R., Okasha, H., & Wang, L. (2022). Bayesian Estimation Using Expected LINEX Loss Function: A Novel Approach with Applications. *Mathematics, 10*. doi:https://doi.org/10.3390/math10030436

Navarro, D. (2023, April 12). *The Metropolis-Hastings algorithm.* Retrieved July 12, 2023, from https://blog.djnavarro.net: https://blog.djnavarro.net/posts/2023-04-12_metropolis-hastings/

Neel. (2014, February 8). *R code for multivariate random-walk Metropolis sampling.* Retrieved July 27, 2023, from https://blog.abhranil.net: https://blog.abhranil.net/2014/02/08/r-code-for-multivariate-random-walk-metropolis-hastings-sampling/

Neyman, J., & Pearson, E. S. (1933). On the problem of the most efficient tests of statistical hypotheses. *Philosophical Transactions of the Royal Society A, 231*(694–706), pp. 289–337. doi:https://doi.org/10.1098/rsta.1933.0009

Nguyen, H. T. (1977). *On Random sets and Belief Functions.* University of California, Berkeley, EECS Department. Retrieved from http://www2.eecs.berkeley.edu/Pubs/TechRpts/1977/28879.html

Nguyen, H. T. (1978). On Random Sets and Belief Functions. *Journal of Mathematical Analysis and Applications, 65*, 531–542. doi:https://doi.org/10.1016/0022-247X(78)90161-0

Nguyen, H., & Wang, T. (1997). Belief Functions and Random Sets. In J. Goutsias, R. Mahler, & H. Nguyen (Eds.), *Random Sets* (pp. 243–255). Springer. doi:https://doi.org/10.1007/978-1-4612-1942-2_11

Ni, S., Lei, Y., & Tang, Y. (2020). Improved Base Belief Function-Based Conflict Data Fusion Approach Considering Belief Entropy in the Evidence Theory. *Entropy, 22*(8). doi:https://doi.org/10.3390/e22080801

O'Hara, R., & Sillanpää, M. J. (2009, March). A Review of Bayesian Variable Selection Methods: What, How and Which. *Bayesian Analysis, 4*(1), 85–118. doi:https://doi.org/10.1214/09-BA403

Ostwald, D., Spitzer, B., Guggenmos, M., Schmidt, T. T., Kiebel, S. J., & Blankenburg, F. (2012). Evidence for neural encoding of Bayesian surprise in human somatosensation. *NeuroImage, 62*(1), 177–188. doi:https://doi.org/10.1016/j.neuroimage.2012.04.050

Palm, G. (2023). *Novelty, Information and Surprise.* Germany: Springer-Verlag. doi:https://doi.org/10.1007/978-3-662-65875-8

Pawar, A., & Warbhe, U. (2021). *Optimizing Bayesian acquisition functions in Gaussian Processes.* Retrieved from https://arxiv.org/abs/2111.04930

Pearson, K. (1900). On the criterion that a given system of deviations from the probable in the case of a correlated system of variables is such that it can be reasonably supposed to have arisen from random sampling. *Philosophical Magazine Series 5, 50*(302), pp. 157–175. doi:https://doi.org/10.1080/14786440009463897

Pearson, K. (1904). On the theory of contingency and its relation to association and normal correlation. In B. Series, *Mathematical Contributions to the Theory of Evolution* (Vol. 1, pp. 1–35). Drapers' Company Research Memoirs . Retrieved July 1, 2023, from https://ia800207.us.archive.org/16/items/cu31924003064833/cu31924003064833.pdf

Planck, M. K. (1901). Über das Gesetz der Energieverteilung im Normalspektrum. *Annalen der Physik, 309*(3), 553–563. doi:https://doi.org/10.1002/andp.19013090310

Planck, M. K. (1914). *The Theory of Heat Radiation.* (M. Masius, Trans.) Philadelphia: P. Blakiston's Son & Co.

Pogu, M., & Souza de Cursi, E. (1994). Global optimization by random perturbation of the gradient method with a fixed parameter. *Journal of Global Optimization, 5*(2), pp. 159–180. doi:https://doi.org/10.1007/BF01100691.

Popov, M. (2019, April 6). *Bayesian Optimization in R.* Retrieved July 25, 2023, from https://mpopov.com/: https://mpopov.com/tutorials/bayesopt-r/

Press, S. J., & Tanur, J. M. (2001). *The Subjectivity of Scientists and the Bayesian Approach.* New York: John Wiley & Sons.

Quiroga-Martinez, D., Hansen, N., Højlund, A., Pearce, M., Brattico, E., & Vuust, P. (2020). Decomposing neural responses to melodic surprise in musicians and non-musicians: Evidence for a hierarchy of predictions in the auditory system. *NeuroImage.* doi:https://doi.org/10.1016/j.neuroimage.2020.116816

Rahman, A., Gao, J., D'Este, C., & Ahmed, S. E. (2016, September). An Assessment of the Effects of Prior Distributions on the Bayesian Predictive Inference. *International Journal of Statistics and Probability, 5*(5), 31–42. doi:https://doi.org/10.5539/ijsp.v5n5p31

Rao, C. R. (1948, January). Large sample tests of statistical hypotheses concerning several parameters with applications to problems of estimation. *Mathematical Proceedings of the Cambridge Philosophical Society, 44*(1), pp. 50–57 . doi:https://doi.org/10.1017/S0305004100023987

Rényi, A. (1961). On Measures of Entropy and Information. In J. Neyman (Ed.), *4th Berkeley Symposium on Mathematics, Statistics and Probability,. 1*, pp. 547–561. University of California Press.

Risti, B., & Smets, P. (2006). Belief function theory on the continuous space with an application to model based classification. In B. Bouchon-Meunier, G. Coletti, & R. R. Yager (Ed.), *Modern Information Processing: From Theory to Applications. IPMU'04* (pp. 11–24). Perugia, Italy.: Elsevier. doi:https://doi.org/10.1016/B978-044452075-3/50002-9

Robert, C. P. (1994). *The Bayesian Choice.* New York, NY, USA: Springer Verlag. doi:https://doi.org/10.1007/978-1-4757-4314-2

Robert, C. P. (2016, January 27). *The Metropolis-Hastings algorithm.* Retrieved July 30, 2023, from arXiv:1504.01896: https://arxiv.org/abs/1504.01896

Robert, C., & Casella, G. (2004). *Monte Carlo Statistical Methods.* New York: Springer Science +Business Media. doi:https://doi.org/10.1007/978-1-4757-4145-2

Robert, C., & Casella, G. (2010). *Introducing Monte Carlo Methods with R.* New York: Springer. doi:https://doi.org/10.1007/978-1-4419-1576-4

Saravanan, R., & Levine, R. (2022). Surprisal analysis of diffusion processes. *Chemical Physics, 556.* doi:https://doi.org/10.1016/j.chemphys.2022.111450

Savchuk, V. P., & Tsokos, C. P. (2011). *Bayesian Theory and Methods with Applications.* Paris: Atlantis Press.

Schwarz, G. E. (1978). Estimating the dimension of a model. *Annals of Statistics,, 6*(2), 461–464. doi:https://doi.org/10.1214/aos/1176344136, MR 0468014

Shafer, G. (1976). *A Mathematical Theory of Evidence* . New Jersey: Princeton University Press.

Shafer, G. (1990). Perspectives on the Theory and Practice of Belief Functions. *International Journal of Approximate Reasoning, 4*(5–6), 323–362. doi:https://doi.org/10.1016/0888-613X(90)90012-Q

Shannon, C. E. (1948a). A Mathematical Theory of Communication. *The Bell System Technical Journal, 27*(3), 379–423. doi:https://doi.org/10.1002/j.1538-7305.1948.tb01338.x

Shannon, C. E. (1948b). A Mathematical Theory of Communication. *Bell System Technical Journal, 27*(4), 623–666. doi:https://doi.org/10.1002/j.1538-7305.1948.tb00917.x

Shannon, C. E. (n.d.). *A mathematical theory of comunication—Nokia Bell Labs*. Retrieved 2 13, 2023, from https://www.bell-labs.com/claude-shannon/assets/images/discoveries: https://www.bell-labs.com/claude-shannon/assets/images/discoveries/1948-04-21-a-mathematical-theory-of-communication-parts-I-and-carousel-01.pdf

Shore, J., & Johnson, R. (1980, January). Axiomatic derivation of the principle of maximum entropy and the principle of minimum cross-entropy. *IEEE Transactions on Information Theory, 26*(1), 26–37. doi:https://doi.org/10.1109/TIT.1980.1056144.

Simons, D. (2006). *Optimal State Estimation.* New Jersey, USA: John Wiley & Sons. doi:https://doi.org/10.1002/0470045345

Smets, P. (1990). Constructing the Pignistic Probability Function in a Context of Uncertainty. *Machine Intelligence and Pattern Recognition, 10*, 29–39. doi:https://doi.org/10.1016/B978-0-444-88738-2.50010-5

Smets, P. (2000). Data fusion in the transferable belief model. *Proceedings of the Third International Conference on Information Fusion. 1*, pp. 21–33. Paris, France: IEEE. doi:https://doi.org/10.1109/IFIC.2000.862713.

Smets, P. (2005). Belief functions on real numbers. *International Journal of Approximate Reasoning, 40*, 181–223. doi:https://doi.org/10.1016/j.ijar.2005.04.001

Sohrab, S. H. (2014). Boltzmann entropy of thermodynamics versus Shannon entropy of information theory. *International Journal of Mechanics, 8*, 73-84. Retrieved February 15, 2023, from https://www.naun.org/main/NAUN/mechanics/2014/a182003-086.pdf

Soize, C. (2005, April). Random matrix theory for modeling uncertainties in computational mechanics. *Computer Methods in Applied Mechanics and Engineering, 194*(12–16), 1333–1366. doi:https://doi.org/10.1016/j.cma.2004.06.038

Soize, C. (2008). Construction of probability distributions in high dimension using the maximum entropy principle: Applications to stochastic processes, random fields and random matrices. *Int. J. Numer. Meth. Engng, 76*, 1583–1611. doi:https://doi.org/10.1002/nme.2385

Souza de Cursi, E. (2023). *Uncertainty Quantification using R.* Springer Cham.

Souza de Cursi, E., & Sampaio, R. (2015). *Uncertainty Quantification and Stochastic Modeling with Matlab.* ISTE/Elsevier. doi:https://doi.org/10.1016/C2014-0-04713-2

Speekenbrink, M. (2016, August). A tutorial on particle filters. *Journal of Mathematical Psychology, 73*, pp. 140–152. doi:https://doi.org/10.1016/j.jmp.2016.05.006

Stigler, S. M. (1982). Thomas Bayes's Bayesian Inference. *Journal of the Royal Statistical Society. Series A (General), 145*(2), pp. .250–258.

Strat, T. (1984). Continuous belief functions for evidential reasoning. *Proceedings of the 4th National Conference on Artificial Intelligence.* Austin, Texas. Retrieved December 18, 2022, from https://www.aaai.org/Papers/AAAI/1984/AAAI84-035.pdf

Strat, T. (1987). The Generation of Explanations within Evidential Reasoning Systems. In J. P. McDermott (Ed.), *10th. International Joint Conference on Artificial Intelligence (IJCAI)*, (pp. 1097–1104). Milan, Italy. Retrieved December 20, 2022, from https://www.ijcai.org/Proceedings/87-2/Papers/104.pdf

Strat, T. (1990). Decision analysis using belief functions. *International Journal of Approximate Reasoning, 4*(5–6), 391–417. doi:https://doi.org/10.1016/0888-613X(90)90014-S

Taillandier, P., & Therond, O. (2011). Use of the Belief Theory to formalize Agent DecisionMaking Processes : Application to cropping Plan Decision Making. *European Simulation and Modelling Conference*, (pp. 138–142). Guimaraes, Portugal. Retrieved 12 17, 2022, from https://hal.archives-ouvertes.fr/hal-00688405

The Metropolis Algorithm. (n.d.). Retrieved July 12, 2023, from The Arcane Algorithm Archive: https://www.algorithm-archive.org/contents/metropolis/metropolis.html

Turing, A. M. (1941). *The Applications of Probability to Cryptography.* Bletchley Park. Available at Archive.org and https://www.nationalarchives.gov.uk/. Retrieved from https://archive.org/details/hw-25-37

Turing, A. M. (2015, May 26). *The Applications of Probability to Cryptography.* doi:https://doi.org/10.48550/arXiv.1505.04714

Tusell, F. (2011, March 1). Kalman Filtering in R. *Journal of Statistical Software, 39*(2), pp. 1–27. doi:https://doi.org/10.18637/jss.v039.i02

Wald, A. (1939). Contributions to the Theory of Statistical Estimation and Testing Hypotheses. *Annals of Mathematical Statistics, 10* (4), pp. 299–326. doi:https://doi.org/10.1214/aoms/1177732144

Wald, A. (1943, November). Tests of Statistical Hypotheses concerning several parameters when the number of observations is large. *Transactions of the American Mathematical Society, 54*(3), pp. 426–482. doi:https://doi.org/10.2307/1990256

Wang, L. (2008). *K arhunen-Loeve E xpansions and their Applications.* The London School of Economics and Political Science. Ann Arbor, MI, USA: ProQuest LLC. Retrieved March 25, 2023, from http://etheses.lse.ac.uk/2950/1/U615901.pdf

Wasserman, L. A. (1990, September). Belief functions and statistical inference. *Canadian Journal of Statistics, 18*(3), 183–196. doi:https://doi.org/10.2307/3315449

Wesner, J. S., & Pomeranz, J. P. (2021). Choosing priors in Bayesian ecological models by simulating from the prior predictive distribution. *Ecosphere, 12*(9). doi:https://doi.org/10.1002/ecs2.3739

Wilks, S. S. (1938, March). The Large-Sample Distribution of the Likelihood Ratio for Testing Composite Hypotheses. *Ann. Math. Statist., 9*(1), pp. 60–62. doi:https://doi.org/10.1214/aoms/1177732360

Wilson, J. T., Hutter, F., & Deisenroth, M. P. (2018). Maximizing acquisition functions for Bayesian optimization. *NIPS'18: 32nd International Conference on Neural Information Processing SystemsDecember 2018Pages 9906–9917*, (pp. 9906–9917). Retrieved from https://dl.acm.org/doi/10.5555/3327546.3327655; https://proceedings.neurips.cc/paper_files/paper/2018/file/498f2c21688f6451d9f5fd09d53edda7-Paper.pdf

Winkler, R. L. (1966). *The assessment of prior distributions in Bayesian analysis.* Chicago, IL, USA: University of Chicago.

Winkler, R. L. (1967, September). The Assessment of Prior Distributions in Bayesian Analysis. *Journal of the American Statistical Association, 62*(319), 776–800. doi:https://doi.org/10.2307/2283671

Wu, W.-Z., & Mi, J.-S. (2008). An Interpretation of Belief Functions on Infinite Universes in the Theory of Rough Sets. In C. Chan, J. Grzymala-Busse, & W. P. Ziarko (Ed.), *Rough Sets and Current Trends in Computing. RSCTC 2008* (pp. 71–80). Akron, OH, USA: Springer. doi: https://doi.org/10.1007/978-3-540-88425-5_8

Yakovenko, V. M. (2010). Statistical Mechanics of Money, Debt, and Energy Comsumption. *Science and Culture, 76*(9–10), 430–436. doi:https://doi.org/10.48550/arXiv.1008.2179

Yakovenko, V. M., & Rosser, J. B. (2009, December 2). Statistical mechanics of money, wealth, and income. *Rev. Mod. Phys., 81*(4), 1703–1725. doi:https://doi.org/10.1103/RevModPhys.81.1703

Yao, Y. Y., & Lingras, P. J. (1998). Interpretations of Belief Functions in the Theory of Rough Sets. *Information Sciences, 104*(1–2), 81–106. doi:https://doi.org/10.1016/S0020-0255(97)00076-5

Yosida, K. (1995). *Functional Analysis.* Berlin Heidelberg: Springer-Verlag . doi:https://doi.org/10.1007/978-3-642-61859-8

Zanchini, E., & Beretta, G. P. (2008). Rigorous Axiomatic Definition of Entropy Valid Also for Non-Equilibrium States. In G. P. Beretta, A. Ghoniem, & G. Hatsopoulos (Ed.), *MEETING THE ENTROPY CHALLENGE: An International Thermodynamics Symposium in Honor and*

Memory of Professor Joseph H. Keenan. 1033. Cambridge, MA, USA: AIP Conference Proceedings. doi:https://doi.org/10.1063/1.2979048

Zellner, A. (1986a, June). Bayesian Estimation and Prediction Using Asymmetric Loss Functions. *Journal of the American Statistical Association, 81*(394), pp. 446–451. doi:https://doi.org/10.2307/2289234

Zellner, A. (1986b). On Assessing Prior Distributions and Bayesian Regression Analysis with g Prior Distributions. In P. Goel, & A. Zellner (Eds.), *Bayesian Inference and Decision Techniques: Essays in Honor of Bruno de Finetti.* (pp. 233–243). Elsevier.

Zhang, H., & Deng, Y. (2020). Weighted belief function of sensor data fusion in engine fault diagnosis. *Soft Comput, 24*, 2329–2339. doi:https://doi.org/10.1007/s00500-019-04063-7

Zhang, Y.-Y. (2020). The Bayesian Posterior Estimators under Six Loss Functions for Unrestricted and Restricted Parameter Spaces. In N. Tang (Ed.), *Bayesian Inference on Complicated Data.* doi:https://doi.org/10.5772/intechopen.88587

Zhou, K., Martin, A., & Pan, Q. (2018). A belief combination rule for a large number of sources. . *Journal of Advances in Information Fusion, 13*(2). Retrieved December 2022, 20, from https://www.researchgate.net/deref/https%3A%2F%2Fhal.archives-ouvertes.fr%2Fhal-01883239

Zondervan-Zwijnenburg, M., Peeters, M., Depaoli, S., & Van de Schoot, R. (2017). Where Do Priors Come From? Applying Guidelines to Construct Informative Priors in Small Sample Research. *Research in Human Development, 14*(4), 305–320. doi:https://doi.org/10.1080/15427609.2017.1370966

Index

E. Souza de Cursi, *Uncertainty Quantification with R*, International Series in Operations Research & Management Science 352,
https://doi.org/10.1007/978-3-031-48208-3

S

T

U

V